MOLECULAR QUANTUM ELECTRODYNAMICS

MOLECULAR QUANTUM ELECTRODYNAMICS

Long-Range Intermolecular Interactions

AKBAR SALAM
Department of Chemistry
Wake Forest University
Winston-Salem, North Carolina

WILEY
A John Wiley & Sons, Inc., Publication

Published by John Wiley & Sons, Inc., Hoboken, New Jersey
Published simultaneously in Canada

For general information on our other products and services or for technical support, please contact our Customer Care Department within the United States at (800) 762-2974, outside the United States at (317) 572-3993 or fax (317) 572-4002.

Wiley also publishes its books in a variety of electronic formats. Some content that appears in print may not be available in electronic formats. For more information about Wiley products, visit our web site at www.wiley.com.

Library of Congress Cataloging-in-Publication Data:

Salam, A. (Akbar)
Molecular quantum electrodynamics : long-range intermolecular interactions / A. Salam.
p. cm.
Includes bibliographical references and index.
ISBN 978-0-470-25930-6 (cloth)
1. Quantum electrodynamics. 2. Intermolecular forces. I. Title.
QC680.S15 2010
530.14′33--dc22

2009035853

Printed in the United States of America

10 9 8 7 6 5 4 3 2 1

To
C, P, and T

CONTENTS

PREFACE

The theory of molecular quantum electrodynamics has its origins in the work carried out by the founding fathers of quantum mechanics in the late 1920s. A complete formulation of quantum theory applicable to nonrelativistic particles was quickly developed, enabling a fundamental description to be given of the chemical and physical properties of matter at the atomic and molecular levels. It was only natural that soon thereafter the new mechanics were applied to electromagnetic radiation, with the consequence that the photon emerged as the particle associated with the quantization of the Maxwell field. This was followed by the construction of a single conceptual and calculational quantum mechanical framework for the study of the interaction of light with matter, which provides a means to probe the structure of atoms and molecules that manifest in numerous forms of spectroscopy.

While the next generation of theoretical physicists focused their attention on formulating a self-consistent and fully covariant theory of electron–photon interaction, an endeavor that was ultimately successful, despite the introduction of the mathematically and physically unsatisfactory technique of renormalization, a device that continues to be used in the present-day calculations to yield results of physically observable quantities that are not divergent, a negative feature that besets all such field theories and is inherent to them, progress in the noncovariant theory of radiation–matter interaction continued, especially in regard to the foundations of the subject and its application to problems of interest in the field of chemical physics. Detailed expositions may be found in texts such as Power's *Introductory Quantum*

Electrodynamics (Longmans, London, 1964), Healy's *Non-Relativistic Quantum Electrodynamics* (Academic Press, London, 1982), Craig and Thirunamachandran's *Molecular Quantum Electrodynamics* (Dover, New York, 1998), and Milonni's *The Quantum Vacuum* (Academic Press, San Diego, 1994). Highlights included the explanation of spontaneous emission and the computation of its rate, the calculation of the Lamb shift—the splitting of the $2S_{1/2}$ and $2P_{1/2}$ levels in atomic hydrogen—and the derivation of the form of the retarded dispersion potential by Casimir and Polder, to name but three historically significant examples. Predictions of these and other phenomena, such as the anomalous magnetic moment of the electron, have been compared with measured values to unparalleled levels of accuracy, in the process providing remarkable agreement with experiment and reassuringly high degrees of confidence in the theory.

Atomic, molecular, and optical physicists and theoretical chemists have extended the domain of application of molecular quantum electrodynamics. These include investigation of single- and multiphoton absorption and emission and scattering of light by matter and the study of chiroptical processes such as optical rotation and circular dichroism. With continuing progress being made in the generation of coherent sources of radiation, numerous and wide-ranging applications have been made to phenomena occurring in the area of nonlinear and quantum optics. Examples include, but are by no means limited to, the laser-induced optical activity, hyper-Rayleigh and Raman scattering, coherent anti-Stokes Raman scattering, optical Kerr effect, second-, third-, and high-harmonic generation, and four-, five-, and six-wave mixing, many of whose theoretical bases may be found in the books by Mukamel (*Principles of Nonlinear Optical Spectroscopy*, Oxford University Press, New York, 1995) and Andrews and Allcock (*Optical Harmonics in Molecular Systems*, Wiley-VCH, Weinheim, 2002).

A topic of widespread interest and fundamental nature in which quantum electrodynamics has made significant contribution is the field of interatomic/intermolecular interactions. The QED formalism lends itself naturally to a description in which coupling between matter occurs via the exchange of one or more virtual photons. Considerable advances have taken place in the last twenty-five years in the quantum electrodynamical theory of intermolecular forces, which mainly constitutes the subject matter of this book. Before giving a standard presentation of the concept of an intermolecular potential and a semiclassical perturbation theory treatment of short- and long-range forces, the latter decomposed into familiar electrostatic, induction, and dispersion terms in Chapter 3, the quantum theory of the nonrelativistic interaction of a charged particle with a radiation field is explicated in Chapter 1. This entails construction of the total Hamiltonian for the system comprising matter, electromagnetic field, and their coupling, beginning with the classical Lagrangian function and applying the canonical quantization procedure. Solutions for the interacting system are developed via perturbation theory expansion. An alternative formulation of the quantized theory of electron–photon interaction is expounded in Chapter 2. The viewpoint adopted and the method developed is slightly less familiar field theoretic approach, in which the Maxwell field interacts with the electron wavefield, whose rigorous construction is found to take place most easily in the Heisenberg picture of quantum mechanics by employing the techniques of second quantization. The time dependence is now contained solely within the dynamical variables. By solving Heisenberg operator equations of motion for the fermion and boson operators, the Maxwell field operators may be obtained in series of powers of the electronic charge. This is in contrast to the theory presented in Chapter 1, for which the Schrödinger picture is advantageous, with the time dependence occurring exclusively within the states of the system. Either formulation may be used to calculate expectation values for quantum mechanical observables for processes involving the interaction of one or more sources of external radiation with a single atom or molecule—the optical phenomena mentioned above, although not covered in the present work. Both theoretical schemes are, however, applied to the calculation of a variety of intermolecular interactions that are especially effective at long range and for which molecular quantum electrodynamics is best suited as it automatically accounts for the finite speed of propagation of electromagnetic signals. Hence, effort is made to elucidate contributions to the intermolecular interaction energy for which quantum electrodynamical predictions are found to differ from results obtained using a semiclassical prescription. This begins in Chapter 4, which is devoted to resonant transfer of energy. Chapter 5 exclusively deals with dispersion forces between pairs of molecules. Energy shifts between ground and/or electronically excited species are evaluated using three different physical viewpoints and calculational schemes. These include diagrammatic time-dependent perturbation theory, a response formalism using Maxwell fields, and a method in which molecules couple via fluctuations in their charge distribution. Discriminatory effects are also studied in excitation energy transfer and van der Waals dispersion occurring between optically active species. The three different computational approaches are then applied to the calculation of many-body forces in Chapter 6. Explicit results are obtained for retardation corrected three-, four-, and N-body dispersion potentials. In Chapter 7, the modification of intermolecular energy shifts by external radiation is investigated. These include changes to pairs of molecules coupled via exchange of one and two virtual photons. Interactions of this type form the basis of optical binding forces and, although small in magnitude, have recently come within the range of experimental detection. Interesting field-induced interaction energies between chiral molecules are also calculated.

While the formalism presented in the first two chapters and its application in the last four chapters may at first sight appear imposing, it is intended that at the very least Chapter 1 will provide a detailed and self-contained account of the theory of molecular quantum electrodynamics derived from the first principles. Mastering this chapter will enable readers to reproduce perturbation theory calculations of intermolecular interactions given in later chapters and their extension to more complex problems. Background material required for a thorough understanding of the content covered is a first course in quantum theory in which the postulates are presented along with standard elementary examples leading to examination of atomic and molecular systems and the common methods of solution by techniques of approximation. This is well within the scope of advanced undergraduate students of chemistry and physics. Advantageous is a prior exposure to mathematical methods. The same could also be said of classical mechanics and classical electrodynamics, which along with quantum mechanics form the foundations of quantum electrodynamics. For this purpose, the Lagrangian function and Hamiltonian formulation are introduced in Chapter 1 along with Maxwell's equations. A similar approach is adopted in the presentation of coupled wavefields in Chapter 2. Understanding of this chapter will allow the response theory calculations of intermolecular forces to be followed easily.

It is hoped that the book will prove useful to both the theorists seeking to acquaint themselves with new methods and experimentalists requiring knowledge of the latest forms of intermolecular potential energy functions. In addition, the book will be accessible to students of the subject as well as to researchers expert in the discipline.

The study of intermolecular forces has been a long and ongoing endeavor and the results obtained over the years have in fact impacted all areas of science. A variety of differing approaches within the framework of molecular quantum electrodynamics will be used in this book to compute the interaction energy of two or more bodies as a function of their separation and orientation. This theory is preferred for treating interactions between atoms and molecules, both for its rigor and for the chemical and physical insights it affords. The emphasis, therefore, is on developing the theory and applying it to fundamental intermolecular processes. The presentation given is general enough to hold for a wide range of situations, yet applicable to specific systems of interest. Despite three differing physical viewpoints being proffered in this work, there remain others within the realm of

Coulomb gauge quantum electrodynamics that can be deployed to explore intermolecular forces. Perhaps the best known among these alternatives is the dressed atom approach. Little mention is made of this methodology, as being superbly dealt with in the treatise by Compagno, Passante, and Persico (*Atom-Field Interactions and Dressed Atoms*, Cambridge University Press, Cambridge, 1995). For this and other reasons, the choice of topics covered has been the authors alone. Hence, there is no discussion of medium effects and methods to treat this important aspect. Likewise, no details are given concerning measurement of intermolecular forces or interactions occurring between atoms or molecules with surfaces. These and other more general aspects of intermolecular forces may be found in Margenau and Kestner's monograph, *Theory of Intermolecular Forces* (Pergamon, Oxford, 1969) and the book by Maitland, Rigby, Smith, and Wakeham (*Intermolecular Forces*, Oxford University Press, Clarendon, 1981). Discussion of experiments is limited to those attempting to directly measure different manifestations of the so-called Casimir effects, including van der Waals dispersion forces. Also excluded are recent proposals for enhancing and suppressing transfer of excitation by external radiation in the so-called laser-assisted resonant energy transfer (LARET) processes. Likewise, no details are given of the dynamical Casimir effect.

No attempt has been made to provide a comprehensive list of references. Only primary texts and key articles are cited. SI units have been used throughout. All errors lay solely with the author.

Words of thanks are due to many. First and foremost, the author owes a great debt of gratitude to "Thiru" Thirunamachandran, who unfortunately did not live to read this book. His influence on the author as teacher, research guide, and friend are all too evident, as is his imprint on this work. The author benefited greatly from numerous discussions with T. Thirunamachandran relating to molecular quantum electrodynamics, as well as with David P. Craig and the late Edwin A. Power. Thanks are also due to David L. Andrews who read Section 1.10 and offered useful insights into state sequence diagrams. Bridget W. Alligood is kindly thanked for skillfully drawing all of the figures contained in this book and for critically reading Chapter 1.

Finally, thanks go to Anita Lekhwani, Senior Acquisitions Editor at John Wiley & Sons, Inc., for her enthusiasm in supporting this project, and to Rebekah Amos, Editorial Assistant at this publishing house, for her prompt and friendly assistance during the writing and production process.

Akbar Salam

Winston-Salem, North Carolina
August 2009

CHAPTER 1

MOLECULAR QUANTUM ELECTRODYNAMICS: BASIC THEORY

One finds then that the Hamiltonian for the interaction of the field with an atom is of the same form as that for the interaction of an assembly of light-quanta with the atom. There is thus a complete formal reconciliation between the wave and light-quantum points of view.

—P. A. M. Dirac, *Proc. R. Soc. Lond. A* **114**, 710 (1927).

1.1 BACKGROUND

Quantum electrodynamics (QED) (Schwinger, 1958) is the physical theory that describes the interaction of electromagnetic radiation with matter. Its characteristic feature is that the radiation field, as well as the material system, is subject to the postulates of quantum mechanics. Therefore, the theoretical framework consists of a single, closed dynamical system comprising both matter and electromagnetic radiation in mutual interaction within which energy is conserved. This is unlike the situation in the so-called semiclassical theory, the historical precursor to QED. In the former construct, only matter obeys quantum mechanical principles, while the radiation field is considered as a prescribed, external perturbation on the

Molecular Quantum Electrodynamics, by Akbar Salam

system and does not form an integral part of it. Even though use of the semiclassical formalism remains widespread, due largely to its physical and computational simplicity relative to QED, its inherent deficiencies lead to critical shortcomings and ultimately limit its scope of availability. This is especially the case in the treatment of electromagnetic fields interacting with atoms and molecules, where continuing progress in the generation of coherent light sources has necessitated a fully quantum mechanical approach to calculating and explaining a variety of optical phenomena (Mukamel, 1995; Andrews and Allcock, 2002). In this regard, QED is the most successful physical theory to date (Feynman, 1985). This statement is justified on two counts. First, development of the QED formalism has provided a rigorous foundation for the understanding of electron–photon interactions at the most fundamental level currently known. Phenomena cover a vast range of length scales, manifested by particles varying in size from the gigameter down to the attometer. Second, and perhaps more convincing, is the ability of the theory to yield numerical values of measurable properties and the unprecedented agreement with experiment in those cases where comparison is possible.

A key step that led to the formulation of QED was the recognition that the mechanical vibrations of a system with infinitely many degrees of freedom could be represented by quantizing a collection of noninteracting harmonic oscillators (Born et al., 1926). This insight prompted Dirac (1927) to quantize the electromagnetic field and to calculate quantum mechanical probabilities for the absorption, stimulated emission, and spontaneous emission of light by atoms. Subsequent advances carried out by many workers, in particular, the significant contributions of Feynman, Schwinger, Tomonaga, and Dyson, resulted in a formulation of QED that satisfied all of the requirements stipulated by the special theory of relativity, which was ultimately explicated in both the particle and field theoretic points of view (Schweber, 1994). This last aspect, for instance, finally enabled the duality of the wave and particle descriptions of radiation and matter to be rationalized on the basis of a single theoretical scheme. Interacting quantum mechanical fermionic matter and bosonic electromagnetic fields are therefore entirely equivalent to a many-body representation of a system of material particles—electrons—interacting with quantized particles of light—the photons.

Early application of both the nonrelativistic and fully covariant versions of the theory was made to outstanding problems. These included calculations of spontaneous decay rates from atoms in electronically excited states, the Lamb shift, and the anomalous magnetic moment of the electron, to select but three historically significant examples. With continuing advances occurring in computational power and experimental procedure during the intervening years, convergence between theory and experiment has been ever closer. More accurate calculations and improvements in measurement capability have yielded for the electronic g-factor, for instance, values in units of Bohr magnetons of $g/2$ (experiment) = 1.00115965218073(28) (Hanneke et al., 2008) and $g/2$ (theory) = 1.00115965218279 (Aoyama et al., 2007). In the case of the Lamb shift in atomic hydrogen, experimental and calculated values for the splitting between the $2S_{1/2}$ and $2P_{1/2}$ levels are 1,057,839(12) kHz (Hagley and Pipkin, 1994) and 1,057,838(6) kHz (Pachucki, 1994), respectively.

While these and other achievements are indeed remarkably impressive, there remain difficulties in the underlying foundations of the theory. Chief among them is the renormalization procedure, without whose aid no finite quantities may be calculated but which even today lacks rigorous mathematical justification. This problem continues to beset other field theories of modern physics, of which QED is but one example. Another limitation, which also applies in general to other field theories, is the approximate nature of solutions generated when systems interact. A common method of solution is a perturbative expansion in series of powers of some appropriate coupling constant, with no *a priori* guarantee of convergence of successively higher order terms, or whether summation of the infinite series is indeed possible. In QED, for example, the eigenstates of one or more charged particles in isolation are taken to be known, and the microscopic Maxwell's equations in free space can be solved and appropriately quantized. By forming a product state, the wavefunctions of this separable system are then used as base states for a perturbation theory solution in series of powers of the electronic charge or the fine structure constant.

Although QED correctly treats the coupling of radiation and matter at high energies, where fermionic pair particle creation and destruction events occur concomitantly with changes in photon number, the emphasis in the presentation to follow will be on conservation of the number of charged particles, which may exchange energy directly or indirectly with the radiation field. Conversely, as there is no conservation of photon number, any integer quantity of real or virtual photons may be created and annihilated during the course of a particular process. Hence, the formalism to be developed and applied will be suitable for systems composed of charged particles such as bound electrons in atoms and molecules possessing energies much less than mc^2, m being the mass of the aggregate and c the speed of light. Since the mass is assumed to be a constant, the system of interest is slow moving, with low velocity v, and automatically satisfies the condition $v \ll c$. When this limiting procedure is applied to covariant

QED, the result is a nonrelativistic version of the theory, which is more commonly known as molecular QED (Power, 1964; Healy, 1982; Craig and Thirunamachandran, 1998a). Its construction from first principles and its application to a variety of long-range intermolecular interactions form the subject of this book.

In addition to providing valuable insight and expressions for forces between particles, the most celebrated being the formula for the retarded van der Waals dispersion potential between a pair of neutral molecules in the ground state, the theory of molecular QED has been employed with considerable success to radiation–molecule interactions. Examples include single- and multiphoton absorption, emission and scattering of light, optical activity and chiroptical spectroscopy, and nonlinear and quantum optical phenomena. Specific processes studied involving the interaction of radiation with one center include calculation of the circular dichroism rate, the angle of rotation of plane polarized light as it traverses a chiral medium, Rayleigh and Raman scattering of linear and circularly polarized radiation and their hyperanalogues, second, third, and higher harmonic generation, four-, five-, and six-wave mixing, and laser-induced optical activity.

Since intermolecular interactions are mediated by electromagnetic forces, coupling of radiation with matter, as well as between two or more particles, may be treated correctly and consistently within the framework provided by the nonrelativistic quantum theory of electron–photon interaction or molecular QED. This is most commonly delineated for the interaction of a charged particle with electromagnetic radiation; it begins with the classical Lagrangian function and ends with the construction of a quantum mechanical Hamiltonian operator that is arrived at via the familiar canonical quantization procedure. Details are presented in this chapter.

More fundamental, though less well known—despite much progress being made in the past 25 years or so—is the field theoretic formulation of molecular QED (Salam, 2008), involving the interaction of second quantized matter and radiation fields, which is the subject of Chapter 2. For a variety of applications, this approach offers a number of advantages over the conventional method in which only the radiation field is second quantized. For instance, properties directly related to electron and photon fields, such as the Maxwell fields in the vicinity of a source of charge, its energy density, and rates of flow of electromagnetic energy, may be evaluated. Moreover, the electromagnetic fields are subsequently employed to calculate energy shifts between molecules using a version of response theory.

The interactions occurring between atomic and molecular systems, especially those operating at large separation distances, form the focus of the remainder of the work. In Chapter 3, the standard presentation of intermolecular forces originating from classical electrostatics is given. The pair potential is first divided into short- and long-range regions, with the exchange–repulsion energy being the major contribution to the interaction energy in the former distance regime. At large separations, the charge distribution between pairs of molecules is expanded in an electric multipole series and quantum mechanical perturbation theory is used to extract the electrostatic, induction, and dispersion terms within the formalism of semiclassical radiation theory in which no account is taken of the photonic character of electromagnetic radiation. Applications of the quantum electrodynamical theory to intermolecular interactions are then given in the chapters to follow. In Chapter 4, the theory of resonance energy transfer is given, a fundamental process taking place in microscopic systems but prototypical in that its interpretation as due to the exchange of a single virtual photon between the pair serves as a basis for the study of other molecular interactions using the formalism of molecular QED. Both perturbation and response theory calculations are presented, followed by their application to the evaluation of the transfer rate between two chiral species. Chapter 5 is devoted to van der Waals dispersion forces. Three different physical viewpoints and calculational techniques are detailed for the computation of the energy shift between two neutral polarizable molecules in the ground state. In addition to perturbation and response theories, the induced multipole moment approach is introduced and shown to simplify calculations considerably. Results are also obtained for electronically excited molecules undergoing dispersive coupling, and the functional form of the discriminatory potential between two optically active molecules is derived. Chapter 6 covers nonadditive and many-body forces. Attention is focused on the effects of retardation on energy transfer and dispersion interactions taking place among three or more particles. In Chapter 7, the effect of an external electromagnetic field in modifying the molecular pair interaction energy is treated. Adoption of the approach whereby molecules couple with each other and to the incident laser via the moments induced by the radiation field is shown to be a more efficient calculational method than the diagrammatic perturbation theory computation. Changes in energy shift induced by an applied field and dependent on the handedness of individual bodies are also studied.

1.2 QUANTUM DESCRIPTION OF MATTER

The continued failure of the application of the laws of classical mechanics to microscopic particles eventually led to the formulation of a new dynamics—quantum mechanics (Dirac, 1958). At its foundation, however, remain a number of key concepts and ideas from classical theory (Goldstein, 1960). One of these fundamental constructs is the Lagrangian function, and another is the physical variational principle and the versatility associated with it. Taken together, they yield the Euler–Lagrange equations of motion and provide an alternative formulation of classical mechanics to Newton's laws of motion. The former may be used to solve any problem within the classical domain and ultimately to provide a rigorous means of quantizing the specific system of interest. As long as a judicious choice of coordinates is made, often the most difficult task at hand, the ensuing equations to be solved are frequently simpler than those obtained by direct application of Newton's laws, and yet contain all of the essential physics. This freedom in the selection of the specific coordinate representation lends itself to the formulation of the classical Lagrangian L for a system of particles in terms of generalized coordinates and velocities q and $\dot{q}$; the Lagrangian is a function of these two variables as well as of the time t. If the Lagrangian does not depend explicitly on the time, then it is defined to be the difference of the kinetic energy T and the potential energy V; the energy of the system is therefore conserved. For a system in which the potential energy is a function of the position only, the Lagrangian has identical definition, namely, $L = T - V$. The equations of motion follow after invoking Hamilton's principle, namely, that of all of the allowed paths the system may pursue between initial and final times t_i and t_f, the actual one taken in configuration space is that for which the variation of the time integral of the Lagrangian, also called the variation of the action S, is an extremum. By application of standard calculus of variations, the Euler–Lagrange equations of motion for a system of particles ξ, with N degrees of freedom, are found to be

$$\frac{\mathrm{d}}{\mathrm{d}t}\left(\frac{\partial L}{\partial \dot{q}_\xi}\right) - \frac{\partial L}{\partial q_\xi} = 0, \quad \xi = 1, 2, \ldots, N. \tag{1.2.1}$$

From the form of the Lagrangian, it is evident that motion in classical mechanics is reversible. Replacing t by $-t$ leaves the Lagrangian as well as the equations of motion unaltered.

To facilitate the transition to quantum mechanics, in which the positions and momenta of the system of particles are canonical, it is convenient to define the Hamiltonian H as

$$H = \sum_{\xi=1}^{N} p_\xi \dot{q}_\xi - L. \tag{1.2.2}$$

This quantity is a function of the generalized coordinates, their canonically conjugate momenta, and time. The momenta are found from

$$p_\xi = \frac{\partial L}{\partial \dot{q}_\xi}, \quad \xi = 1, 2, \ldots, N. \tag{1.2.3}$$

When the Hamiltonian is time independent, it is equal to the sum of the kinetic and potential energy, $H = T + V$. Evaluating the total differential of (1.2.2) and noting from equations (1.2.1) and (1.2.3) that

$$\dot{p}_\xi = \frac{\partial L}{\partial q_\xi} \tag{1.2.4}$$

yields Hamilton's canonical equations

$$\dot{q}_\xi = \frac{\partial H}{\partial p_\xi}; \quad \dot{p}_\xi = -\frac{\partial H}{\partial q_\xi}, \quad \xi = 1, 2, \ldots, N, \tag{1.2.5}$$

and

$$\frac{\partial H}{\partial t} = -\frac{\partial L}{\partial t}, \tag{1.2.6}$$

which are now the equations of motion. A consequence of the resulting dynamics is that $2N$ first-order equations (1.2.5) have to be solved, rather than N second-order equations (1.2.1). After writing the classical Hamiltonian function in terms of the canonically conjugate dynamical variables, the quantum mechanical form of the Hamiltonian operator is obtained straightforwardly by promoting the classical variables to quantum operators, the latter obeying the canonical commutation relations for particles ξ and ξ',

$$[\vec{q}_\xi, \vec{q}_{\xi'}] = 0; \quad [\vec{p}_\xi, \vec{p}_{\xi'}] = 0; \quad [\vec{q}_\xi, \vec{p}_{\xi'}] = i\hbar\delta_{\xi\xi'}. \tag{1.2.7}$$

By way of illustration of the development above, consider a system of particles ξ with charges e_ξ, masses m_ξ, and position and velocity vectors $\vec{q}_\xi$ and $\dot{\vec{q}}_\xi$, whose classical Lagrangian function is (Landau and Lifshitz, 1972)

$$L(\vec{q}, \dot{\vec{q}}) = \frac{1}{2}\sum_\xi m_\xi \dot{\vec{q}}_\xi^{\,2} - \frac{1}{4\pi\varepsilon_0}\sum_{\xi<\xi'} \frac{e_\xi e_{\xi'}}{|\vec{q}_\xi - \vec{q}_{\xi'}|}. \tag{1.2.8}$$

This leads to the quantum mechanical Hamiltonian operator

$$H(\vec{q},\vec{p}) = \sum_{\xi} \frac{\vec{p}_{\xi}^{\,2}}{2m_{\xi}} + \frac{1}{4\pi\varepsilon_0} \sum_{\xi<\xi'} \frac{e_{\xi} e_{\xi'}}{|\vec{q}_{\xi} - \vec{q}_{\xi'}|}. \tag{1.2.9}$$

It is easily verified that the Lagrangian (1.2.8) gives rise to the correct equations of motion: applying (1.2.1) produces Newton's second law equation for particle ξ,

$$m_{\xi} \ddot{\vec{q}}_{\xi} = \frac{1}{4\pi\varepsilon_0} \sum_{\xi<\xi'} \frac{e_{\xi} e_{\xi'}}{|\vec{q}_{\xi} - \vec{q}_{\xi'}|^2}, \tag{1.2.10}$$

whose right-hand side is recognizable as the generalized Coulomb force, obtained from the potential energy term in the Lagrangian via $\vec{F} = -\partial V(\vec{q})/\partial \vec{q}$.

1.3 ELECTRODYNAMICS AND MAXWELL EQUATIONS

The unification of electricity and magnetism with classical optics occurred with the formulation of Maxwell's equations (Jackson, 1963)—the basic laws underlying the behavior of electromagnetic radiation. As a consequence, light was understood to be an electromagnetic wave in which radiation of all frequencies could, in principle, be generated. Maxwell's quartet of equations is expressed as

$$\vec{\nabla} \cdot \vec{D} = \rho, \tag{1.3.1}$$

$$\vec{\nabla} \cdot \vec{B} = 0, \tag{1.3.2}$$

$$\vec{\nabla} \times \vec{E} + \frac{\partial \vec{B}}{\partial t} = 0, \tag{1.3.3}$$

$$\vec{\nabla} \times \vec{H} = \frac{\partial \vec{D}}{\partial t} + \vec{J}. \tag{1.3.4}$$

These fundamental equations completely determine the electromagnetic field, and the electrodynamic nature of such radiation is clearly evident from them. The fields $\vec{E}$ and $\vec{B}$ shall be termed the electric and magnetic fields, respectively, even though the latter is more properly called the magnetic flux density or the magnetic induction. The auxiliary fields $\vec{D}$

and $\vec{H}$ are designated the electric displacement and magnetic field, respectively. No confusion shall result from the use of the descriptor "magnetic field," as the appropriate symbol shall be used for the quantity concerned, either $\vec{B}$ or $\vec{H}$. The $\vec{E}$ and $\vec{B}$ fields are fundamental in that they propagate in regions of space that contain no sources of charge. Further, if the charge density ρ and the current density $\vec{J}$ account for all charged entities, then $\vec{E}$ and $\vec{B}$ describe the radiation field in its totality. On the other hand, including the contributions to ρ and $\vec{J}$ of the elementary charges and their currents, which are manifested in the form of polarization fields and currents, necessitates the introduction of the two auxiliary fields $\vec{D}$ and $\vec{H}$. The bound charged particles are viewed as forming a medium that contains the fields. These in turn describe the response of the material system to the applied fields via the electric polarization and magnetization fields. The connection between the derived and fundamental fields is arrived at through constitutive relations and the introduction of the electric permittivity ε and the magnetic permeability μ to describe the properties of the medium.

Although the interaction of radiation and matter is to be applied to bound systems moving at a very small fraction of the velocity of light, the equations of electrodynamics are themselves invariant in form under Lorentz transformations and are compatible with Einstein's relativity theories. When Maxwell's equations are combined with Newton's second law of motion and the Lorentz force equation, a complete description of the nonrelativistic classical dynamics of charged particles interacting with electromagnetic fields results.

For the subsequent development of the quantum theory of electron–photon interaction, it is advantageous to work with Maxwell's equations in microscopic form rather than employ relations (1.3.1)–(1.3.4) that are applicable when the distribution of charge is taken to be continuous. In place of the macroscopic Maxwell equations, their microscopic equivalents can be expressed solely in terms of the microscopic forms of the electric and magnetic field vectors $\vec{e}$ and $\vec{b}$ and the sources of charge and are given by

$$\vec{\nabla} \cdot \vec{e} = \rho/\varepsilon_0, \tag{1.3.5}$$

$$\vec{\nabla} \cdot \vec{b} = 0, \tag{1.3.6}$$

$$\vec{\nabla} \times \vec{e} + \frac{\partial \vec{b}}{\partial t} = 0, \tag{1.3.7}$$

$$\vec{\nabla} \times \vec{b} = \frac{1}{c^2}\frac{\partial \vec{e}}{\partial t} + \frac{1}{\varepsilon_0 c^2}\vec{j}. \tag{1.3.8}$$

To facilitate the microscopic treatment of matter, a discrete particle description is adopted for a collection of charged particles α possessing electrical charge e_α, situated at $\vec{q}_\alpha$ and moving with velocity $\dot{\vec{q}}_\alpha$, for which the charge and current density are defined to be

$$\rho(\vec{r}) = \sum_\alpha e_\alpha \delta(\vec{r}-\vec{q}_\alpha) \tag{1.3.9}$$

and

$$\vec{j}(\vec{r}) = \sum_\alpha e_\alpha \dot{\vec{q}}_\alpha \delta(\vec{r}-\vec{q}_\alpha), \tag{1.3.10}$$

where $\delta(\vec{r})$ is the Dirac delta function, which is strongly localized at the origin of the charge. By carrying out an average of the microscopic field over the molecular volume, the macroscopic Maxwell equations can be obtained from the microscopic Maxwell–Lorentz equations (1.3.5)–(1.3.8). In Section 1.2, it was remarked that forward and reverse motions are identical in classical mechanics. The same is true for the electromagnetic field in relativity theory, with the additional requirement that the sign of the magnetic field is reversed as well as $t \to -t$. It is easy to see that the equations of motion of a charged particle in a field are unchanged on transforming $t \to -t$, $\vec{e} \to \vec{e}$, and $\vec{b} \to -\vec{b}$ in the Lorentz force expression

$$\frac{d\vec{p}_\alpha}{dt} = -e_\alpha(\vec{e} + \vec{v}_\alpha \times \vec{b}), \tag{1.3.11}$$

where $\vec{v}_\alpha = d\vec{q}_\alpha/dt$ is the velocity and $\vec{p}_\alpha = m_\alpha \vec{v}_\alpha$ is the momentum of particle α.

The coupled first-order partial differential equations of Maxwell can be solved for the fields $\vec{e}$ and $\vec{b}$ for a variety of simple cases in electromagnetic theory. However, for many other situations and for the eventual quantization of the radiation field via the canonical quantization scheme, it is convenient to introduce two electromagnetic potentials and to rewrite Maxwell's equations in terms of them. One is the scalar potential ϕ and the other is the vector potential $\vec{a}$. The definition of the latter readily follows from the second Maxwell equation (1.3.6),

$$\vec{b} = \vec{\nabla} \times \vec{a}, \tag{1.3.12}$$

on noting that the divergence of the curl of a vector field vanishes. Inserting (1.3.12) into the third Maxwell equation (1.3.7) yields

$$\vec{\nabla} \times \left(\vec{e} + \frac{\partial \vec{a}}{\partial t} \right) = 0, \tag{1.3.13}$$

so that the factor within parentheses can be defined in terms of the gradient of a scalar function, in this case the scalar potential, since the curl of the gradient of a scalar field is zero,

$$-\vec{\nabla}\phi = \vec{e} + \frac{\partial \vec{a}}{\partial t}. \tag{1.3.14}$$

Substituting relation (1.3.14) into the first inhomogeneous microscopic Maxwell equation (1.3.5) produces

$$\vec{\nabla}^2 \phi + \frac{\partial}{\partial t}(\vec{\nabla} \cdot \vec{a}) = -\frac{\rho}{\varepsilon_0}, \tag{1.3.15}$$

while using definitions (1.3.12) and (1.3.14) in the last Maxwell equation (1.3.8) gives

$$\vec{\nabla}^2 \vec{a} - \frac{1}{c^2}\frac{\partial^2 \vec{a}}{\partial t^2} - \vec{\nabla}\left(\vec{\nabla} \cdot \vec{a} + \frac{1}{c^2}\frac{\partial \phi}{\partial t} \right) = -\frac{1}{\varepsilon_0 c^2}\vec{j} \tag{1.3.16}$$

on using the vector identity

$$\vec{\nabla} \times (\vec{\nabla} \times \vec{a}) = -\vec{\nabla}^2 \vec{a} + \vec{\nabla}(\vec{\nabla} \cdot \vec{a}). \tag{1.3.17}$$

Maxwell's equations have now been reduced to two coupled equations instead of four, with the potentials related directly to the sources. The equations (1.3.15) and (1.3.16) can be further simplified into two separate inhomogeneous wave equations, one dependent on ϕ only and the other on $\vec{a}$ only. This may be achieved by taking advantage of the gauge freedom associated with the potentials. From relation (1.3.12), it can be seen that $\vec{a}$ is undetermined to the extent that the gradient of a scalar function of the position and time, f, can be added to it,

$$\vec{a} \rightarrow \vec{a}' = \vec{a} + \vec{\nabla} f, \tag{1.3.18}$$

a transformation that leaves $\vec{b}$ invariant. Substituting (1.3.18) into (1.3.14) enables the form of the transformation that must simultaneously be made to the scalar potential such that $\vec{e}$ is unchanged to be derived:

$$-\vec{e} = \frac{\partial \vec{a}}{\partial t} + \vec{\nabla}\phi = \frac{\partial}{\partial t}(\vec{a} + \vec{\nabla} f) + \vec{\nabla}\phi', \tag{1.3.19}$$

from which

$$\vec{\nabla}\left(\phi'-\phi+\frac{\partial f}{\partial t}\right)=0 \tag{1.3.20}$$

or

$$\phi\rightarrow\phi'=\phi-\frac{\partial f}{\partial t}. \tag{1.3.21}$$

The scalar potential is therefore determined to within the time derivative of the same function *f*. The two relations (1.3.18) and (1.3.21) constitute the gauge transformation. From them, a set of potentials $(\vec{a},\phi)$ can always be chosen such that the Lorentz condition (1.3.22) is satisfied:

$$\vec{\nabla}\cdot\vec{a}+\frac{1}{c^2}\frac{\partial\phi}{\partial t}=0, \tag{1.3.22}$$

which when inserted into (1.3.15) and (1.3.16) results in the wave equations

$$\left(\vec{\nabla}^2-\frac{1}{c^2}\frac{\partial^2}{\partial t^2}\right)\phi=-\frac{\rho}{\varepsilon_0} \tag{1.3.23}$$

and

$$\left(\vec{\nabla}^2-\frac{1}{c^2}\frac{\partial^2}{\partial t^2}\right)\vec{a}=-\frac{1}{\varepsilon_0 c^2}\vec{j}. \tag{1.3.24}$$

The most convenient choice of gauge, from the point of view of nonrelativistic theory, is the one in which the vector potential is solenoidal, that is, $\vec{\nabla}\cdot\vec{a}=0$, also known as the Coulomb, radiation, or transverse gauge. In this gauge, ϕ is seen from equations (1.3.22) and (1.3.23) to obey Poisson's equation

$$\vec{\nabla}^2\phi=-\frac{\rho}{\varepsilon_0}, \tag{1.3.25}$$

with solution

$$\phi(\vec{r},t)=\frac{1}{4\pi\varepsilon_0}\int\frac{\rho(\vec{r}\,',t)}{|\vec{r}-\vec{r}\,'|}\mathrm{d}^3\vec{r}\,', \tag{1.3.26}$$

which represents the instantaneous Coulomb potential due to the charge density and from which this gauge takes its primary name. Fixing the gauge in equation (1.3.16) yields the inhomogeneous wave equation satisfied by the vector potential,

$$\left(\vec{\nabla}^2 - \frac{1}{c^2}\frac{\partial^2}{\partial t^2}\right)\vec{a} - \frac{1}{c^2}\vec{\nabla}\left(\frac{\partial \phi}{\partial t}\right) = -\frac{1}{\varepsilon_0 c^2}\vec{j}. \tag{1.3.27}$$

A further advance is made through the explicit decomposition of the vector fields $\vec{e}$ and $\vec{b}$ into their parallel (||) and perpendicular ($\perp$) components; this is known as Helmholtz's theorem, which holds for any vector field and gives rise to irrotational and solenoidal vectors, respectively. From (1.3.6), $\vec{b}$ is purely transverse and (1.3.5) and (1.3.7) become

$$\vec{\nabla}\cdot\vec{e}^{\,||} = \frac{\rho}{\varepsilon_0} \tag{1.3.28}$$

and

$$\vec{\nabla}\times\vec{e}^{\,\perp} = -\frac{\partial \vec{b}}{\partial t}, \tag{1.3.29}$$

while the fourth microscopic Maxwell equation (1.3.8) separates into

$$0 = \frac{1}{c^2}\frac{\partial \vec{e}^{\,||}}{\partial t} + \frac{1}{\varepsilon_0 c^2}\vec{j}^{\,||} \tag{1.3.30}$$

and

$$\vec{\nabla}\times\vec{b} = \frac{1}{c^2}\frac{\partial \vec{e}^{\,\perp}}{\partial t} + \frac{1}{\varepsilon_0 c^2}\vec{j}^{\,\perp}. \tag{1.3.31}$$

The equation of continuity,

$$\vec{\nabla}\cdot\vec{j}^{\,||} + \frac{\partial \rho}{\partial t} = 0, \tag{1.3.32}$$

follows immediately on taking the divergence of (1.3.30) and using (1.3.28). Similarly, equation (1.3.14) divides as

$$\vec{e}^{\,||} = -\vec{\nabla}\phi \tag{1.3.33}$$

and

$$\vec{e}^{\,\perp} = -\frac{\partial \vec{a}}{\partial t} \tag{1.3.34}$$

in the Coulomb gauge, enabling the inhomogeneous wave equation for the vector potential (1.3.27) to be expressed exclusively in terms of transverse variables,

$$\left(\vec{\nabla}^2 - \frac{1}{c^2}\frac{\partial^2}{\partial t^2}\right)\vec{a} = -\frac{1}{\varepsilon_0 c^2}\vec{j}^{\perp}, \qquad (1.3.35)$$

with ϕ continuing to satisfy Poisson's equation (1.3.25). The solution of the equation (1.3.35) for the vector potential is given by

$$\vec{a}(\vec{r},t) = \frac{1}{4\pi\varepsilon_0 c^2}\int \frac{\vec{j}^{\perp}(\vec{r}',t-|\vec{r}-\vec{r}'|/c)}{|\vec{r}-\vec{r}'|}\mathrm{d}^3\vec{r}', \qquad (1.3.36)$$

which appears to be retarded, but is in fact not so. This is because its source is the transverse rather than the total current, the former being nonlocal, resulting in $\vec{a}^{\perp}$ having identical characteristics also. Causality is recovered in the Coulomb gauge by including both transverse and longitudinal components, thereby ensuring that all static contributions cancel one another. No such difficulty arises in the Lorentz gauge (1.3.22), the solutions to the wave equations (1.3.23) and (1.3.24) being properly retarded, recognizing that the total current appears as the source in the equation for $\vec{a}$,

$$\phi(\vec{r},t) = \frac{1}{4\pi\varepsilon_0}\int \frac{\rho(\vec{r}',t-|\vec{r}-\vec{r}'|/c)}{|\vec{r}-\vec{r}'|}\mathrm{d}^3\vec{r}', \qquad (1.3.37)$$

$$\vec{a}(\vec{r},t) = \frac{1}{4\pi\varepsilon_0 c^2}\int \frac{\vec{j}(\vec{r}',t-|\vec{r}-\vec{r}'|/c)}{|\vec{r}-\vec{r}'|}\mathrm{d}^3\vec{r}'. \qquad (1.3.38)$$

It is worth noting that $\vec{a}^{\perp}$ is gauge invariant, since from the transformation (1.3.18) only the longitudinal component of $\vec{a}$ can change. Consequently, from (1.3.34), it is seen that $\vec{e}^{\perp} = -\dot{\vec{a}}^{\perp}$ in all gauges, and the effect of a gauge transformation is to change the contributions from $\vec{a}$ and ϕ to $\vec{e}^{\parallel}$.

1.4 QUANTIZATION OF THE FREE ELECTROMAGNETIC FIELD

In Section 1.2, it was shown how the quantum mechanics of a system of particles is rigorously built up from a classical mechanics in canonical form. In this section, it is shown how the same principles may be applied and suitably adapted to the radiation field propagating *in vacuo*. Ultimately, this will lead to the quantum mechanical Hamiltonian for the electromagnetic field and its corresponding eigenvalues and eigenfunctions, the latter in a form convenient for its later adoption as a basis set in the perturbation theory solution to the interacting matter–radiation problem.

When there are no sources present, both the charge and current density are zero so that the microscopic Maxwell equations (1.3.5)–(1.3.8) applicable to the electromagnetic field in free space become

$$\vec{\nabla}\cdot\vec{e}=0, \tag{1.4.1}$$

$$\vec{\nabla}\cdot\vec{b}=0, \tag{1.4.2}$$

$$\vec{\nabla}\times\vec{e}+\frac{\partial\vec{b}}{\partial t}=0, \tag{1.4.3}$$

$$\vec{\nabla}\times\vec{b}-\frac{1}{c^2}\frac{\partial\vec{e}}{\partial t}=0. \tag{1.4.4}$$

Solutions are easily found for the fields $\vec{e}$ and $\vec{b}$, which describe electromagnetic waves in a vacuum, as well as for $\vec{a}$. Continuing the development in the Coulomb gauge, clearly from equation (1.4.1), $\vec{e}$ is purely transverse, which from (1.3.14) means that ϕ can be taken to vanish so that $\vec{e}=-\dot{\vec{a}}$. Substituting this relation and (1.3.12) into (1.4.4) and using identity (1.3.17) leads to d'Alembert's equation for the vector potential,

$$\left(\vec{\nabla}^2-\frac{1}{c^2}\frac{\partial^2}{\partial t^2}\right)\vec{a}=0. \tag{1.4.5}$$

$\vec{e}$ and $\vec{b}$ satisfy identical wave equations. By taking the curl of equation (1.4.3) and substituting for $\vec{\nabla}\times\vec{b}$ from (1.4.4) yields the wave equation for $\vec{e}$, and carrying out a similar procedure on (1.4.4) first, results in the equation for $\vec{b}$. One form of solution to the wave equation for each of the three fields is in terms of plane waves,

$$\vec{a}=\vec{a}_0\,\mathrm{e}^{i\vec{k}\cdot\vec{r}-i\omega t}, \tag{1.4.6}$$

$$\vec{e}=e_0\,\vec{e}_0\,\mathrm{e}^{i\vec{k}\cdot\vec{r}-i\omega t}, \tag{1.4.7}$$

and

$$\vec{b}=b_0\,\vec{b}_0\,\mathrm{e}^{i\vec{k}\cdot\vec{r}-i\omega t}, \tag{1.4.8}$$

where in the last two solutions the pre-exponential factors, respectively, denote the scalar amplitude and the polarization vector of the respective quantity, with $\vec{e}_0 = ick\vec{a}_0$ on using $\vec{e} = -\dot{\vec{a}}$. The direction of propagation of the electromagnetic wave is described by the wavevector $\vec{k}$, whose magnitude is obtained after insertion of the appropriate solution into the wave equation from which it is found. Thus, $|\vec{k}| = k = \omega/c$, where ω is the circular frequency. The relation between the amplitudes of the two electromagnetic fields is found on inserting (1.4.7) and (1.4.8) into (1.4.3) and (1.4.4), producing

$$\hat{k} \times \vec{e}_0 = c\vec{b}_0 \tag{1.4.9}$$

and

$$\hat{k} \times \vec{b}_0 = -\frac{1}{c}\vec{e}_0, \tag{1.4.10}$$

respectively, where the circumflex designates a unit vector. From the last two relations, it may be inferred that the three vectors $\vec{e}_0$, $\vec{b}_0$, and $\hat{k}$ are mutually perpendicular and form a right-handed set, at the same time illustrating the transverse nature of electromagnetic waves. Transversality also follows on substituting the harmonic solutions (1.4.7) and (1.4.8) into the first two source-free Maxwell equations (1.4.1) and (1.4.2), respectively.

Because the respective polarization vectors in the plane wave solutions for $\vec{e}$ (1.4.7) and $\vec{b}$ (1.4.8) always point in the same direction, the waves are described as being linearly polarized. A wave with a more general state of polarization may be formed by combining two such independent waves. An example is the case of two different electric fields, each possessing a phase δ_1 and δ_2, whose superposition produces an elliptically polarized wave

$$\vec{e}(\vec{r}, t) = (e_1\vec{e}_1\,\mathrm{e}^{i\delta_1} + e_2\vec{e}_2\,\mathrm{e}^{i\delta_2})\mathrm{e}^{i\vec{k}\cdot\vec{r}-i\omega t}. \tag{1.4.11}$$

If both waves have identical amplitudes $e_1 = e_2 = e$, but a phase difference of $\pm\pi/2$, a circularly polarized wave results. For orthogonal unit vectors $\hat{e}_1$ and $\hat{e}_2$, the left- and right-handed circular polarizations are defined to be

$$\hat{e}^{(\mathrm{L/R})} = \frac{1}{\sqrt{2}}(\hat{e}_1 \pm i\hat{e}_2). \tag{1.4.12}$$

If, however, the phase difference between the two waves in (1.4.11) is $\delta_1 - \delta_2 = 0, \pm\pi$, then linear polarization results, where the tangent of the angle of the polarization vector with respect to $\vec{e}_1$ is given by the ratio of the amplitudes e_2 to e_1, that is, $\tan\theta = (e_2/e_1)$, and with a modulus $e = \sqrt{e_1^2 + e_2^2}$.

In free space, the wavevector $\vec{k}$ associated with the plane waves is unrestricted in value. To enumerate and normalize the allowed states in the quantum theory, the field is described in terms of modes as first carried out by Rayleigh (1900) and later by Jeans (1905). With $\vec{a}$, $\vec{e}$, and $\vec{b}$ satisfying vector Helmholtz equations of the type (1.4.5), a complete set of states is readily obtained by expansion in Cartesian coordinates over a parallelepiped of volume $V = L_xL_yL_z$, where L_x, L_y, and L_z are the dimensions along the three axes of the box, x, y, and z. This corresponds to a multiple Fourier series, which for the vector potential, when subject to the periodic boundary condition that $\vec{a}$ has identical values on opposite sides of the box, gives for the number of allowed modes

$$(n_x, n_y, n_z) = \frac{1}{(2\pi)^3}(k_x, k_y, k_z)L_xL_yL_z, \tag{1.4.13}$$

where n_x, n_y, and n_z are integers and k_i, $i = x, y, z$, are the wavevector components, with the two modes of the field characterized by the three values of n. Monochromatic solutions to the wave equation for the vector potential (1.4.5) are easily found via separation of variables $\vec{a}^{(\lambda)}(\vec{k}, \vec{r}, t) = \vec{a}^{(\lambda)}(\vec{k}, \vec{r})a(t)$, with the spatial part obeying the Helmholtz equation

$$\vec{\nabla}^2\vec{a}^{(\lambda)}(\vec{k}, \vec{r}) + k^2\vec{a}^{(\lambda)}(\vec{k}, \vec{r}) = 0 \tag{1.4.14}$$

and the temporal part satisfying

$$\frac{\partial^2 a(t)}{\partial t^2} + \omega^2 a(t) = 0, \tag{1.4.15}$$

with $\omega = ck$ the circular frequency. As a Fourier series expansion in the plane waves subject to (1.4.13), the vector potential is

$$\vec{a}(\vec{r}, t) = \sum_{\vec{k}, \lambda}[\vec{e}^{(\lambda)}(\vec{k})a^{(\lambda)}(\vec{k})e^{i\vec{k}\cdot\vec{r} - i\omega t} + \bar{\vec{e}}^{(\lambda)}(\vec{k})\bar{a}^{(\lambda)}(\vec{k})e^{-i\vec{k}\cdot\vec{r} + i\omega t}], \tag{1.4.16}$$

applicable to a wave propagating along $\hat{k}$ at speed c. In expansion (1.4.16), $\vec{e}^{(\lambda)}(\vec{k})$ is the unit electric polarization vector of mode $(\vec{k}, \lambda)$, λ being the index of polarization, and $a^{(\lambda)}(\vec{k})$ is the Fourier amplitude, with the overbar denoting the complex conjugate. From the Coulomb gauge condition, the transverse property of the modes is readily apparent, namely, $\hat{k} \cdot \vec{a}^{(\lambda)}(\vec{k}) = 0$. With $\vec{e}$ and $\vec{b}$ also transverse, their unit vectors are resolved parallel and perpendicular to the vector potential for each mode $\vec{k}$, so that $\vec{e}^{(\lambda_1)}(\vec{k}) = \vec{e}^{(\lambda)}(\vec{k})$, $\vec{e}^{(\lambda_2)}(\vec{k}) = \vec{b}^{(\lambda)}(\vec{k})$, and $\hat{k}$ form a set of mutually orthogonal unit vectors. The mode expansions for the electric fields are easily obtained from (1.4.16) via $\vec{e}(\vec{r},t) = -\partial\vec{a}(\vec{r},t)/\partial t$ and $\vec{b}(\vec{r},t) = \vec{\nabla} \times \vec{a}(\vec{r},t)$. Once field quantization is carried out, their explicit normalized forms will be given.

Having outlined the essential characteristics of the free classical radiation field, and its description in terms of modes when confined to a box of volume V, as a precursor to quantization, electrodynamics is presented in terms of the Lagrangian formulation. To account for the infinite number of degrees of freedom possessed by the radiation field, the Euler–Lagrange equations (1.2.1) for a system of particles require modification. To accurately describe the smooth and continuous variation of the field, a Lagrangian density $\mathcal{L}$ is introduced, which is a functional of the field and the variables that define the latter. Integrating $\mathcal{L}$ over all space yields the Lagrangian function, L. Analogous to functions, which enable a variable to be converted to a number, a functional provides a means for going from a function to a number, in this case assigning a number to the field. It may be recalled that in the analytical dynamics of particles, the Lagrangian was a function of the generalized positions and velocities. For the electromagnetic field, however, instead of $\vec{q}$, the generalized coordinate is chosen to be the vector potential, while the velocity analogous to $\dot{\vec{q}}$ is taken to be $\dot{\vec{a}}$. In addition, $\mathcal{L}$ is a function of the gradient of $\vec{a}$, thereby ensuring that spatial variations are properly included. Thus,

$$L = \int \mathcal{L}(\vec{a}, \vec{\nabla}\vec{a}, \dot{\vec{a}}, t)\mathrm{d}^3\vec{r}. \tag{1.4.17}$$

Applying the variational calculus along with Hamilton's principle as in the case of particles earlier, but with variation now performed over the new variables, the Euler–Lagrange equations of motion for the electromagnetic field are modified to

$$\frac{\partial}{\partial t}\left(\frac{\partial\mathcal{L}}{\partial\dot{a}_i}\right) + \frac{\partial}{\partial x_j}\frac{\partial\mathcal{L}}{\partial(\partial a_i/\partial x_j)} - \frac{\partial\mathcal{L}}{\partial a_i} = 0, \tag{1.4.18}$$

in which the second term is new relative to equation (1.2.1) and expresses the rate of change with respect to position of the variation of the Lagrangian density with the spatial derivative of the vector potential.

On taking, as is common, the square of the electric field to be proportional to the kinetic energy of the field and the square of the magnetic field as contributing to the electromagnetic potential energy, the Lagrangian density for the free field in analogy with point particles is written as the difference in kinetic and potential energy,

$$\mathcal{L} = \frac{\varepsilon_0}{2}\left\{\dot{\vec{a}}^2 - c^2(\vec{\nabla}\times\vec{a})^2\right\}. \tag{1.4.19}$$

Using (1.4.19) in (1.4.18) leads to Maxwell's equations, the appropriate equations of motion for the radiation field. Specifically, the wave equation for $\vec{a}$ (1.4.5) results,

$$\left(\vec{\nabla}^2 - \frac{1}{c^2}\frac{\partial^2}{\partial t^2}\right)a_i = 0, \tag{1.4.20}$$

as originally obtained from Maxwell's equations.

Like the Lagrangian for the field (1.4.17), the Hamiltonian H is a functional and is defined in terms of a density functional $\mathcal{H}$,

$$H = \int \mathcal{H}(\vec{a}, \vec{\Pi}, \vec{\nabla}\vec{a}, t)\mathrm{d}^3\vec{r}, \tag{1.4.21}$$

with $\mathcal{H}$ itself found from

$$\mathcal{H} = \vec{\Pi}\cdot\dot{\vec{a}} - \mathcal{L}. \tag{1.4.22}$$

$\vec{\Pi}(\vec{r})$ is the field momentum canonically conjugate to the vector potential, defined as

$$\vec{\Pi}(\vec{r}) = \frac{\partial\mathcal{L}}{\partial\dot{\vec{a}}}, \tag{1.4.23}$$

which from (1.4.19) is seen to be

$$\vec{\Pi}(\vec{r}) = \varepsilon_0\dot{\vec{a}}, \tag{1.4.24}$$

being proportional to the electric field $\vec{e}$. When expressed in terms of canonically conjugate variables, the Hamiltonian density (1.4.22) is written as

$$\mathcal{H} = \frac{1}{2\varepsilon_0}\left\{\vec{\Pi}^2 + \varepsilon_0^2 c^2(\vec{\nabla}\times\vec{a})^2\right\}, \tag{1.4.25}$$

which is equivalent to the electromagnetic energy density $(\varepsilon_0/2)(\vec{e}^2 + c^2\vec{b}^2)$. From the preceding development, in particular, the description of electromagnetic radiation in terms of modes of the field, it is now a simple matter to show that the radiation field enclosed in a fixed volume is equivalent to a mechanical oscillator. This theorem was first proved by Jeans (1905). On substituting the mode expansions for $\vec{a}$ and $\vec{\Pi}$—the latter obtained from (1.4.24)—into the Hamiltonian density (1.4.25), or alternatively using the derived mode expansions for $\vec{e}$ and $\vec{b}$ in the energy density, the radiation field Hamiltonian (1.4.21) can be written as

$$H = 2\varepsilon_0 c^2 V \sum_{\vec{k},\lambda} k^2 a^{(\lambda)}(\vec{k})\bar{a}^{(\lambda)}(\vec{k}), \tag{1.4.26}$$

after the field modes have been normalized. Two new, real canonically conjugate variables are now defined according to

$$q_{\vec{k},\lambda} = (\varepsilon_0 V)^{1/2}\left(a_{\vec{k}}^{(\lambda)} + \bar{a}_{\vec{k}}^{(\lambda)}\right) \tag{1.4.27}$$

and

$$p_{\vec{k},\lambda} = -ick(\varepsilon_0 V)^{1/2}\left(a_{\vec{k}}^{(\lambda)} - \bar{a}_{\vec{k}}^{(\lambda)}\right), \tag{1.4.28}$$

which yields for the Hamiltonian (1.4.26) the expression

$$H = \sum_{\vec{k},\lambda} H_{\vec{k},\lambda} = \sum_{\vec{k},\lambda} \frac{1}{2}(p_{\vec{k},\lambda}^2 + \omega^2 q_{\vec{k},\lambda}^2), \tag{1.4.29}$$

which is seen to be a mode sum over classical harmonic oscillator Hamiltonians in mass-weighted coordinates. Hamilton's canonical equations (1.2.5) are easily seen to be satisfied by the choice of conjugate variables. Recalling (1.4.6),

$$\frac{\mathrm{d}q_{\vec{k},\lambda}}{\mathrm{d}t} = p_{\vec{k},\lambda}; \quad \frac{\mathrm{d}p_{\vec{k},\lambda}}{\mathrm{d}t} = -\omega^2 q_{\vec{k},\lambda}, \tag{1.4.30}$$

and from (1.4.29)

$$\frac{\partial H}{\partial p_{\vec{k},\lambda}} = p_{\vec{k},\lambda} = \dot{q}_{\vec{k},\lambda}; \quad \frac{\partial H}{\partial q_{\vec{k},\lambda}} = \omega^2 q_{\vec{k},\lambda} = -\dot{p}_{\vec{k},\lambda}, \tag{1.4.31}$$

which provides desired confirmation of the result.

From (1.4.29), it may be concluded that quantization of the free field may be accomplished by quantizing a collection of noninteracting harmonic oscillators. The solution of the latter problem is familiar from quantum mechanics. A brief summary is given of the method of solution and of the resulting energy eigenvalues and eigenfunctions in a form that makes it readily applicable to quantization of electromagnetic radiation (Dirac, 1958). For an individual oscillator α of mass m_α and angular frequency ω_α representing a single mode of the radiation field, the Hamiltonian for the electromagnetic field is given by

$$H = \sum_\alpha \frac{1}{2m_\alpha}(p_\alpha^2 + m_\alpha^2\omega_\alpha^2 q_\alpha^2), \tag{1.4.32}$$

where the dynamical variables q_α and p_α are the coordinate and canonically conjugate momentum.

Consider a one-dimensional harmonic oscillator, whose classical Hamiltonian function is

$$H = \frac{1}{2m}(p^2 + m^2\omega^2 q^2), \tag{1.4.33}$$

there now being no need for the subscript α. The corresponding quantum mechanical Hamiltonian is taken to be of the same form as (1.4.33), with q and p represented by their respective operator equivalents subject to the fundamental commutator

$$[q, p] = i\hbar. \tag{1.4.34}$$

By introducing two mutually adjoint operators a and $a^\dagger$ in terms of p and q,

$$a = \frac{1}{\sqrt{2}}\left(\sqrt{\frac{m\omega}{\hbar}}\, q + i\sqrt{\frac{1}{m\omega\hbar}}\, p\right) \tag{1.4.35}$$

and

$$a^\dagger = \frac{1}{\sqrt{2}}\left(\sqrt{\frac{m\omega}{\hbar}}\, q - i\sqrt{\frac{1}{m\omega\hbar}}\, p\right), \tag{1.4.36}$$

the quantum mechanical Hamiltonian can be expressed as

$$H = \frac{\hbar\omega}{2}(aa^\dagger + a^\dagger a). \tag{1.4.37}$$

Although both a and $a^\dagger$ are real, they are not symmetric and hence not Hermitian unlike q and p. Using the fundamental commutator (1.4.34), it is easily verified that

$$[a, a^\dagger] = 1. \tag{1.4.38}$$

Hence, the Hamiltonian can be written in two other ways equivalent to (1.4.37), namely,

$$H = \left(a^\dagger a + \frac{1}{2}\right)\hbar\omega = \left(aa^\dagger - \frac{1}{2}\right)\hbar\omega. \tag{1.4.39}$$

Its characteristic solutions are then given by the eigenvalues and eigenfunctions of the operator $a^\dagger a$. This operator is called the number operator n. Its eigenvalues are the positive integers and zero, representing the numbers of quantized particles in the allowed eigenstates $|n\rangle$. In the case of the electromagnetic field, these particles are called photons. They satisfy Bose–Einstein statistics, with the wavefunction for n identical such particles being totally symmetric. Thus,

$$a^\dagger a|n\rangle = n|n\rangle, \quad n = 0, 1, 2, \ldots, \tag{1.4.40}$$

the ground state ket $|0\rangle$, for example, having an eigenvalue of zero. From (1.4.39), it is easily seen that the eigenvalues of the harmonic oscillator are

$$\left(n + \frac{1}{2}\right)\hbar\omega, \quad n = 0, 1, 2, \ldots, \tag{1.4.41}$$

with the lowest energy corresponding to $(\hbar\omega/2)$, the zero-point energy of the field (Milonni,1994). A ladder of states separated by a quantum of energy $\hbar\omega$ is generated in accord with Planck's quantum hypothesis. The individual operators a and $a^\dagger$ are annihilation and creation operators, acting on the occupation number state and, respectively, decreasing and increasing the number of particles by unity. This aspect of being able to tackle changes in particle number together with the correct statistical laws that the particles obey is called second quantization. It provides the link between quantum field theory and the many-body formulation (Mandl, 1959). For a normalized state $|n\rangle$, the operator equations are

$$a|n\rangle = \begin{cases} 0, & n = 0, \\ n^{1/2}|n-1\rangle, & n = 1, 2, 3, \ldots \end{cases} \tag{1.4.42}$$

and

$$a^{\dagger}|n\rangle = (n+1)^{1/2}|n+1\rangle, \quad n = 0, 1, 2, \ldots. \tag{1.4.43}$$

The wavefunction formed from the projection of the states in the Hilbert space of the system is now taken to be an operator instead of a classical number and is interpreted as a quantized field. This generalized many-particle occupation number space is called a Fock space and applies to both fermions and bosons (Fock, 1932).

The eigenvalues and eigenfunctions found for a single harmonic oscillator are easily adapted to the solution of the many-particle uncoupled harmonic oscillator Hamiltonian (1.4.32), which has been shown to be equivalent to a sum over Hamiltonians for each mode of the radiation field. In terms of the destruction and creation operators (1.4.35) and (1.4.36) for a $(\vec{k}, \lambda)$-mode photon, which are subject to the commutation relations

$$\begin{aligned} &[a^{(\lambda)}(\vec{k}), a^{(\lambda')}(\vec{k}')] = 0, \\ &[a^{\dagger(\lambda)}(\vec{k}), a^{\dagger(\lambda')}(\vec{k}')] = 0, \\ &[a^{(\lambda)}(\vec{k}), a^{\dagger(\lambda')}(\vec{k}')] = \delta_{\lambda\lambda'}\delta_{\vec{k}\vec{k}'}, \end{aligned} \tag{1.4.44}$$

the analogues of (1.4.37) and (1.4.39) are

$$\begin{aligned} H &= \sum_{\vec{k},\lambda} \frac{1}{2}\left[a^{(\lambda)}(\vec{k})a^{\dagger(\lambda)}(\vec{k}) + a^{\dagger(\lambda)}(\vec{k})a^{(\lambda)}(\vec{k})\right]\hbar ck \\ &= \sum_{\vec{k},\lambda}\left[a^{\dagger(\lambda)}(\vec{k})a^{(\lambda)}(\vec{k}) + \frac{1}{2}\right]\hbar ck \\ &= \sum_{\vec{k},\lambda}\left[a^{(\lambda)}(\vec{k})a^{\dagger(\lambda)}(\vec{k}) - \frac{1}{2}\right]\hbar ck, \end{aligned} \tag{1.4.45}$$

whose eigenenergy is the sum over all oscillators α of the energy of a single oscillator,

$$\begin{aligned} H|n_1(\vec{k}_1,\lambda_1), n_2(\vec{k}_2,\lambda_2),\ldots\rangle &= \sum_{\alpha}\left(n_{\alpha}(\vec{k}_{\alpha},\lambda_{\alpha}) + \frac{1}{2}\right)\hbar\omega_{\alpha}|n_1(\vec{k}_1,\lambda_1), \\ &\quad n_2(\vec{k}_2,\lambda_2),\ldots\rangle, \end{aligned} \tag{1.4.46}$$

where n_{α} denotes the occupation number of oscillator α. From (1.4.42), it is seen that when $n = 0$, it is not possible to absorb a particle from the ground state of the system. For the electromagnetic field, the state in which all single particle states are empty, that is, $n_\alpha = 0$ for all α, corresponds to the electromagnetic vacuum. By successive application of the creation operators on the vacuum state, all other basis states of the field may be generated, as in

$$|n_1(\vec{k}_1, \lambda_1), n_2(\vec{k}_2, \lambda_2), \ldots\rangle = \prod_\alpha \frac{[a^{\dagger(\lambda_\alpha)}(\vec{k}_\alpha)]^{n_\alpha}}{(n_\alpha!)^{1/2}} |0(\vec{k}_1, \lambda_1), 0(\vec{k}_2, \lambda_2), \ldots\rangle, \tag{1.4.47}$$

and are known as number states, and they form an orthonormal basis set. It is customary to specify only nonzero occupation numbers of the field. The expression (1.4.47) is the analogue of the wavefunction in the one-particle theory. Hence for a $(\vec{k}, \lambda)$-mode photon,

$$a^{\dagger(\lambda)}(\vec{k})a^{(\lambda)}(\vec{k})|n(\vec{k}, \lambda)\rangle = n|n(\vec{k}, \lambda)\rangle, \quad n = 0, 1, 2, \ldots, \tag{1.4.48}$$

$$a^{(\lambda)}(\vec{k})|n(\vec{k}, \lambda)\rangle = n^{1/2}|(n-1)(\vec{k}, \lambda)\rangle, \quad n = 1, 2, \ldots, \tag{1.4.49}$$

$$a^{\dagger(\lambda)}(\vec{k})|n(\vec{k}, \lambda)\rangle = (n+1)^{1/2}|(n+1)(\vec{k}, \lambda)\rangle, \quad n = 0, 1, 2, \ldots. \tag{1.4.50}$$

The quantum mechanical counterpart to the classical mode expansion for the vector potential (1.4.16) at $t = 0$ is of the form

$$\vec{a}(\vec{r}) = \sum_{\vec{k}, \lambda} \left(\frac{\hbar}{2\varepsilon_0 ckV}\right)^{1/2} \left[\vec{e}^{(\lambda)}(\vec{k})a^{(\lambda)}(\vec{k})e^{i\vec{k}\cdot\vec{r}} + \bar{\vec{e}}^{(\lambda)}(\vec{k})a^{\dagger(\lambda)}(\vec{k})e^{-i\vec{k}\cdot\vec{r}}\right], \tag{1.4.51}$$

where in the quantum theory the Fourier amplitudes a and $a^\dagger$ are understood to be annihilation and creation operators obeying the rules of commutation (1.4.44). The normalizing factor appearing in (1.4.51) is obtained on evaluating the expectation value of the energy of the radiation field for a number state $|n(\vec{k}, \lambda)\rangle$, which is known to be $(n + 1/2)\hbar\omega$. The mode expansions for the quantum electric, magnetic, and canonically conjugate momentum field $\vec{\Pi}(\vec{r})$ follow from (1.4.51) and their defining equations given earlier. They are

$$\vec{e}(\vec{r}) = i\sum_{\vec{k},\lambda}\left(\frac{\hbar ck}{2\varepsilon_0 V}\right)^{1/2}\left[\vec{e}^{(\lambda)}(\vec{k})a^{(\lambda)}(\vec{k})e^{i\vec{k}\cdot\vec{r}} - \bar{\vec{e}}^{(\lambda)}(\vec{k})a^{\dagger(\lambda)}(\vec{k})e^{-i\vec{k}\cdot\vec{r}}\right], \tag{1.4.52}$$

$$\vec{b}(\vec{r}) = i\sum_{\vec{k},\lambda}\left(\frac{\hbar k}{2\varepsilon_0 cV}\right)^{1/2}\left[\vec{b}^{(\lambda)}(\vec{k})a^{(\lambda)}(\vec{k})e^{i\vec{k}\cdot\vec{r}} - \bar{\vec{b}}^{(\lambda)}(\vec{k})a^{\dagger(\lambda)}(\vec{k})e^{-i\vec{k}\cdot\vec{r}}\right], \tag{1.4.53}$$

$$\vec{\Pi}(\vec{r}) = -i\sum_{\vec{k},\lambda}\left(\frac{\hbar ck\varepsilon_0}{2V}\right)^{1/2}\left[\vec{e}^{(\lambda)}(\vec{k})a^{(\lambda)}(\vec{k})e^{i\vec{k}\cdot\vec{r}} - \bar{\vec{e}}^{(\lambda)}(\vec{k})a^{\dagger(\lambda)}(\vec{k})e^{-i\vec{k}\cdot\vec{r}}\right]. \tag{1.4.54}$$

In the expressions for the mode expansions, the quantization volume appears explicitly. Quantum mechanical observables, however, are of course independent of this quantity. For systems normalized in finite but large volumes, the summation over the allowed wavevectors $\vec{k}$, which are restricted by (1.4.13), may be replaced by an integral through the correspondence

$$\frac{1}{V}\sum_{\vec{k}} \xrightarrow[V\to\infty]{} \frac{1}{(2\pi)^3}\int d^3\vec{k}, \tag{1.4.55}$$

where $d^3\vec{k}$ is the volume element in momentum space.

Often when calculating transfer rates, scattering cross sections, and energy shifts involving photons of a particular polarization—be they real or virtual photons, a sum over the two polarizations is required. Recalling that $\vec{e}^{(1)}(\vec{k}), \vec{e}^{(2)}(\vec{k})$, which can in general be complex, and $\hat{k}$ form a set of mutually perpendicular unit vectors, it follows that

$$\sum_{\lambda=1,2} e_i^{(\lambda)}(\vec{k})\bar{e}_j^{(\lambda)}(\vec{k}) = \delta_{ij} - \hat{k}_i\hat{k}_j. \tag{1.4.56}$$

Using the definition for the unit magnetic polarization vector $\vec{b}^{(\lambda)}(\vec{k}) = \hat{k}\times\vec{e}^{(\lambda)}(\vec{k})$ allows two additional sum rules to be obtained for electric–magnetic and magnetic–magnetic combinations. They are

$$\sum_{\lambda=1,2} e_i^{(\lambda)}(\vec{k})\bar{b}_j^{(\lambda)}(\vec{k}) = \varepsilon_{ijk}\hat{k}_k, \tag{1.4.57}$$

where ε_{ijk} is the Levi-Civita third rank antisymmetric tensor, and

$$\sum_{\lambda=1,2} b_i^{(\lambda)}(\vec{k})\bar{b}_j^{(\lambda)}(\vec{k}) = \delta_{ij} - \hat{k}_i\hat{k}_j. \qquad (1.4.58)$$

The imposition of quantum mechanical principles to the vibrational modes of a classical electromagnetic wave led to the automatic emergence of the quantized particle of light—the photon—from the formalism. In the process, the underlying duality of the wave and particle pictures of light has been revealed. This complementary description can also be found in reverse. Beginning instead with the photon, application of quantum mechanics to assemblies of such particles yields quantization of a set of classical mode oscillators. This second viewpoint is applicable to bosons in general, as well as to fermions, and forms the basis of quantum field theory. The presentation of the latter in a form applicable to interacting matter and electromagnetic wavefields is the subject of Chapter 2.

1.5 INTERACTING PARTICLE–RADIATION FIELD SYSTEM

Thus far, the variational calculus and Hamilton's principle of least action have been applied first to a system of isolated charged particles and then to the free radiation field. In each case, the equations of motion were obtained from the classical Lagrangian function expressed in terms of generalized coordinates and velocities. For material particles undergoing nonrelativistic kinematics, the equations of motion lead directly to Newton's dynamical laws, while for electromagnetic radiation, Maxwell's equations resulted. The classical Hamiltonian function was then constructed for each noninteracting system by defining the momentum canonically conjugate to the generalized coordinate variable and eliminating the generalized velocity in favor of this new quantity. The respective Hamiltonian was then converted to its quantum mechanical form by elevating the dynamical particle and field variables to operators, and the ensuing Schrödinger equation was solved for radiation and matter eigenvalues and eigenfunctions. An analogous procedure is now followed for a system of charged particles and radiation field in mutual interaction (Heitler, 1954; Power, 1964; Healy, 1982; Craig and Thirunamachandran, 1998a). It will be seen that this problem is no longer separable. Particle and field are inextricably linked—the dynamics of one affecting the other, and vice versa. Overall, however, energy is conserved as that given up by matter is gained by the field and that lost by radiation is acquired by the system of charged particles. Ultimately, this leads to perturbation theory solutions of the coupled matter–field system. As before, there is considerable freedom in the specific choice of classical Lagrangian function. Its only limitation is that it must lead to the correct equations of motion. Since the noninteracting matter–field system constitutes a completely separable case, it is sensible to partition the total Lagrangian into a sum of molecule, field, and interaction Lagrangians,

$$L = L_{\mathrm{mol}} + L_{\mathrm{rad}} + L_{\mathrm{int}}, \tag{1.5.1}$$

where

$$L_{\mathrm{mol}} = \frac{1}{2}\sum_{\alpha} m_{\alpha}\dot{\vec{q}}_{\alpha}^{2} - V(\vec{q}), \tag{1.5.2}$$

$$L_{\mathrm{rad}} = \int \mathcal{L}_{\mathrm{rad}}\, \mathrm{d}^3\vec{r} = \frac{1}{2}\varepsilon_0 \int \{\dot{\vec{a}}^2 - c^2(\vec{\nabla}\times\vec{a})^2\}\mathrm{d}^3\vec{r}, \tag{1.5.3}$$

$$L_{\mathrm{int}} = \int \mathcal{L}_{\mathrm{int}}(\vec{r})\mathrm{d}^3\vec{r} = \int \vec{j}^{\perp}(\vec{r})\cdot a(\vec{r})\mathrm{d}^3\vec{r}. \tag{1.5.4}$$

Unsurprisingly, the molecular and radiation field Lagrangians (1.5.2) and (1.5.3) are identical to the Lagrangians (1.2.8) and (1.4.19), respectively, when the quantum mechanics of a collection of charged particles, and the electromagnetic field in the absence of sources, was studied. The form of the interaction Lagrangian (1.5.4) is a direct result of working in the Coulomb gauge. The scalar potential, describing the electrostatic Coulomb potential, is replaced by the electrostatic potential energy with the transverse vector potential describing the radiation field. In (1.5.4), $\vec{j}^{\perp}(\vec{r})$ is the transverse part of the current density, obtained by projecting the total current density onto the transverse delta function dyadic $\delta_{ij}^{\perp}(\vec{r})$ (Belinfante, 1946) so that

$$j_i^{\perp}(\vec{r}) = \sum_{\alpha} e_{\alpha}\dot{q}_{j(\alpha)}\delta_{ij}^{\perp}(\vec{r}-\vec{q}_{\alpha}). \tag{1.5.5}$$

It is straightforward to demonstrate that the Lagrangian (1.5.1) is of the appropriate form. Using (1.5.1) in the Euler–Lagrange equations for the field (1.4.18), the vector potential is seen to obey the inhomogeneous wave equation,

$$\left(\vec{\nabla}^2 - \frac{1}{c^2}\frac{\partial^2}{\partial t^2}\right)\vec{a}(\vec{r}) = -\frac{1}{\varepsilon_0 c^2}\vec{j}^{\perp}(\vec{r}), \tag{1.5.6}$$

instead of its source-free counterpart (1.4.20). Note that (1.5.6) is identical to equation (1.3.35), the latter following directly from the Maxwell–Lorentz equations in the Coulomb gauge. Substituting the total Lagrangian (1.5.1) into the Euler–Lagrange equations for an assembly of particles (1.2.1) yields for the ith component

$$m_\alpha \ddot{q}_{i(\alpha)} = -\frac{\partial V}{\partial q_{i(\alpha)}} + e_\alpha e_i^\perp(\vec{q}_\alpha) + e_\alpha[\dot{\vec{q}}_\alpha \times \vec{b}(\vec{q}_\alpha)]_i, \tag{1.5.7}$$

where relations (1.3.12) and (1.3.34) have been used, and which is immediately recognizable as Newton's equation of motion modified by the addition of the Lorentz force law terms representing the interaction of charged particles with the transverse radiation field.

The total Hamiltonian may be evaluated from the total Lagrangian in the usual way according to

$$H = \sum_\alpha \vec{p}_\alpha \cdot \dot{\vec{q}}_\alpha + \int \vec{\Pi} \cdot \dot{\vec{a}}\, d^3\vec{r} - L, \tag{1.5.8}$$

after calculating the momenta canonically conjugate to the generalized position and vector potential. The former is now given by

$$\vec{p}_\alpha = \frac{\partial L}{\partial \dot{\vec{q}}_\alpha} = m_\alpha \dot{\vec{q}}_\alpha + e_\alpha \vec{a}(\vec{q}_\alpha), \tag{1.5.9}$$

while the latter is identical to that obtained using the free field,

$$\vec{\Pi}(\vec{r}) = \frac{\partial \mathcal{L}}{\partial \dot{\vec{a}}} = \varepsilon_0 \dot{\vec{a}}(\vec{r}) = -\varepsilon_0 \vec{e}^\perp(\vec{r}). \tag{1.5.10}$$

Substituting for $\dot{\vec{q}}_\alpha$ and $\dot{\vec{a}}$ into (1.5.8) produces what is universally known as the minimal-coupling Hamiltonian (Craig and Thirunamachandran, 1998a),

$$H^{\min} = \sum_\alpha \frac{1}{2m_\alpha}\{\vec{p}_\alpha - e_\alpha \vec{a}(\vec{q}_\alpha)\}^2 + V(\vec{q}) + \frac{1}{2\varepsilon_0}\int \{\vec{\Pi}^2(\vec{r}) + \varepsilon_0^2 c^2 (\vec{\nabla} \times \vec{a}(\vec{r}))^2\} d^3\vec{r}. \tag{1.5.11}$$

At this stage, it is advantageous to collect the charged particles α together to form atoms and molecules ξ. Further, it is approximated that the nuclei are located at fixed positions in space relative to electrons, which are allowed to move. Hence, the dynamical variables of the charged particle system are the electronic coordinates and momenta. The clamped nuclei approximation is justified on the grounds that the nucleons are significantly more massive than electrons. This is a simplification that is frequently made in chemical physics and is adopted in what follows. For many situations, however, such as the treatment of molecular vibrations and the dynamics of chemical reactions, nuclear motions cannot be ignored. For such applications, it is convenient to describe the interaction of radiation with quantum mechanical electrons, but with the nuclei moving classically in a specific version of a semiclassical formulation of the theory. This has been achieved by coupling the Maxwell and Schrödinger equations in a canonical prescription (Masiello et al., 2005). The time evolution is followed by integrating the first-order Hamilton's equations subject to well-defined initial conditions for the dynamical variables. The solutions are formally exact in the limit of infinite basis sets, though in practice computations are carried out with a set of truncated functions. The other, more difficult option, is a fully quantum mechanical treatment of electronic and nuclear degrees of freedom coupled to radiation.

By dividing the total electrostatic potential energy into a sum of one-particle and two-center terms,

$$V = \sum_{\xi} V(\xi) + \sum_{\xi<\xi'} V(\xi, \xi'), \tag{1.5.12}$$

the Hamiltonian (1.5.11) can be written as

$$H^{\text{min}} = H^{\text{min}}_{\text{mol}} + H^{\text{min}}_{\text{rad}} + H^{\text{min}}_{\text{int}}, \tag{1.5.13}$$

where the molecular Hamiltonian is

$$H^{\text{min}}_{\text{mol}} = \sum_{\xi} \left\{ \frac{1}{2m} \sum_{\alpha} \vec{p}^{\,2}_{\alpha}(\xi) + V(\xi) \right\}, \tag{1.5.14}$$

in which $\vec{p}_\alpha$ is the momentum of electron α with position vector $\vec{q}_\alpha$ and $V(\xi)$ is the intramolecular potential energy of molecule ξ. The radiation field Hamiltonian is

$$H^{\text{min}}_{\text{rad}} = \frac{1}{2\varepsilon_0} \int \left\{ \vec{\Pi}^2 + \varepsilon_0^2 c^2 (\vec{\nabla} \times \vec{a})^2 \right\} \mathrm{d}^3\vec{r} = \frac{\varepsilon_0}{2} \int (\vec{e}^{\perp 2} + c^2 \vec{b}^2) \mathrm{d}^3\vec{r}, \tag{1.5.15}$$

expressed in terms of the vector potential and its canonically conjugate momentum, or in terms of electric and magnetic fields. The third term of (1.5.13) accounts for the interaction between radiation and matter and is explicitly given by

$$H^{\text{min}}_{\text{int}} = \frac{e}{m} \sum_{\xi} \sum_{\alpha} \vec{p}_\alpha(\xi) \cdot \vec{a}(\vec{q}_\alpha(\xi)) + \frac{e^2}{2m} \sum_{\xi} \sum_{\alpha} \vec{a}^2(\vec{q}_\alpha(\xi)) + V_{\text{inter}}, \tag{1.5.16}$$

while V_{inter} is given by the second term of (1.5.12) and describes the intermolecular potential energy between molecules ξ and ξ'. The superscript min is inserted because the total and individual Hamiltonians are constructed from dynamical variables specific to this particular formulation.

The quantum mechanical analogue of the classical Hamiltonian (1.5.13) is obtained on promoting the particle and field coordinates and canonically conjugate momenta to quantum operators subject to the following commutation relations valid at equal time:

$$[q_{i(\alpha)}(\xi), p_{j(\beta)}(\xi')] = i\hbar\delta_{ij}\,\delta_{\alpha\beta}\,\delta_{\xi\xi'} \tag{1.5.17}$$

and

$$[a_i(\vec{r}), \Pi_j(\vec{r}\,')] = i\hbar\delta_{ij}^{\perp}(\vec{r}-\vec{r}\,'). \tag{1.5.18}$$

The commutator between field operators (1.5.18) was expressed alternatively by (1.4.44) in terms of the annihilation and destruction operators for a mode of the radiation field in free space.

While the minimal-coupling Hamiltonian (1.5.13) rigorously describes the interaction of a charged particle with the electromagnetic field, it proves to be awkward when it is applied to radiation–molecule and molecule–molecule processes. This is due to the appearance of the particle momentum, the vector potential of the radiation field, and the intermolecular coupling term in the interaction component of the Hamiltonian (1.5.16). The first of these variables is not the most appropriate for a chemical species, the second is more often expressed in terms of the fundamental electric and magnetic fields, while the third term must always be included when treating two or more particles. In the following two sections, an alternative Lagrangian and Hamiltonian are obtained that have proved to be more suitable for application to atomic and molecular systems interacting with the radiation field.

1.6 MULTIPOLAR LAGRANGIAN

The Lagrangian for the interacting charged particle–electromagnetic field system (1.5.1) is a function of particle coordinates and velocities $\vec{q}_\alpha$ and $\dot{\vec{q}}_\alpha$ as well as a functional of the analogous field dynamical variables $\vec{a}(\vec{r})$ and $\dot{\vec{a}}(\vec{r})$. It was shown to lead to the correct Euler–Lagrange equations of motion. Lagrangians that differ in the total time derivative of a function or functional of the coordinates and the time, $f(\vec{q}, \vec{a}, t)$, are said to be equivalent. Thus,

$$L^{\text{new}} = L^{\text{old}} - \frac{\mathrm{d}}{\mathrm{d}t} f(\vec{q}, \vec{a}, t). \tag{1.6.1}$$

Because the variations in the path correct to first order between the initial and final times vanish,

$$\delta q(t_i) = \delta q(t_f) = 0, \tag{1.6.2a}$$

$$\delta a_j(t_i) = \delta a_j(t_f) = 0, \tag{1.6.2b}$$

the variations of the action integral involving old and new Lagrangians are then identical,

$$\delta S^{\text{new}} = \delta \int_{t_i}^{t_f} L^{\text{new}}\, \mathrm{d}t = \delta \int_{t_i}^{t_f} L^{\text{old}}\, \mathrm{d}t = \delta S^{\text{old}}, \tag{1.6.3}$$

where the action S is the time integral of the Lagrangian. Hence, identical Euler–Lagrange equations of motion follow from L^{new} as derived from L^{old}.

A Lagrangian equivalent to that of (1.5.1), one that leads to a Hamiltonian that is better suited to deal with atomic and molecular systems, is obtained by adding a function of the form

$$\int \vec{p}^{\perp}(\vec{r}) \cdot \vec{a}(\vec{r}) \mathrm{d}^3\vec{r}, \tag{1.6.4}$$

as was first suggested by Göppert-Mayer (1931). If $\vec{p}^{\perp}(\vec{r})$ in (1.6.4) is taken to be the transverse component of the electric polarization field, the resulting new Lagrangian is of the multipolar form, whose explicit structure will be given below. Before going on to this, the decomposition of charge and current densities in terms of electric and magnetic polarization fields is carried out and their multipole expanded forms are given.

In a medium, the electric and magnetic polarizations result from charge and current densities. The former is separated into contributions from free and bound charges, while the current density is composed of terms arising from electric polarization and magnetization currents due to the relative motions of bound charges, and the contributions from convective and Röntgen currents. For a neutral system, the convective current vanishes, while for stationary nuclei there is no Röntgen term.

The charge density (1.3.9) may be partitioned as

$$\rho(\vec{r}) = \sum_{\alpha} e_\alpha \delta(\vec{r} - \vec{q}_\alpha) = \sum_{\alpha} e_\alpha \delta(\vec{r} - \vec{R}) - \vec{\nabla} \cdot \vec{p}(\vec{r}) = \rho^{\text{true}} - \vec{\nabla} \cdot \vec{p}(\vec{r}), \tag{1.6.5}$$

in which $\rho^{\text{true}} = \sum_\alpha e_\alpha \delta(\vec{r}-\vec{R})$ is the net charge density of the distribution. This division of the source necessitates the introduction of the vector $\vec{R}_\xi$, an expansion point about which multipole moments are defined and which may be taken as the center of mass, an inversion center, or the origin of a local chromophore center. The electric polarization field $\vec{p}(\vec{r})$ can be written in closed form as the parametric integral (Woolley, 1971)

$$\begin{aligned}\vec{p}(\xi,\vec{r}) &= \sum_\alpha e_\alpha(\vec{q}_\alpha(\xi)-\vec{R}_\xi)\int_0^1 \delta(\vec{r}-\vec{R}_\xi-\lambda(\vec{q}_\alpha(\xi)-\vec{R}_\xi))\mathrm{d}\lambda \\ &+ \sum_a e_a Z_a(\xi)(\vec{Q}_a(\xi)-\vec{R}_\xi)\int_0^1 \delta(\vec{r}-\vec{R}_\xi-\lambda(\vec{Q}_a(\xi)-\vec{R}_\xi))\mathrm{d}\lambda,\end{aligned} \tag{1.6.6}$$

a sum of electronic and nuclear contributions, with $Z_a(\xi)$ and $\vec{Q}_a(\xi)$ the atomic number and position of nucleus a of molecule ξ, and

$$\vec{p}(\vec{r}) = \sum_\xi \vec{p}(\xi,\vec{r}). \tag{1.6.7}$$

Concentrating on the electronic term and expanding the delta function produces

$$\begin{aligned}\vec{p}(\xi,\vec{r}) &= \sum_\alpha e_\alpha(\vec{q}_\alpha(\xi)-\vec{R}_\xi)\int_0^1 \Big[1-\Big\{\lambda(\vec{q}_\alpha(\xi)-\vec{R}_\xi)\cdot\vec{\nabla}\Big\} \\ &+ \frac{1}{2!}\{\lambda(\vec{q}_\alpha(\xi)-\vec{R}_\xi)\cdot\vec{\nabla}\}^2 - \cdots\Big]\delta(\vec{r}-\vec{R}_\xi)\mathrm{d}\lambda,\end{aligned} \tag{1.6.8}$$

which after carrying out the λ-integral results in the familiar electric multipole series expansion of the polarization distribution; the dipole term is given by

$$\vec{p}^{(1)}(\xi,\vec{r}) = \sum_\alpha e_\alpha(\vec{q}_\alpha(\xi)-\vec{R}_\xi)\delta(\vec{r}-\vec{R}_\xi) \tag{1.6.9}$$

and the quadrupole polarization by

$$\vec{p}^{(2)}(\xi,\vec{r}) = -\sum_\alpha \frac{1}{2!} e_\alpha(\vec{q}_\alpha(\xi)-\vec{R}_\xi)(\vec{q}_\alpha(\xi)-\vec{R}_\xi)\cdot\vec{\nabla}\delta(\vec{r}-\vec{R}_\xi). \tag{1.6.10}$$

The superscript indicates the order of the moment, with the familiar electric dipole defined as

$$\mu_i(\xi) = \sum_\alpha e_\alpha(\vec{q}_\alpha(\xi)-\vec{R}_\xi)_i \tag{1.6.11}$$

and the electric quadrupole moment tensor as

$$Q_{ij}(\xi) = \frac{1}{2!}\sum_{\alpha} e_{\alpha}(\vec{q}_{\alpha}(\xi)-\vec{R}_{\xi})_i(\vec{q}_{\alpha}(\xi)-\vec{R}_{\xi})_j. \tag{1.6.12}$$

With the polarization field defined by (1.6.8), the multipolar Lagrangian is obtained from the minimal-coupling Lagrangian (1.5.1) on using (1.6.4) in (1.6.1); that is,

$$\begin{aligned} L^{\text{mult}} &= L^{\text{min}} - \frac{\mathrm{d}}{\mathrm{d}t}\int \vec{p}^{\perp}(\vec{r})\cdot\vec{a}(\vec{r})\mathrm{d}^3\vec{r} \\ &= \sum_{\xi}\left\{\frac{m}{2}\sum_{\alpha}\dot{\vec{q}}_{\alpha}^{2}(\xi) - V(\xi)\right\} + \frac{\varepsilon_0}{2}\int\left\{\dot{\vec{a}}^{2}(\vec{r}) - c^2(\vec{\nabla}\times\vec{a}(\vec{r}))^2\right\}\mathrm{d}^3\vec{r} \\ &\quad + \int\left(\vec{j}^{\perp}(\vec{r}) - \frac{\mathrm{d}\vec{p}^{\perp}(\vec{r})}{\mathrm{d}t}\right)\cdot\vec{a}(\vec{r})\mathrm{d}^3\vec{r} - \int \vec{p}^{\perp}(\vec{r})\cdot\dot{\vec{a}}(\vec{r})\mathrm{d}^3\vec{r} - \sum_{\xi<\xi'} V_{\text{inter}}(\xi,\xi'), \end{aligned} \tag{1.6.13}$$

where the partitioning of the total electrostatic potential energy into intra- and intermolecular terms according to (1.5.12) has been used explicitly.

From the definition of the transverse current density and the electric dipole polarization distribution (1.6.9), it is seen that in the electric dipole approximation,

$$\vec{j}^{\perp}(\vec{r}) = \frac{\mathrm{d}\vec{p}^{\perp}(\vec{r})}{\mathrm{d}t}, \tag{1.6.14}$$

and the particle momentum canonically conjugate to the position,

$$\vec{p}_{\alpha} = \frac{\partial L^{\text{mult}}}{\partial\dot{\vec{q}}_{\alpha}} = m\dot{\vec{q}}_{\alpha}, \tag{1.6.15}$$

is equal to the kinetic momentum.

To proceed further, the third term of L^{mult} is examined in detail. Employing the identity (Power and Thirunamachandran, 1971)

$$\begin{aligned} \dot{\vec{q}}\delta(\vec{r}-\vec{q}) &= \frac{\mathrm{d}}{\mathrm{d}t}(\vec{q}-\vec{R})\int_0^1 \delta(\vec{r}-\vec{R}-\lambda(\vec{q}-\vec{R}))\mathrm{d}\lambda \\ &\quad + \vec{\nabla}\times\left[(\vec{q}-\vec{R})\times\dot{\vec{q}}\int_0^1 \lambda\delta(\vec{r}-\vec{R}-\lambda(\vec{q}-\vec{R}))\mathrm{d}\lambda\right] \end{aligned} \tag{1.6.16}$$

enables the difference between the transverse current and polarization densities to be written as

$$\vec{j}^{\perp}(\vec{r}) - \frac{\mathrm{d}\vec{p}^{\perp}(\vec{r})}{\mathrm{d}t} = \vec{\nabla} \times \vec{m}(\vec{r}), \tag{1.6.17}$$

showing that the current density for a system of neutral molecules moving with zero velocity is composed of electric and magnetic polarization contributions. On the right-hand side of (1.6.17) is the curl of the magnetization field, $\vec{m}(\vec{r})$,

$$\vec{m}(\xi,\vec{r}) = \sum_{\alpha} e_{\alpha}\{(\vec{q}_{\alpha}(\xi)-\vec{R}_{\xi}) \times \dot{\vec{q}}_{\alpha}(\xi)\} \int_0^1 \lambda\delta(\vec{r}-\vec{R}_{\xi}-\lambda(\vec{q}_{\alpha}(\xi)-\vec{R}_{\xi}))\mathrm{d}\lambda, \tag{1.6.18}$$

with the contribution from nuclei given by

$$+\sum_{a} e_a Z_a(\xi)\{(\vec{Q}_a(\xi)-\vec{R}_{\xi}) \times \dot{\vec{Q}}_a(\xi)\} \int_0^1 \lambda\delta(\vec{r}-\vec{R}_{\xi}-\lambda(\vec{Q}_a(\xi)-\vec{R}_{\xi}))\mathrm{d}\lambda, \tag{1.6.19}$$

so that

$$\vec{m}(\vec{r}) = \sum_{\xi} \vec{m}(\xi,\vec{r}). \tag{1.6.20}$$

On again expanding the δ-function in a Taylor series and performing the λ-integral, the ith component of the electronic contribution to the magnetization (1.6.18) is given by

$$\begin{aligned} m_i(\xi,\vec{r}) &= \sum_{\alpha} e_{\alpha}\{(\vec{q}_{\alpha}(\xi)-\vec{R}_{\xi}) \times \dot{\vec{q}}_{\alpha}(\xi)\}_i \int_0^1 \lambda\Big[1-\{\lambda(\vec{q}_{\alpha}(\xi)-\vec{R}_{\xi})_j\vec{\nabla}_j\} \\ &\quad +\frac{1}{2!}\{\lambda(\vec{q}_{\alpha}(\xi)-\vec{R}_{\xi})_j(\vec{q}_{\alpha}(\xi)-\vec{R}_{\xi})_k\vec{\nabla}_j\vec{\nabla}_k\}^2-\cdots\Big]\delta(\vec{r}-\vec{R}_{\xi})\mathrm{d}\lambda \\ &= \sum_{\alpha} e_{\alpha}\{(\vec{q}_{\alpha}(\xi)-\vec{R}_{\xi})\times\dot{\vec{q}}_{\alpha}(\xi)\}_i \\ &\quad \times\left\{\begin{aligned} &\frac{1}{2!}-\frac{2}{3!}(\vec{q}_{\alpha}(\xi)-\vec{R}_{\xi})_j\vec{\nabla}_j \\ &+\frac{3}{4!}(\vec{q}_{\alpha}(\xi)-\vec{R}_{\xi})_j(\vec{q}_{\alpha}(\xi)-\vec{R}_{\xi})_k\vec{\nabla}_j\vec{\nabla}_k-\cdots \end{aligned}\right\}\delta(\vec{r}-\vec{R}_{\xi}) \\ &= \Big\{m_i^{(1)}-m_{ij}^{(2)}\vec{\nabla}_j+m_{ijk}^{(3)}\vec{\nabla}_j\vec{\nabla}_k-\cdots\Big\}\delta(\vec{r}-\vec{R}_{\xi}), \end{aligned} \tag{1.6.21}$$

where the first term, including $\delta(\vec{r}-\vec{R}_\xi)$, is the magnetic dipole magnetization, the second is the magnetic quadrupole magnetization, and so on, with the magnetic multipole moments defined as

$$m_i^{(1)}(\xi)=\sum_\alpha \frac{1}{2!}e_\alpha\{(\vec{q}_\alpha(\xi)-\vec{R}_\xi)\times\dot{\vec{q}}_\alpha(\xi)\}_i, \tag{1.6.22}$$

$$m_{ij}^{(2)}(\xi)=\sum_\alpha \frac{2}{3!}e_\alpha\{(\vec{q}_\alpha(\xi)-\vec{R}_\xi)\times\dot{\vec{q}}_\alpha(\xi)\}_i(\vec{q}_\alpha(\xi)-\vec{R}_\xi)_j. \tag{1.6.23}$$

Of course, the right-hand side of (1.6.17) and the form of (1.6.18) are obtained by explicitly evaluating the left-hand side of the former, the two terms being given by

$$\begin{aligned} j_i(\vec{r})=\sum_\alpha e_\alpha\dot{q}_{i(\alpha)}(\xi)\Big[1-(\vec{q}_\alpha(\xi)-\vec{R}_\xi)_j\vec{\nabla}_j \\ +\frac{1}{2!}(\vec{q}_\alpha(\xi)-\vec{R}_\xi)_j(\vec{q}_\alpha(\xi)-\vec{R}_\xi)_k\vec{\nabla}_j\vec{\nabla}_k-\cdots\Big]\delta(\vec{r}-\vec{R}_\xi) \end{aligned} \tag{1.6.24}$$

and

$$\begin{aligned} \frac{\mathrm{d}p_i(\vec{r})}{\mathrm{d}t}=\sum_\alpha e_\alpha\dot{q}_{i(\alpha)}(\xi)\Big[1-\frac{1}{2!}(\vec{q}_\alpha(\xi)-\vec{R}_\xi)_j\vec{\nabla}_j \\ +\frac{1}{3!}(\vec{q}_\alpha(\xi)-\vec{R}_\xi)_j(\vec{q}_\alpha(\xi)-\vec{R}_\xi)_k\vec{\nabla}_j\vec{\nabla}_k-\cdots\Big]\delta(\vec{r}-\vec{R}_\xi) \\ -\sum_\alpha e_\alpha(\vec{q}_\alpha(\xi)-\vec{R}_\xi)_i\dot{q}_{j(\alpha)}(\xi)\vec{\nabla}_j\Big[\frac{1}{2!}-\frac{2}{3!}(\vec{q}_\alpha(\xi)-\vec{R}_\xi)_k\vec{\nabla}_k+\cdots\Big]\delta(\vec{r}-\vec{R}_\xi). \end{aligned} \tag{1.6.25}$$

Returning to the multipolar Lagrangian (1.6.13), inserting (1.6.17) produces for L^{mult}

$$L^{\mathrm{mult}}=L_{\mathrm{mol}}+L_{\mathrm{rad}}+\int[\vec{\nabla}\times\vec{m}(\vec{r})]\cdot\vec{a}(\vec{r})\mathrm{d}^3\vec{r}-\int\vec{p}^{\perp}(\vec{r})\cdot\dot{\vec{a}}(\vec{r})\mathrm{d}^3\vec{r}-\sum_{\xi<\xi'}V_{\mathrm{inter}}(\xi,\xi'), \tag{1.6.26}$$

where

$$L_{\mathrm{mol}}=\sum_\xi\left\{\frac{1}{2}m\sum_\alpha\dot{\vec{q}}_\alpha^2(\xi)-V(\xi)\right\} \tag{1.6.27}$$

and

$$L_{\mathrm{rad}}=\frac{\varepsilon_0}{2}\int\{\dot{\vec{a}}^2(\vec{r})-c^2(\vec{\nabla}\times\vec{a}(\vec{r}))^2\}\mathrm{d}^3\vec{r}. \tag{1.6.28}$$

The last three terms of (1.6.26) now constitute the new interaction terms with coupling now occurring through the polarization and magnetization fields instead of through the transverse current density as in (1.5.4) (Power and Thirunamachandran, 1978).

Even though the formal equations of motion are identical for equivalent Lagrangians, their actual expressions will differ due to the introduction of new variables. The result of the addition of the time derivative of (1.6.4) to the minimal-coupling Lagrangian to form L^{mult} led to the appearance of polarization and magnetization fields. The equations of motion for the radiation field, expressed in terms of $\vec{p}(\vec{r})$ and $\vec{m}(\vec{r})$, are known as the atomic field equations. They are obtained from the Euler–Lagrange equations of motion for the field and (1.6.26) as follows. Defining the appropriate Lagrangian density $\mathcal{L}^{\mathrm{mult}}$ from (1.6.26), the three terms of the equations of motion

$$\frac{\partial}{\partial t}\left(\frac{\partial\mathcal{L}^{\mathrm{mult}}}{\partial\dot{a}_i}\right)+\frac{\partial}{\partial x_j}\frac{\partial\mathcal{L}^{\mathrm{mult}}}{\partial(\partial a_i/\partial x_j)}-\frac{\partial\mathcal{L}^{\mathrm{mult}}}{\partial a_i}=0 \tag{1.6.29}$$

are

$$\frac{\partial\mathcal{L}^{\mathrm{mult}}}{\partial\dot{a}_i}=\varepsilon_0\dot{a}_i(\vec{r})-p_i^{\perp}(\vec{r})=-d_i^{\perp}(\vec{r}), \tag{1.6.30}$$

where $d_i^{\perp}(\vec{r})$ is the transverse component of the electric displacement field,

$$\vec{d}(\vec{r})=\varepsilon_0\vec{e}(\vec{r})+\vec{p}(\vec{r}), \tag{1.6.31}$$

$$\frac{\partial\mathcal{L}^{\mathrm{mult}}}{\partial(\partial a_i/\partial x_j)}=-c^2\varepsilon_0\left(\frac{\partial a_i}{\partial x_j}-\frac{\partial a_j}{\partial x_i}\right), \tag{1.6.32}$$

and

$$\frac{\partial\mathcal{L}^{\mathrm{mult}}}{\partial a_i}=[\vec{\nabla}\times\vec{m}(\vec{r})]_i. \tag{1.6.33}$$

Taking the time derivative of (1.6.30) and adding to (1.6.32) and (1.6.33) produces for (1.6.29)

$$\vec{\nabla}\times\vec{b}(\vec{r})=\frac{1}{\varepsilon_0 c^2}\left\{\frac{\partial\vec{d}^{\perp}(\vec{r})}{\partial t}+\left[\vec{\nabla}\times\vec{m}(\vec{r})\right]\right\}, \tag{1.6.34}$$

which on substituting (1.6.17) for $\vec{j}^{\perp}(\vec{r})$ and using (1.6.31) yields the source-dependent Maxwell–Lorentz equation

$$\vec{\nabla} \times \vec{b}(\vec{r}) = \frac{1}{c^2}\frac{\partial \vec{e}^{\perp}(\vec{r})}{\partial t} + \frac{1}{\varepsilon_0 c^2}\vec{j}^{\perp}(\vec{r}). \tag{1.6.35}$$

By defining the magnetic analogue $\vec{h}(\vec{r})$ of the displacement field,

$$\vec{h}(\vec{r}) = \varepsilon_0 c^2 \vec{b}(\vec{r}) - \vec{m}(\vec{r}), \tag{1.6.36}$$

(1.6.34) can be written in terms of both auxiliary fields as

$$\vec{\nabla} \times \vec{h}(\vec{r}) = \frac{\partial \vec{d}^{\perp}(\vec{r})}{\partial t}, \tag{1.6.37}$$

which expresses the Maxwell–Lorentz equation (1.6.35) in terms of $\vec{d}(\vec{r})$ and $\vec{h}(\vec{r})$.

As expected, Newton's force law with Lorentz term—the equation of motion for the particle (1.5.7)—results when L^{mult} (1.6.26) is used in the Euler–Lagrange equation

$$\frac{\mathrm{d}}{\mathrm{d}t}\frac{\partial L^{\text{mult}}}{\partial \dot{\vec{q}}_{\alpha}(\xi)} - \frac{\partial L^{\text{mult}}}{\partial \vec{q}_{\alpha}(\xi)} = 0. \tag{1.6.38}$$

1.7 MULTIPOLAR HAMILTONIAN

Having obtained in the previous section the equations of motion satisfied by the multipolar Lagrangian for the interaction of a system of charged particles and electromagnetic radiation, it remains to apply the canonical quantization scheme to L^{mult} given by (1.6.26) to derive the multipolar form of the Hamiltonian operator. Before the latter can be constructed, the momenta canonically conjugate to the particle coordinate and the vector potential of the field are first evaluated using (1.6.26). The first of these is given by

$$\begin{aligned}\vec{p}_{\alpha}(\xi) &= \frac{\partial L^{\text{mult}}}{\partial \dot{\vec{q}}_{\alpha}(\xi)} = m\dot{\vec{q}}_{\alpha}(\xi) + \frac{\partial}{\partial \dot{\vec{q}}_{\alpha}(\xi)}\int \{\vec{\nabla} \times \vec{m}_{\alpha}(\vec{r})\}\cdot \vec{a}(\vec{r})\mathrm{d}^3\vec{r} \\ &= m\dot{\vec{q}}_{\alpha}(\xi) + e(\vec{q}_{\alpha}(\xi) - \vec{R}_{\xi})\int\int_0^1 \lambda\delta(\vec{r} - \vec{R}_{\xi} - \lambda(\vec{q}_{\alpha}(\xi) - \vec{R}_{\xi}))\mathrm{d}\lambda \times \vec{b}(\vec{r})\mathrm{d}^3\vec{r}\end{aligned} \tag{1.7.1}$$

after inserting the expression for the magnetization field applicable to a single charged particle (1.6.18). Defining a further vector field,

$$\vec{n}_\alpha(\xi;\vec{r}) = -e(\vec{q}_\alpha(\xi)-\vec{R}_\xi)\int_0^1 \lambda\delta(\vec{r}-\vec{R}_\xi-\lambda(\vec{q}_\alpha(\xi)-\vec{R}_\xi))\mathrm{d}\lambda, \tag{1.7.2}$$

which is seen to be λ times the electronic part of the polarization field $\vec{p}(\xi,\vec{r})$ (1.6.6), with

$$\vec{n}(\vec{r}) = \sum_{\xi,\alpha} \vec{n}_\alpha(\xi,\vec{r}), \tag{1.7.3}$$

the conjugate momentum (1.7.1) can be written more succinctly as

$$\vec{p}_\alpha(\xi) = m\dot{\vec{q}}_\alpha(\xi) - \int \vec{n}_\alpha(\xi;\vec{r})\times\vec{b}(\vec{r})\mathrm{d}^3\vec{r}. \tag{1.7.4}$$

Note that the kinetic and canonical momenta are no longer identical, but differ in the second term on the right-hand side of the last relation. The momentum conjugate to the field has been evaluated previously and is given by

$$\Pi_i(\vec{r}) = \frac{\partial\mathcal{L}^{\mathrm{mult}}}{\partial\dot{a}_i(\vec{r})} = \varepsilon_0\dot{a}_i(\vec{r}) - p_i^\perp(\vec{r}) = -d_i^\perp(\vec{r}), \tag{1.7.5}$$

being the negative of the transverse displacement field; this is unlike the situation in minimal coupling, where $\vec{\Pi}(\vec{r})$ was found to be proportional to the negative transverse electric field (1.5.10). The multipolar Hamiltonian is obtained from (1.5.8) on inserting L^{mult} and summing over particles α and aggregates ξ,

$$H^{\mathrm{mult}} = \sum_{\xi,\alpha}\vec{p}_\alpha(\xi)\cdot\dot{\vec{q}}_\alpha(\xi) + \int\vec{\Pi}(\vec{r})\cdot\dot{\vec{a}}(\vec{r})\mathrm{d}^3\vec{r} - L^{\mathrm{mult}}, \tag{1.7.6}$$

and replacing $\dot{\vec{q}}$ and $\dot{\vec{a}}$ by using (1.7.4) and (1.7.5). This results in

$$\begin{aligned}H^{\mathrm{mult}} = {} & \frac{1}{2m}\sum_{\xi,\alpha}\vec{p}_\alpha^{\,2}(\xi) + \sum_\xi V(\xi) + \frac{\varepsilon_0}{2}\int\left\{\left(\frac{\vec{\Pi}(\vec{r})}{\varepsilon_0}\right)^2 + c^2\vec{b}^{\,2}(\vec{r})\right\}\mathrm{d}^3\vec{r} \\ & + \varepsilon_0^{-1}\int\vec{p}^{\,\perp}(\vec{r})\cdot\vec{\Pi}(\vec{r})\mathrm{d}^3\vec{r} - \int\vec{m}(\vec{r})\cdot\vec{b}(\vec{r})\mathrm{d}^3\vec{r} \\ & + \frac{1}{2m}\sum_{\xi,\alpha}\left\{\int\vec{n}_\alpha(\xi,\vec{r})\times\vec{b}(\vec{r})\mathrm{d}^3\vec{r}\right\}^2 \\ & + \frac{1}{2\varepsilon_0}\int|\vec{p}^{\,\perp}(\vec{r})|^2\mathrm{d}^3\vec{r} + \sum_{\xi<\xi'}V_{\mathrm{inter}}(\xi,\xi'),\end{aligned} \tag{1.7.7}$$

where the magnetization field $\vec{m}(\vec{r})$ is given in symmetrized form by

$$\vec{m}(\vec{r})=\frac{1}{2m}\sum_{\xi,\alpha}\{\vec{n}_\alpha(\xi,\vec{r})\times\vec{p}_\alpha(\xi)-\vec{p}_\alpha(\xi)\times\vec{n}_\alpha(\xi,\vec{r})\}, \tag{1.7.8}$$

and which differs from (1.6.18) because the kinetic and canonical momenta are now no longer equal to each other.

The penultimate term of (1.7.7), proportional to the square of the transverse polarization, is composed of intra- and intermolecular terms,

$$\varepsilon_0^{-1}\int\sum_{\xi}|\vec{p}^{\perp}(\xi,\vec{r})|^2\mathrm{d}^3\vec{r}+\varepsilon_0^{-1}\int\sum_{\xi\neq\xi'}\vec{p}^{\perp}(\xi,\vec{r})\cdot\vec{p}^{\perp}(\xi',\vec{r})\mathrm{d}^3\vec{r}. \tag{1.7.9}$$

Noting that $\vec{p}(\vec{r})=\vec{p}^{\perp}(\vec{r})+\vec{p}^{||}(\vec{r})$, the total intermolecular polarization product vanishes for nonoverlapping charge distributions due to the rapid fall off with r of the polarization field outside the source molecule. Hence,

$$\varepsilon_0^{-1}\int\sum_{\xi\neq\xi'}\vec{p}^{\perp}(\xi,\vec{r})\cdot\vec{p}^{\perp}(\xi',\vec{r})\mathrm{d}^3\vec{r}=-\varepsilon_0^{-1}\int\sum_{\xi\neq\xi'}\vec{p}^{||}(\xi,\vec{r})\cdot\vec{p}^{||}(\xi',\vec{r})\mathrm{d}^3\vec{r}. \tag{1.7.10}$$

The right-hand side of (1.7.10) can be shown to be equal to $-\sum_{\xi<\xi'}V_{\text{inter}}(\xi,\xi')$. An explicit demonstration within the context of electron wavefields is given in Section 2.3. Therefore, in the multipolar Hamiltonian, the intermolecular Coulomb interaction energy is canceled by the intermolecular part of the transverse polarization, leaving an intramolecular self-energy term $(1/2\varepsilon_0)\int\sum_{\xi}|\vec{p}^{\perp}(\xi,\vec{r})|^2\mathrm{d}^3\vec{r}$. The multipolar Hamiltonian (1.7.7) can now be written as

$$H^{\text{mult}}=H^{\text{mult}}_{\text{mol}}+H^{\text{mult}}_{\text{rad}}+H^{\text{mult}}_{\text{int}}+\frac{1}{2\varepsilon_0}\int\sum_{\xi}|\vec{p}^{\perp}(\xi,\vec{r})|^2\mathrm{d}^3\vec{r}, \tag{1.7.11}$$

where

$$H^{\text{mult}}_{\text{mol}}=\sum_{\xi}\left\{\frac{1}{2m}\sum_{\alpha}\vec{p}_\alpha^2(\xi)+V(\xi)\right\}, \tag{1.7.12}$$

$$\begin{aligned}H^{\text{mult}}_{\text{rad}}&=\frac{1}{2}\int\left\{\frac{\vec{\Pi}^2(\vec{r})}{\varepsilon_0}+\varepsilon_0c^2(\vec{\nabla}\times\vec{a}(\vec{r}))^2\right\}\mathrm{d}^3\vec{r}\\&=\frac{1}{2\varepsilon_0}\int\{\vec{d}^{\perp 2}(\vec{r})+\varepsilon_0^2c^2\vec{b}^2(\vec{r})\}\mathrm{d}^3\vec{r},\end{aligned} \tag{1.7.13}$$

and

$$H_{\text{int}}^{\text{mult}} = \varepsilon_0^{-1}\int \vec{p}^{\perp}(\vec{r})\cdot\vec{\Pi}(\vec{r})\mathrm{d}^3\vec{r} - \int \vec{m}(\vec{r})\cdot(\vec{\nabla}\times\vec{a}(\vec{r}))\mathrm{d}^3\vec{r} + \frac{1}{2m}\left\{\int \vec{n}(\vec{r})\right.$$
$$\left.\times(\vec{\nabla}\times\vec{a}(\vec{r}))\mathrm{d}^3\vec{r}\right\}^2 = -\varepsilon_0^{-1}\int \vec{p}(\vec{r})\cdot\vec{d}^{\perp}(\vec{r})\mathrm{d}^3\vec{r} - \int \vec{m}(\vec{r})\cdot\vec{b}(\vec{r})\mathrm{d}^3\vec{r}$$
$$+\frac{1}{2}\int O_{ij}(\vec{r},\vec{r}\,')b_i(\vec{r})b_j(\vec{r}\,')\mathrm{d}^3\vec{r}\,\mathrm{d}^3\vec{r}\,'. \tag{1.7.14}$$

The last term in (1.7.11) is the aforementioned one-center contribution from the transverse electric polarization field; it is independent of the radiation field and does not contribute to processes that involve a change in the state of the electromagnetic field, although it must be included when calculating self-energy corrections. The last term of (1.7.14) represents the diamagnetic interaction, with the field $O_{ij}(\vec{r},\vec{r}\,')$ given by

$$O_{ij}(\vec{r},\vec{r}\,') = \sum_{\xi,\xi'}\frac{1}{m}\varepsilon_{ikl}\varepsilon_{jml}n_k(\xi,\vec{r})n_m(\xi',\vec{r}\,')$$
$$= \frac{e^2}{m}\varepsilon_{ikl}\varepsilon_{jml}\sum_{\xi,\xi'}\sum_{\alpha,\beta}(\vec{q}_\alpha(\xi)-\vec{R}_\xi)_k(\vec{q}_\beta(\xi')-\vec{R}_{\xi'})_m$$
$$\times\int_0^1\!\!\int_0^1 \lambda\lambda'\delta(\vec{r}-\vec{R}_\xi-\lambda(\vec{q}_\alpha(\xi)-\vec{R}_\xi))\delta(\vec{r}\,'-\vec{R}_{\xi'}-\lambda'(\vec{q}_\beta(\xi')-\vec{R}_{\xi'}))\mathrm{d}\lambda\,\mathrm{d}\lambda'. \tag{1.7.15}$$

Noteworthy features of the multipolar Hamiltonian include the fact that the Maxwell fields $\vec{d}^{\perp}(\vec{r})$ and $\vec{b}(\vec{r})$ appear explicitly in the radiation field and interaction Hamiltonians. The dependence of $H_{\text{int}}^{\text{mult}}$ on the electromagnetic fields only, rather than on the electromagnetic potentials clearly has the advantage of making equation (1.7.14) independent of gauge. Molecules couple directly to the radiation fields through the electric polarization, magnetization, and diamagnetization fields. Absent from H^{mult} is the intermolecular electrostatic interaction term. Interaction between molecules is now mediated by the field via the exchange of transverse photons that propagate with speed c.

The interaction Hamiltonian may be conveniently expanded in terms of multipole moments using the relations (1.6.8), (1.6.21), and (1.7.15) for the fields $\vec{p}(\xi,\vec{r})$, $\vec{m}(\xi,\vec{r})$, and $O_{ij}(\vec{r},\vec{r}\,')$, respectively, so as to simplify its use in subsequent applications that depend only on specific multipole moments. After carrying out the volume integral, the first few terms of (1.7.14) are therefore

$$H_{\text{int}}^{\text{mult}} = \sum_{\xi} \{ -\varepsilon_0^{-1} \vec{\mu}(\xi) \cdot \vec{d}^{\perp}(\vec{R}_{\xi}) - \varepsilon_0^{-1} Q_{ij}(\xi) \vec{\nabla}_j d_i^{\perp}(\vec{R}_{\xi}) - \vec{m}(\xi) \cdot \vec{b}(\vec{R}_{\xi}) \} + \frac{e^2}{8m} \sum_{\xi,\alpha} \{ (\vec{q}_{\alpha}(\xi) - \vec{R}_{\xi}) \times \vec{b}(\vec{R}_{\xi}) \}^2 + \cdots, \tag{1.7.16}$$

comprising electric dipole $\vec{\mu}(\xi)$, electric quadrupole $Q_{ij}(\xi)$, magnetic dipole $\vec{m}(\xi)$, and lowest order diamagnetic coupling of species ξ, with this last term of (1.7.16) being proportional to the square of the magnetic field.

From the definition of the electric displacement field (1.7.5) and the mode expansion (1.4.54) for the canonically conjugate momentum $\vec{\Pi}(\vec{r})$, the mode expansion for the transverse electric displacement field operator is

$$\vec{d}^{\perp}(\vec{r}) = i \sum_{\vec{k},\lambda} \left(\frac{\hbar c k \varepsilon_0}{2V} \right)^{1/2} \left[\vec{e}^{(\lambda)}(\vec{k}) a^{(\lambda)}(\vec{k}) e^{i\vec{k}\cdot\vec{r}} - \bar{\vec{e}}^{(\lambda)}(\vec{k}) a^{\dagger(\lambda)}(\vec{k}) e^{-i\vec{k}\cdot\vec{r}} \right]. \tag{1.7.17}$$

It is of interest to point out that from the definition of the transverse current density (1.5.5) and the electric dipole polarization distribution (1.6.9)

$$\vec{j}^{\perp}(\vec{r}) = \frac{d\vec{p}^{\perp}(\vec{r})}{dt} \tag{1.7.18}$$

in the electric dipole approximation and that the particle momentum canonically conjugate to the position operator,

$$\vec{p}_{\alpha} = \frac{\partial L^{\text{mult}}}{\partial \dot{\vec{q}}_{\alpha}} = m \dot{\vec{q}}_{\alpha}, \tag{1.7.19}$$

is equal to the kinetic momentum. Relation (1.7.5) still holds for the momentum canonically conjugate to the vector potential, but now also in the long-wavelength approximation. The effect of (1.7.18) is to replace coupling to $\vec{j}^{\perp}(\vec{r})$ by the transverse polarization in the multipolar Lagrangian (1.6.13) so that

$$L_{\text{int}}^{\text{mult}} = -\int \vec{p}^{\perp}(\vec{r}) \cdot \dot{\vec{a}}(\vec{r}) d^3\vec{r} - \sum_{\xi<\xi'} V_{\text{inter}}(\xi, \xi'). \tag{1.7.20}$$

As before, H^{mult} equation (1.7.11) results when (1.7.20) is used instead of the last three terms of (1.6.13), but with $H_{\text{int}}^{\text{mult}}$ now given by the first term of equation (1.7.16) only, which is the electric dipole form of the interaction operator.

Converting the dynamical variables to quantum mechanical operators, which obey the commutation rules

$$[q_{i(\alpha)}(\xi), p_{j(\beta)}(\xi')] = i\hbar\delta_{ij}\,\delta_{\alpha\beta}\,\delta_{\xi\xi'} \tag{1.7.21}$$

and

$$[a_i(\vec{r}), \Pi_j(\vec{r}\,')] = i\hbar\delta_{ij}^{\perp}(\vec{r}-\vec{r}\,'), \tag{1.7.22}$$

produces the quantum analogue of the multipolar Hamiltonian (1.7.11). This Hamiltonian is frequently used as the starting point in the calculation of processes involving the interaction of radiation with matter as well as for the study of long-range intermolecular forces. A completely second quantized form of H^{mult} proves to be advantageous for many of the applications to be detailed in subsequent chapters. The presentation of this field theoretic viewpoint is left for Chapter 2. Finally, it is interesting to note that the commutation relations (1.7.21) and (1.7.22) are identical to those occurring for minimal-coupling variables. The preservation of commutation rules is a direct consequence of carrying out a transformation that yielded an equivalent Lagrangian.

1.8 CANONICAL TRANSFORMATION

In the Hamiltonian formulation of classical mechanics, the equations of motion satisfied by the generalized coordinates q_i and momenta p_i, $i = 1, \ldots, 3N$, for a system of N particles are Hamilton's canonical equations (1.2.5). Alternatively, these relations may be expressed in terms of the Poisson bracket, which for two differentiable functions f and g that are functions of p and q,

$$\{f, g\} = \left\{\frac{\partial f}{\partial q}\frac{\partial g}{\partial p} - \frac{\partial f}{\partial p}\frac{\partial g}{\partial q}\right\}, \tag{1.8.1}$$

so that

$$\dot{q} = \{q, H\} \tag{1.8.2}$$

and

$$\dot{p} = \{p, H\}. \tag{1.8.3}$$

From the definition (1.8.1), it is seen that the Poisson bracket for the canonical pair q and p is unity,

$$\{q, p\} = 1. \tag{1.8.4}$$

If f is explicitly a function of time, namely, $f = f(q, p, t)$, then the time evolution of f is given by

$$\dot{f} = \{f, H\} + \frac{\partial f}{\partial t}. \tag{1.8.5}$$

In Section 1.6, it was shown how equivalent Lagrangians could be generated by the addition of the total time derivative of a function of the coordinates, which amounted to a canonical transformation of the dynamical variables, with construction of the corresponding Hamiltonian function requiring the evaluation of the canonically conjugate momenta. In general, the momenta obtained from two equivalent Lagrangians differ. Under a canonical transformation, however, the Poisson bracket (1.8.1) and Hamilton's canonical equations of motion (1.8.2) and (1.8.3) are preserved.

When f is a quantum mechanical operator, the counterpart to (1.8.5) is the Heisenberg equation of motion

$$\dot{f} = \frac{1}{i\hbar}[f, H] + \frac{\partial f}{\partial t}, \tag{1.8.6}$$

in which the commutator bracket now appears instead of the Poisson bracket. Analogously, the quantum versions of Hamilton's canonical equations take the form

$$i\hbar\dot{q} = [q, H] \tag{1.8.7}$$

and

$$i\hbar\dot{p} = [p, H], \tag{1.8.8}$$

with the fundamental commutator between position and momentum given by

$$[q, p] = i\hbar. \tag{1.8.9}$$

Application of a quantum canonical transformation leaves the commutation relation (1.8.9) and the operator equations of motion (1.8.7) and (1.8.8) unchanged. One type of transformation that possesses these properties is given by

$$q' = e^{iS} q e^{-iS} \tag{1.8.10}$$

and

$$p' = e^{iS} p e^{-iS}, \tag{1.8.11}$$

where S is an Hermitian operator and is called the generating function of the transformation. The transformed Hamiltonian is then found from the starting one by expressing it in terms of the transformed variables. In effect, this amounts to transforming the original Hamiltonian by the application of the canonical transformation; that is,

$$H' = e^{iS} H e^{-iS}. \tag{1.8.12}$$

Hence, the multipolar Hamiltonian can be obtained from the minimal-coupling Hamiltonian via the application of a canonical transformation of the form (1.8.12) with a suitable choice of the generator S (Power and Zienau, 1959). It now remains to find the connection between the function f, whose time derivative when taken and added to a Lagrangian produces equivalent Lagrangians, and the function S, which when utilized according to (1.8.12) gives rise to equivalent Hamiltonians.

Beginning with the Lagrangian L, an equivalent Lagrangian L' is given by

$$L'(q, \dot{q}, t) = L(q, \dot{q}, t) + \frac{d}{dt} f(q, t) = L(q, \dot{q}, t) + \frac{\partial}{\partial q} f(q, t)\dot{q} + \frac{\partial}{\partial t} f(q, t), \tag{1.8.13}$$

where f does not depend on the velocities. The Hamiltonian obtained from L' is

$$H' = p'\dot{q} - L', \tag{1.8.14}$$

with

$$p' = \frac{\partial L'}{\partial \dot{q}} = \frac{\partial L}{\partial \dot{q}} + \frac{\partial f(q, t)}{\partial q}, \tag{1.8.15}$$

since $q' = q$.

Employing the identity (Craig and Thirunamachandran, 1998a)

$$e^{A} B e^{-A} = B + [A, B] + \frac{1}{2!}[A, [A, B]] + \frac{1}{3!}[A, [A, [A, B]]] + \cdots, \tag{1.8.16}$$

in which A and B are two noncommuting operators, the transformed momentum (1.8.11) becomes

$$\begin{aligned} p' &= p + i[S,p] + \frac{i^2}{2!}[S,[S,p]] + \cdots \\ &= p - \hbar\frac{\partial S}{\partial q}, \end{aligned} \tag{1.8.17}$$

where S is taken to be a function of q only, and making use of the commutation relation

$$[S(q),p] = i\hbar\frac{\partial S}{\partial q}, \tag{1.8.18}$$

which results in only the first two terms of (1.8.17) surviving and all successive commutator expressions vanishing. From (1.8.15) and (1.8.17), the relation between the function f and the generator S is easily seen to be

$$f = -\hbar S. \tag{1.8.19}$$

Recalling (1.6.4), the generator that transforms H^{min} to give the same H^{mult} as that calculated from L^{mult} when starting with L^{min} is therefore

$$S = \frac{1}{\hbar}\int \vec{p}^{\perp}(\vec{r})\cdot\vec{a}(\vec{r})\mathrm{d}^3\vec{r}. \tag{1.8.20}$$

Because S is a function only of the generalized coordinates, only the canonically conjugate particle and field momenta are affected by the canonical transformation. Thus, the former is calculated from

$$\vec{p}^{\,\text{mult}}_{(\alpha)}(\xi) = \mathrm{e}^{iS}\vec{p}^{\,\text{min}}_{(\alpha)}(\xi)\mathrm{e}^{-iS} = \vec{p}^{\,\text{min}}_{(\alpha)}(\xi) + i[S,\vec{p}^{\,\text{min}}_{(\alpha)}(\xi)] + \cdots, \tag{1.8.21}$$

on using the identity (1.8.16). Inserting the expression for the electronic part of the electric polarization field (1.6.6), it can be shown that

$$\vec{p}^{\,\text{mult}}_{(\alpha)}(\xi) = \vec{p}^{\,\text{min}}_{(\alpha)}(\xi) + e\vec{a}(\vec{q}_{\alpha}(\xi)) - \int \vec{n}_{\alpha}(\xi,\vec{r})\times\vec{b}(\vec{r})\mathrm{d}^3\vec{r}. \tag{1.8.22}$$

Similarly, the transformed field momentum is

$$\begin{aligned} \vec{\Pi}^{\text{mult}}(\vec{r}) &= \mathrm{e}^{iS}\vec{\Pi}^{\text{min}}(\vec{r})\mathrm{e}^{-iS} = \vec{\Pi}^{\text{min}}(\vec{r}) + i[S,\vec{\Pi}^{\text{min}}(\vec{r})] + \cdots \\ &= \vec{\Pi}^{\text{min}}(\vec{r}) + \frac{i}{\hbar}\left[\int \vec{p}^{\perp}(\vec{r}\,')\cdot\vec{a}(\vec{r}\,')\mathrm{d}^3\vec{r}\,', \vec{\Pi}^{\text{min}}(\vec{r})\right] + \cdots. \end{aligned} \tag{1.8.23}$$

The commutator bracket is evaluated using the relation (1.7.22), giving

$$\vec{\Pi}^{\text{mult}}(\vec{r}) = \vec{\Pi}^{\text{min}}(\vec{r}) - \vec{p}^{\perp}(\vec{r}), \tag{1.8.24}$$

with all subsequent terms in the expansion (1.8.23) vanishing due to the fact that S commutes with the first commutator.

Substituting for $\vec{p}_{\alpha}^{\text{mult}}(\xi)$ and $\vec{\Pi}^{\text{mult}}(\vec{r})$ from (1.8.22) and (1.8.24) into the minimal-coupling Hamiltonian (1.5.11) results in

$$H^{\text{mult}} = \sum_{\xi}\left[\frac{1}{2m}\sum_{\alpha}\left\{\vec{p}_{\alpha}(\xi) + \int \vec{n}_{\alpha}(\xi,\vec{r}) \times \vec{b}(\vec{r})\mathrm{d}^3\vec{r}\right\}^2 + V(\xi)\right]$$
$$+ \frac{1}{2\varepsilon_0}\int\{[\vec{\Pi}(\vec{r}) + \vec{p}^{\perp}(\vec{r})]^2 + \varepsilon_0^2 c^2(\vec{\nabla} \times \vec{a}(\vec{r}))^2\}\mathrm{d}^3\vec{r} + \sum_{\xi<\xi'} V_{\text{inter}}(\xi,\xi'), \tag{1.8.25}$$

which is identical to the multipolar Hamiltonian (1.7.7) constructed from the multipolar Lagrangian.

It is straightforward to demonstrate that the time derivatives of the generalized coordinates are unchanged by the canonical transformation and are in fact equal to each other in the two formalisms. For this purpose, it is convenient to use the Heisenberg equation of motion (1.8.7). Thus,

$$\dot{\vec{q}}_{\alpha}^{\text{min}}(\xi) = \frac{1}{i\hbar}\left[\vec{q}_{\alpha}(\xi), H^{\text{min}}\right] = \frac{1}{m}\{\vec{p}_{\alpha}^{\text{min}}(\xi) + e\vec{a}(\vec{q}_{\alpha}(\xi))\} \tag{1.8.26}$$

and

$$\dot{\vec{q}}_{\alpha}^{\text{mult}}(\xi) = \frac{1}{i\hbar}\left[\vec{q}_{\alpha}(\xi), H^{\text{mult}}\right] = \frac{1}{m}\left\{\vec{p}_{\alpha}^{\text{mult}}(\xi) + \int \vec{n}(\xi,\vec{r}) \times \vec{b}(\vec{r})\mathrm{d}^3\vec{r}\right\}. \tag{1.8.27}$$

Clearly,

$$\dot{\vec{q}}_{\alpha}^{\text{min}}(\xi) = \dot{\vec{q}}_{\alpha}^{\text{mult}}(\xi) \tag{1.8.28}$$

on using (1.8.22). In a similar fashion,

$$\dot{\vec{a}}^{\text{min}}(\vec{r}) = \frac{1}{i\hbar}\left[\vec{a}(\vec{r}), H^{\text{min}}\right] = \varepsilon_0^{-1}\vec{\Pi}^{\text{min}}(\vec{r}) \tag{1.8.29}$$

and

$$\dot{\vec{a}}^{\text{mult}}(\vec{r}) = \frac{1}{i\hbar}\left[\vec{a}(\vec{r}), H^{\text{mult}}\right] = \varepsilon_0^{-1}\left\{\vec{\Pi}^{\text{mult}}(\vec{r}) + \vec{p}^{\perp}(\vec{r})\right\}, \tag{1.8.30}$$

which with the aid of (1.8.24) prove that

$$\dot{\vec{a}}^{\mathrm{min}}(\vec{r}) = \dot{\vec{a}}^{\mathrm{mult}}(\vec{r}). \tag{1.8.31}$$

Finally, it should be mentioned that the transformation discussed in this section is more precisely known as a Lagrangian-induced quantum completely canonical transformation (Power, 1978). This corresponds to a quantum canonical transformation with time-independent generator S. The transformation is therefore unitary, and the eigenspectra resulting from the use of either H^{min} or H^{mult} are identical.

1.9 PERTURBATION THEORY SOLUTION

In Section 1.7, it was shown how the multipolar Hamiltonian in Coulomb gauge for a system of charged particles in interaction with electromagnetic radiation could be written as a sum of molecular, radiation field, and interaction Hamiltonians, as well as including a term involving the square of the intramolecular transverse polarization field (1.7.11). It now remains to discuss how such a system of quantum mechanical equations is solved in general.

In the absence of any interaction between radiation and matter, the total Hamiltonian is simply a sum of molecular and radiation field Hamiltonians, $H_{\mathrm{mol}} + H_{\mathrm{rad}}$. Such a Hamiltonian is separable, with eigenenergy being the sum of the molecule and radiation field energies and eigenfunctions being the product states of molecule and radiation field wavefunctions. The quantization of the free electromagnetic field was carried out in Section 1.4, where an occupation number representation was used to specify the state of the radiation field, with n quanta of frequency ω having energy $n\hbar\omega$. Earlier in Section 1.2 it was shown how the application of the variational calculus to the classical Lagrangian function for a system of charged particles and the subsequent application of the canonical quantization prescription led to the familiar quantum mechanical molecular Hamiltonian H_{mol}, whose solution using a vast array of quantum chemical techniques is formally taken to be known, yielding eigenfunctions $|E_m\rangle$ for a molecular state of the system characterized by quantum number m with eigenenergy E_m. The form of the total Hamiltonian for the coupled matter–field system, be it in the minimal-coupling or multipolar frameworks (1.5.13) and (1.7.11), respectively, naturally lends itself to a perturbation theory solution.

The total quantum electrodynamical Hamiltonian is divided according to

$$H = H_0 + H_{\text{int}}, \tag{1.9.1}$$

where

$$H_0 = H_{\text{mol}} + H_{\text{rad}} \tag{1.9.2}$$

constitutes the unperturbed Hamiltonian, with

$$H_{\text{mol}} = \sum_{\xi} \left\{ \frac{1}{2m} \sum_{\alpha} \vec{p}_{\alpha}^{\,2}(\xi) + V(\xi) \right\} \tag{1.9.3}$$

and

$$H_{\text{rad}} = \frac{1}{2\varepsilon_0} \int \{ \vec{\Pi}^2(\vec{r}) + \varepsilon_0^2 c^2 (\vec{\nabla} \times \vec{a}(\vec{r}))^2 \} \mathrm{d}^3\vec{r}. \tag{1.9.4}$$

The eigenstates of H_0 are written as product states of molecular and radiation field eigenfunctions $|E_m^{\xi}; n(\vec{k}, \lambda)\rangle = |E_m^{\xi}\rangle |n(\vec{k}, \lambda)\rangle$ corresponding to molecule ξ in electronic state $|m\rangle$ and the electromagnetic field characterized by n photons of mode $(\vec{k}, \lambda)$. These orthonormal functions form a basis set that is employed in the perturbation theory solution. The justification for such a treatment is that H_0 represents the solution to a known problem, in the case of (1.9.2) the noninteracting system of molecule(s) and the radiation field. Further, the coupling of radiation and matter, given by the second term of (1.9.1), is viewed as a small perturbation on the total system. This is based on the fact that the particle–field interaction terms are considerably smaller in magnitude than the strengths of Coulombic fields present within an atomic or molecular system. Except for very intense fields, of the order of 10^{12} V m^{-1}, the perturbation approximation holds true for both the minimal- and multipolar-coupling schemes. The interaction Hamiltonians for these two formalisms are, respectively, given by

$$H_{\text{int}}^{\text{min}} = \frac{e}{m} \sum_{\xi} \sum_{\alpha} \vec{p}_{\alpha}(\xi) \cdot \vec{a}(\vec{q}_{\alpha}(\xi)) + \frac{e^2}{2m} \sum_{\xi} \sum_{\alpha} \vec{a}^2(\vec{q}_{\alpha}(\xi)) + V_{\text{inter}} \tag{1.9.5}$$

and

$$H_{\text{int}}^{\text{mult}} = -\varepsilon_0^{-1} \int \vec{p}(\vec{r}) \cdot \vec{d}^{\perp}(\vec{r}) \mathrm{d}^3\vec{r} - \int \vec{m}(\vec{r}) \cdot \vec{b}(\vec{r}) \mathrm{d}^3\vec{r} + \frac{1}{2} \int O_{ij}(\vec{r}, \vec{r}\,') b_i(\vec{r}) b_j(\vec{r}\,') \mathrm{d}^3\vec{r}\, \mathrm{d}^3\vec{r}\,'. \tag{1.9.6}$$

The expansion of the three terms of (1.9.6) in lowest order multipole moments was given by equation (1.7.16).

The two most common perturbative approaches for the solution of equations of the type (1.9.1) differ in their dependence on time. In time-independent perturbation theory, the equation to be solved is the time-independent Schrödinger equation,

$$(H_0 + \lambda H_{\text{int}})|i\rangle = E_i|i\rangle, \tag{1.9.7}$$

for the energy of the perturbed system E_i, which is taken to be nondegenerate, and its corresponding perturbed state function $|i\rangle$ in a power series expansion in the perturbation operator H_{int}. This is most frequently done using the method of Rayleigh and Schrödinger (Levine, 2000) for Hamiltonians that do not depend on time and is most useful for calculating shifts in energy levels of the perturbed system along with its perturbed wavefunction. The parameter λ, which lies between 0 and 1, ensures the perturbation operator, in the present case the interaction Hamiltonian, is applied smoothly. Ultimately, λ is eliminated by setting it equal to unity, corresponding to the situation in which the perturbation is acting fully. When $\lambda = 0$, equation (1.9.7) reduces to the unperturbed problem, represented by the Hamiltonian H_0, which satisfies the eigenvalue equation

$$H_0|i^{(0)}\rangle = E_i^{(0)}|i^{(0)}\rangle, \tag{1.9.8}$$

whose eigenfunction $|i^{(0)}\rangle$ and eigenenergy $E_i^{(0)}$ are taken to be known. The perturbed state and energy are expanded in series of powers of λ,

$$|i\rangle = |i^{(0)}\rangle + \lambda|i^{(1)}\rangle + \lambda^2|i^{(2)}\rangle + \cdots \tag{1.9.9}$$

and

$$E_i = E_i^{(0)} + \lambda E_i^{(1)} + \lambda E_i^{(2)} + \cdots; \tag{1.9.10}$$

$|i^{(1)}\rangle, |i^{(2)}\rangle, \ldots$ and $E^{(1)}, E^{(2)}, \ldots$ are successive perturbative corrections to the state function and energy, respectively, of the perturbed problem. Well-known formulas result for the perturbed state $|i\rangle$ and energy E_i. In terms of the unperturbed state $|i^{(0)}\rangle$ and energy $E_i^{(0)}$, they are

$$|i\rangle = |i^{(0)}\rangle + \sum_{j^{(0)} \neq i^{(0)}} \frac{\langle j^{(0)}|H_{\text{int}}|i^{(0)}\rangle}{E_i^{(0)} - E_j^{(0)}} |j^{(0)}\rangle + \cdots \tag{1.9.11}$$

and

$$E_i = E_i^{(0)} + \langle i^{(0)}|H_{\text{int}}|i^{(0)}\rangle + \sum_{j^{(0)}\neq i^{(0)}} \frac{\langle i^{(0)}|H_{\text{int}}|j^{(0)}\rangle\langle j^{(0)}|H_{\text{int}}|i^{(0)}\rangle}{E_i^{(0)} - E_j^{(0)}} + \cdots. \tag{1.9.12}$$

In the term corresponding to the first-order correction to the wavefunction and to the second-order correction to the energy, the sum is executed over all unperturbed states except $|i^{(0)}\rangle$.

When the interaction Hamiltonian or the total Hamiltonian is time dependent, the dynamics involves the time evolution of the stationary states of the unperturbed system, which can now make transitions from one state to another under the influence of the perturbation. The dynamics is governed by the time-dependent Schrödinger equation

$$i\hbar \frac{\partial}{\partial t}|\Psi(t)\rangle = H|\Psi(t)\rangle, \tag{1.9.13}$$

in which the states $|\Psi(t)\rangle$ are explicitly time dependent. This is characteristic of the Schrödinger picture of quantum mechanics. It is convenient to view the time variation of the state function $|\Psi(t)\rangle$ as due to the action of a transformation operator $U(t)$ on a fixed state of the system at some initial time t_0,

$$|\Psi(t)\rangle = U(t)|\Psi(t_0)\rangle. \tag{1.9.14}$$

$U(t)$ is more commonly called the time evolution operator, and it is unitary so that the normalization properties associated with $|\Psi(t)\rangle$ are retained for all t. Substituting (1.9.14) into (1.9.13) shows that $U(t)$ satisfies the equation of motion

$$i\hbar \frac{\partial}{\partial t}U(t) = HU(t), \tag{1.9.15}$$

whose formal solution is simply

$$U(t) = \mathrm{e}^{-iH(t-t_0)/\hbar}, \tag{1.9.16}$$

when subject to the initial condition $U(t_0) = 1$, enabling $|\Psi(t)\rangle$ to be evaluated at any time via (1.9.14).

In the standard treatment of time-dependent perturbation theory, all system states are written in terms of the eigenfunctions $|i\rangle$ of the unperturbed Hamiltonian H_0 with energy E_i, $H_0|i\rangle = E_i|i\rangle$, along with their time-dependent factors, as in

$$|\Psi(t)\rangle = \sum_i a_i(t)\mathrm{e}^{-iE_it/\hbar}|i\rangle, \tag{1.9.17}$$

and the problem amounts to determining the time-dependent coefficients $a_i(t)$ from which state-to-state transition probabilities may be calculated. This is known as Dirac's variation of constants method. Equivalent results for the time evolution may be obtained in terms of the operator $U(t)$ introduced above (Ziman, 1969). Further calculational advantages ensue, however, if the time factor is removed from each basis state and absorbed into the operator itself, so that using (1.9.7)

$$|\Psi(t)\rangle = \mathrm{e}^{-iH_0(t-t_0)/\hbar}U_I(t, t_0)|\Psi(t_0)\rangle. \tag{1.9.18}$$

The perturbation operator becomes

$$H^I_{\mathrm{int}}(t) = \mathrm{e}^{iH_0(t-t_0)/\hbar}H_{\mathrm{int}}\mathrm{e}^{-iH_0(t-t_0)/\hbar}. \tag{1.9.19}$$

Both the states and the operators are time dependent, and the focus is on the effect of the perturbation on the system. This representation lies in between the Schrödinger picture and the Heisenberg formalism. Recall that in the latter viewpoint, the dynamical variables are all time dependent and the states are time independent. The new picture symbolized by equations (1.9.18) and (1.9.19) is called the interaction representation or the Dirac representation, as indicated by the additional label I. Hence, $U_I(t, t_0)$ is the interaction picture time-evolution operator. It satisfies the operator equation of motion

$$i\hbar\frac{\partial}{\partial t}U_I(t, t_0) = H^I_{\mathrm{int}}(t)U_I(t, t_0) \tag{1.9.20}$$

on using (1.9.18) and (1.9.19) in the time-dependent Schrödinger equation (1.9.13).

Continuing in the interaction representation, but now dropping the symbol I, the state function at time t is given by

$$|\Psi(t)\rangle = U(t, t_0)|\Psi(t_0)\rangle. \tag{1.9.21}$$

If H_{int} is time independent, then analogously to (1.9.16), the formal solution to (1.9.20) is

$$U(t, t_0) = \mathrm{e}^{-iH_{\mathrm{int}}(t-t_0)/\hbar}. \tag{1.9.22}$$

On the other hand, if $H_{\text{int}} = H_{\text{int}}(t)$, but is taken to be a c-number, the differential equation (1.9.20) can be integrated to give for the time-evolution operator the solution

$$U(t, t_0) = \exp\left\{ -\frac{i}{\hbar} \int_{t_0}^{t} H_{\text{int}}(t')\mathrm{d}t' \right\}. \tag{1.9.23}$$

This form of $U(t, t_0)$, however, cannot be used in the quantum theory, since H_{int} is an operator and does not, in general, commute with itself at two different times t_1 and t_2. A formal series solution for $U(t, t_0)$ may be constructed for application in quantum mechanical problems when H_{int} is an operator. This is accomplished by integrating (1.9.20) with respect to time, subject to the initial condition

$$U(t_0, t_0) = 1, \tag{1.9.24}$$

to give

$$U(t, t_0) = 1 + \frac{1}{i\hbar} \int_{t_0}^{t} \mathrm{d}t_1\, H_{\text{int}}(t_1) U(t_1, t_0). \tag{1.9.25}$$

The right-hand side of (1.9.25) is reinserted as an expression for $U(t_1, t_0)$ under the integral sign and successively iterated in this way, eventually leading to a power series for $U(t, t_0)$ in terms of H_{int}. Thus,

$$U(t,t_0) = 1 + \sum_{n=1}^{\infty} \left(\frac{1}{i\hbar}\right)^n \int_{t_0}^{t} \mathrm{d}t_1 \int_{t_0}^{t_1} \mathrm{d}t_2 \ldots \int_{t_0}^{t_{n-1}} \mathrm{d}t_n H_{\text{int}}(t_1) \ldots H_{\text{int}}(t_n). \tag{1.9.26}$$

The time ordering of the operators is explicit and the series result (1.9.26) is exact. In applying perturbation theory, the central problem is to compute successive terms, and if at all possible, the series sum.

It is now a simple matter to use (1.9.26) to calculate the probability amplitude of finding the system in the final state $|f\rangle$ at time t as a result of a perturbation H_{int}, which began to act at a time t_0 when the system was in the initial state $|i\rangle$. The matrix elements of the time evolution operator are

$$\langle f|U(t, t_0)|i\rangle = \delta_{if} - \sum_{\xi} M_{fi}(\xi) \frac{[\mathrm{e}^{i(E_f - E_i)(t-t_0)/\hbar} - 1]}{E_f - E_i}, \tag{1.9.27}$$

where the matrix element for the process is denoted by M_{fi}. Clearly, there is no probability of finding the system in state $|f\rangle$ in zeroth order of H_{int}

because the initial and final states are orthogonal to one another. In powers of the perturbation operator, the matrix element is expanded as

$$M_{fi} = \langle f|H_{\text{int}}|i\rangle + \sum_{I} \frac{\langle f|H_{\text{int}}|I\rangle\langle I|H_{\text{int}}|i\rangle}{(E_i - E_I)} + \sum_{I,II} \frac{\langle f|H_{\text{int}}|II\rangle\langle II|H_{\text{int}}|I\rangle\langle I|H_{\text{int}}|i\rangle}{(E_i - E_I)(E_i - E_{II})}$$
$$+ \sum_{I,II,III} \frac{\langle f|H_{\text{int}}|III\rangle\langle III|H_{\text{int}}|II\rangle\langle II|H_{\text{int}}|I\rangle\langle I|H_{\text{int}}|i\rangle}{(E_i - E_I)(E_i - E_{II})(E_i - E_{III})} + \cdots, \tag{1.9.28}$$

where as before summation is carried out over all intermediate states that connect initial to final excluding the system states $|i\rangle$ and $|f\rangle$.

Ignoring the Kronecker delta term of (1.9.27), the time-dependent probability of finding the system in the final state is

$$P_{f\leftarrow i}(t) = |\langle f|U(t,t_0)|i\rangle|^2 = \sum_{\xi} 4|M_{fi}(\xi)|^2 \frac{\sin^2 \omega_{fi}(t-t_0)/2}{\hbar^2\omega_{fi}^2}, \tag{1.9.29}$$

where energy is conserved subject to

$$\hbar\omega_{fi} = (E_f - E_i). \tag{1.9.30}$$

In atomic and molecular systems, transitions can occur between discrete bound states that form part of an energy manifold. If the final state belongs to a continuum of levels centered at some frequency ω with range $\Delta\omega$, the total probability $P_{\text{tot}}(t)$ is a sum of all individual probabilities,

$$P_{\text{tot}}(t) = \sum_{f} P_{f\leftarrow i}(t), \tag{1.9.31}$$

with $P_{f\leftarrow i}(t)$ given by (1.9.29). This picture of a transition from a discrete state to a continuum of levels holds if each transition is taken to be independent of every other one. This is true for small $t - t_0$. Thus,

$$P_{\text{tot}}(t) = \sum_{\xi} \frac{2\pi}{\hbar} |M_{fi}(\xi)|^2 (t-t_0)\rho_f, \tag{1.9.32}$$

where ρ_f is the density of final states, namely, the number of levels per unit energy, $\mathrm{d}n_f/\mathrm{d}E_f$. Taking the time derivative of (1.9.32) leads to the familiar Fermi golden rule transition rate expression

$$\Gamma = \frac{\mathrm{d}}{\mathrm{d}t} P_{\text{tot}}(t) = \sum_{\xi} \frac{2\pi}{\hbar} |M_{fi}(\xi)|^2 \rho_f. \tag{1.9.33}$$

The evaluation of level shifts and spectroscopic rates using the perturbative expansions (1.9.12) and (1.9.28) is greatly facilitated by the introduction of diagrammatic techniques, the most well known being the time-ordered graphs of Feynman (Feynman, 1948, 1949a, 1949b; Mattuck, 1976). These diagrams have the additional advantage of yielding further valuable insight into the underlying physical process by providing a visual representation of individual electron–photon interactions in time-ordered sequences of emission and absorption events. For a process involving the interaction of n photons, the leading contribution to the quantum amplitude is given by the nth-order term in H_{int} when the coupling between radiation and matter is linear in the electric and magnetic fields, as is the case for the electric and magnetic multipolar series. At a specific order in perturbation theory, the summation over all intermediate states that link the same initial and final states $|i\rangle$ and $|f\rangle$ amounts to the drawing of all possible topologically distinct time orderings of photon creation and destruction operations. Therefore, an individual time-ordered diagram is isomorphic to one term in the time-dependent perturbation theory sum of the probability amplitude for a given nth-order process.

Feynman diagrams have proved to be extremely versatile in depicting, and beneficial in computing, electron–photon interactions occurring in elementary particle physics, many-body perturbation theory, atomic, molecular, and optical physics, and theoretical chemistry, though originally developed to be applied in quantum electrodynamics. Soon after their initial deployment, Dyson (1949) immediately recognized the power of the newly proposed visual representation, remarking that "In Feynman's theory the graph corresponding to a particular matrix element is regarded, not merely as an aid to calculation, but as a picture of the physical process which gives rise to that matrix element." As such, an adapted version of Feynman's originally proposed diagrammatic technique aids in the calculation of matrix elements within nonrelativistic theory. Numerous radiation–molecule and molecule–molecule interactions have been calculated and understood at a fundamental level, through the employment of time-dependent perturbation theory together with time-ordered diagrams.

There are, however, a number of drawbacks associated with this particular pictorial representation. Chief among them is that for higher order processes, for example, those involving emission/absorption of a significant number of real and/or virtual photons at either single and/or multiple molecular centers, the number of possible time orderings describing evolution from the same initial to final state for a specific process can very quickly grow in number. Obviously, this limits the actual drawing of all contributory diagrams required in the perturbation theory summation over all possible intermediate states to ensure that the resulting amplitude or energy shift correctly accounts for all terms. Illustration of a process by one graph on its own or by a finite subset of the total number of time orderings therefore provides an incomplete picture and is devoid of physical meaning. A further deficiency is that common features associated with distinct processes are not brought to the fore in the time-ordering approach. To overcome some of these shortcomings, a new visual representation of laser–matter and intermolecular interactions—a state sequence diagram—has recently been formulated. A summary of the principles underlying their formal construction is given in the next section.

1.10 STATE SEQUENCE DIAGRAMS

An alternative visual representation to time-ordered diagrams has been developed in which one picture, termed a state sequence diagram, is employed to illustrate and aid in the calculation of a specific laser–molecule or intermolecular process (Jenkins et al., 2002). Drawing of these latter type of diagrams is made possible by transforming the representation of photon–matter couplings in hyperspace, whose dimension is determined by the particular process under investigation, to a coordinate network existing in one plane; application of linkage rules allows valid connections to be forged between initial, intermediate, and final states in a systematic manner without explicit reference to individual photon emission and/or absorption events. The treatment given is general enough to enable both unique and indistinguishable electron–photon interactions to be examined in one formalism, thereby accommodating the effects of possible photon degeneracy.

The first step in the construction of a state sequence diagram involves initial identification of the total number of electron–photon interactions, n, occurring in some process. This quantifies the hyperspace dimension of the interacting system. Next, an index is assigned to each photon interaction. This labeling conceals any physical significance associated with photon creation and annihilation, with the procedure to be followed in the drawing of network planes and state sequence diagrams essentially involving index manipulation. Next, an orthonormal basis set that spans the n-space is introduced by representing each index by a vector $\vec{i}_i$,

$$I = \{c_1\vec{i}_1, c_2\vec{i}_2, \ldots, c_n\vec{i}_n\}, \tag{1.10.1}$$

where the coefficient c_i of the ith vector denotes the photon multiplicity. Clearly,

$$\sum_{i=1}^{j} c_i = n, \tag{1.10.2}$$

where the sum is executed over the number of distinct indices, j. Also evident is that when each of the photonic events is unique, $n=j$ and all of the vector coefficients are unity. When more than one photon of a particular mode is emitted or absorbed, however, the particular coefficient will no longer be one so that $j \neq n$, but since n is fixed in value, $n-j$ indices will have vanishing coefficients and are therefore superfluous; hence, no vector representation is necessary in this case. Thus, the truncated set of vectors $I = \{c_1\vec{i}_1, \ldots, c_j\vec{i}_j\}$ represents a subspace of dimension j in the full n-dimensional hyperspace, with the coefficients arranged subject to

$$c_j \geq c_{j-1} \geq \cdots \geq c_1 \geq 1. \tag{1.10.3}$$

The initial, final, and intermediate states for a process are then denoted by points in the hyperspace through generation of coordinates $(C_1, C_2, \ldots, C_j)$ from the vectors (1.10.1). The scalars C_j take on values $C_j = 0, 1, \ldots, c_j$ to accommodate the possible sequencing of I. Because the initial and final states are well defined, they correspond to fixed points on the planar interaction network. The intermediate states, on the other hand, in general require sequence ordering. Taken together, a two-dimensional array of points results, their joining indicating an allowed ordering of photon creation and destruction events.

The points occurring in the interaction plane are designated by the coordinates (k, h). Those points lying on the vertical axis are termed hyperspace numbers h and are obtained as follows. From the vector coefficient with the highest value is found the numerical base of the space, B, through the relation

$$B = c_j + 1. \tag{1.10.4}$$

The numbers h are found by first converting the hyperspace coordinates to base B and subsequent re-expression in base 10, namely,

$$(C_1, C_2, \ldots, C_j) \to C_1C_2\ldots C_j|_B \equiv \text{hyperspace number}|_{10} = h. \tag{1.10.5}$$

Meanwhile, the points lying on the horizontal axis, k, are found from

$$k = \sum_{i=1}^{j} C_i, \tag{1.10.6}$$

with $k = 0, 1, \ldots, n$ comprising $n + 1$ subsets of the complete set of $\{(k, h)\}$ points. Each subset is composed of m elements denoted by the vertex r_k^m with the values of m restricted to

$$m = 1, \ldots, |r_k|, \tag{1.10.7}$$

from which k can be construed as accounting for the number of steps between point r_k^m and the point representing the initial state $|i\rangle$. Because $|i\rangle$ represents the state of the system before any interaction has occurred, its network coordinate is taken to be the origin (0, 0), since $k = 0$ and $C_i = 0$, and can also be denoted by r_0^1. In the interaction plane, the coordinate of the final state is denoted by (n, h_f) or r_n^1 since k takes on its maximum value of n. The hyperspace number for f, h_f, corresponds to the situation in which the number of components for each member of the set I is a maximum and is obtained from

$$(c_1, c_2, \ldots, c_j) \rightarrow c_1 c_2 \ldots c_j|_{B=c_j+1} \equiv h_f|_{10}. \tag{1.10.8}$$

Finally, rules that connect two points in the interaction plane must be formulated, from which all allowed paths between the $|i\rangle$ and $|f\rangle$ termini can be constructed so that the state sequence diagram can ultimately be sketched. For any two vertices r_k^m and $r_{k'}^{m'}$ mapping coordinates $(C_1, C_2, \ldots, C_i)$ to $(C'_1, C'_2, \ldots, C'_i)$

$$\sum_{i=1}^{j} |C'_i - C_i| = 1, \tag{1.10.9}$$

which together with equation (1.10.6) results in the linkage rule

$$k' = k \pm 1. \tag{1.10.10}$$

The total number of paths $\mathcal{P}$ permitted between $|i\rangle$ and $|f\rangle$ is given by

$$\mathcal{P} = \frac{n!}{\prod_i (c_i!)} \tag{1.10.11}$$

and represents permutations of interaction indices. Application of this procedure enables the construction of an interaction plane network that serves as a template for the eventual state sequence depiction of all processes of order n involving distinguishable electron–photon coupling events. The number of permutations calculated from (1.10.11) is equal to the number of possible time-ordered sequences of photon absorption and emission events as drawn in separate time-ordered diagrams when a particular process is considered. Hence, it is apparent that the advantage offered by the state sequence approach lies in the information embodied by the interaction plane network and that one state sequence diagram contains all time orderings.

It proves useful to introduce additional properties associated with the interaction vertices to further aid the drawing and understanding of network planes. One of these is the structure coefficient, defined by

$$^{F}T_{k}^{n,j} = |r_k|. \tag{1.10.12}$$

To aid in the arrangement of vertex sets and to specify the value of m, a partition function F is introduced through

$$F = \{c_1; \ldots; c_j\}, \tag{1.10.13}$$

which relates the basis set I (1.10.1) to the structure coefficients. For the specific case in which all of the photon creation and destruction events are distinguishable, namely, all of the indices $\vec{i}_j$ are unique, $F = \{1; \ldots; 1\}$, and $n = j$, the values of the structure coefficients (1.10.12) are the binomial coefficients,

$$^{\{1;\ldots;1\}}T_{k}^{n,n} = \begin{cases} \binom{n}{k} = \dfrac{n!}{(n-k)!k!}, & 0 \leq k \leq n, \\ 0, & k < 0, k > n. \end{cases} \tag{1.10.14}$$

Hence, a process containing n distinct photonic events will give rise to an interaction plane network and an eventual state sequence diagram partitioned with coefficients generated by the nth row of Pascal's triangle. Furthermore, knowledge of the structure coefficient (1.10.12) for a given value of k enables the structure of an interaction network plane to be predicted via the recursion relation

$$^{\{c_1;\ldots;c_j\}}T_{k}^{n,j} = \sum_{k'=k-c_j}^{k} {}^{\{c_1;\ldots;c_{j'}\}}T_{k'}^{n',j'}, \tag{1.10.15}$$

where $n' = n - c_j$ and $j' = j - 1$. Iteration of (1.10.15) creates partitions that contain n-space coefficients greater than unity, enabling degenerate photons also to be treated. Successive application decreases the value of the coefficient, finally bottoming out at $F = \{1; \ldots; 1\}$ or the null set, in the process reducing to the result (1.10.14). For the case when $C_j > 1$, the partition function (1.10.13) can be expressed as

$$F = \{c_1; \ldots; c_u; \ldots c_j\}, \tag{1.10.16}$$

after the introduction of an index u, with $c_j \geq \cdots \geq c_u > 1$, and so describes degenerate cases where $n > j$. All structure coefficients can therefore be expressed through nested sums of binomial coefficients,

$$
{}^{\{c_1;\ldots;c_u;\ldots c_j\}}T_k^{n,j} = \sum_{k_1=k-c_j}^{k} \sum_{k_2=k_1-c_{j-1}}^{k_1} \cdots \sum_{k_{j-u+1}=k_{j-u}-c_u}^{k_{j-u}} \binom{u-1}{k_{j-u+1}}. \tag{1.10.17}
$$

Formal application of the prescription presented herein to enable drawing of state sequence diagrams for a variety of intermolecular interactions will be undertaken in the chapters to follow as appropriate; these will be compared and contrasted with the commonly used Feynman diagram approach. State sequence diagrams have been employed with success to processes such as second harmonic generation, six-wave mixing, two-photon distributed absorption, energy transfer and pooling in two-, three-, and four-center systems, and laser-induced intermolecular forces.

CHAPTER 2

MOLECULAR QUANTUM ELECTRODYNAMICS: FIELD THEORETIC TREATMENT

> *... the natural formulation of the quantum theory of electrons is obtained by simultaneously conceiving radiation and matter as waves in interaction in three dimensional space...*
>
> —P. Jordan, *Z. Phys.* **44**, 473 (1927).

2.1 INTRODUCTION

Two common formulations and representations of quantum mechanics are the Schrödinger and Heisenberg pictures (Dirac, 1958). In the former description, the time evolution of the system is determined by the time-dependent Schrödinger equation. It has the characteristic feature that the state vector of the system changes with time while the operators are time independent. In the Heisenberg point of view, on the other hand, the dynamical variables correspond to moving operators and the states to fixed vectors. Now the dynamics is dictated by the Heisenberg equations of motion for the dynamical variables. The two pictures are related by the transformation theory, in which a unitary transformation is applied to both the state vectors and the linear operators. This ensures that the expectation

---

value of an observable quantity is identical in either representation. Clearly, at the initial time, the Schrödinger and Heisenberg picture operators are the same. Because the Heisenberg equation of motion for an operator is the quantum mechanical analogue of the classical mechanical equation defining the time variation of a dynamical variable subject to Hamilton's principle, Heisenberg's formulation is often beneficial in that it enables various aspects of classical mechanics to be easily converted to quantum theory. Hence, for many applications, adoption of the Heisenberg picture offers calculational advantages as well as additional physical insight, even though the Schrödinger picture is the more commonly chosen viewpoint that typically leads to simpler equations to be solved. As a result, quantum electrodynamics can be developed and applied according to either viewpoint.

Formulation of the theory in the Heisenberg picture is most easily carried out using the techniques of second quantization on adopting a field theoretic point of view, in which electron and photon fields interact. While a well-defined classical limit exists for the quantized electromagnetic field, no such limit exists for the quantized electron wavefield. Picturing matter and radiation as fields is, however, entirely equivalent to a many-body description of a system of electrons interacting with the photons of the radiation field, bringing to the fore the complementary nature of the wave and particle points of view. Even though electron–positron pair production can be treated easily using the formalism of second quantization, such an effect does not occur for the systems to be examined in what follows. Hence, the fermion operator simply causes a change in the excitation energy level of a bound electron.

In Section 2.2, the second quantized minimal-coupling Hamiltonian is derived from the Lagrangian for the interacting electron and photon wavefields by following the standard canonical quantization procedure. It is then shown how L_{min} may be transformed to L_{mult} by changing the generalized coordinate of the electron wavefield, from which the second quantized multipolar Hamiltonian then results. Section 2.3 demonstrates how application of a canonical transformation on H_{min} yields H_{mult}. Exact expressions are obtained in Section 2.4 for the time-dependent vector potential and Maxwell fields within the electric dipole approximation. Comparison and contrast are made between minimal- and multipolar-coupling radiation fields in Section 2.5. In the following section, the electric displacement and magnetic field operators linear and quadratic in the electric dipole moment are computed. It is found that for many applications it is necessary to go beyond the first-order fields. Similarly, a number of problems require, for their solution, knowledge of the higher multipole moment Maxwell fields. Sections 2.7 and 2.8 contain expressions for the Maxwell field dependent upon magnetic dipole, electric quadrupole, and diamagnetic couplings. In the final two sections of this chapter, observables associated with the radiation fields are calculated. These include the electromagnetic energy density in the vicinity of a source and the Poynting vector.

2.2 NONRELATIVISTIC QUANTUM FIELD THEORY

In Chapter 1, the key steps were given for obtaining the quantum electrodynamical Hamiltonian operator in either the minimal-coupling or multipolar form for a charged particle interacting with a radiation field starting from the classical minimal-coupling Lagrangian function. In this section, a detailed presentation is given of the alternative field theoretic approach in which both matter and radiation are considered as wavefields, leading ultimately to the multipolar Hamiltonian in second quantized form.

If the electron is viewed as wavelike, the Lagrangian for the situation in which no radiation field is present, and self-interaction energy terms are neglected, gives rise to the Schrödinger equation and its complex conjugate when the Euler–Lagrange equations are applied to the electron wavefields $\psi(\vec{q})$ and $\bar{\psi}(\vec{q})$ (Schiff, 1955). It was shown in Section 1.5 that to correctly include the effects of electromagnetic radiation, and its subsequent interaction with matter, the principle of minimal electromagnetic coupling could be invoked, that is, to simply substitute $-i\hbar\vec{\nabla}^{(\vec{q})}$ (the gradient form of the momentum operator) by $-i\hbar\vec{\nabla}^{(\vec{q})} + e\vec{a}^{\perp(\vec{q})}$. Including now the self-interaction energy, as well as the external potential $V(\vec{q})$, the minimal-coupling Lagrangian for the system is written as (Power and Thirunamachandran, 1983a)

$$
\begin{aligned}
L_{\text{min}}(\vec{a}^{\perp},\psi,\bar{\psi};\dot{\vec{a}}^{\perp},\dot{\psi},\dot{\bar{\psi}}) &= \int \mathcal{L}_{\text{min}}\,\mathrm{d}\tau \\
&= -\int \bar{\psi}(\vec{q})\left\{\frac{1}{2m}\left[-i\hbar\vec{\nabla}^{(\vec{q})} + e\vec{a}^{\perp}(\vec{q})\right]^2 + V(\vec{q})\right. \\
&\quad \left. + \frac{e^2}{8\pi\varepsilon_0}\int \frac{\bar{\psi}(\vec{q}\,')\psi(\vec{q}\,')}{|\vec{q}-\vec{q}\,'|}\,\mathrm{d}^3\vec{q}\,'\right\}\psi(\vec{q})\mathrm{d}^3\vec{q} \\
&\quad + \frac{i\hbar}{2}\int\left[\bar{\psi}(\vec{q})\dot{\psi}(\vec{q}) - \dot{\bar{\psi}}(\vec{q})\psi(\vec{q})\right]\mathrm{d}^3\vec{q} + \frac{\varepsilon_0}{2}\int\left\{\dot{\vec{a}}^{\perp 2}(\vec{r}) - c^2(\vec{\nabla}\times\vec{a}(\vec{r}))^2\right\}\mathrm{d}^3\vec{r},
\end{aligned}
\tag{2.2.1}
$$

where $\mathcal{L}_{\text{min}}$ is the Lagrangian density and $\mathrm{d}\tau$ is the complete volume element. It may be verified that L_{min} leads to the correct equations of motion. Effecting variation of the electron wavefield $\psi(\vec{q})$, using the Euler–Lagrange equation

$$\frac{\partial}{\partial t}\frac{\partial \mathcal{L}_{\text{min}}}{\partial \dot{\bar{\psi}}}+\frac{\partial}{\partial x_j}\frac{\partial \mathcal{L}_{\text{min}}}{\partial(\partial\bar{\psi}/\partial x_j)}-\frac{\partial \mathcal{L}_{\text{min}}}{\partial \bar{\psi}}=0, \tag{2.2.2}$$

produces

$$i\hbar\dot{\psi}(\vec{q})=\left\{\frac{1}{2m}\left[-i\hbar\vec{\nabla}^{(\vec{q})}+e\vec{a}^{\perp}(\vec{q})\right]^2+V(\vec{q})+\frac{e^2}{4\pi\varepsilon_0}\int\frac{\bar{\psi}(\vec{q}\,')\psi(\vec{q}\,')}{|\vec{q}-\vec{q}\,'|}\mathrm{d}^3\vec{q}\,'\right\}\psi(\vec{q}), \tag{2.2.3}$$

which is the Schrödinger equation for an electron coupled to a radiation field. On making use of equation (1.4.18) to carry out the variation with respect to the coordinate variable of the electromagnetic field, namely, the transverse vector potential yields, after eliminating the vector potential and its time derivative in terms of the magnetic field and the electric field, respectively,

$$\vec{\nabla}\times\vec{b}(\vec{r})=\frac{1}{c^2}\frac{\partial\vec{e}^{\perp}(\vec{r})}{\partial t}+\frac{1}{\varepsilon_0 c^2}\vec{j}^{\perp}(\vec{r}), \tag{2.2.4}$$

which is the transverse component of the fourth Maxwell equation (1.3.31). The transverse current is expressed as

$$j_i^{\perp}(\vec{r})=-e\int\bar{\psi}(\vec{q})\left[-i\hbar\vec{\nabla}^{(\vec{q})}+e\vec{a}^{\perp}(\vec{q})\right]_j\delta_{ij}^{\perp}(\vec{r}-\vec{q})\psi(\vec{q})\mathrm{d}^3\vec{q}, \tag{2.2.5}$$

where $\delta_{ij}^{\perp}(\vec{r}-\vec{q})$ is the transverse delta function dyadic (Belinfante, 1946). By following the procedure for constructing the dynamics in canonical form, the minimal-coupling Hamiltonian may be obtained from the minimal-coupling Lagrangian (2.2.1) by first calculating the fields canonically conjugate to $\psi(\vec{q})$, $\bar{\psi}(\vec{q})$, and $\vec{a}^{\perp}(\vec{q})$. These are

$$\Pi^{\text{min}}(\vec{q})=\frac{\partial\mathcal{L}_{\text{min}}}{\partial\dot{\psi}(\vec{q})}=\frac{i\hbar}{2}\bar{\psi}(\vec{q}), \tag{2.2.6}$$

$$\bar{\Pi}^{\text{min}}(\vec{q})=\frac{\partial\mathcal{L}_{\text{min}}}{\partial\dot{\bar{\psi}}(\vec{q})}=-\frac{i\hbar}{2}\psi(\vec{q}), \tag{2.2.7}$$

and

$$\pi_i^{\text{min}}(\vec{r})=\frac{\partial\mathcal{L}_{\text{min}}}{\partial\dot{a}_i^{\perp}(\vec{r})}=\varepsilon_0\dot{a}_i^{\perp}(\vec{r})=-\varepsilon_0 e_i^{\perp}(\vec{r}). \tag{2.2.8}$$

Note that because the equations of motion for the electron fields are first order in time, the canonical conjugates $\Pi^{\min}(\vec{q})$ and $\bar{\Pi}^{\min}(\vec{q})$ are proportional to the Hermitian conjugates of the wavefields $\bar{\psi}(\vec{q})$ and $\psi(\vec{q})$. As found in the treatment for a charged particle, the momentum conjugate to the vector potential $\vec{\pi}^{\min}(\vec{r})$ is equal to $-\varepsilon_0\vec{e}^{\perp}(\vec{r})$. The minimal-coupling Hamiltonian is then calculated in the usual manner via

$$\begin{aligned} H_{\min} &= \int \Pi^{\min}(\vec{q})\dot{\psi}(\vec{q})\mathrm{d}^3\vec{q} + \int \bar{\Pi}^{\min}(\vec{q})\dot{\bar{\psi}}(\vec{q})\mathrm{d}^3\vec{q} \\ &\quad + \int \vec{\pi}^{\min}(\vec{r})\cdot\dot{\vec{a}}^{\perp}(\vec{r})\mathrm{d}^3\vec{r} - L_{\min} \\ &= \int \bar{\psi}(\vec{q})\left\{\frac{1}{2m}\left[-i\hbar\vec{\nabla}^{(\vec{q})} + e\vec{a}^{\perp}(\vec{q})\right]^2 + V(\vec{q}) \right. \\ &\quad \left. + \frac{e^2}{8\pi\varepsilon_0}\int \frac{\bar{\psi}(\vec{q}\,')\psi(\vec{q}\,')}{|\vec{q}-\vec{q}\,'|}\mathrm{d}^3\vec{q}\,'\right\}\psi(\vec{q})\mathrm{d}^3\vec{q} \\ &\quad + \frac{1}{2\varepsilon_0}\int \{\vec{\pi}^2(\vec{r}) + \varepsilon_0^2 c^2(\vec{\nabla}\times\vec{a}(\vec{r}))^2\}\mathrm{d}^3\vec{r}, \end{aligned} \tag{2.2.9}$$

on using the last three relations and Lagrangian (2.2.1). It is usual to decompose the Hamiltonian into a sum of molecular, radiation field, and interaction terms as follows:

$$H_{\min} = H_{\mathrm{mol}}^{\min} + H_{\mathrm{rad}}^{\min} + H_{\mathrm{int}}^{\min}, \tag{2.2.10}$$

where

$$H_{\mathrm{mol}}^{\min} = \int \bar{\psi}(\vec{q})\left\{-\frac{\hbar^2}{2m}(\vec{\nabla}^{(\vec{q})})^2 + V(\vec{q}) + \frac{e^2}{8\pi\varepsilon_0}\int \frac{\bar{\psi}(\vec{q}\,')\psi(\vec{q}\,')}{|\vec{q}-\vec{q}\,'|}\mathrm{d}^3\vec{q}\,'\right\}\psi(\vec{q})\mathrm{d}^3\vec{q}, \tag{2.2.11}$$

$$H_{\mathrm{rad}}^{\min} = \frac{1}{2\varepsilon_0}\int \{\vec{\pi}^2(\vec{r}) + \varepsilon_0^2 c^2(\vec{\nabla}\times\vec{a}(\vec{r}))^2\}\mathrm{d}^3\vec{r} = \frac{\varepsilon_0}{2}\int \{\vec{e}^{\perp 2}(\vec{r}) + c^2\vec{b}^2(\vec{r})\}\mathrm{d}^3\vec{r}, \tag{2.2.12}$$

and

$$H_{\mathrm{int}}^{\min} = \frac{e}{m}\int \bar{\psi}(\vec{q})\left[-i\hbar\vec{\nabla}^{(\vec{q})}\cdot\vec{a}^{\perp}(\vec{q})\right]\psi(\vec{q})\mathrm{d}^3\vec{q} + \frac{e^2}{2m}\int \bar{\psi}(\vec{q})\vec{a}^{\perp 2}(\vec{q})\psi(\vec{q})\mathrm{d}^3\vec{q}. \tag{2.2.13}$$

Note that no interparticle interaction terms appear in equation (2.2.10) because the treatment presented thus far in this section is applicable to a single electron. The extension to many charges will be presented in Section 2.3.

The quantum mechanical version of minimal-coupling Hamiltonian (2.2.10) is obtained by promoting the canonical variables to quantum operators. The appropriate pairs satisfy the equal-time anticommutation or commutation relations

$$[\psi(\vec{q}),\bar{\psi}(\vec{q}\,')]_+ = \delta(\vec{q}-\vec{q}\,'), \tag{2.2.14}$$

$$[a_i^\perp(\vec{r}),\pi_j(\vec{r}\,')]_- = i\hbar\delta_{ij}^\perp(\vec{r}-\vec{r}\,'). \tag{2.2.15}$$

Note that two other useful commutation relations ensue from (2.2.15) on using (2.2.8) and $\vec{b} = \vec{\nabla}\times\vec{a}(\vec{r})$, respectively. They are

$$[a_i^\perp(\vec{r}),e_j^\perp(\vec{r}\,')]_- = -\frac{i\hbar}{\varepsilon_0}\delta_{ij}^\perp(\vec{r}-\vec{r}\,') \tag{2.2.16}$$

and

$$[e_i^\perp(\vec{r}),b_j(\vec{r}\,')]_- = \frac{i\hbar}{\varepsilon_0}\varepsilon_{ijk}\vec{\nabla}_k^{(\vec{r}\,')}\delta(\vec{r}-\vec{r}\,'), \tag{2.2.17}$$

where ε_{ijk} is the Levi-Civita alternating tensor, and the gradient acts on variable $\vec{r}\,'$.

It was shown in Section 1.6 that by adding a total time derivative of a function of the coordinates of a particle and the time to the minimal-coupling Lagrangian, L_{min} could be converted to L_{mult}, from which H_{mult} could be constructed directly. A number of advantages were given for the use of the multipolar formalism in studying radiation–molecule and molecule–molecule processes, especially those associated with $H_{\text{int}}^{\text{mult}}$. For interacting electron and Maxwell fields, the Lagrangian functions in the two frameworks are also related, but this time by a change in the generalized coordinate of the electron field, which is transformed through the relation

$$\psi(\vec{q}) = \mathrm{e}^{-iS(\vec{q})}\phi(\vec{q}), \tag{2.2.18}$$

where

$$S(\vec{q}) = \frac{1}{\hbar}\int \vec{p}(\vec{r},\vec{q})\cdot\vec{a}^\perp(\vec{r})\mathrm{d}^3\vec{r}. \tag{2.2.19}$$

The coordinate describing the electromagnetic field, the transverse vector potential $\vec{a}^\perp(\vec{r})$, however, remains unchanged. In equation (2.2.19), $\vec{p}(\vec{r},\vec{q})$

is the electric polarization field of a single bound electron, expressed in the form of a parametric integral and implicitly containing the complete electric multipole series. It is given by

$$\vec{p}(\vec{r},\vec{q}) = -e(\vec{q}-\vec{R})\int_0^1 \delta(\vec{r}-\vec{R}-\lambda(\vec{q}-\vec{R}))\mathrm{d}\lambda. \tag{2.2.20}$$

If the first term in the expansion of $\vec{p}(\vec{r},\vec{q})$ is retained, for instance, giving the electric dipole polarization field $-e(\vec{q}-\vec{R})\delta(\vec{r}-\vec{R})$, the electric dipole approximated multipolar Lagrangian and Hamiltonian result. This approximation corresponds to the neglect of any spatial variation of the vector potential over the extent of the species. When the first spatial derivative of $\vec{a}^{\perp}(\vec{r})$ is included along with the electric quadrupole polarization field, three new interaction terms appear after the electric dipole coupling term in the multipolar Hamiltonian. They are the magnetic dipole, electric quadrupole, and lowest order diamagnetic interactions. In what follows, the total electric polarization field (2.2.20) is used.

By effecting (2.2.18), the minimal-coupling Lagrangian transforms to

$$\begin{aligned} L_{\min}(\vec{a}^{\perp},\psi,\bar{\psi};\dot{\vec{a}}^{\perp},\dot{\psi},\dot{\bar{\psi}}) &= L_{\min}\left(\vec{a}^{\perp},\mathrm{e}^{iS}\phi,\bar{\phi}\mathrm{e}^{-iS};\dot{\vec{a}}^{\perp},\frac{\mathrm{d}}{\mathrm{d}t}(\mathrm{e}^{iS}\phi),\frac{\mathrm{d}}{\mathrm{d}t}(\bar{\phi}\mathrm{e}^{-iS})\right) \\ &= L_{\mathrm{mult}}(\vec{a}^{\perp},\phi,\bar{\phi};\dot{\vec{a}}^{\perp},\dot{\phi},\dot{\bar{\phi}}), \end{aligned} \tag{2.2.21}$$

with L_{mult} for the complete multipolar series given explicitly by

$$\begin{aligned} L_{\mathrm{mult}} &= \int \mathcal{L}_{\mathrm{mult}}\mathrm{d}\tau \\ &= -\int \bar{\phi}(\vec{q})\left\{\frac{1}{2m}\left[-i\hbar\vec{\nabla}^{(\vec{q})}+e\vec{a}^{\perp}(\vec{q})+\hbar\vec{\nabla}^{(\vec{q})}S(\vec{q})\right]^2+V(\vec{q})\right. \\ &\quad \left.+\frac{e^2}{8\pi\varepsilon_0}\int\frac{\bar{\phi}(\vec{q}\,')\phi(\vec{q}\,')}{|\vec{q}-\vec{q}\,'|}\mathrm{d}^3\vec{q}\,'\right\}\phi(\vec{q})\mathrm{d}^3\vec{q} \\ &\quad +\frac{i\hbar}{2}\int[\bar{\phi}(\vec{q})\dot{\phi}(\vec{q})-\dot{\bar{\phi}}(\vec{q})\phi(\vec{q})]\mathrm{d}^3\vec{q}+\frac{\varepsilon_0}{2}\int\{\dot{\vec{a}}^{\perp 2}(\vec{r})-c^2(\vec{\nabla}\times\vec{a}(\vec{r}))^2\}\mathrm{d}^3\vec{r} \\ &\quad -\int\int\bar{\phi}(\vec{q})\vec{p}(\vec{r},\vec{q})\cdot\dot{\vec{a}}^{\perp}(\vec{r})\phi(\vec{q})\mathrm{d}^3\vec{r}\,\mathrm{d}^3\vec{q}. \end{aligned} \tag{2.2.22}$$

It is convenient for later use to replace the last two terms occurring within square brackets of the first term in equation (2.2.22) by the identity

$$e\vec{a}^{\perp}(\vec{q})+\hbar\vec{\nabla}^{(\vec{q})}S(\vec{q}) = e\vec{a}^{\perp}(\vec{q})+\vec{\nabla}^{(\vec{q})}\int\vec{p}(\vec{r},\vec{q})\cdot\vec{a}^{\perp}(\vec{r})\mathrm{d}^3\vec{r}$$
$$= \int\vec{n}(\vec{r},\vec{q})\times\vec{b}(\vec{r})\mathrm{d}^3\vec{r}, \tag{2.2.23}$$

where

$$\vec{n}(\vec{r},\vec{q}) = -e(\vec{q}-\vec{R})\int_0^1\lambda\delta(\vec{r}-\vec{R}-\lambda(\vec{q}-\vec{R}))\mathrm{d}\lambda \tag{2.2.24}$$

is a polarization field that differs from the electric polarization field (2.2.20) by a multiple of λ.

The equations of motion for the generalized coordinates of the multipolar Lagrangian density follow from equations (2.2.22) and (2.2.23). Using these equations in the analogue of equation (2.2.2), it is found that for the electron wavefield $\bar{\phi}(\vec{q})$, the Euler–Lagrange equation $\frac{\partial}{\partial t}\frac{\partial\mathcal{L}_{\mathrm{mult}}}{\partial\dot{\bar{\phi}}}+\frac{\partial}{\partial x_j}\frac{\partial\mathcal{L}_{\mathrm{mult}}}{\partial(\partial\bar{\phi}/\partial x_j)}-\frac{\partial\mathcal{L}_{\mathrm{mult}}}{\partial\bar{\phi}}=0$ produces

$$i\hbar\dot{\phi}(\vec{q}) = \left\{\begin{array}{l} -\dfrac{\hbar^2}{2m}(\vec{\nabla}^{(\vec{q})})^2+V(\vec{q})+\dfrac{e^2}{4\pi\varepsilon_0}\displaystyle\int\frac{\bar{\phi}(\vec{q}')\phi(\vec{q}')}{|\vec{q}-\vec{q}'|}\mathrm{d}^3\vec{q}' \\ -\displaystyle\int\vec{p}(\vec{r},\vec{q})\cdot\vec{e}^{\perp}(\vec{r})\mathrm{d}^3\vec{r} \\ -\displaystyle\int\vec{m}(\vec{r},\vec{q})\cdot\vec{b}(\vec{r})\mathrm{d}^3\vec{r}+\frac{1}{2m}\left[\int\vec{n}(\vec{r},\vec{q})\times\vec{b}(\vec{r})\mathrm{d}^3\vec{r}\right]^2 \end{array}\right\}\phi(\vec{q}). \tag{2.2.25}$$

Note the appearance in (2.2.25) of the symmetrized magnetization field $\vec{m}(\vec{r},\vec{q})$, defined by

$$\vec{m}(\vec{r},\vec{q}) = \frac{1}{2m}\left\{\vec{n}(\vec{r},\vec{q})\times\left(-i\hbar\vec{\nabla}^{(\vec{q})}+\int\vec{n}(\vec{r}',\vec{q})\times\vec{b}(\vec{r}')\mathrm{d}^3\vec{r}'\right)\right\}$$
$$-\frac{1}{2m}\left\{\left(-i\hbar\vec{\nabla}^{(\vec{q})}+\int\vec{n}(\vec{r}',\vec{q})\times\vec{b}(\vec{r}')\mathrm{d}^3\vec{r}'\right)\times\vec{n}(\vec{r},\vec{q})\right\}. \tag{2.2.26}$$

The first terms in the expansion of each of the fields $\vec{m}(\vec{r},\vec{q})$ and $\vec{n}(\vec{r},\vec{q})$ are

$$\vec{m}(\vec{r},\vec{q}) = -\frac{e}{2m}(\vec{q}-\vec{R})\times(-i\hbar\vec{\nabla}^{(\vec{q})})\delta(\vec{r}-\vec{R}) \tag{2.2.27}$$

and

$$\vec{n}(\vec{r},\vec{q}) = -\frac{e}{2}(\vec{q}-\vec{R})\delta(\vec{r}-\vec{R}), \tag{2.2.28}$$

which lead to the recognizable magnetic dipole and lowest order diamagnetic interaction terms in (2.2.25), respectively,

$$\frac{e}{2m}\left[(\vec{q}-\vec{R}) \times -i\hbar\vec{\nabla}^{(\vec{q})}\right]\cdot\vec{b}(\vec{R}) \tag{2.2.29}$$

and

$$\frac{e^2}{8m}\left[(\vec{q}-\vec{R}) \times \vec{b}(\vec{R})\right]^2. \tag{2.2.30}$$

Variation of the multipolar Lagrangian density (2.2.22) with respect to the vector potential gives contributions

$$\frac{\partial}{\partial t}\frac{\partial\mathcal{L}_{\text{mult}}}{\partial\dot{a}_i} = \varepsilon_0\ddot{a}_i^{\perp}(\vec{r}) - \frac{\mathrm{d}}{\mathrm{d}t}\int\bar{\phi}(\vec{q})p_i^{\perp}(\vec{r},\vec{q})\phi(\vec{q})\mathrm{d}^3\vec{q} \tag{2.2.31}$$

and

$$\begin{aligned}
&\frac{\partial}{\partial x_j}\frac{\partial\mathcal{L}_{\text{mult}}}{\partial(\partial a_i/\partial x_j)} - \frac{\partial\mathcal{L}_{\text{mult}}}{\partial a_i}\\
&= -c^2\varepsilon_0[\vec{\nabla}\times\vec{\nabla}\times\vec{a}(\vec{r})]_i + \frac{1}{2m}\varepsilon_{ilm}\varepsilon_{jkm}\int\bar{\phi}(\vec{q})\\
&\times\left\{\begin{array}{l}\left(-i\hbar\vec{\nabla}_j^{(\vec{q})} + \int[\vec{n}(\vec{r}\,',\vec{q})\times\vec{b}(\vec{r}\,')]_j\,\mathrm{d}^3\vec{r}\,'\right)\vec{\nabla}_l n_k(\vec{r},\vec{q})\\ +\vec{\nabla}_l n_k(\vec{r},\vec{q})\left(-i\hbar\vec{\nabla}_j^{(\vec{q})} + \int[\vec{n}(\vec{r}\,',\vec{q})\times\vec{b}(\vec{r}\,')]_j\,\mathrm{d}^3\vec{r}\,'\right)\end{array}\right\}\phi(\vec{q})\mathrm{d}^3\vec{q}.
\end{aligned} \tag{2.2.32}$$

On using the definition for $\vec{m}(\vec{r},\vec{q})$ given by equation (2.2.26), the equation of motion for the photon field obtained from the sum of expressions (2.2.31) and (2.2.32) can be written as

$$\begin{aligned}
&c^2\varepsilon_0\vec{\nabla}\times\left\{\vec{b}(\vec{r}) - \frac{1}{c^2\varepsilon_0}\int\bar{\phi}(\vec{q})\vec{m}(\vec{r},\vec{q})\phi(\vec{q})\mathrm{d}^3\vec{q}\right\}\\
&\quad= \frac{\mathrm{d}}{\mathrm{d}t}\left\{-\varepsilon_0\dot{\vec{a}}^{\perp}(\vec{r}) + \int\bar{\phi}(\vec{q})\vec{p}^{\perp}(\vec{r},\vec{q})\phi(\vec{q})\mathrm{d}^3\vec{q}\right\}.
\end{aligned} \tag{2.2.33}$$

At this stage of the development, it is beneficial to introduce two auxiliary fields, the electric displacement field

$$\vec{d}(\vec{r}) = \varepsilon_0\vec{e}(\vec{r}) + \vec{p}(\vec{r}) \tag{2.2.34}$$

and its magnetic analogue

$$\vec{h}(\vec{r}) = \varepsilon_0 c^2\vec{b}(\vec{r}) - \vec{m}(\vec{r}), \tag{2.2.35}$$

where

$$\vec{p}(\vec{r}) = \int \bar{\phi}(\vec{q})\vec{p}(\vec{r},\vec{q})\phi(\vec{q})\mathrm{d}^3\vec{q} \tag{2.2.36}$$

and

$$\vec{m}(\vec{r}) = \int \bar{\phi}(\vec{q})\vec{m}(\vec{r},\vec{q})\phi(\vec{q})\mathrm{d}^3\vec{q}. \tag{2.2.37}$$

The equation of motion (2.2.33) then simply becomes

$$\vec{\nabla} \times \vec{h}(\vec{r}) = \frac{\partial \vec{d}^{\perp}(\vec{r})}{\partial t}, \tag{2.2.38}$$

in which the currents are contained implicitly within $\vec{d}(\vec{r})$ and $\vec{b}(\vec{r})$. Equation (2.2.38) is an alternative expression of the source-dependent Maxwell–Lorentz equation

$$\vec{\nabla} \times \vec{b}(\vec{r}) = \frac{1}{c^2}\frac{\partial \vec{e}^{\perp}(\vec{r})}{\partial t} + \frac{1}{\varepsilon_0 c^2}\vec{j}^{\perp}(\vec{r}), \tag{2.2.39}$$

which can be obtained from (2.2.33) on inserting

$$\vec{j}^{\perp}(\vec{r}) = \frac{\partial \vec{p}^{\perp}(\vec{r})}{\partial t} + \vec{\nabla} \times \vec{m}(\vec{r}). \tag{2.2.40}$$

The two auxiliary fields $\vec{d}(\vec{r})$ and $\vec{h}(\vec{r})$ describe the effects of a medium formed by the bound charges within which they operate and are subsequently modified by.

In a manner similar to that used to derive H_{min}, the multipolar Hamiltonian H_{mult} is now obtained from L_{mult} given by (2.2.22). First, the canonically conjugate momenta are evaluated. They are found to be

$$\Pi^{\text{mult}}(\vec{q}) = \frac{\partial \mathcal{L}_{\text{mult}}}{\partial \dot{\phi}(\vec{q})} = \frac{i\hbar}{2}\bar{\phi}(\vec{q}), \tag{2.2.41}$$

$$\bar{\Pi}^{\text{mult}}(\vec{q}) = \frac{\partial \mathcal{L}_{\text{mult}}}{\partial \dot{\bar{\phi}}(\vec{q})} = -\frac{i\hbar}{2}\phi(\vec{q}), \tag{2.2.42}$$

and

$$\pi_i^{\text{mult}}(\vec{r}) = \frac{\partial \mathcal{L}_{\text{mult}}}{\partial \dot{a}_i^{\perp}(\vec{r})} = \varepsilon_0 \dot{a}_i^{\perp}(\vec{r}) - \int \bar{\phi}(\vec{q}) p_i^{\perp}(\vec{r},\vec{q})\phi(\vec{q})\mathrm{d}^3\vec{q} = -d_i^{\perp}(\vec{r}). \tag{2.2.43}$$

Again the momenta canonically conjugate to the electron fields are their Hermitian conjugates. In contrast to the minimal-coupling scheme, in which the momentum conjugate to the vector potential was proportional to the transverse electric field, equation (2.2.8), in the multipolar formalism, the conjugate momentum is equal to the negative of the transverse electric displacement field. Analogous to (2.2.9), the multipolar Hamiltonian is

$$\begin{aligned}
H_{\text{mult}} &= \int \Pi^{\text{mult}}(\vec{q})\dot{\phi}(\vec{q})\mathrm{d}^3\vec{q} + \int \bar{\Pi}^{\text{mult}}(\vec{q})\dot{\bar{\phi}}(\vec{q})\mathrm{d}^3\vec{q} + \int \vec{\pi}^{\text{mult}}(\vec{r})\cdot\dot{\vec{a}}^{\perp}(\vec{r})\mathrm{d}^3\vec{r} - L_{\text{mult}} \\
&= \int \bar{\phi}(\vec{q})\Bigg\{\frac{1}{2m}\left(-i\hbar\vec{\nabla}^{(\vec{q})} + \int \vec{n}(\vec{r},\vec{q})\times\vec{b}(\vec{r})\mathrm{d}^3\vec{r}\right)^2 \\
&\quad + V(\vec{q}) + \frac{e^2}{8\pi\varepsilon_0}\int \frac{\bar{\phi}(\vec{q}')\phi(\vec{q}')}{|\vec{q}-\vec{q}'|}\mathrm{d}^3q'\Bigg\}\phi(\vec{q})\mathrm{d}^3\vec{q} \\
&\quad + \frac{\varepsilon_0}{2}\int\left\{\varepsilon_0^{-2}\left(\vec{\pi}(\vec{r}) + \int \bar{\phi}(\vec{q})\vec{p}^{\perp}(\vec{r},\vec{q})\phi(\vec{q})\mathrm{d}^3\vec{q}\right)^2 + c^2\left[\vec{\nabla}\times\vec{a}(\vec{r})\right]^2\right\}\mathrm{d}^3\vec{r},
\end{aligned} \tag{2.2.44}$$

which can also be partitioned into three terms consisting of molecular, radiation field, and interaction contributions,

$$H_{\text{mult}} = H_{\text{mol}}^{\text{mult}} + H_{\text{rad}}^{\text{mult}} + H_{\text{int}}^{\text{mult}}. \tag{2.2.45}$$

Each term is given individually by

$$\begin{aligned}
H_{\text{mol}}^{\text{mult}} &= \int \bar{\phi}(\vec{q})\left\{-\frac{\hbar^2}{2m}(\vec{\nabla}^{(\vec{q})})^2 + V(\vec{q}) + \frac{e^2}{8\pi\varepsilon_0}\int \frac{\bar{\phi}(\vec{q}')\phi(\vec{q}')}{|\vec{q}-\vec{q}'|}\mathrm{d}^3\vec{q}'\right\}\phi(\vec{q})\mathrm{d}^3\vec{q} \\
&\quad + \frac{1}{2\varepsilon_0}\int\left(\int \bar{\phi}(\vec{q})\vec{p}^{\perp}(\vec{r},\vec{q})\phi(\vec{q})\mathrm{d}^3\vec{q}\right)^2\mathrm{d}^3\vec{r},
\end{aligned} \tag{2.2.46}$$

$$H_{\rm rad}^{\rm mult} = \frac{1}{2}\int\left\{\frac{\vec{\pi}^2(\vec{r})}{\varepsilon_0} + \varepsilon_0 c^2(\vec{\nabla}\times\vec{a}(\vec{r}))^2\right\}\mathrm{d}^3\vec{r} = \frac{1}{2}\int\left\{\frac{\vec{d}^{\perp 2}(\vec{r})}{\varepsilon_0} + \varepsilon_0 c^2\vec{b}^2(\vec{r})\right\}\mathrm{d}^3\vec{r}, \tag{2.2.47}$$

$$\begin{aligned} H_{\rm int}^{\rm mult} = & -\varepsilon_0^{-1}\int\bar{\phi}(\vec{q})\vec{p}(\vec{r},\vec{q})\cdot\vec{d}^{\perp}(\vec{r})\phi(\vec{q})\mathrm{d}^3\vec{q}\,\mathrm{d}^3\vec{r} \\ & -\int\bar{\phi}(\vec{q})\vec{m}(\vec{r},\vec{q})\cdot\vec{b}(\vec{r})\phi(\vec{q})\mathrm{d}^3\vec{q}\,\mathrm{d}^3\vec{r} \\ & +\frac{1}{2m}\int\bar{\phi}(\vec{q})\left(\int\vec{n}(\vec{r},\vec{q})\times\vec{b}(\vec{r})\mathrm{d}^3\vec{r}\right)^2\phi(\vec{q})\mathrm{d}^3\vec{q}. \end{aligned} \tag{2.2.48}$$

The interaction Hamiltonian (2.2.48) contains the complete multipolar expansion of the charge, comprising electric, magnetic, and diamagnetic terms. It should be noted that expression (2.2.44) for the multipolar Hamiltonian may be considered as a sum of molecule and radiation field energy terms, with each in turn being a sum of kinetic and potential energy contributions. By promoting the canonically conjugate pairs of variables associated with the electron wavefield to operators subject to the equal-time anticommutation relation

$$[\phi(\vec{q}),\bar{\phi}(\vec{q}\,')]_+ = \delta(\vec{q}-\vec{q}\,') \tag{2.2.49}$$

and using commutator (2.2.15) for $a_i^{\perp}(\vec{r})$ and its conjugate momentum, $H_{\rm mult}$ of equation (2.2.44) takes on quantum mechanical form. All pairs of canonical operators other than those represented by relations (2.2.14), (2.2.15), and (2.2.49) either commute or anticommute.

2.3 QUANTUM CANONICAL TRANSFORMATION

The quantum electrodynamical multipolar Hamiltonian (2.2.45) may be obtained directly from the quantum mechanical minimal-coupling Hamiltonian (2.2.9) by applying a quantum canonical transformation on the latter (Power and Thirunamachandran, 1983a) instead of transforming $L_{\rm min}$ to $L_{\rm mult}$ via the application of the classical coordinate transformation (2.2.18) and constructing $H_{\rm mult}$ from $L_{\rm mult}$. A characteristic feature of a quantum canonical transformation is that the commutator relation (2.2.15) and anticommutators (2.2.14) and (2.2.49) are preserved along with the equations of motion for the operator. This is directly analogous to canonical or contact transformations in classical mechanics (Goldstein, 1960; Power, 1978), which leave the Poisson bracket between two canonical variables and their respective dynamical equations invariant. The relation between a multipolar canonical variable $O_{\rm mult}$ and its minimal-coupling counterpart

O_{min}, when a quantum canonical transformation is effected with a generator $\mathcal{S}$ that is Hermitian, is given by

$$O_{\text{mult}} = \mathrm{e}^{-i\mathcal{S}} O_{\text{min}} \mathrm{e}^{i\mathcal{S}}. \tag{2.3.1}$$

Hence,

$$\begin{aligned} H_{\text{min}}&(\vec{a}^{\perp}_{\text{min}}(\vec{r}), \vec{\pi}_{\text{min}}(\vec{r}); \psi(\vec{q}), \bar{\psi}(\vec{q})) \\ &= H_{\text{min}}(\mathrm{e}^{-i\mathcal{S}}\vec{a}^{\perp}_{\text{mult}}(\vec{r})\mathrm{e}^{i\mathcal{S}}, \mathrm{e}^{-i\mathcal{S}}\vec{\pi}_{\text{mult}}(\vec{r})\mathrm{e}^{i\mathcal{S}}; \mathrm{e}^{-i\mathcal{S}}\phi(\vec{q})\mathrm{e}^{i\mathcal{S}}, \mathrm{e}^{-i\mathcal{S}}\bar{\phi}(\vec{q})\mathrm{e}^{i\mathcal{S}}) \\ &= \mathrm{e}^{-i\mathcal{S}} H_{\text{min}}(\vec{a}^{\perp}_{\text{mult}}(\vec{r}), \vec{\pi}_{\text{mult}}(\vec{r}); \phi(\vec{q}), \bar{\phi}(\vec{q}))\mathrm{e}^{i\mathcal{S}} \\ &= H_{\text{mult}}(\vec{a}^{\perp}_{\text{mult}}(\vec{r}), \vec{\pi}_{\text{mult}}(\vec{r}); \phi(\vec{q}), \bar{\phi}(\vec{q})), \end{aligned} \tag{2.3.2}$$

enabling the transformation to be interpreted as a unitary rotation in Hilbert space. The choice of $\mathcal{S}$ is dictated by the fact that the same H_{mult} must result as that obtained by application of the coordinate transformation (2.2.18). Therefore,

$$\mathcal{S} = \int \bar{\phi}(\vec{q}) S(\vec{q}) \phi(\vec{q}) \mathrm{d}^3\vec{q} = \frac{1}{\hbar}\int \bar{\phi}(\vec{q})\vec{p}(\vec{r}, \vec{q}) \cdot \vec{a}^{\perp}(\vec{r})\phi(\vec{q})\mathrm{d}^3\vec{q}\mathrm{d}^3\vec{r}. \tag{2.3.3}$$

With this choice of $\mathcal{S}$, the vector potential stays the same, while its canonically conjugate momentum $\vec{\pi}_{\text{min}}$ changes to

$$\begin{aligned} \vec{\pi}_{\text{mult}}(\vec{r}) &= \vec{\pi}_{\text{min}}(\vec{r}) + i[\mathcal{S}, \vec{\pi}_{\text{min}}(\vec{r})]_- + \cdots \\ &= \vec{\pi}_{\text{min}}(\vec{r}) + \frac{i}{\hbar}\int \bar{\phi}(\vec{q}) p_i(\vec{r}, \vec{q})\phi(\vec{q})\mathrm{d}^3\vec{q}[a_i^{\perp}(\vec{r}\,'), \vec{\pi}_{\text{min}}(\vec{r})]_-\mathrm{d}^3\vec{r}\,' \\ &= \vec{\pi}_{\text{min}}(\vec{r}) - \vec{p}^{\perp}(\vec{r}), \end{aligned} \tag{2.3.4}$$

on using the commutator (2.2.15) and the identity (1.8.16). Because the commutator in (2.3.4) commutes with the generator (2.3.3), only the first bracket in the expansion (1.8.16) is nonvanishing.

In similar fashion, for the operator electron field,

$$\begin{aligned} \phi(\vec{q}) &= \psi(\vec{q}) + i[\mathcal{S}, \psi(\vec{q})]_- + \frac{1}{2!}[\mathcal{S}, [\mathcal{S}, \psi(\vec{q})]_-]_- + \cdots \\ &= \psi(\vec{q}) + i\int [\bar{\psi}(\vec{q}\,')S(\vec{q}\,')\psi(\vec{q}\,'), \psi(\vec{q})]_-\mathrm{d}^3\vec{q}\,' \\ &\quad + \frac{1}{2!}\int [\bar{\psi}(\vec{q}\,')S(\vec{q}\,')\psi(\vec{q}\,'), [\bar{\psi}(\vec{q}\,'')S(\vec{q}\,'')\psi(\vec{q}\,''), \psi(\vec{q})]_-]_-\mathrm{d}^3\vec{q}\,'\mathrm{d}^3\vec{q}\,'' + \cdots \\ &= \psi(\vec{q}) - i\int S(\vec{q}\,')\psi(\vec{q})\delta(\vec{q} - \vec{q}\,')\mathrm{d}^3\vec{q}\,' \\ &\quad + \frac{1}{2!}\int S(\vec{q}\,')S(\vec{q}\,'')\delta(\vec{q} - \vec{q}\,')\delta(\vec{q} - \vec{q}\,'')\psi(\vec{q})\mathrm{d}^3\vec{q}\,'\mathrm{d}^3\vec{q}\,'' + \cdots \\ &= \mathrm{e}^{-iS(\vec{q})}\psi(\vec{q}), \end{aligned} \tag{2.3.5}$$

and its Hermitian-conjugate field $\bar{\phi}(\vec{q}) = \bar{\psi}(\vec{q})e^{iS(\vec{q})}$; all terms in the series are seen to contribute. It is worth noting that the second quantized operator transform $\phi(\vec{q}) = e^{iS}\psi(\vec{q})e^{-iS}$ that leads to (2.3.5) is the quantum mechanical analogue of the transformation for first quantized fields (2.2.18). A consequence of the transformation of the electron fields in the two cases is the changes produced in the equations of motion. From the Heisenberg operator equation of motion, these are found to be

$$
\begin{aligned}
i\hbar\dot{\psi}(\vec{q}) &= [\psi(\vec{q}), H_{\text{min}}]_- \\
&= \left\{\frac{1}{2m}\left[-i\hbar\vec{\nabla}^{(\vec{q})} + e\vec{a}^{\perp}(\vec{q})\right]^2 + V(\vec{q}) + \frac{e^2}{4\pi\varepsilon_0}\int \frac{\bar{\psi}(\vec{q}')\psi(\vec{q}')}{|\vec{q}-\vec{q}'|}\,\mathrm{d}^3q'\right\}\psi(\vec{q})
\end{aligned}
\tag{2.3.6}
$$

and

$$
\begin{aligned}
i\hbar\dot{\phi}(\vec{q}) &= [\phi(\vec{q}), H_{\text{mult}}]_- \\
&= \left\{
\begin{aligned}
&\frac{1}{2m}\left[-i\hbar\vec{\nabla}^{(\vec{q})} + \int \vec{n}(\vec{r},\vec{q}) \times \vec{b}(\vec{r})\mathrm{d}^3\vec{r}\right]^2 + V(\vec{q}) \\
&+ \frac{e^2}{4\pi\varepsilon_0}\int \frac{\bar{\phi}(\vec{q}')\phi(\vec{q}')}{|\vec{q}-\vec{q}'|}\,\mathrm{d}^3q' \\
&+ \frac{1}{\varepsilon_0}\int \vec{p}(\vec{r},\vec{q}) \cdot \vec{\pi}_{\text{mult}}(\vec{r})\mathrm{d}^3\vec{r} \\
&+ \frac{1}{2\varepsilon_0}\int \bar{\phi}(\vec{q}')\vec{p}^{\perp}(\vec{r},\vec{q}')\phi(\vec{q}')\mathrm{d}^3q' \cdot \vec{p}^{\perp}(\vec{r},\vec{q})\mathrm{d}^3\vec{r}
\end{aligned}
\right\}\phi(\vec{q}),
\end{aligned}
\tag{2.3.7}
$$

which are the second quantized versions of equations (2.2.3) and (2.2.25), respectively. Because the electron wavefields in the two formalisms are nonidentical, they satisfy different Schrödinger equations. Further, it is worthy of remark that Schrödinger equation (2.3.7) may also be obtained by starting with the equation of motion for $\psi(\vec{q})$(2.3.6) and carrying out the transformation (2.3.5).

The second quantized formulation of the minimal- and multipolar-coupling Hamiltonians presented thus far has been limited to the description of a single charge interacting with a radiation field. For later application of the theory of interacting electron and Maxwell fields to atomic and molecular systems, and especially to intermolecular interactions, a suitable many-body Hamiltonian needs to be developed. This extension is made by considering each atom, molecule, functional unit, chromophore, to give rise to a single-electron field with the assumption that the fermion fields of different centers are noninteracting. Hence, there is no exchange of electrons between distinct bodies, although within a particular unit, the Pauli exclusion principle is obeyed. Ignoring terms due to the kinetic energy of the nuclei and denoting the electron wavefield of center ξ by $\phi_\xi(\vec{q})$, whose total nuclear charge is $Z_\xi e$, the generalization of the multipolar Lagrangian (2.2.22) is

$$
\begin{aligned}
L_{\text{mult}} = & -\sum_\xi \int \bar{\phi}_\xi(\vec{q}) \left\{ \begin{array}{c} \dfrac{1}{2m}\left[-i\hbar\vec{\nabla}^{(\vec{q})} + \displaystyle\int \vec{n}_\xi(\vec{r},\vec{q}) \times \vec{b}(\vec{r}) \mathrm{d}^3\vec{r} \right]^2 \\ + V_\xi(\vec{q}) + \dfrac{e^2}{8\pi\varepsilon_0} \displaystyle\int \dfrac{\bar{\phi}_\xi(\vec{q}')\phi_\xi(\vec{q}')}{|\vec{q}-\vec{q}'|} \mathrm{d}^3\vec{q}' \end{array} \right\} \phi_\xi(\vec{q}) \mathrm{d}^3\vec{q} \\
& -\frac{1}{4\pi\varepsilon_0} \sum_{\substack{\xi,\xi' \\ \xi\neq\xi'}} \bar{\phi}_\xi(\vec{q}) \left\{ -\frac{Z_{\xi'} e^2}{|\vec{q}-\vec{R}_{\xi'}|} + \frac{e^2}{2} \int \frac{\bar{\phi}_{\xi'}(\vec{q}')\phi_{\xi'}(\vec{q}')}{|\vec{q}-\vec{q}'|} \mathrm{d}^3\vec{q}' \right\} \phi_\xi(\vec{q}) \mathrm{d}^3\vec{q} \\
& + \frac{e^2}{8\pi\varepsilon_0} \sum_{\substack{\xi,\xi' \\ \xi\neq\xi'}} \frac{Z_\xi Z_{\xi'}}{|\vec{R}_\xi - \vec{R}_{\xi'}|} + \frac{i\hbar}{2} \sum_\xi \int [\bar{\phi}_\xi(\vec{q})\dot{\phi}_\xi(\vec{q}) - \dot{\bar{\phi}}_\xi(\vec{q})\phi_\xi(\vec{q})] \mathrm{d}^3\vec{q} \\
& + \frac{\varepsilon_0}{2} \int \{ \dot{\vec{a}}^{\perp 2}(\vec{r}) - c^2(\vec{\nabla} \times \vec{a}(\vec{r}))^2 \} \mathrm{d}^3\vec{r} \\
& - \sum_\xi \int\int \bar{\phi}_\xi(\vec{q}) \vec{p}_\xi(\vec{r},\vec{q}) \cdot \dot{\vec{a}}^{\perp}(\vec{r}) \phi_\xi(\vec{q}) \mathrm{d}^3\vec{r}\, \mathrm{d}^3\vec{q}, \qquad (2.3.8)
\end{aligned}
$$

where $\vec{R}_\xi$ is the location of center ξ. In the above Lagrangian, $V_\xi(\vec{q})$ is the Coulomb potential energy between the nuclei of aggregate ξ and between the electron field $\phi_\xi(\vec{q})$ and these nuclei, while the second and third terms describe the intermolecular Coulombic contributions.

It is now shown how both the inter- and intramolecular Coulomb terms are contained within the longitudinal components of the total electric polarization field. First, it is noted that because of the Coulomb gauge condition, the Coulombic terms may be expressed as the following field energy,

$$\frac{\varepsilon_0}{2} \int \vec{e}^{\parallel}(\vec{r}) \cdot \vec{e}^{\parallel}(\vec{r}) \mathrm{d}^3\vec{r}, \qquad (2.3.9)$$

where the longitudinal electric field $\vec{e}^{\parallel}(\vec{r}) = -\vec{\nabla}\varphi(\vec{r})$, where $\varphi(\vec{r})$ is the scalar potential, which is a solution of Poisson's equation. In the neutral system being considered, $\vec{\nabla} \cdot \vec{d}(\vec{r}) = 0$ because the true charges are the only sources contributing to the displacement field. Hence, $\vec{d}^{\parallel}(\vec{r}) = 0$ when there are no residual charges, and from the definition of the electric displacement field (2.2.34), $\vec{\nabla}\cdot\vec{e}^{||}(\vec{r}) = -(1/\varepsilon_0)\vec{\nabla}\cdot\vec{p}^{||}(\vec{r})$, so that equation (2.3.9) can be written as

$$\frac{1}{2\varepsilon_0}\int|\vec{p}^{||}(\vec{r})|^2\mathrm{d}^3\vec{r}, \tag{2.3.10}$$

where the longitudinal component of the electric polarization field is defined by

$$\vec{p}^{||}(\vec{r}) = \sum_{\xi}\vec{p}_{\xi}^{||}(\vec{r}) = \sum_{\xi}\int\bar{\phi}_{\xi}(\vec{q})\vec{p}_{\xi}^{||}(\vec{r},\vec{q})\phi_{\xi}(\vec{q})\mathrm{d}^3\vec{q}, \tag{2.3.11}$$

with

$$p_{i(\xi)}^{||}(\vec{r},\vec{q}) = -e(\vec{q}-\vec{R}_{\xi})_j\int_0^1\delta_{ij}^{||}(\vec{r}-\vec{R}_{\xi}-\lambda(\vec{q}-\vec{R}_{\xi}))\mathrm{d}\lambda, \tag{2.3.12}$$

where $\delta_{ij}^{||}(\vec{r})$ is the longitudinal delta function dyadic (Belinfante, 1946). Using the relation

$$\delta_{ij}^{||}(\vec{r}) = -\vec{\nabla}_i\vec{\nabla}_j\frac{1}{4\pi r} \tag{2.3.13}$$

and the identity

$$a_i\vec{\nabla}_i f(\vec{r}-\lambda\vec{a}) = -\frac{\mathrm{d}}{\mathrm{d}\lambda}f(\vec{r}-\lambda\vec{a}) \tag{2.3.14}$$

produces for expression (2.3.12) the formula

$$p_{i(\xi)}^{||}(\vec{r},\vec{q}) = \frac{e}{4\pi}\vec{\nabla}_i\left[\frac{1}{|\vec{r}-\vec{q}|}-\frac{1}{|\vec{r}-\vec{R}_{\xi}|}\right]. \tag{2.3.15}$$

Inserting (2.3.15) into equation (2.3.11), which in turn is substituted into (2.3.10), and using conservation of charge,

$$\int\bar{\phi}_{\xi}(\vec{q})\phi_{\xi}(\vec{q})\mathrm{d}^3\vec{q} = Z_{\xi}, \tag{2.3.16}$$

the intra- and intermolecular Coulombic interactions are

$$\begin{aligned} V_{\text{intra}} &= \frac{1}{2\varepsilon_0}\sum_{\xi}\int p_{\xi}^{||}(\vec{r})\cdot p_{\xi}^{||}(\vec{r})\mathrm{d}^3\vec{r} \\ &= \sum_{\xi}\int\bar{\phi}_{\xi}(\vec{q})\left\{V_{\xi}(\vec{q}) + \frac{e^2}{8\pi\varepsilon_0}\int\frac{\bar{\phi}_{\xi}(\vec{q}')\phi_{\xi}(\vec{q}')}{|\vec{q}-\vec{q}'|}\mathrm{d}^3\vec{q}'\right\}\phi_{\xi}(\vec{q})\mathrm{d}^3\vec{q} \end{aligned} \tag{2.3.17}$$

and

$$
\begin{aligned}
V_{\text{inter}} &= \sum_{\substack{\xi,\xi' \\ \xi \neq \xi'}} \frac{1}{2\varepsilon_0} \int \vec{p}_{\xi}^{||}(\vec{r}) \cdot \vec{p}_{\xi'}^{||}(\vec{r}) \mathrm{d}^3\vec{r} \\
&= \frac{1}{4\pi\varepsilon_0} \sum_{\substack{\xi,\xi' \\ \xi \neq \xi'}} \int \bar{\phi}_{\xi}(\vec{q}) \left\{ -\frac{Z_{\xi'} e^2}{|\vec{q} - \vec{R}_{\xi'}|} + \frac{e^2}{2} \int \frac{\bar{\phi}_{\xi'}(\vec{q}\,')\phi_{\xi'}(\vec{q}\,')}{|\vec{q} - \vec{q}\,'|} \mathrm{d}^3\vec{q}\,' \right\} \phi_{\xi}(\vec{q}) \mathrm{d}^3\vec{q} \\
&\quad + \frac{e^2}{8\pi\varepsilon_0} \sum_{\substack{\xi,\xi' \\ \xi \neq \xi'}} \frac{Z_{\xi} Z_{\xi'}}{|\vec{R}_{\xi} - \vec{R}_{\xi'}|}.
\end{aligned} \tag{2.3.18}
$$

In an analogous manner, the corresponding transverse polarization field contribution $\frac{1}{2\varepsilon_0}\int |\vec{p}^{\perp}(\vec{r})|^2 \mathrm{d}^3\vec{r}$ can be partitioned into intra- and intermolecular terms. Adding the intermolecular part of this contribution to equation (2.3.18) produces

$$
\begin{aligned}
&\frac{1}{2\varepsilon_0} \int |\vec{p}^{\perp}(\vec{r})|^2 \mathrm{d}^3\vec{r} + \frac{1}{2\varepsilon_0} \sum_{\substack{\xi,\xi' \\ \xi \neq \xi'}} \int \vec{p}_{\xi}^{||}(\vec{r}) \cdot \vec{p}_{\xi'}^{||}(\vec{r}) \mathrm{d}^3\vec{r} \\
&\quad = \frac{1}{2\varepsilon_0} \sum_{\xi} \int |\vec{p}_{\xi}^{\perp}(\vec{r})|^2 \mathrm{d}^3\vec{r} + \frac{1}{2\varepsilon_0} \sum_{\substack{\xi,\xi' \\ \xi \neq \xi'}} \int \vec{p}_{\xi}(\vec{r}) \cdot \vec{p}_{\xi'}(\vec{r}) \mathrm{d}^3\vec{r}.
\end{aligned} \tag{2.3.19}
$$

The rightmost term of expression (2.3.19) represents the total polarization field, which is localized at a specific center; hence, this term vanishes for each pairwise contribution. An effective intramolecular potential energy function composed of the one-center terms of equations (2.3.17) and (2.3.19) may be written as

$$
\begin{aligned}
V'_{\xi} &= \frac{1}{2\varepsilon_0} \int |p_{\xi}^{\perp}(\vec{r})|^2 \mathrm{d}^3\vec{r} \\
&\quad + \int \bar{\phi}_{\xi}(\vec{q}) \left\{ V_{\xi}(\vec{q}) + \frac{e^2}{8\pi\varepsilon_0} \int \frac{\bar{\phi}_{\xi}(\vec{q}\,')\phi_{\xi}(\vec{q}\,')}{|\vec{q} - \vec{q}\,'|} \mathrm{d}^3\vec{q}\,' \right\} \phi_{\xi}(\vec{q}) \mathrm{d}^3\vec{q}.
\end{aligned} \tag{2.3.20}
$$

This finally allows the multipolar Lagrangian to be expressed as

$$\begin{aligned}L_{\text{mult}} = &-\sum_{\xi}\left[\int \bar{\phi}_{\xi}(\vec{q})\left\{\frac{1}{2m}\left[-i\hbar\vec{\nabla}^{(\vec{q})} + \int \vec{n}_{\xi}(\vec{r},\vec{q}) \times \vec{b}(\vec{r})\mathrm{d}^3\vec{r}\right]^2\right\}\right.\\ &\qquad\left.\times\, \phi_{\xi}(\vec{q})\mathrm{d}^3\vec{q} + V'_{\xi}\right]\\ &+\frac{i\hbar}{2}\sum_{\xi}\int\left[\bar{\phi}_{\xi}(\vec{q})\dot{\phi}_{\xi}(\vec{q}) - \dot{\bar{\phi}}_{\xi}(\vec{q})\phi_{\xi}(\vec{q})\right]\mathrm{d}^3\vec{q}\\ &+\frac{\varepsilon_0}{2}\int\left\{\left[\varepsilon_0\dot{\vec{a}}^{\perp}(\vec{r}) - \vec{p}^{\perp}(\vec{r})\right]^2 - c^2(\vec{\nabla}\times\vec{a}(\vec{r}))^2\right\}\mathrm{d}^3\vec{r}. \qquad (2.3.21)\end{aligned}$$

This Lagrangian contains no intermolecular terms; all of the molecular dependent contributions are of one center in nature. As for charged particle theory, the intermolecular interaction occurs via the transverse electromagnetic field. This is in direct contrast to the minimal-coupling framework, in which both the Lagrangian and the Hamiltonian contain pair electrostatic interaction terms. By following the standard canonical quantization procedure, the multipolar Hamiltonian for a molecular assembly may be obtained from equation (2.3.21) and partitioned as

$$H_{\text{mult}} = \sum_{\xi} H_{\text{mol}}^{\text{mult}}(\xi) + H_{\text{rad}}^{\text{mult}} + \sum_{\xi} H_{\text{int}}^{\text{mult}}(\xi). \qquad (2.3.22)$$

Individual contributions are given in the next section.

2.4 MULTIPOLAR MAXWELL FIELDS

In the previous two sections, it has been shown how the multipolar Hamiltonian in second quantized form could be obtained from the classical minimal-coupling Hamiltonian by changing the generalized coordinates of the system or by applying a quantum canonical transformation to the minimal-coupling Hamiltonian operator. In the electric dipole approximation, the multipolar Hamiltonian for a single molecule is given explicitly by

$$H_{\text{mult}} = H_{\text{mol}}^{\text{mult}} + H_{\text{rad}}^{\text{mult}} + H_{\text{int}}^{\text{mult}}, \qquad (2.4.1)$$

where

$$H_{\text{mol}}^{\text{mult}} = \int \bar{\phi}(\vec{q})\left\{-\frac{\hbar^2}{2m}\left(\vec{\nabla}^{(\vec{q})}\right)^2 + V(\vec{q})\right\}\phi(\vec{q})\mathrm{d}^3\vec{q}, \qquad (2.4.2)$$

$$H_{\mathrm{rad}}^{\mathrm{mult}} = \frac{\varepsilon_0}{2} \int \left\{ \frac{\vec{d}^{\perp 2}(\vec{r})}{\varepsilon_0^2} + c^2 \vec{b}^2(\vec{r}) \right\} \mathrm{d}^3\vec{r}, \tag{2.4.3}$$

and

$$H_{\mathrm{int}}^{\mathrm{mult}} = -\varepsilon_0^{-1} \int \bar{\phi}(\vec{q}) \vec{\mu} \cdot \vec{d}^{\perp}(\vec{R}) \phi(\vec{q}) \mathrm{d}^3\vec{q}, \tag{2.4.4}$$

for a source electric dipole moment $\vec{\mu}$ located at $\vec{R}$, with self-energy terms being ignored since they are independent of the Maxwell fields and do not affect their equations of motion. In the second quantized formalism, the electron wavefield is expressed as

$$\phi(\vec{q}, t) = \sum_n b_n(t) \phi_n(\vec{q}), \tag{2.4.5}$$

where $\phi_n(\vec{q})$ is the orthonormal electron field mode and $b_n(t)$ is the time-dependent fermion annihilation operator for the state $|n\rangle$ of energy E_n. The boson creation and destruction operators $a^{(\lambda)}(\vec{k})$ and $a^{\dagger(\lambda)}(\vec{k})$ for a $(\vec{k}, \lambda)$-mode photon are contained implicitly in the last two terms of the total Hamiltonian (2.4.1) through the mode expansions for the radiation fields $\vec{d}^{\perp}(\vec{r})$ and $\vec{b}(\vec{r})$.

It is instructive to examine the time dependence of dynamical variables in the multipolar- and minimal-coupling frameworks and to calculate the time-dependent field operators, including that for the vector potential (Power, 1993; Power and Thirunamachandran, 1999a,1999b). The time-dependent mode expansions for the vector potential, electric displacement field, and magnetic field are

$$\vec{a}(\vec{r}, t) = \sum_{\vec{k},\lambda} \left(\frac{\hbar}{2\varepsilon_0 ckV} \right)^{1/2} \left[\vec{e}^{(\lambda)}(\vec{k}) a^{(\lambda)}(\vec{k}, t) \mathrm{e}^{i\vec{k}\cdot\vec{r}} + \bar{\vec{e}}^{(\lambda)}(\vec{k}) a^{\dagger(\lambda)}(\vec{k}, t) \mathrm{e}^{-i\vec{k}\cdot\vec{r}} \right], \tag{2.4.6}$$

$$\vec{d}^{\perp}(\vec{r}, t) = i \sum_{\vec{k},\lambda} \left(\frac{\hbar ck\varepsilon_0}{2V} \right)^{1/2} \left[\vec{e}^{(\lambda)}(\vec{k}) a^{(\lambda)}(\vec{k}, t) \mathrm{e}^{i\vec{k}\cdot\vec{r}} - \bar{\vec{e}}^{(\lambda)}(\vec{k}) a^{\dagger(\lambda)}(\vec{k}, t) \mathrm{e}^{-i\vec{k}\cdot\vec{r}} \right], \tag{2.4.7}$$

and

$$\vec{b}(\vec{r}, t) = i \sum_{\vec{k},\lambda} \left(\frac{\hbar k}{2\varepsilon_0 cV} \right)^{1/2} \left[\vec{b}^{(\lambda)}(\vec{k}) a^{(\lambda)}(\vec{k}, t) \mathrm{e}^{i\vec{k}\cdot\vec{r}} - \bar{\vec{b}}^{(\lambda)}(\vec{k}) a^{\dagger(\lambda)}(\vec{k}, t) \mathrm{e}^{-i\vec{k}\cdot\vec{r}} \right]. \tag{2.4.8}$$

The time evolution of the boson annihilation operator $a^{(\lambda)}(\vec{k},t)$, for example, is obtained from the Heisenberg equation of motion

$$i\hbar\dot{a}^{\mathrm{mult}}(t) = [a^{\mathrm{mult}}(t), H^{\mathrm{mult}}]_- \tag{2.4.9}$$

on inserting (2.4.1) and using the commutation relation at equal time,

$$[a^{\mathrm{mult}(\lambda)}(\vec{k},t), a^{\mathrm{mult}\dagger(\lambda')}(\vec{k}',t)]_- = \delta_{\vec{k}\vec{k}'}\delta_{\lambda\lambda'}, \tag{2.4.10}$$

where the mode dependence has been suppressed in formula (2.4.9). This leads to

$$a^{\mathrm{mult}}(t) = a^{\mathrm{mult}}(0)\mathrm{e}^{-i\omega t} + \frac{1}{\hbar}\left(\frac{\hbar ck}{2\varepsilon_0 V}\right)^{1/2} \mathrm{e}^{-i(\vec{k}\cdot\vec{R}+\omega t)}\bar{\vec{e}}^{(\lambda)}(\vec{k})\cdot\int_0^t \mathrm{d}t'\vec{\mu}(t')\mathrm{e}^{i\omega t'} \tag{2.4.11}$$

and consists of a source-independent and a source-dependent term. It is convenient to partition the fields (2.4.6)–(2.4.8) into free- and source-field contributions. Hence, for the vector potential,

$$\vec{a}(\vec{r},t) = \vec{a}^{(0)}(\vec{r},t) + \vec{a}^{(s)}(\vec{r},t), \tag{2.4.12}$$

where the source-free term is obtained from the first term of expression (2.4.11),

$$\begin{aligned}\vec{a}^{(0)\mathrm{mult}}(\vec{r},t) = \sum_{\vec{k},\lambda}\left(\frac{\hbar}{2\varepsilon_0 ckV}\right)^{1/2} \Big[&\vec{e}^{(\lambda)}(\vec{k})a^{(\lambda)\mathrm{mult}}(\vec{k},0)\mathrm{e}^{i\vec{k}\cdot\vec{r}-i\omega t} \\ &+ \bar{\vec{e}}^{(\lambda)}(\vec{k})a^{\dagger(\lambda)\mathrm{mult}}(\vec{k},0)\mathrm{e}^{-i\vec{k}\cdot\vec{r}+i\omega t}\Big].\end{aligned} \tag{2.4.13}$$

Substituting the second term of (2.4.11) into the mode expansion (2.4.6) gives for the source-dependent contribution to the vector potential,

$$a_i^{(s)\mathrm{mult}}(\vec{r},t) = \sum_{\vec{k},\lambda}\left(\frac{1}{2\varepsilon_0 V}\right)\left\{e_i^{(\lambda)}(\vec{k})\bar{e}_j^{(\lambda)}(\vec{k})\mathrm{e}^{i\vec{k}\cdot(\vec{r}-\vec{R})}\int_0^t \mathrm{d}t'\mu_j(t')\mathrm{e}^{-i\omega(t-t')} + \mathrm{H.C.}\right\}, \tag{2.4.14}$$

where H.C. is the Hermitian-conjugate term. Two necessary and key steps used repeatedly in the computation of Maxwell fields, and in subsequent applications to follow, are the summations over polarization and wavevector. The former may be carried out with the identities given in Section 1.4. In the present case, use is made of relation (1.4.56). For the wavevector sum, use is made of the continuum approximation as exemplified in the prescription (1.4.55), with $\mathrm{d}^3\vec{k}=k^2\mathrm{d}k\mathrm{d}\Omega$ in spherical polar coordinates, where $\mathrm{d}\Omega$ is an infinitesimal element of solid angle. The angular integration is performed by noting that

$$\frac{1}{4\pi}\int \mathrm{e}^{\pm i\vec{k}\cdot\vec{r}}\mathrm{d}\Omega=\frac{\sin kr}{kr}, \tag{2.4.15}$$

from which

$$\begin{aligned}\frac{1}{4\pi}\int(\delta_{ij}-\hat{k}_i\hat{k}_j)\mathrm{e}^{\pm i\vec{k}\cdot\vec{r}}\mathrm{d}\Omega&=\frac{1}{k^3}\left(-\vec{\nabla}^2\delta_{ij}+\vec{\nabla}_i\vec{\nabla}_j\right)\frac{\sin kr}{r}\\&=\left\{(\delta_{ij}-\hat{r}_i\hat{r}_j)\frac{\sin kr}{kr}+(\delta_{ij}-3\hat{r}_i\hat{r}_j)\left(\frac{\cos kr}{k^2r^2}-\frac{\sin kr}{k^3r^3}\right)\right\}.\end{aligned} \tag{2.4.16}$$

Hence, equation (2.4.14) becomes

$$\begin{aligned}a_i^{(s)\mathrm{mult}}(\vec{r},t)=&\frac{1}{4\pi^2\varepsilon_0}\left(-\vec{\nabla}^2\delta_{ij}+\vec{\nabla}_i\vec{\nabla}_j\right)\frac{1}{|\vec{r}-\vec{R}|}\int_0^t\mathrm{d}t'\mu_j(t')\\&\times\int_0^\infty\mathrm{d}k\frac{2}{k}\sin(k|\vec{r}-\vec{R}|)\cos(kc(t-t')).\end{aligned} \tag{2.4.17}$$

Using the result

$$\frac{2}{\pi}\int_0^\infty\mathrm{d}k\frac{1}{k}\sin(kr)\cos(kc(t-t'))=\frac{1}{2}\{\mathrm{sgn}[r-c(t-t')]+\mathrm{sgn}[r+c(t-t')]\}, \tag{2.4.18}$$

equation (2.4.17) becomes

$$a_i^{(s)\mathrm{mult}}(\vec{r},t)=\begin{cases}\dfrac{1}{4\pi\varepsilon_0}(-\vec{\nabla}^2\delta_{ij}+\vec{\nabla}_i\vec{\nabla}_j)\dfrac{1}{|\vec{r}-\vec{R}|}\displaystyle\int_{t-|\vec{r}-\vec{R}|/c}^{t}\mathrm{d}t'\mu_j(t'), & t>|\vec{r}-\vec{R}|/c>0,\\[2ex]\dfrac{1}{4\pi\varepsilon_0}(-\vec{\nabla}^2\delta_{ij}+\vec{\nabla}_i\vec{\nabla}_j)\dfrac{1}{|\vec{r}-\vec{R}|}\displaystyle\int_0^t\mathrm{d}t'\mu_j(t'), & t<|\vec{r}-\vec{R}|/c.\end{cases} \tag{2.4.19}$$

In expression (2.4.19), the gradient operators act on the retarded time as well as on $1/(|\vec{r}-\vec{R}|)$. Noteworthy is the fact that for $t<|\vec{r}-\vec{R}|/c$, the vector potential is nonzero.

In a similar fashion, the transverse displacement field may be partitioned into vacuum and source fields and computed using the relations (2.4.11), (2.4.16), and (2.4.18). The free displacement field is given by

$$\vec{d}^{(0)\text{mult}}(\vec{r},t) = i\sum_{\vec{k},\lambda}\left(\frac{\hbar ck\varepsilon_0}{2V}\right)^{1/2}\left[\vec{e}^{(\lambda)}(\vec{k})a^{(\lambda)\text{mult}}(\vec{k},0)\mathrm{e}^{i\vec{k}\cdot\vec{r}-i\omega t}\right.$$
$$\left.-\bar{\vec{e}}^{(\lambda)}(\vec{k})a^{\dagger(\lambda)\text{mult}}(\vec{k},0)\mathrm{e}^{-i\vec{k}\cdot\vec{r}+i\omega t}\right]. \tag{2.4.20}$$

The free field operates entirely in the boson space, changing the number of photons by one. Substituting the second term of (2.4.11) and its adjoint into the mode expansion (2.4.7) gives for the electric dipole-dependent vector operator the form

$$d_i^{(s)\text{mult}}(\vec{r},t) = i\sum_{\vec{k},\lambda}\left(\frac{ck}{2V}\right)\left\{e_i^{(\lambda)}(\vec{k})\bar{e}_j^{(\lambda)}(\vec{k})\mathrm{e}^{i\vec{k}\cdot(\vec{r}-\vec{R})}\int_0^t \mathrm{d}t'\mu_j(t')\mathrm{e}^{-i\omega(t-t')} - \text{H.C.}\right\}. \tag{2.4.21}$$

Carrying out the mode sum yields

$$\begin{aligned} d_i^{(s)\text{mult}}(\vec{r},t) &= \frac{c}{4\pi^2}(-\vec{\nabla}^2\delta_{ij}+\vec{\nabla}_i\vec{\nabla}_j)\frac{1}{|\vec{r}-\vec{R}|}\int_0^t \mathrm{d}t'\mu_j(t') \\ &\quad\times\int_0^\infty \mathrm{d}k\, 2\sin(k|\vec{r}-\vec{R}|)\sin(kc(t-t')) \\ &= \frac{c}{4\pi}(-\vec{\nabla}^2\delta_{ij}+\vec{\nabla}_i\vec{\nabla}_j)\frac{1}{|\vec{r}-\vec{R}|} \\ &\quad\times\int_0^t \mathrm{d}t'\mu_j(t')\{\delta(|\vec{r}-\vec{R}|-c(t-t'))-\delta(|\vec{r}-\vec{R}|+c(t-t'))\}. \end{aligned} \tag{2.4.22}$$

Only the first δ-function contributes in (2.4.22) because $t' = t-|\vec{r}-\vec{R}|/c$ lies between 0 and t; the second term vanishes as it lies outside the range of integration over t'. The source field is thus strictly causal, and the displacement field must be evaluated at the retarded time. Finally,

$$d_i^{(s)}(\vec{r},t)=\begin{cases}\dfrac{1}{4\pi}\left(-\vec{\nabla}^2\delta_{ij}+\vec{\nabla}_i\vec{\nabla}_j\right)\dfrac{\mu_j(t-|\vec{r}-\vec{R}|/c)}{|\vec{r}-\vec{R}|}, & t>|\vec{r}-\vec{R}|/c>0\\[2ex] 0, & t<|\vec{r}-\vec{R}|/c,\end{cases} \tag{2.4.23}$$

with the gradient operators acting on the retarded time as well as on $1/(|\vec{r}-\vec{R}|)$. From the source-dependent vector potential (2.4.19), it is a simple matter to evaluate the source-dependent magnetic field and the transverse and total electric field operators, even though the last two fields do not appear in the multipolar formalism. Nevertheless, their evaluation will help compare the fields in the two frameworks. The magnetic field is obtained from

$$\begin{aligned}b_i^{(s)\mathrm{mult}}(\vec{r},t)&=[\vec{\nabla}\times\vec{a}^{(s)\mathrm{mult}}(\vec{r},t)]_i\\ &=\begin{cases}-\dfrac{1}{4\pi\varepsilon_0 c}\varepsilon_{ijk}\vec{\nabla}_k\dfrac{1}{|\vec{r}-\vec{R}|}\dfrac{\mathrm{d}}{\mathrm{d}t}\mu_j(t-|\vec{r}-\vec{R}|/c), & t>|\vec{r}-\vec{R}|/c>0,\\[2ex] 0, & t<|\vec{r}-\vec{R}|/c,\end{cases}\end{aligned} \tag{2.4.24}$$

which also vanishes for $t<|\vec{r}-\vec{R}|/c$, while the source-dependent transverse electric field is

$$\begin{aligned}e_i^{\perp(s)\mathrm{mult}}(\vec{r},t)&=-\frac{\mathrm{d}}{\mathrm{d}t}a_i^{(s)\mathrm{mult}}(\vec{r},t)\\ &=\begin{cases}\dfrac{1}{4\pi\varepsilon_0}\left(-\vec{\nabla}^2\delta_{ij}+\vec{\nabla}_i\vec{\nabla}_j\right)\dfrac{1}{|\vec{r}-\vec{R}|}\left[\mu_j(t-|\vec{r}-\vec{R}|/c)-\mu_j(t)\right], & t>|\vec{r}-\vec{R}|/c>0,\\[2ex] -\dfrac{1}{4\pi\varepsilon_0}(-\vec{\nabla}^2\delta_{ij}+\vec{\nabla}_i\vec{\nabla}_j)\dfrac{1}{|\vec{r}-\vec{R}|}\mu_j(t), & t<|\vec{r}-\vec{R}|/c,\end{cases}\end{aligned} \tag{2.4.25}$$

which like the source-dependent vector potential contains a nonvanishing term for $t<|\vec{r}-\vec{R}|/c$. In the Coulomb gauge, the longitudinal electric field is related to a static distribution of charges. For an electric dipole source,

$$e_i^{||\mathrm{mult}}(\vec{r},t)=-\varepsilon_0^{-1}\mu_j(t)\delta_{ij}^{\perp}(\vec{r})=-\frac{1}{4\pi\varepsilon_0 r^3}(\delta_{ij}-3\hat{r}_i\hat{r}_j)\mu_j(t),\quad r>0, \tag{2.4.26}$$

which is related to the electric polarization field via

$$e_i^{\|\text{mult}}(\vec{r},t) = -\varepsilon_0^{-1} p_i^{\|}(\vec{r},t) = \varepsilon_0^{-1} p_i^{\perp}(\vec{r},t). \tag{2.4.27}$$

Thus, expression (2.4.25) can be written as

$$e_i^{\perp(s)\text{mult}}(\vec{r},t) = \begin{cases} \dfrac{1}{4\pi\varepsilon_0}\left(-\vec{\nabla}^2\delta_{ij}+\vec{\nabla}_i\vec{\nabla}_j\right)\dfrac{1}{|\vec{r}-\vec{R}|}\mu_j(t-|\vec{r}-\vec{R}|/c)-\varepsilon_0^{-1}p_i^{\perp}(\vec{r},t), & t>|\vec{r}-\vec{R}|/c>0, \\ -\varepsilon_0^{-1}p_i^{\perp}(\vec{r},t), & t<|\vec{r}-\vec{R}|/c, \end{cases} \tag{2.4.28}$$

with the total electric field given by

$$e_i^{\text{tot}(s)\text{mult}}(\vec{r},t) = \begin{cases} \dfrac{1}{4\pi\varepsilon_0}\left(-\vec{\nabla}^2\delta_{ij}+\vec{\nabla}_i\vec{\nabla}_j\right)\dfrac{1}{|\vec{r}-\vec{R}|}\mu_j(t-|\vec{r}-\vec{R}|/c), & t>|\vec{r}-\vec{R}|/c>0, \\ 0, & t<|\vec{r}-\vec{R}|/c, \end{cases} \tag{2.4.29}$$

which is seen to be equal to $\varepsilon_0^{-1} d_i^{(s)\text{mult}}(\vec{r},t)$ expressed by (2.4.23).

2.5 MINIMAL-COUPLING MAXWELL FIELDS

In the previous section, explicit formulas were obtained for the time-dependent vector potential, transverse and total electric fields, and electric displacement and magnetic field operators due to an electric dipole source moment by starting from the second quantized multipolar Hamiltonian and calculating the time evolution of boson operators from the Heisenberg equations of motion. To further examine the similarities and differences between multipolar- and minimal-coupling frameworks (Salam, 2008), the corresponding fields are evaluated in the latter scheme. Now the starting point is the minimal-coupling Hamiltonian. For a single particle, this may be written as

$$H^{\text{min}} = H_{\text{mol}}^{\text{min}} + H_{\text{rad}}^{\text{min}} + \frac{e}{m}\vec{p}\cdot\vec{a}(\vec{q},t) + \frac{e^2}{2m}\vec{a}^2(\vec{q},t), \tag{2.5.1}$$

where the first two terms of (2.5.1) are the molecular and radiation field Hamiltonians (2.2.11) and (2.2.12) and the last two terms are the interaction terms. It should be noted that the total Hamiltonian (2.5.1) is expressed in terms of canonically conjugate variables derived within the minimal-coupling formalism, which differ from their multipolar counterparts. From equation (2.2.8), for instance, the conjugate momentum of the radiation field coordinate is equal to $-\varepsilon_0\vec{e}^{\perp}(\vec{r})$. For radiation field wavelengths larger than molecular dimensions, the spatial variations of the vector potential may be neglected to a first approximation, thereby leading to the electric dipole approximated form. Situating this dipole at $\vec{R}$ as before, the interaction Hamiltonian is then

$$H_{\text{int}}^{\text{min}} = \frac{e}{m}\vec{p}\cdot\vec{a}(\vec{R},t) + \frac{e^2}{2m}\vec{a}^2(\vec{R},t). \tag{2.5.2}$$

The time evolution of the photon annihilation operator in minimal coupling is obtained from the Heisenberg equation of motion

$$i\hbar\dot{a}^{\text{min}}(t) = [a^{\text{min}}(t), H^{\text{min}}]_- = \hbar\omega a + \frac{e}{m}\left(\frac{\hbar}{2\varepsilon_0 ckV}\right)^{1/2}\left[\vec{p}+e\vec{a}(\vec{R},t)\right]\cdot\bar{\vec{e}}^{(\lambda)}(\vec{k}), \tag{2.5.3}$$

on using the commutation rule (2.4.10), which remains valid since the two Hamiltonians are related by a quantum canonical transformation. Performing the time integral in (2.5.3) leads to

$$a^{\text{min}}(t) = a^{\text{min}}(0)\mathrm{e}^{-i\omega t} + \frac{i}{\hbar}\left(\frac{\hbar}{2\varepsilon_0 ckV}\right)^{1/2}\mathrm{e}^{-i\omega t}\bar{\vec{e}}^{(\lambda)}(\vec{k})\cdot\int_0^t \mathrm{d}t'\dot{\vec{\mu}}(t')\mathrm{e}^{-i\vec{k}\cdot\vec{R}+i\omega t'}, \tag{2.5.4}$$

where use has been made of the relation between the kinetic and canonical momentum of the material system,

$$e\frac{\mathrm{d}}{\mathrm{d}t}(\vec{q}-\vec{R}) = \frac{e}{m}\left[\vec{p}+e\vec{a}(\vec{R})\right]. \tag{2.5.5}$$

As in the case of the source-dependent fields and vector potential in the multipolar framework, the second term of (2.5.4) is substituted into the respective mode expansions and the wavevector and polarization sums carried out. From (2.4.6), the source-dependent minimal-coupling vector potential is found to be

$$
\begin{aligned}
a_i^{(s)\min}(\vec{r},t) &= i\sum_{\vec{k},\lambda}\left(\frac{1}{2\varepsilon_0 ckV}\right) \\
&\quad\times\left\{e_i^{(\lambda)}(\vec{k})\bar{e}_j^{(\lambda)}(\vec{k})\mathrm{e}^{i\vec{k}\cdot(\vec{r}-\vec{R})}\int_0^t \mathrm{d}t'\,\dot{\mu}_j(t')\mathrm{e}^{i\omega(t-t')}+\mathrm{H.C.}\right\} \\
&= \frac{1}{4\pi^2\varepsilon_0 c}(-\vec{\nabla}^2\delta_{ij}+\vec{\nabla}_i\vec{\nabla}_j)\frac{1}{|\vec{r}-\vec{R}|}\int_0^t \mathrm{d}t'\,\dot{\mu}_j(t') \\
&\quad\times\int_0^\infty \mathrm{d}k\frac{2}{k^2}\sin\left[k|\vec{r}-\vec{R}|\right]\sin[kc(t-t')],
\end{aligned}
\tag{2.5.6}
$$

which after carrying out the k-integral results in

$$
a_i^{(s)\min}(\vec{r},t)=\begin{cases}
\dfrac{1}{4\pi\varepsilon_0}(-\vec{\nabla}^2\delta_{ij}+\vec{\nabla}_i\vec{\nabla}_j)\dfrac{1}{|\vec{r}-\vec{R}|}\displaystyle\int_{t-|\vec{r}-\vec{R}|/c}^{t}\mathrm{d}t'\,\mu_j(t'), & t>|\vec{r}-\vec{R}|/c>0,\\[2ex]
\dfrac{1}{4\pi\varepsilon_0}(-\vec{\nabla}^2\delta_{ij}+\vec{\nabla}_i\vec{\nabla}_j)\dfrac{1}{|\vec{r}-\vec{R}|}\displaystyle\int_0^t\mathrm{d}t'[\mu_j(t')-\mu_j(0)], & t<|\vec{r}-\vec{R}|/c.
\end{cases}
\tag{2.5.7}
$$

The transverse electric field can be derived in a manner similar to that used to obtain the vector potential in the multipolar formalism. Alternatively, it is given straightforwardly by

$$
\begin{aligned}
e_i^{\perp(s)\min}(\vec{r},t) &= -\dot{a}_i^{(s)\min}(\vec{r},t) \\
&= \begin{cases}
\dfrac{1}{4\pi\varepsilon_0}(-\vec{\nabla}^2\delta_{ij}+\vec{\nabla}_i\vec{\nabla}_j)\dfrac{1}{|\vec{r}-\vec{R}|}\left[\mu_j(t-|\vec{r}-\vec{R}|/c)-\mu_j(t)\right], & t>|\vec{r}-\vec{R}|/c>0,\\[2ex]
\dfrac{1}{4\pi\varepsilon_0}(-\vec{\nabla}^2\delta_{ij}+\vec{\nabla}_i\vec{\nabla}_j)\dfrac{1}{|\vec{r}-\vec{R}|}\left[\mu_j(0)-\mu_j(t)\right], & t<|\vec{r}-\vec{R}|/c.
\end{cases}
\end{aligned}
\tag{2.5.8}
$$

Since the longitudinal electric field is the same in both frameworks, the dipole-dependent term (2.4.26) can be added to equation (2.5.8) to give the total electric field in minimal coupling,

$$
e_i^{\mathrm{tot}(s)\min}(\vec{r},t)=\begin{cases}
\dfrac{1}{4\pi\varepsilon_0}(-\vec{\nabla}^2\delta_{ij}+\vec{\nabla}_i\vec{\nabla}_j)\dfrac{1}{|\vec{r}-\vec{R}|}\mu_j\left(t-|\vec{r}-\vec{R}|/c\right), & t>|\vec{r}-\vec{R}|/c>0,\\[2ex]
\dfrac{1}{4\pi\varepsilon_0}(-\vec{\nabla}^2\delta_{ij}+\vec{\nabla}_i\vec{\nabla}_j)\dfrac{1}{|\vec{r}-\vec{R}|}\mu_j(0), & t<|\vec{r}-\vec{R}|/c.
\end{cases}
\tag{2.5.9}
$$

Interestingly, the source-dependent transverse and total electric fields for $t>|\vec{r}-\vec{R}|/c>0$ are identical to their multipolar analogues (2.4.28) and (2.4.29). In contrast, the functional forms differ for $t<|\vec{r}-\vec{R}|/c$. Neither the transverse nor the total minimal-coupling electric field vanishes in this time interval, unlike the multipolar total electric field (2.4.29), which is causal.

Once again the source-dependent vector potential (2.5.7) may be used to calculate the time-dependent magnetic field operator due to an electric dipole. Thus,

$$b_i^{(s)\min}(\vec{r},t) = \left[\vec{\nabla}\times\vec{a}^{(s)\min}(\vec{r},t)\right]_i = \begin{cases} -\dfrac{1}{4\pi\varepsilon_0 c}\varepsilon_{ijk}\vec{\nabla}_k \dfrac{1}{|\vec{r}-\vec{R}|}\dfrac{\mathrm{d}}{\mathrm{d}t}\mu_j(t-|\vec{r}-\vec{R}|/c), & t>|\vec{r}-\vec{R}|/c>0, \\ 0, & t<|\vec{r}-\vec{R}|/c, \end{cases} \tag{2.5.10}$$

which is equivalent to the result (2.4.24) obtained using multipolar equations of motion. From the analysis of this and the previous section, it is found that the field operators are dependent on the time-dependent source electric dipole moment to which no label min or mult has been attached, despite the differing time evolution of the two types of Hamiltonian operator. No distinction is necessary due to the fact that $\vec{\mu}(t)$ is independent of any canonically conjugate momenta and therefore remains invariant in both constructions. The same is true of the magnetic field, $\vec{b}(\vec{r},t)$.

To further explore the relationship between the various radiation fields in the two formalisms, it is useful to find the connection between the minimal- and multipolar-coupling photon creation and annihilation operators. By applying the transformation

$$a^{\mathrm{mult}} = \mathrm{e}^{iS}a^{\min}\mathrm{e}^{-iS}, \tag{2.5.11}$$

where in the electric dipole approximation the generator is

$$S = \frac{1}{\hbar}\vec{\mu}\cdot\vec{a}(\vec{R}), \tag{2.5.12}$$

it is found that

$$a^{\mathrm{mult}}(t) = a^{\min}(t) - i\left(\frac{1}{2\hbar ck\varepsilon_0 V}\right)^{1/2}\mu_j(t)\bar{e}_j\mathrm{e}^{-i\vec{k}\cdot\vec{R}}. \tag{2.5.13}$$

New features arise in the functional forms of the vacuum radiation fields as a result of the difference above, as witnessed by the substitution of minimal boson operators into the multipolar mode expansions. Illustrating first for the vector potential,

$$
\begin{aligned}
a_i^{(0)\text{mult}}(\vec{r},t) &= \sum_{\vec{k},\lambda}\left(\frac{\hbar}{2ck\varepsilon_0 V}\right)^{1/2} \\
&\quad\times\left[e_i^{(\lambda)}(\vec{k})a^{\text{min}}(0)\mathrm{e}^{i\vec{k}\cdot\vec{r}-i\omega t}+\bar{e}_i^{(\lambda)}(\vec{k})a^{\dagger\text{min}}(0)\mathrm{e}^{-i\vec{k}\cdot\vec{r}+i\omega t}\right] \\
&\quad -i\sum_{\vec{k},\lambda}\left(\frac{1}{2ck\varepsilon_0 V}\right)\mu_j(0) \\
&\quad\times\left[e_i^{(\lambda)}(\vec{k})\bar{e}_j^{(\lambda)}(\vec{k})\mathrm{e}^{i\vec{k}\cdot(\vec{r}-\vec{R})-i\omega t}-\bar{e}_i^{(\lambda)}(\vec{k})e_j^{(\lambda)}(\vec{k})\mathrm{e}^{-i\vec{k}\cdot(\vec{r}-\vec{R})+i\omega t}\right] \\
&= a_i^{(0)\text{min}}(\vec{r},t)+\begin{cases} 0, & t>|\vec{r}-\vec{R}|/c>0, \\ -\dfrac{t}{4\pi}(-\vec{\nabla}^2\delta_{ij}+\vec{\nabla}_i\vec{\nabla}_j)\dfrac{1}{|\vec{r}-\vec{R}|}\mu_j(0), & t<|\vec{r}-\vec{R}|/c, \end{cases}
\end{aligned}
\tag{2.5.14}
$$

showing that the free vector potential is identical in both frameworks for $t>|\vec{r}-\vec{R}|/c$, but differs for $t<|\vec{r}-\vec{R}|/c$. Similarly, for the free electric displacement and magnetic fields,

$$
\begin{aligned}
d_i^{(0)\text{mult}}(\vec{r},t) &= i\sum_{\vec{k},\lambda}\left(\frac{\hbar ck\varepsilon_0}{2V}\right)^{1/2} \\
&\qquad\times\left[e_i^{(\lambda)}(\vec{k})a^{\text{min}}(0)\mathrm{e}^{i\vec{k}\cdot\vec{r}-i\omega t}-\bar{e}_i^{(\lambda)}(\vec{k})a^{\dagger\text{min}}(0)\mathrm{e}^{-i\vec{k}\cdot\vec{r}+i\omega t}\right] \\
&\qquad+\sum_{\vec{k},\lambda}\frac{1}{2V}\mu_j(0)\left[e_i^{(\lambda)}(\vec{k})\bar{e}_j^{(\lambda)}(\vec{k})\mathrm{e}^{i\vec{k}\cdot(\vec{r}-\vec{R})-i\omega t}\right. \\
&\qquad\left.+\bar{e}_i^{(\lambda)}(\vec{k})e_j^{(\lambda)}(\vec{k})\mathrm{e}^{-i\vec{k}\cdot(\vec{r}-\vec{R})+i\omega t}\right] \\
&\quad=\varepsilon_0 e_i^{\perp(0)\text{min}}(\vec{r},t) \\
&\qquad+\begin{cases} 0, & t>|\vec{r}-\vec{R}|/c>0, \\ \dfrac{1}{4\pi}(-\vec{\nabla}^2\delta_{ij}+\vec{\nabla}_i\vec{\nabla}_j)\dfrac{1}{|\vec{r}-\vec{R}|}\mu_j(0), & t<|\vec{r}-\vec{R}|/c, \end{cases}
\end{aligned}
\tag{2.5.15}
$$

and

$$\begin{aligned} b_i^{(0)\mathrm{mult}}(\vec{r},t) = {} & i\sum_{\vec{k},\lambda}\left(\frac{\hbar k}{2\varepsilon_0 cV}\right)^{1/2} \\ & \times\left[b_i^{(\lambda)}(\vec{k})a^{\mathrm{min}}(0)\mathrm{e}^{i\vec{k}\cdot\vec{r}-i\omega t}-\bar{b}_i^{(\lambda)}(\vec{k})a^{\dagger\mathrm{min}}(0)\mathrm{e}^{-i\vec{k}\cdot\vec{r}+i\omega t}\right] \\ & +\sum_{\vec{k},\lambda}\left(\frac{1}{2\varepsilon_0 cV}\right)\mu_j(0)\left[b_i^{(\lambda)}(\vec{k})\bar{e}_j^{(\lambda)}(\vec{k})\mathrm{e}^{i\vec{k}\cdot(\vec{r}-\vec{R})-i\omega t}\right. \\ & \left.+\bar{b}_i^{(\lambda)}(\vec{k})e_j^{(\lambda)}(\vec{k})\mathrm{e}^{-i\vec{k}\cdot(\vec{r}-\vec{R})+i\omega t}\right]. \end{aligned} \tag{2.5.16}$$

After carrying out the mode sum, the second term in expression (2.5.16) is found to vanish for $|\vec{r}-\vec{R}|\neq ct$ so that the free magnetic field is equivalent in both formalisms,

$$b_i^{(0)\mathrm{mult}}(\vec{r},t) = b_i^{(0)\mathrm{min}}(\vec{r},t), \quad |\vec{r}-\vec{R}|\neq ct. \tag{2.5.17}$$

Finally, the source-dependent multipolar total electric field at time t outside the source, when expressed in terms of minimal-coupling boson operators using the relation (2.5.13), is

$$\begin{aligned} e_i^{(\mathrm{tot})\mathrm{mult}}(\vec{r},t) = \varepsilon_0^{-1}d_i(\vec{r},t) = {} & \frac{i}{\varepsilon_0}\sum_{\vec{k},\lambda}\left(\frac{\hbar ck\varepsilon_0}{2V}\right)^{1/2} \\ & \times\left[e_i a^{\mathrm{mult}}(t)\mathrm{e}^{i\vec{k}\cdot\vec{r}}-\bar{e}_i a^{\dagger\mathrm{mult}}(t)\mathrm{e}^{-i\vec{k}\cdot\vec{r}}\right] \\ = \frac{i}{\varepsilon_0}\sum_{\vec{k},\lambda}\left(\frac{\hbar ck\varepsilon_0}{2V}\right)^{1/2} & \left[e_i a^{\mathrm{min}}(t)\mathrm{e}^{i\vec{k}\cdot\vec{r}}-\bar{e}_i a^{\dagger\mathrm{min}}(t)\mathrm{e}^{-i\vec{k}\cdot\vec{r}}\right] \\ & +\varepsilon_0^{-1}\sum_{\vec{k},\lambda}\frac{1}{2V}\mu_j(t)\left[e_i^{(\lambda)}(\vec{k})\bar{e}_j^{(\lambda)}(\vec{k})\mathrm{e}^{i\vec{k}\cdot(\vec{r}-\vec{R})}+\bar{e}_i^{(\lambda)}(\vec{k})e_j^{(\lambda)}(\vec{k})\mathrm{e}^{-i\vec{k}\cdot(\vec{r}-\vec{R})}\right] \\ = e_i^{\perp\mathrm{min}}(\vec{r},t)-\frac{1}{4\pi\varepsilon_0|\vec{r}-\vec{R}|^3} & \mu_j(t)\left[\delta_{ij}-3(\hat{r}-\hat{R})_i(\hat{r}-\hat{R})_j\right] \\ = e_i^{\perp\mathrm{min}}(\vec{r},t)+\varepsilon_0^{-1}p_i^{\|}(\vec{r},t) & = e_i^{\perp(\mathrm{min})}(\vec{r},t)+e_i^{\|}(\vec{r},t) = e_i^{(\mathrm{tot})\mathrm{min}}(\vec{r},t), \end{aligned} \tag{2.5.18}$$

demonstrating their equivalence in both schemes. Thus, from (2.5.18),

$$\vec{d}^{\mathrm{mult}}(\vec{r},t) = \varepsilon_0\vec{e}^{\perp\mathrm{min}}(\vec{r},t)+\vec{p}^{\perp}(\vec{r},t), \quad \text{for all } t, \tag{2.5.19}$$

from which it is seen that the total fields are identical for all times. Meanwhile from equation (2.5.15),

$$\vec{d}^{(0)\text{mult}}(\vec{r},t) = \varepsilon_0 \vec{e}^{\perp(0)\text{min}}(\vec{r},t) + \begin{cases} 0, & t > |\vec{r}-\vec{R}|/c > 0, \\ \vec{p}^{\perp}(\vec{r},0), & t < |\vec{r}-\vec{R}|/c, \end{cases} \tag{2.5.20}$$

showing that the free electric fields are the same for $t > |\vec{r}-\vec{R}|/c$. Interestingly, the source-dependent polarization field is time dependent, but in the absence of sources is independent of t. Noting that from (2.4.23) $\vec{d}^{(s)\text{mult}} = \vec{d}^{\text{mult}} - \vec{d}^{(0)\text{mult}}$, the separation of the total minimal-coupling electric field using equations (2.5.18) and (2.5.19) is

$$e_i^{(\text{tot})\text{min}}(\vec{r},t) = e_i^{\perp(0)\text{min}}(\vec{r},t) + \begin{cases} \dfrac{1}{4\pi\varepsilon_0}(-\vec{\nabla}^2\delta_{ij} + \vec{\nabla}_i\vec{\nabla}_j)\dfrac{1}{|\vec{r}-\vec{R}|}\mu_j(t-|\vec{r}-\vec{R}|/c), & t > |\vec{r}-\vec{R}|/c > 0, \\ \dfrac{1}{4\pi\varepsilon_0}(-\vec{\nabla}^2\delta_{ij} + \vec{\nabla}_i\vec{\nabla}_j)\dfrac{1}{|\vec{r}-\vec{R}|}\mu_j(0), & t < |\vec{r}-\vec{R}|/c, \end{cases} \tag{2.5.21}$$

with the division of the magnetic field being the same in both formalisms.

In summary, the vacuum field operators differ in the minimal- and multipolar-coupling frameworks. This is due to the boson creation and annihilation operators being nonidentical in the two schemes as their time evolution is governed by two different Hamiltonians. The source-dependent fields at positive retarded time on the other hand are the same. In addition to evaluating the time evolution of photon operators, the time dependence of fermion creation and annihilation operators in the two formalisms can be calculated analogously using Heisenberg's equation of motion. Like their boson counterparts, $b_n(t)$ and $b_n^\dagger(t)$ are found to differ in the multipolar- and minimal-coupling versions of the theory (Power and Thirunamachandran, 1999b). A consequence of the boson operators being different is that expectation values of observables involving these operators will differ. To carry out such a calculation requires the selection of a set of unperturbed matter and radiation field states. The ground state of such a system is identical in the two coupling schemes and corresponds to a true no-particle state. Other states, however, will in general be different as they are generated by the action of nonidentical operators. Examples of observables that differ include the expectation value of the photon number operator $n^{\mathrm{min/mult}}(t) = a^{\dagger\mathrm{min/mult}}(t)a^{\mathrm{min/mult}}(t)$ and the occupation number for the matter field $N^{\mathrm{min/mult}}(t) = b_m^{\dagger\mathrm{min/mult}}(t)b_m^{\mathrm{min/mult}}(t)$, as well as non-identical line-shape functions.

The distinction between $a^{(\lambda)}(\vec{k}, t)$ and $a^{\dagger(\lambda)}(\vec{k}, t)$ in the minimal- and multipolar-coupling formalisms is due to the description of the quantized electromagnetic field in terms of photons. If the radiation field is instead viewed as primary, these differences vanish. It should be borne in mind that both the minimal- and multipolar-coupling Hamiltonians give rise to identical equations for the radiation field operators, these being Maxwell's equations. An analogous viewpoint applies to the fermion creation and destruction operators in the two versions of the theory. Now the equation of motion for the electron wavefield is Schrödinger's equation including the effects of electromagnetic radiation.

2.6 MULTIPOLAR MAXWELL FIELDS IN THE VICINITY OF A SOURCE

Results from the analysis of the two previous sections of the forms of the quantum electrodynamical radiation field operators in minimal- and multipolar-coupling frameworks showed that the fields independent of the source are the same for positive retarded times $t > |\vec{r}-\vec{R}|/c$ and that the total fields—source plus free field—are identical for all t, with the source dipole moment operator evaluated at the delayed time $t-|\vec{r}-\vec{R}|/c$. To be able to apply the second quantized Maxwell field operators to the calculation of quantum mechanical observables, for which matrix elements of the field operators are required, it is necessary to express the Heisenberg fields at the initial time $t = 0$. One way this may be achieved is by iterating the coupled integro-differential equations for the boson and fermion creation and annihilation operators, which are obtained from the Heisenberg operator equation of motion. This enables a series solution to be found for the Maxwell fields in successive powers of the multipole moments in terms of photon and electron operators evaluated at the initial time. This is carried out for the multipolar framework electric displacement and magnetic fields only (Power and Thirunamachandran, 1983b; Salam, 2008) by calculating the time evolution of $a^{\mathrm{mult}}(t)$ and $b_n^{\mathrm{mult}}(t)$ and their Hermitian adjoints by starting from the second quantized multipolar Hamiltonian. In this case, the source-dependent Maxwell fields are strictly causal, vanishing for $t < |\vec{r}-\vec{R}|/c$.

In the electric dipole approximation, the multipolar Hamiltonian is given by

$$\begin{aligned} H^{\text{mult}} = & \sum_n b_n^\dagger(t) b_n(t) E_n + \sum_{\vec{k},\lambda} a^{\dagger(\lambda)}(\vec{k},t) a^{(\lambda)}(\vec{k},t)\hbar\omega \\ & -\varepsilon_0^{-1} \sum_{m,n} b_m^\dagger(t) b_n(t) \vec{\mu}^{mn} \cdot \vec{d}^{\perp}(\vec{R},t), \end{aligned} \tag{2.6.1}$$

where the self-energy term has been neglected. In equation (2.6.1), the transition electric dipole moment matrix element is given by

$$\vec{\mu}^{mn} = \int \bar{\phi}_m(\vec{q}) \vec{\mu} \phi_n(\vec{q}) \mathrm{d}^3\vec{q}. \tag{2.6.2}$$

The time development of the photon and electron creation and annihilation operators is found from the Heisenberg equations of motion

$$i\hbar \dot{a}^{(\lambda)}(\vec{k},t) = \left[a^{(\lambda)}(\vec{k},t), H_{\text{mult}}\right]_- \tag{2.6.3}$$

and

$$i\hbar \dot{b}_n(t) = [b_n(t), H_{\text{mult}}]_+ \tag{2.6.4}$$

on using the standard equal-time boson commutator and fermion anticommutator relations

$$\left[a^{(\lambda)}(\vec{k},t), a^{\dagger(\lambda')}(\vec{k}',t)\right]_- = \delta_{\vec{k}\vec{k}'}\delta_{\lambda\lambda'} \tag{2.6.5}$$

and

$$\left[b_m(t), b_n^\dagger(t)\right]_+ = \delta_{mn}. \tag{2.6.6}$$

Transforming to new variables $\alpha(t)$ and $\beta_n(t)$ via the substitutions $a(t) = \alpha(t)\mathrm{e}^{-i\omega t}$ and $b_n(t) = \beta_n(t)\mathrm{e}^{-i\omega_n t}$ yields

$$\alpha(t) = \alpha(0) + \left(\frac{ck}{2\varepsilon_0\hbar V}\right)^{1/2} \sum_{m,n} \bar{e}_j \mathrm{e}^{-i\vec{k}\cdot\vec{R}} \int_0^t \mathrm{d}t'\, \mu_j^{mn}(t') \mathrm{e}^{i(\omega_{mn}+\omega)t'} \beta_m^\dagger(t') \beta_n(t') \tag{2.6.7}$$

and

$$\begin{aligned} \beta_n(t) = \beta_n(0) - & \sum_{\vec{k},\lambda} \sum_m \left(\frac{ck}{2\varepsilon_0\hbar V}\right)^{1/2} \int_0^t \mathrm{d}t'\, \beta_m(t') \mu_j^{nm}(t') [e_j \mathrm{e}^{i\vec{k}\cdot\vec{R} - i(\omega_{mn}+\omega)t'} \alpha(t') \\ & - \bar{e}_j \mathrm{e}^{-i\vec{k}\cdot\vec{R} - i(\omega_{mn}-\omega)t'} \alpha^\dagger(t')], \end{aligned} \tag{2.6.8}$$

where the mode dependence of the radiation field operators and associated polarization vectors has been dropped from the last two equations. These coupled equations may be solved by iteration, in the process generating a solution in series of powers of the transition dipole moment. The boson operators are then inserted into the mode expansions for the displacement and magnetic fields, allowing formulas for the Maxwell fields in the neighborhood of the source to be found. In the Heisenberg picture, the mode expansions for the displacement and magnetic fields are

$$d_i^{\perp}(\vec{r},t) = i\sum_{\vec{k},\lambda}\left(\frac{\hbar ck\varepsilon_0}{2V}\right)^{1/2}\left[e_i\alpha(t)e^{i\vec{k}\cdot\vec{r}-i\omega t} - \bar{e}_i\alpha^{\dagger}(t)e^{-i\vec{k}\cdot\vec{r}+i\omega t}\right] \quad (2.6.9)$$

and

$$b_i(\vec{r},t) = i\sum_{\vec{k},\lambda}\left(\frac{\hbar k}{2\varepsilon_0 cV}\right)^{1/2}\left[b_i\alpha(t)e^{i\vec{k}\cdot\vec{r}-i\omega t} - \bar{b}_i\alpha^{\dagger}(t)e^{-i\vec{k}\cdot\vec{r}+i\omega t}\right], \quad (2.6.10)$$

which may be written as a series expansion in terms of the order of iteration, n,

$$d_i^{\perp}(\vec{r},t) = \sum_{n=0}^{\infty} d_i^{(n)}(\vec{r},t) = d_i^{(0)}(\vec{r},t) + d_i^{(1)}(\vec{r},t) + d_i^{(2)}(\vec{r},t) + \cdots \quad (2.6.11)$$

and

$$b_i(\vec{r},t) = \sum_{n=0}^{\infty} b_i^{(n)}(\vec{r},t) = b_i^{(0)}(\vec{r},t) + b_i^{(1)}(\vec{r},t) + b_i^{(2)}(\vec{r},t) + \cdots. \quad (2.6.12)$$

The first terms of equations (2.6.7) and (2.6.8) correspond to boson and fermion operators at the initial time and are simply $\alpha(0)$ and $\beta_n(0)$, respectively, and are clearly source independent. They are used to obtain the vacuum fields, which are

$$d_i^{(0)}(\vec{r},t) = i\sum_{\vec{k},\lambda}\left(\frac{\hbar ck\varepsilon_0}{2V}\right)^{1/2}\left[e_i\alpha(0)e^{i\vec{k}\cdot\vec{r}-i\omega t} - \bar{e}_i\alpha^{\dagger}(0)e^{-i\vec{k}\cdot\vec{r}+i\omega t}\right] \quad (2.6.13)$$

and

$$b_i^{(0)}(\vec{r},t) = i\sum_{\vec{k},\lambda}\left(\frac{\hbar k}{2\varepsilon_0 cV}\right)^{1/2}\left[b_i\alpha(0)\mathrm{e}^{i\vec{k}\cdot\vec{r}-i\omega t} - \bar{b}_i\alpha^{\dagger}(0)\mathrm{e}^{-i\vec{k}\cdot\vec{r}+i\omega t}\right]. \tag{2.6.14}$$

The zeroth-order fields are seen to operate exclusively in the boson space, increasing or decreasing the number of photons by unity.

Substituting $\beta_n(0)$ and its Hermitian conjugate into the right-hand side of equation (2.6.7) and integrating produces an expression for the photon annihilation operator that is linearly dependent on the transition electric dipole moment, which is used to compute the source-dependent terms of the Maxwell fields (2.6.11) and (2.6.12). Thus,

$$\alpha^{(1)}(t) = \left(\frac{ck}{2\hbar\varepsilon_0 V}\right)^{1/2}\sum_{m,n}\mu_j^{mn}\bar{e}_j\mathrm{e}^{-i\vec{k}\cdot\vec{R}}\beta_m^{\dagger}(0)\beta_n(0)\left(\frac{\mathrm{e}^{i(\omega_{mn}+\omega)t}-1}{i(\omega_{mn}+\omega)}\right), \tag{2.6.15}$$

where μ_j^{mn} is a time-independent matrix element of the transition electric dipole moment. Equation (2.6.15) is used to obtain the first-order fields, since

$$d_i^{(1)}(\vec{r},t) = i\sum_{\vec{k},\lambda}\left(\frac{\hbar ck\varepsilon_0}{2V}\right)^{1/2}\left[e_i\alpha^{(1)}(t)\mathrm{e}^{i\vec{k}\cdot\vec{r}-i\omega t} - \bar{e}_i\alpha^{\dagger(1)}(t)\mathrm{e}^{-i\vec{k}\cdot\vec{r}+i\omega t}\right] \tag{2.6.16}$$

and

$$b_i^{(1)}(\vec{r},t) = i\sum_{\vec{k},\lambda}\left(\frac{\hbar k}{2\varepsilon_0 cV}\right)^{1/2}\left[b_i\alpha^{(1)}(t)\mathrm{e}^{i\vec{k}\cdot\vec{r}-i\omega t} - \bar{b}_i\alpha^{\dagger(1)}(t)\mathrm{e}^{-i\vec{k}\cdot\vec{r}+i\omega t}\right]. \tag{2.6.17}$$

Illustrating explicitly for the linear displacement field, inserting expression (2.6.15) into expansion (2.6.16) produces

$$\begin{aligned} d_i^{(1)}(\vec{r},t) = &\sum_{\vec{k},\lambda}\sum_{m,n}\left(\frac{ck}{2V}\right) \\ &\times\left\{\beta_m^{\dagger}(0)\beta_n(0)\mu_j^{mn}e_i\bar{e}_j\mathrm{e}^{i\vec{k}\cdot(\vec{r}-\vec{R})}\left[\frac{\mathrm{e}^{i\omega_{mn}t}-\mathrm{e}^{-i\omega t}}{(\omega_{mn}+\omega)}\right] + \mathrm{H.C.}\right\}, \end{aligned} \tag{2.6.18}$$

where H.C. stands for the Hermitian-conjugate term. The mode sum is performed using the relations (1.4.55) and (1.4.56), while the angular integral makes use of the result (2.4.16), giving for equation (2.6.18),

$$d_i^{(1)}(\vec{r},t) = \frac{1}{8\pi^2 i}\sum_{m,n} \beta_m^\dagger(0)\beta_n(0) \times \int_0^\infty \frac{dk}{k-k_{nm}} \mu_j^{mn}(-\vec{\nabla}^2\delta_{ij} + \vec{\nabla}_i\vec{\nabla}_j) \times \left(\frac{e^{ik|\vec{r}-\vec{R}|} - e^{-ik|\vec{r}-\vec{R}|}}{|\vec{r}-\vec{R}|}\right)\left[e^{-ik_{nm}ct} - e^{-ikct}\right] + \text{H.C.} \tag{2.6.19}$$

Because the replacement of k by its negative in the Hermitian-conjugate terms produces essentially the same contribution as the first term, but with integration limits $(-\infty, 0)$, the range of integration in equation (2.6.19) can be extended to $(-\infty, \infty)$, so that

$$d_i^{(1)}(\vec{r},t) = \frac{1}{8\pi^2 i}\sum_{m,n} \beta_m^\dagger(0)\beta_n(0)\mu_j^{mn}(-\vec{\nabla}^2\delta_{ij} + \vec{\nabla}_i\vec{\nabla}_j)\frac{1}{|\vec{r}-\vec{R}|} \times \int_{-\infty}^{\infty} \frac{dk}{k-k_{nm}}\left[e^{ik|\vec{r}-\vec{R}|}e^{-ik_{nm}ct} - e^{ik(|\vec{r}-\vec{R}|-ct)} - e^{-ik|\vec{r}-\vec{R}|}e^{-ik_{nm}ct} + e^{-ik(|\vec{r}-\vec{R}|+ct)}\right], \tag{2.6.20}$$

which on integrating yields, irrespective of the way in which the pole is displaced, the first-order electric displacement field

$$d_i^{(1)}(\vec{r},t) = \begin{cases} \dfrac{1}{4\pi}\displaystyle\sum_{m,n} \beta_m^\dagger(0)\beta_n(0)\mu_j^{mn}(-\vec{\nabla}^2\delta_{ij} + \vec{\nabla}_i\vec{\nabla}_j)\dfrac{e^{ik_{nm}(|\vec{r}-\vec{R}|-ct)}}{|\vec{r}-\vec{R}|}, & t > |\vec{r}-\vec{R}|/c > 0, \\ 0, & t < |\vec{r}-\vec{R}|/c, \end{cases} \tag{2.6.21}$$

which is seen to obey Einstein causality. For a source dipole situated at the origin in which $\vec{R} = 0$, evaluating the gradients results in the expression

$$d_i^{(1)}(\vec{r},t) = \frac{1}{4\pi}\sum_{m,n}\beta_m^{\dagger}(0)\beta_n(0)\mu_j^{mn}\mathrm{e}^{-ik_{nm}ct}k_{nm}^3 \times\left[-\frac{(\delta_{ij}-\hat{r}_i\hat{r}_j)}{k_{nm}r}+(\delta_{ij}-3\hat{r}_i\hat{r}_j)\left(\frac{-i}{k_{nm}^2r^2}+\frac{1}{k_{nm}^3r^3}\right)\right]\mathrm{e}^{ik_{nm}r}, \tag{2.6.22}$$

which is recognizable as the quantum electrodynamical analogue of the classical electric displacement field of an oscillating electric dipole. It operates entirely in the electron Fock space, changing only the molecular state of the system. Further, for all $\vec{r}$, the diagonal matrix element of expression (2.6.22) is the electrostatic field of a permanent electric dipole moment.

The magnetic field due to an electric dipole source is derived in a manner similar to that used to obtain the first-order displacement field. Equation (2.6.15) is substituted into the expansion for the linear magnetic field (2.6.17). Now the polarization sum is executed using the identity (1.4.57), and the wavevector sum converted to an integral via the relation (1.4.55). For the evaluation of the angular integral, use is made of the result

$$\frac{1}{4\pi}\int \varepsilon_{ijk}\hat{k}_k\mathrm{e}^{\pm i\vec{k}\cdot\vec{r}}\mathrm{d}\Omega = \mp\frac{i}{k^2}\varepsilon_{ijk}\vec{\nabla}_k\frac{\sin kr}{r} = \mp i\varepsilon_{ijk}\hat{r}_k\left(\frac{\cos kr}{kr}-\frac{\sin kr}{k^2r^2}\right). \tag{2.6.23}$$

Integrating over wavevector then gives

$$b_i^{(1)}(\vec{r},t) = \begin{cases}\dfrac{i}{4\pi\varepsilon_0 c}\displaystyle\sum_{m,n}\beta_m^{\dagger}(0)\beta_n(0)\mu_j^{mn}k_{nm}\varepsilon_{ijk}\vec{\nabla}_k\dfrac{\mathrm{e}^{ik_{nm}(|\vec{r}-\vec{R}|-ct)}}{|\vec{r}-\vec{R}|}, & t>|\vec{r}-\vec{R}|/c>0,\\ 0, & t<|\vec{r}-\vec{R}|/c,\end{cases} \tag{2.6.24}$$

which on taking the gradient after letting $\vec{R}=0$ becomes

$$b_i^{(1)}(\vec{r},t) = \frac{-1}{4\pi\varepsilon_0 c}\sum_{m,n}\beta_m^{\dagger}(0)\beta_n(0)\mu_j^{mn}\mathrm{e}^{-ik_{nm}ct}k_{nm}^3\varepsilon_{ijk}\hat{r}_k\left(\frac{1}{k_{nm}r}+\frac{i}{k_{nm}^2r^2}\right)\mathrm{e}^{ik_{nm}r}, \tag{2.6.25}$$

the familiar form for the magnetic field of an electric dipole. The properties of the first-order magnetic field (2.6.25) are identical to those discussed previously for the linear displacement field operator. In addition, in the near zone, the displacement field has inverse cube distance dependence while the magnetic field exhibits r^{-2} behavior; both fields have an inverse power law form in the radiation zone.

Higher order contributions to the Maxwell field operators may also be evaluated, the technical procedure becoming progressively more complicated as the order of iteration increases. The next term in the expansions of the fields (2.6.11) and (2.6.12) is the one in which operators depend quadratically upon the electric dipole source. For their evaluation, formulas for $\alpha^{(2)}(t)$ and $\beta_n^{(1)}(t)$ and their Hermitian conjugates are required. Illustrating explicitly for the second-order displacement field, its mode expansion is given by

$$d_i^{(2)}(\vec{r},t) = i\sum_{\vec{k},\lambda}\left(\frac{\hbar ck\varepsilon_0}{2V}\right)^{1/2}\left[e_i\alpha^{(2)}(t)\mathrm{e}^{i\vec{k}\cdot\vec{r}-i\omega t}-\bar{e}_i\alpha^{\dagger(2)}(t)\mathrm{e}^{-i\vec{k}\cdot\vec{r}+i\omega t}\right], \tag{2.6.26}$$

where $\alpha^{(2)}(t)$ is obtained from (2.6.7) on inserting $\beta_n^{(1)}(t')$ and its adjoint from (2.6.8). Thus,

$$\alpha^{(2)}(t) = \sum_{m,n}\left(\frac{ck}{2\varepsilon_0\hbar V}\right)^{1/2}\mu_j^{mn}\bar{e}_j\,\mathrm{e}^{-i\vec{k}\cdot\vec{R}} \times\int_0^t \mathrm{d}t'\,\mathrm{e}^{i(\omega_{mn}+\omega)t'}[\beta_m^{\dagger(0)}(t')\beta_n^{(1)}(t')+\beta_m^{\dagger(1)}(t')\beta_n^{(0)}(t')] \tag{2.6.27}$$

and

$$\beta_n^{(1)}(t) = \frac{1}{i\hbar}\sum_{\vec{k},\lambda}\sum_p\left(\frac{\hbar ck}{2\varepsilon_0 V}\right)^{1/2}\beta_p(0)\left[\mu_j^{np}e_j\mathrm{e}^{i\vec{k}\cdot\vec{R}}\alpha(0)\left(\frac{\mathrm{e}^{-i(\omega_{pn}+\omega)t}-1}{(\omega_{pn}+\omega)}\right) -\mu_j^{np}\bar{e}_j\mathrm{e}^{-i\vec{k}\cdot\vec{R}}\alpha^{\dagger}(0)\left(\frac{\mathrm{e}^{-i(\omega_{pn}-\omega)t}-1}{(\omega_{pn}-\omega)}\right)\right], \tag{2.6.28}$$

giving for equation (2.6.27), after performing the time integral,

$$\alpha^{(2)}(t) = -\sum_{\vec{k}',\lambda'}\sum_{m,n,p}\left(\frac{ck}{2\varepsilon_0\hbar V}\right)^{1/2}\left(\frac{ck'}{2\varepsilon_0\hbar V}\right)^{1/2}\mu_j^{mn}\bar{e}_j e^{-i\vec{k}\cdot\vec{R}}$$
$$\times\left[\begin{array}{l}\beta_m^\dagger(0)\beta_p(0)\left\{\begin{array}{l}\mu_l^{np}e'_l e^{i\vec{k}'\cdot\vec{R}}\alpha'(0)\left(\dfrac{e^{i(\omega_{mp}+\omega-\omega')t}-1}{(\omega_{pn}+\omega')(\omega_{mp}+\omega-\omega')}-\dfrac{e^{i(\omega_{mn}+\omega)t}-1}{(\omega_{pn}+\omega')(\omega_{mn}+\omega)}\right)\\ -\mu_l^{np}\bar{e}'_l e^{-i\vec{k}'\cdot\vec{R}}\alpha'^\dagger(0)\left(\dfrac{e^{i(\omega_{mp}+\omega+\omega')t}-1}{(\omega_{pn}-\omega')(\omega_{mp}+\omega+\omega')}-\dfrac{e^{i(\omega_{mn}+\omega)t}-1}{(\omega_{pn}-\omega')(\omega_{mn}+\omega)}\right)\end{array}\right\}\\ +\beta_p^\dagger(0)\beta_n(0)\left\{\begin{array}{l}\mu_l^{pm}\bar{e}'_l e^{-i\vec{k}'\cdot\vec{R}}\alpha'^\dagger(0)\left(\dfrac{e^{i(\omega_{pn}+\omega+\omega')t}-1}{-(\omega_{pm}+\omega')(\omega_{pn}+\omega+\omega')}-\dfrac{e^{i(\omega_{mn}+\omega)t}-1}{-(\omega_{pm}+\omega')(\omega_{mn}+\omega)}\right)\\ -\mu_l^{pm}e'_l e^{i\vec{k}'\cdot\vec{R}}\alpha'(0)\left(\dfrac{e^{i(\omega_{pn}+\omega-\omega')t}-1}{-(\omega_{pm}-\omega')(\omega_{pn}+\omega-\omega')}-\dfrac{e^{i(\omega_{mn}+\omega)t}-1}{-(\omega_{pm}-\omega')(\omega_{mn}+\omega)}\right)\end{array}\right\}\end{array}\right], \tag{2.6.29}$$

where the prime denotes the photon mode $(\vec{k}',\lambda')$. Equation (2.6.29) and its Hermitian conjugate are substituted into the mode expansion (2.6.26) and the $(\vec{k},\lambda)$-mode sum is carried out, producing for the quadratic displacement field the result

$$d_i^{(2)}(\vec{r},t) = \frac{i}{4\pi}\sum_{\vec{k},\lambda}\sum_{m,p}\left(\frac{\hbar ck}{2\varepsilon_0 V}\right)^{1/2}$$
$$\times\left\{\begin{array}{l}e_k\alpha(0)\beta_m^\dagger(0)\beta_p(0)e^{i\vec{k}\cdot\vec{R}}(-\vec{\nabla}^2\delta_{ij}+\vec{\nabla}_i\vec{\nabla}_j)\\ \times\left[\begin{array}{l}\displaystyle\sum_n\left\{\frac{\mu_j^{mn}\mu_k^{np}}{E_{np}-\hbar\omega}+\frac{\mu_k^{mn}\mu_j^{np}}{E_{nm}+\hbar\omega}\right\}\frac{e^{i(k_{pm}+k)(|\vec{r}-\vec{R}|-ct)}}{|\vec{r}-\vec{R}|}\\ \displaystyle-\sum_n\frac{\mu_j^{mn}\mu_k^{np}}{E_{np}-\hbar\omega}\frac{e^{ik_{nm}(|\vec{r}-\vec{R}|-ct)}}{|\vec{r}-\vec{R}|}-\sum_n\frac{\mu_k^{mn}\mu_j^{np}}{E_{nm}+\hbar\omega}\frac{e^{-ik_{np}(|\vec{r}-\vec{R}|-ct)}}{|\vec{r}-\vec{R}|}\end{array}\right]\end{array}\right\}+\text{H.C.} \tag{2.6.30}$$

In near-identical fashion, the magnetic field analogue to (2.6.30) is derived by substituting the expression for $\alpha^{(2)}(t)$ and its Hermitian conjugate into the mode expansion for the second-order magnetic field

$$b_i^{(2)}(\vec{r},t) = i\sum_{\vec{k},\lambda}\left(\frac{\hbar k}{2\varepsilon_0 cV}\right)^{1/2}\left[b_i\alpha^{(2)}(t)e^{i\vec{k}\cdot\vec{r}-i\omega t}-\bar{b}_i\alpha^{\dagger(2)}(t)e^{-i\vec{k}\cdot\vec{r}+i\omega t}\right] \tag{2.6.31}$$

and executing the mode sum. This leads to the formula

$$
\begin{aligned}
b_i^{(2)}(\vec{r},t) = & \frac{i}{4\pi\varepsilon_0}\sum_{\vec{k},\lambda}\sum_{m,p}\left(\frac{\hbar k}{2\varepsilon_0 cV}\right)^{1/2} \\
& \times\left\{\begin{array}{l}
e_k\alpha(0)\beta_m^\dagger(0)\beta_p(0)\mathrm{e}^{i\vec{k}\cdot\vec{R}}\left(i\varepsilon_{ijl}\vec{\nabla}_l\right) \\
\times\left[\begin{array}{l}
\sum_n\left\{\dfrac{\mu_j^{mn}\mu_k^{np}}{E_{np}-\hbar\omega}+\dfrac{\mu_k^{mn}\mu_j^{np}}{E_{nm}+\hbar\omega}\right\}(k_{pm}+k)\dfrac{\mathrm{e}^{i(k_{pm}+k)(|\vec{r}-\vec{R}|-ct)}}{|\vec{r}-\vec{R}|} \\
-\sum_n\dfrac{\mu_j^{mn}\mu_k^{np}}{E_{np}-\hbar\omega}k_{nm}\dfrac{\mathrm{e}^{ik_{nm}(|\vec{r}-\vec{R}|-ct)}}{|\vec{r}-\vec{R}|}-\sum_n\dfrac{\mu_k^{mn}\mu_j^{np}}{E_{nm}+\hbar\omega}k_{pn}\dfrac{\mathrm{e}^{-ik_{np}(|\vec{r}-\vec{R}|-ct)}}{|\vec{r}-\vec{R}|}
\end{array}\right]
\end{array}\right\}+\mathrm{H.C.}
\end{aligned}
\tag{2.6.32}
$$

A noteworthy characteristic of the quadratic fields (2.6.30) and (2.6.32) is that in contrast to the vacuum and first-order fields, they act in both the fermion and boson spaces, changing the state of the electron as well as creating or destroying a quantum of electromagnetic radiation. Expressions for cubic and higher order electric dipole-dependent Maxwell fields may be calculated by iterating (2.6.7) and (2.6.8) to third and higher orders, yielding complicated functions of the dynamical variables.

2.7 HIGHER MULTIPOLE MOMENT MAXWELL FIELDS

For many applications, especially those involving optically active molecules, the electric dipole approximation is no longer valid and the contribution due to higher multipole moments needs to be included (Thirunamachandran, 1988; Salam and Thirunamachandran, 1994). Since individual multipole moment terms are additive in the multipolar framework, it is facile to extract a specific contribution, be it a unique coupling term or a collection of terms that are of a similar order of magnitude. By accounting for the spatial variations of the vector potential to first order and retaining terms linear in the electric charge, the interaction Hamiltonian now includes electric quadrupole and magnetic dipole couplings, as well as the leading electric dipole contribution. Hence, the interaction Hamiltonian in multipolar formalism correct to this order of approximation is written as

$$
\begin{aligned}
H_{\mathrm{int}} = & -\sum_{m,n} b_m^\dagger(t)b_n(t)\left[\varepsilon_0^{-1}\vec{\mu}^{mn}\cdot\vec{d}^\perp(\vec{R},t)\right. \\
& \left.+\vec{m}^{mn}\cdot\vec{b}(\vec{R},t)+\varepsilon_0^{-1}Q_{ij}^{mn}\vec{\nabla}_j d_i^\perp(\vec{R},t)\right],
\end{aligned}
\tag{2.7.1}
$$

where $\vec{m}^{mn}$ and Q_{ij}^{mn} are matrix elements of the magnetic dipole and electric quadrupole moment operators and are defined analogously to that given previously for $\vec{\mu}^{mn}$ by equation (2.6.2). The basic forms of the time-dependent boson and fermion operator equations remain essentially the same, as given by formulas (2.6.7) and (2.6.8), but are now modified slightly due to the effects of including higher multipole terms. By retaining all three coupling terms in expansion (2.7.1), the boson and fermion operators linear in the electronic charge are

$$\begin{aligned}\alpha^{(1)}(t) = \left(\frac{ck}{2\hbar\varepsilon_0 V}\right)^{1/2} \sum_{m,n} \left(\mu_j^{mn}\bar{e}_j + \frac{1}{c}m_j^{mn}\bar{b}_j + (-ik_k)Q_{jk}^{mn}\bar{e}_j\right) \\ \times \mathrm{e}^{-i\vec{k}\cdot\vec{R}}\beta_m^{\dagger}(0)\beta_n(0)\left(\frac{\mathrm{e}^{i(\omega_{mn}+\omega)t}-1}{i(\omega_{mn}+\omega)}\right)\end{aligned} \tag{2.7.2}$$

and

$$\begin{aligned}\beta_n^{(1)}(t) = \frac{1}{i\hbar}\sum_{\vec{k},\lambda}\sum_{p}\left(\frac{\hbar ck}{2\varepsilon_0 V}\right)^{1/2}\beta_p(0)\left[\left(\mu_j^{np}e_j + \frac{1}{c}m_j^{np}b_j + (ik_k)Q_{jk}^{np}e_j\right)\right. \\ \times \mathrm{e}^{i\vec{k}\cdot\vec{R}}\alpha(0)\left(\frac{\mathrm{e}^{-i(\omega_{pn}+\omega)t}-1}{(\omega_{pn}+\omega)}\right) - \left(\mu_j^{np}\bar{e}_j + \frac{1}{c}m_j^{np}\bar{b}_j + (-ik_k)Q_{jk}^{np}\bar{e}_j\right) \\ \left.\times \mathrm{e}^{-i\vec{k}\cdot\vec{R}}\alpha^{\dagger}(0)\left(\frac{\mathrm{e}^{-i(\omega_{pn}-\omega)t}-1}{(\omega_{pn}-\omega)}\right)\right].\end{aligned} \tag{2.7.3}$$

The Maxwell fields in the neighborhood of these additional sources are evaluated in identical fashion to that demonstrated in the previous section in the electric dipole approximation. In the present case, use is also made of the polarization sum (1.4.58) and the following angular averages,

$$\begin{aligned}\frac{1}{4\pi}\int(\delta_{ij}-\hat{k}_i\hat{k}_j)\hat{k}_k\mathrm{e}^{\pm i\vec{k}\cdot\vec{r}}\mathrm{d}\Omega = \mp\frac{i}{k^4}(-\vec{\nabla}^2\delta_{ij}+\vec{\nabla}_i\vec{\nabla}_j)\vec{\nabla}_k\frac{\sin kr}{r} \\ = \mp i\left\{(\delta_{ij}-\hat{r}_i\hat{r}_j)\hat{r}_k\left(\frac{\cos kr}{kr}-\frac{\sin kr}{k^2r^2}\right) + (\delta_{ij}\hat{r}_k+\delta_{ik}\hat{r}_j+\delta_{jk}\hat{r}_i-5\hat{r}_i\hat{r}_j\hat{r}_k)\right. \\ \left.\times\left(-\frac{\sin kr}{k^2r^2}-\frac{3\cos kr}{k^3r^3}+\frac{3\sin kr}{k^4r^4}\right)\right\}\end{aligned} \tag{2.7.4}$$

and

$$\frac{1}{4\pi}\int \varepsilon_{ijk}\hat{k}_k\hat{k}_l e^{\pm i\vec{k}\cdot\vec{r}} d\Omega = -\frac{1}{k^3}\varepsilon_{ijk}\vec{\nabla}_k\vec{\nabla}_l\frac{\sin kr}{r}$$
$$= \varepsilon_{ijk}\left\{\hat{r}_k\hat{r}_l\frac{\sin kr}{kr} - (\delta_{kl} - 3\hat{r}_k\hat{r}_l)\left(\frac{\cos kr}{k^2r^2} - \frac{\sin kr}{k^3r^3}\right)\right\}. \tag{2.7.5}$$

Including the electric dipole term, the first-order electric displacement and magnetic fields correct up to the electric quadrupole coupling term are

$$d_i^{(1)}(\vec{r},t) = \begin{cases} \dfrac{1}{4\pi}\sum_{m,n}\beta_m^\dagger(0)\beta_n(0)\Big[\mu_j^{mn}(-\vec{\nabla}^2\delta_{ij} + \vec{\nabla}_i\vec{\nabla}_j) - \dfrac{i}{c}m_j^{mn}k_{nm}\varepsilon_{ijk}\vec{\nabla}_k \\ \quad -Q_{jk}^{mn}(-\vec{\nabla}^2\delta_{ij} + \vec{\nabla}_i\vec{\nabla}_j)\vec{\nabla}_k\Big]\dfrac{e^{ik_{nm}(|\vec{r}-\vec{R}|-ct)}}{|\vec{r}-\vec{R}|}, & t > |\vec{r}-\vec{R}|/c, \\ 0, & t < |\vec{r}-\vec{R}|/c, \end{cases} \tag{2.7.6}$$

and

$$b_i^{(1)}(\vec{r},t) = \begin{cases} \dfrac{1}{4\pi\varepsilon_0 c}\sum_{m,n}\beta_m^\dagger(0)\beta_n(0)\Big[i\mu_j^{mn}k_{nm}\varepsilon_{ijk}\vec{\nabla}_k + \dfrac{1}{c}m_j^{mn}(-\vec{\nabla}^2\delta_{ij} + \vec{\nabla}_i\vec{\nabla}_j) \\ \quad +iQ_{jk}^{mn}k_{nm}\varepsilon_{ijk}\vec{\nabla}_k\vec{\nabla}_l\Big]\dfrac{e^{ik_{nm}(|\vec{r}-\vec{R}|-ct)}}{|\vec{r}-\vec{R}|}, & t > |\vec{r}-\vec{R}|/c, \\ 0, & t < |\vec{r}-\vec{R}|/c. \end{cases} \tag{2.7.7}$$

In addition to properties listed in Section 2.6 associated with the dipole-dependent first-order Maxwell fields, which also apply to equations (2.7.6) and (2.7.7), it is interesting to note that the electric field of a magnetic dipole is the negative of the magnetic field of an electric dipole, and the electric field of an electric dipole is the same as the magnetic field of a magnetic dipole, with $\vec{\mu}^{mn}$ replaced by $\vec{m}^{mn}$ in both cases.

Higher multipole contributions to the fields quadratic in the sources are obtained as for the electric dipole case, using the extension of equation (2.6.27),

$$\alpha^{(2)}(t) = \sum_{m,n}\left(\frac{ck}{2\varepsilon_0\hbar V}\right)^{1/2}\left(\mu_j^{mn}\bar{e}_j + \frac{1}{c}m_j^{mn}\bar{b}_j + (-ik_k)Q_{jk}^{mn}\bar{e}_j\right)e^{-i\vec{k}\cdot\vec{R}}$$
$$\times\int_0^t dt'\, e^{i(\omega_{mn}+\omega)t'}[\beta_m^{\dagger(0)}(t')\beta_n^{(1)}(t') + \beta_m^{\dagger(1)}(t')\beta_n^{(0)}(t')], \tag{2.7.8}$$

and formula (2.7.3). Explicit expressions for these higher multipole-dependent second-order Maxwell fields are given in Appendix A.

2.8 MAXWELL FIELDS OF A DIAMAGNETIC SOURCE

In many situations, the observable of a quantum mechanical operator is proportional to the square of the magnetic field $\vec{b}(\vec{r})$. In such cases, the Maxwell fields of the leading order diamagnetic coupling term should be included for the sake of consistency. To lowest order, this interaction term is also quadratic in $\vec{b}(\vec{r})$. Its evaluation is carried out in what follows. It extends work of the previous section and the results of Appendix A, in which the electric displacement field and the magnetic field in the neighborhood of an electric dipole, quadrupole, and magnetic dipole moment were computed to second order in the source. Demanding that the diamagnetic coupling term be accounted for is also justified by the fact that it is of a similar order of magnitude to electric quadrupole and magnetic dipole couplings and can also be shown to arise when spatial variations of the vector potential are taken to higher orders.

In a second quantized representation, the lowest order diamagnetic interaction for a source located at $\vec{R}$ is

$$H_{\text{int}}^{\text{dia}} = \frac{e^2}{8m}\int \bar{\phi}(\vec{q})\left\{(\vec{q}-\vec{R})\times\vec{b}(\vec{R})\right\}^2\phi(\vec{q})\mathrm{d}^3\vec{q}. \tag{2.8.1}$$

With this as the sole coupling term and on rewriting the vector cross-product between operators $\vec{q}$ and $\vec{b}(\vec{r})$ as

$$\left\{(\vec{q}-\vec{R})\times\vec{b}(\vec{R})\right\}_i = \varepsilon_{ijk}(\vec{q}-\vec{R})_j b_k(\vec{R}), \tag{2.8.2}$$

the total Hamiltonian after inserting the mode expansion for $\vec{b}(\vec{r})$ is

$$\begin{aligned} H^{\text{mult}} = {} & \sum_n b_n^\dagger b_n E_n + \sum_{\vec{k},\lambda} a^\dagger a\hbar\omega - \frac{e^2}{8m}\varepsilon_{ijp}\varepsilon_{klp}\sum_{\substack{\vec{k},\lambda \\ \vec{k}',\lambda'}}\sum_{m,n}\left(\frac{\hbar k}{2\varepsilon_0 cV}\right)^{1/2} \\ & \times\left(\frac{\hbar k'}{2\varepsilon_0 cV}\right)^{1/2} b_m^\dagger b_n\left[(\vec{q}-\vec{R})_i(\vec{q}-\vec{R})_k\right]^{mn} \\ & \times\left(b_j a\mathrm{e}^{i\vec{k}\cdot\vec{R}} - \bar{b}_j a^\dagger\mathrm{e}^{-i\vec{k}\cdot\vec{R}}\right)\left(b_l' a'\mathrm{e}^{i\vec{k}'\cdot\vec{R}} - \bar{b}_l' a'^\dagger\mathrm{e}^{-i\vec{k}'\cdot\vec{R}}\right). \end{aligned} \tag{2.8.3}$$

In equation (2.8.3), $[(\vec{q}-\vec{R})_i(\vec{q}-\vec{R})_k]^{mn}$ is the mnth matrix element of the product of position operators. All photon mode dependence of the magnetic polarization vectors and boson creation and annihilation operators has been suppressed for notational brevity and all time dependence is implicit. As in Section 2.6, the Heisenberg operator equations of motion are calculated using the total Hamiltonian (2.8.3) and the commutation relation (2.6.5) and the anticommutator (2.6.6). For the boson and fermion operators, respectively, these are found to be

$$\alpha^{\text{dia}}(t) = -\frac{ie^2}{4\hbar m}\varepsilon_{ijp}\varepsilon_{klp}\sum_{\vec{k}',\lambda'}\sum_{m,n}\left(\frac{\hbar k}{2\varepsilon_0 cV}\right)^{1/2}\left(\frac{\hbar k'}{2\varepsilon_0 cV}\right)^{1/2}$$
$$\times[(\vec{q}-\vec{R})_i(\vec{q}-\vec{R})_k]^{mn}\bar{b}_j$$
$$\times\int_0^t dt'\,\mathrm{e}^{-i\vec{k}\cdot\vec{R}+i(\omega_{mn}+\omega)t'}\beta_m^\dagger(t')\beta_n(t')(b_l'\alpha'(t')\mathrm{e}^{i\vec{k}'\cdot\vec{R}-i\omega't'}$$
$$-\bar{b}_l'\alpha'^\dagger(t')\mathrm{e}^{-i\vec{k}'\cdot\vec{R}+i\omega't'}) \tag{2.8.4}$$

and

$$\beta_n^{\text{dia}}(t) = \frac{ie^2}{8\hbar m}\varepsilon_{ijp}\varepsilon_{klp}\sum_{\substack{\vec{k},\lambda\\ \vec{k}',\lambda'}}\sum_{m}\left(\frac{\hbar k}{2\varepsilon_0 cV}\right)^{1/2}\left(\frac{\hbar k'}{2\varepsilon_0 cV}\right)^{1/2}$$
$$\times[(\vec{q}-\vec{R})_i(\vec{q}-\vec{R})_k]^{nm}\int_0^t dt'\,\beta_m(t')$$
$$\times\,(b_j\alpha(t')\mathrm{e}^{i\vec{k}\cdot\vec{R}-i(\omega_{mn}+\omega)t'}-\bar{b}_j\alpha^\dagger(t')\mathrm{e}^{-i\vec{k}\cdot\vec{R}-i(\omega_{mn}-\omega)t'})$$
$$\times\,(b_l'\alpha'(t')\mathrm{e}^{i\vec{k}'\cdot\vec{R}-i(\omega_{mn}+\omega')t'}-\bar{b}_l'\alpha'^\dagger(t')\mathrm{e}^{-i\vec{k}'\cdot\vec{R}-i(\omega_{mn}-\omega')t'}). \tag{2.8.5}$$

For the solution of the electric displacement and magnetic fields due to a diamagnetic source, it is clear that only the zeroth-order term of $\alpha^{\text{dia}}(t)$, equation (2.8.4), is needed since in this approximation the time-dependent boson operator is second order in electronic charge. This leading term is obtained on inserting fermion and boson raising and lowering operators at their initial time $t=0$, that is, $\beta_m^\dagger(0)$, $\beta_n(0)$, $\alpha'^\dagger(0)$, and $\alpha'(0)$ are substituted into expression (2.8.4). Integrating with respect to time and letting $\vec{R}=0$ results in

$$\alpha^{\text{dia}}(t) = -\frac{ie^2}{4\hbar m}\varepsilon_{ijp}\varepsilon_{klp}\sum_{\vec{k}',\lambda'}\sum_{m,n}\left(\frac{\hbar k}{2\varepsilon_0 cV}\right)^{1/2}\left(\frac{\hbar k'}{2\varepsilon_0 cV}\right)^{1/2}$$
$$\times[q_iq_k]^{mn}\beta_m^\dagger(0)\beta_n(0)\bar{b}_j$$
$$\times\left\{b_l'\alpha'(0)\left[\frac{\mathrm{e}^{i(\omega_{mn}+\omega-\omega')t}-1}{i(\omega_{mn}+\omega-\omega')}\right]-\bar{b}_l'\alpha'^\dagger(0)\left[\frac{\mathrm{e}^{i(\omega_{mn}+\omega+\omega')t}-1}{i(\omega_{mn}+\omega+\omega')}\right]\right\}. \tag{2.8.6}$$

Substituting equation (2.8.6) into the mode expansion for the transverse electric displacement field (2.6.9), the diamagnetic contribution is

$$d_i^{\perp(\text{dia})}(\vec{r},t) = \frac{e^2}{4i\hbar m}\varepsilon_{jkp}\varepsilon_{lmp}\sum_{\substack{\vec{k},\lambda\\ \vec{k}',\lambda'}}\sum_{m,n}\left(\frac{\hbar k}{2V}\right)\left(\frac{\hbar k'}{2\varepsilon_0 cV}\right)^{1/2} \times e_i\bar{b}_k e^{i\vec{k}\cdot\vec{r}}[q_j q_l]^{mn}\beta_m^\dagger(0)\beta_n(0)\alpha'(0)b'_m \times\left[\frac{e^{i(\omega_{mn}-\omega')t}-e^{-i\omega t}}{(\omega_{mn}+\omega-\omega')}\right]+\text{H.C.} \tag{2.8.7}$$

Performing the usual sum over polarizations and angular average leads to

$$d_i^{\perp(\text{dia})}(\vec{r},t) = \frac{e^2}{16\pi^2 mc}\varepsilon_{jkp}\varepsilon_{lmp}\sum_{\vec{k}',\lambda'}\sum_{m,n}\left(\frac{\hbar k'}{2\varepsilon_0 cV}\right)^{1/2}[q_j q_l]^{mn}\beta_m^\dagger(0)\beta_n(0) \times\alpha'(0)b'_m\left(i\varepsilon_{iks}\vec{\nabla}_s\frac{1}{r}\right)\int_0^\infty dk\,k(e^{ikr}-e^{-ikr}) \times\left[\frac{e^{i(k_{mn}-k')ct}-e^{-ikct}}{(k_{mn}+k-k')}\right]+\text{H.C.} \tag{2.8.8}$$

Integrating subject to the causality requirement that the field vanish for $r > ct$ and taking the diagonal matrix element of the electronic operators results in the field

$$d_i^{\perp(\text{dia})}(\vec{r},t) = -\frac{ie^2}{8\pi mc}\varepsilon_{jkp}\,\varepsilon_{lmp}\sum_{\vec{k},\lambda}\sum_{m}\left(\frac{\hbar k}{2\varepsilon_0 cV}\right)^{1/2} \times[q_j q_l]^{mm}\beta_m^\dagger(0)\beta_m(0)\alpha(0)b_m \times k^3\varepsilon_{iks}\hat{r}_s\left[\frac{1}{kr}+\frac{i}{k^2r^2}\right]e^{ik(r-ct)}+\text{H.C.} \tag{2.8.9}$$

Similarly, for the diamagnetic contribution to the magnetic field, after substituting relation (2.8.6) into the mode expansion for the $\vec{b}$-field (2.6.10), carrying out the mode sum yields

$$b_i^{(\text{dia})}(\vec{r},t) = \frac{ie^2}{8\pi\varepsilon_0 mc^2}\varepsilon_{jkp}\varepsilon_{lmp}\sum_{\vec{k},\lambda}\sum_{m}\left(\frac{\hbar k}{2\varepsilon_0 cV}\right)^{1/2} \times[q_j q_l]^{mm}\beta_m^\dagger(0)\beta_m(0)\alpha(0)b_m \times k^3\left[\frac{(\delta_{ik}-\hat{r}_i\hat{r}_k)}{kr}-(\delta_{ik}-3\hat{r}_i\hat{r}_k)\left(\frac{-i}{k^2r^2}+\frac{1}{k^3r^3}\right)\right]e^{ik(r-ct)}+\text{H.C.} \tag{2.8.10}$$

This, along with results in the two previous sections and in Appendix A, completes the formal derivation of the Maxwell field operators correct up to second order in the electronic charge.

2.9 ELECTROMAGNETIC ENERGY DENSITY

It was shown in Section 1.4 that the Hamiltonian density for the free radiation field equation (1.4.25) expressed in terms of the vector potential and its canonically conjugate momentum is equivalent to the energy density of the electromagnetic field, $(\varepsilon_0/2)(\vec{e}^2 + c^2\vec{b}^2)$. Having now calculated the multipolar formalism Maxwell field operators in the vicinity of an electric dipole source moment, the electric and magnetic contributions to the Thomson energy density due to such a source may be evaluated (Power and Thirunamachandran, 1992). The importance of the energy density lies in its relation to the intermolecular potential between a pair of polarizable molecules, which arises when a test body is placed in the radiation field of the source.

Since the electric displacement field $\vec{d}(\vec{r})$ is purely transverse in a neutral system, because the divergence of $\vec{d}(\vec{r})$ vanishes, $\varepsilon_0\vec{e}^{\text{tot}}(\vec{r}) = \vec{d}^{\perp}(\vec{r})$ outside the sources due to the fact that the total electric polarization field $\vec{p}(\vec{r})$ is local. Hence, the electric energy density operator is given by

$$
\begin{aligned}
&\frac{1}{2\varepsilon_0}\left[\vec{d}^{\perp}(\vec{r},t)\right]^2 \\
&\quad = \frac{1}{2\varepsilon_0}\left[\vec{d}^{(0)}(\vec{r},t) + \vec{d}^{(1)}(\vec{r},t) + \vec{d}^{(2)}(\vec{r},t) + \cdots\right]^2 \\
&\quad = \frac{1}{2\varepsilon_0}\left[d_i^{(0)}d_i^{(0)} + d_i^{(0)}d_i^{(1)} + d_i^{(1)}d_i^{(0)} + d_i^{(1)}d_i^{(1)} + d_i^{(0)}d_i^{(2)} + d_i^{(2)}d_i^{(0)} + \cdots\right]
\end{aligned}
\tag{2.9.1}
$$

on expanding the field in successive powers of the transition dipole correct up to second order in the source moment. The important contributions to the electric energy density arise from the last three terms of the product written in equation (2.9.1). It is clear that the first term cannot contribute since the free field is independent of the source. The two terms involving the product of the zeroth- and first-order fields likewise do not contribute to the expectation value for a state with a constant number of photons. The first term to be retained in any calculation of the energy density is that arising from the product of the field linear in the electric dipole moment. To this is added the term due to the interference of the vacuum and quadratic field, since this is also proportional to the square of the source dipole moment.

For a molecular state $|p\rangle$ and the radiation field in the vacuum state, the expectation value of the electric energy density involves calculating

$$\frac{1}{2\varepsilon_0}\langle 0(\vec{k},\lambda);p|(d_i^{(1)}d_i^{(1)}+d_i^{(0)}d_i^{(2)}+d_i^{(2)}d_i^{(0)})|p;0(\vec{k},\lambda)\rangle. \tag{2.9.2}$$

Recalling that the first-order field operates only in the electron Fock space, use of expression (2.6.21) for a dipole source situated at the origin gives for the first term of (2.9.2)

$$\frac{1}{2\varepsilon_0}\sum_n \langle p|d_i^{(1)}|n\rangle\langle n|d_i^{(1)}|p\rangle = \frac{1}{32\pi^2\varepsilon_0}\sum_n \mu_j^{pn}\mu_k^{np}k_{pn}^6\bar{F}_{ij}(k_{pn}r)F_{ik}(k_{pn}r), \tag{2.9.3}$$

where $F_{ij}(kr)$ is the tensor field

$$\begin{aligned} F_{ij}(kr) &= \frac{1}{k^3}(-\vec{\nabla}^2\delta_{ij}+\vec{\nabla}_i\vec{\nabla}_j)\frac{\mathrm{e}^{ikr}}{r} \\ &= \left[-(\delta_{ij}-\hat{r}_i\hat{r}_j)\frac{1}{kr}+(\delta_{ij}-3\hat{r}_i\hat{r}_j)\left(\frac{-i}{k^2r^2}+\frac{1}{k^3r^3}\right)\right]\mathrm{e}^{ikr} = f_{ij}(kr)\mathrm{e}^{ikr}. \end{aligned} \tag{2.9.4}$$

For the evaluation of the last two terms of formula (2.9.2), advantage is taken of the fact that the vacuum field does not change the electronic state of the molecule so that the diagonal matrix element over the fermion state may be taken for the quadratic field (2.6.30). For the molecular state $|p\rangle$, the matrix element can be expressed as

$$\langle p|d_i^{(2)}(\vec{r},t)|p\rangle = \frac{i}{4\pi}\sum_{\vec{k},\lambda}\left(\frac{\hbar ck}{2\varepsilon_0 V}\right)^{1/2}\left[e_k\alpha(0)\mathrm{e}^{-i\omega t}\mathcal{F}_{ki}-\bar{e}_k\alpha^{\dagger}(0)\mathrm{e}^{i\omega t}\bar{\mathcal{F}}_{ki}\right], \tag{2.9.5}$$

where

$$\begin{aligned} \mathcal{F}_{ki} = {} & \sum_n \frac{\mu_j^{pn}\mu_k^{np}}{E_{np}-\hbar\omega}k^3F_{ij}(kr) - \sum_n \frac{\mu_j^{pn}\mu_k^{np}}{E_{np}-\hbar\omega}k_{np}^3F_{ij}(k_{np}r)\mathrm{e}^{i(k_{pn}+k)ct} \\ & + \sum_n \frac{\mu_k^{pn}\mu_j^{np}}{E_{np}+\hbar\omega}k^3F_{ij}(kr) - \sum_n \frac{\mu_k^{pn}\mu_j^{np}}{E_{np}+\hbar\omega}k_{pn}^3F_{ij}(k_{pn}r)\mathrm{e}^{-i(k_{pn}-k)ct}. \end{aligned} \tag{2.9.6}$$

As a result, the second-order displacement field, like the free field, operates exclusively in the boson space, so that the contribution to the electric energy density from the interference of these fields is

$$\begin{aligned}&\frac{1}{2\varepsilon_0}\sum_{\vec{k},\lambda}[\langle 0(\vec{k},\lambda);p|d_i^{(0)}|p;1(\vec{k},\lambda)\rangle\langle 1(\vec{k},\lambda);p|d_i^{(2)}|p;0(\vec{k},\lambda)\rangle \\ &\quad+\langle 0(\vec{k},\lambda);p|d_i^{(2)}|p;1(\vec{k},\lambda)\rangle\langle 1(\vec{k},\lambda);p|d_i^{(0)}|p;0(\vec{k},\lambda)\rangle] \\ &=\frac{1}{8\pi\varepsilon_0}\sum_{\vec{k},\lambda}\left(\frac{\hbar ck}{2V}\right)\left[e_i e^{i\vec{k}\cdot\vec{r}}\bar{e}_k\bar{\mathcal{F}}_{ki}+e_k\mathcal{F}_{ki}\bar{e}_i e^{-i\vec{k}\cdot\vec{r}}\right].\end{aligned} \tag{2.9.7}$$

Concentrating on the first term of (2.9.7), the polarization sum is carried out using the identity (1.4.56), the $\vec{k}$-sum is converted to an integral via relation (1.4.55), and the angular average is done using equations (2.4.16) and (2.9.4) to give

$$\frac{\hbar c}{32\pi^2\varepsilon_0}\frac{1}{2\pi i}\int_0^\infty \mathrm{d}k\, k^3[F_{ik}(kr)-\bar{F}_{ik}(kr)]\bar{\mathcal{F}}_{ki}. \tag{2.9.8}$$

Substituting the tensor field (2.9.6) and collecting terms with identical denominator, (2.9.8) can be written as

$$\begin{aligned}&\frac{1}{32\pi^2\varepsilon_0}\sum_n \mu_j^{pn}\mu_k^{np}\frac{\mathrm{PV}}{2\pi i}\int_0^\infty \mathrm{d}k\, k^3 \\ &\times\left\{\begin{aligned}&\frac{[F_{ik}(kr)-\bar{F}_{ik}(kr)][k^3\bar{F}_{ij}(kr)-k_{np}^3\bar{F}_{ij}(k_{np}r)e^{i(k_{np}-k)ct}]}{(k_{np}-k)} \\ &+\frac{[F_{ik}(kr)-\bar{F}_{ik}(kr)][k^3\bar{F}_{ij}(kr)-k_{pn}^3\bar{F}_{ij}(k_{pn}r)e^{i(k_{pn}-k)ct}]}{(k_{np}+k)}\end{aligned}\right\} \\ &=-\frac{1}{32\pi^2\varepsilon_0}\sum_n \mu_j^{pn}\mu_k^{np}\frac{\mathrm{PV}}{2\pi i}\int_0^\infty\frac{\mathrm{d}k\,k^3}{k-k_{np}}\left[-k^3\bar{f}_{ik}(kr)\bar{f}_{ij}(kr)e^{-2ikr}\right. \\ &\left.-k_{np}^3 f_{ik}(kr)\bar{f}_{ij}(k_{np}r)e^{ik(r-ct)}e^{ik_{pn}(r-ct)}+k_{np}^3\bar{f}_{ik}(kr)\bar{f}_{ij}(k_{np}r)e^{-ik(r+ct)}e^{ik_{pn}(r-ct)}\right] \\ &+\frac{1}{32\pi^2\varepsilon_0}\sum_n \mu_j^{pn}\mu_k^{np}\frac{\mathrm{PV}}{2\pi i}\int_0^\infty\frac{\mathrm{d}k\,k^3}{k-k_{pn}}\left[-k^3\bar{f}_{ik}(kr)\bar{f}_{ij}(kr)e^{-2ikr}\right. \\ &\left.-k_{pn}^3 f_{ik}(kr)\bar{f}_{ij}(k_{pn}r)e^{ik(r-ct)}e^{-ik_{pn}(r-ct)}+k_{pn}^3\bar{f}_{ik}(kr)\bar{f}_{ij}(k_{pn}r)e^{-ik(r+ct)}e^{-ik_{pn}(r-ct)}\right],\end{aligned} \tag{2.9.9}$$

where use has been made of the j,k-index symmetry to eliminate the term without an exponential dependence. Expression (2.9.9) contains both time-independent and time-dependent terms. The abbreviation PV denotes the Cauchy principal value, which is taken since exact resonances are excluded in the k-integral when making the continuum approximation to the mode sum. The evaluation of the integral depends on the sign of k_{pn} and is carried out by transforming the integral from one along the real axis to one along the imaginary axis in the complex plane. For a state for which $E_p > E_n$, after making the substitution $k = -iu$, the time-independent part of (2.9.9) is

$$\frac{1}{64\pi^2\varepsilon_0}\sum_n \mu_j^{pn}\mu_k^{np}k_{pn}^6\bar{f}_{ij}(k_{pn}r)f_{ik}(k_{pn}r) + \frac{1}{32\pi^3\varepsilon_0}\sum_n \mu_j^{pn}\mu_k^{np}\int_0^\infty \frac{du\,u^6 e^{-2ur}}{u^2+k_{pn}^2}k_{pn}f_{ij}(iur)f_{ik}(iur). \tag{2.9.10}$$

The time-dependent part is given by

$$\frac{1}{64\pi^3\varepsilon_0}\sum_n \mu_j^{pn}\mu_k^{np}(-k_{pn})^3\int_0^\infty du\,u^3\left[\bar{f}_{ij}(k_{np}r)e^{ik_{pn}(r-ct)}f_{ik}(-iur)\frac{e^{-uc(t-r/c)}}{u+ik_{pn}}\right.$$

$$-\bar{f}_{ij}(k_{np}r)e^{ik_{pn}(r-ct)}\bar{f}_{ik}(-iur)\frac{e^{-uc(t+r/c)}}{u+ik_{pn}} + \bar{f}_{ij}(k_{pn}r)e^{ik_{np}(r-ct)}f_{ik}(-iur)$$

$$\left.\times\frac{e^{-uc(t-r/c)}}{u-ik_{pn}} - \bar{f}_{ij}(k_{pn}r)e^{ik_{np}(r-ct)}\bar{f}_{ik}(-iur)\frac{e^{-uc(t+r/c)}}{u-ik_{pn}}\right]. \tag{2.9.11}$$

Because the exponents decrease for large times, the integrals in (2.9.11) tend to zero for $t \gg r/c$. Also, over a fixed time period, the average of the time-dependent contribution vanishes due to the modulation factors $e^{\pm ik_{pn}ct}$. These oscillatory terms are ignored henceforth. Returning to expression (2.9.9) and evaluating the integral for the case $k_{np} > 0$, the pole contribution is

$$-\frac{1}{64\pi^2\varepsilon_0}\sum_n \mu_j^{pn}\mu_k^{np}k_{pn}^6\bar{f}_{ik}(k_{pn}r)f_{ij}(k_{pn}r), \tag{2.9.12}$$

on using the fact that

$$f_{ij}(\pm kr) = -\bar{f}_{ij}(\mp kr), \tag{2.9.13}$$

a relation easily obtainable from definition (2.9.4). Further, from the definition of the tensor field $F_{ij}(kr)$, it is a simple matter to obtain this quantity for the complex variable $k = iu$, which occurs in equations (2.9.10) and (2.9.11) and will be convenient for future use. Thus,

$$F_{ij}(iur) = \frac{i}{u^3}(-\vec{\nabla}^2\delta_{ij} + \vec{\nabla}_i\vec{\nabla}_j)\frac{\mathrm{e}^{-ur}}{r} = f_{ij}(iur)\mathrm{e}^{-ur}, \tag{2.9.14}$$

where

$$f_{ij}(iur) = i\left[(\delta_{ij} - \hat{r}_i\hat{r}_j)\frac{1}{ur} + (\delta_{ij} - 3\hat{r}_i\hat{r}_j)\left(\frac{1}{u^2r^2} + \frac{1}{u^3r^3}\right)\right], \tag{2.9.15}$$

from which the following useful identities easily ensue:

$$f_{ij}(\pm iur) = \bar{f}_{ij}(\mp iur) = -\bar{f}_{ij}(\pm iur). \tag{2.9.16}$$

For $k_{np} > 0$, the u-integral part is identical to that appearing in the second term of (2.9.10), in which the relations given in (2.9.16) have been applied. It is important to note that the first term of (2.9.10) has the same sign as the corresponding term arising from the first-order fields (2.9.3). For those states n with $E_n > E_p$, however, the pole contribution has opposite sign as indicated by (2.9.12). It is interesting to note the cancellation of the pole term arising when $k_{np} > 0$ with the contribution (2.9.3). The reinforcing and canceling of pole contributions from the zeroth- and second-order fields with terms from the product of the first-order fields is a striking characteristic and is a direct consequence of the inclusion of the second-order field.

Since the second term of (2.9.7) is the complex conjugate of the first, the total contribution to the electric energy density due to an electric dipole source is

$$\frac{1}{16\pi^2\varepsilon_0}\sum_{\substack{n\\E_p>E_n}} \mu_j^{pn}\mu_k^{np}k_{pn}^6\bar{f}_{ik}(k_{pn}r)f_{ij}(k_{pn}r) + \frac{1}{16\pi^3\varepsilon_0}\sum_{\substack{n\\\text{All } E_n}} \mu_j^{pn}\mu_k^{np}$$

$$\times\int_0^\infty \frac{\mathrm{d}u\, u^6\mathrm{e}^{-2ur}}{u^2 + k_{pn}^2}k_{pn}\, f_{ij}(iur)f_{ik}(iur). \tag{2.9.17}$$

This result holds for transitions from the initial state $|p\rangle$ with summation carried out over a complete set of intermediate states $|n\rangle$ for an oriented dipole source. When the initial state is the ground state, only the second term of (2.9.17) survives,

$$\frac{\hbar c}{32\pi^3\varepsilon_0}\int_0^\infty du\, u^6 e^{-2ur}\alpha_{jk}(iu)\left[(\delta_{jk}-\hat{r}_j\hat{r}_k)\left(\frac{1}{u^2r^2}+\frac{2}{u^3r^3}+\frac{2}{u^4r^4}\right)\right.$$
$$\left.+(\delta_{jk}+3\hat{r}_j\hat{r}_k)\left(\frac{1}{u^4r^4}+\frac{2}{u^5r^5}+\frac{1}{u^6r^6}\right)\right], \tag{2.9.18}$$

where

$$\alpha_{ij}(iu) = 2\sum_n \frac{k_{n0}\mu_i^{0n}\mu_j^{n0}}{\hbar c(k_{n0}^2+u^2)} \tag{2.9.19}$$

is the ground-state dynamic polarizability expressed in terms of imaginary wavevector. For an isotropic source, the ground-state energy density is

$$\frac{\hbar c}{16\pi^3\varepsilon_0}\int_0^\infty du\, u^6 e^{-2ur}\alpha(iu)\left[\frac{1}{u^2r^2}+\frac{2}{u^3r^3}+\frac{5}{u^4r^4}+\frac{6}{u^5r^5}+\frac{3}{u^6r^6}\right], \tag{2.9.20}$$

where $\alpha(iu) = (1/3)\delta_{ij}\alpha_{ij}(iu)$ is the rotationally averaged polarizability, obtained using result (B.4) of Appendix B. It is instructive to examine the asymptotic behavior of result (2.9.20) in the limits of large and small distances r. In the far-zone limit, r is much larger than the characteristic wavelength of molecular transitions, that is, $k_{n0}r \gg 1$. After performing the u-integral using the result

$$\int_0^\infty x^n e^{-\eta x}dx = n!\eta^{-n-1}, \quad Re\,\eta > 0, \tag{2.9.21}$$

the far-zone asymptote is

$$\frac{23\hbar c\alpha(0)}{64\pi^3\varepsilon_0 r^7}, \tag{2.9.22}$$

where $\alpha(0)$ is the static polarizability and corresponds to the $\omega \to 0$ limit of expression (2.9.19). In the near zone, r is much smaller than the reduced transition wavelength, that is, $k_{n0}r \ll 1$. Retaining the leading term after setting the exponential factor to unity in equation (2.9.20) yields, for the short-range asymptote, the limit

$$\frac{1}{16\pi^2\varepsilon_0 r^6}\sum_n |\vec{\mu}^{0n}|^2, \tag{2.9.23}$$

which is the familiar electric energy density of a static electric dipole source.

Returning to the result (2.9.17) applicable to an excited-state molecule, the first term is interpreted as an additional contribution to the energy density arising from real photon emission. After multiplying the geometric tensors, this first term is

$$\frac{1}{16\pi^2\varepsilon_0}\sum_{\substack{n\\E_p>E_n}} \mu_j^{pn}\mu_k^{np}k_{pn}^6\left[(\delta_{jk}-\hat{r}_j\hat{r}_k)\left(\frac{1}{k_{pn}^2r^2}-\frac{2}{k_{pn}^4r^4}\right)\right.$$
$$\left.+(\delta_{jk}+3\hat{r}_j\hat{r}_k)\left(\frac{1}{k_{pn}^4r^4}+\frac{1}{k_{pn}^6r^6}\right)\right], \tag{2.9.24}$$

which on orientational averaging using result (B.4) produces

$$\frac{1}{24\pi^2\varepsilon_0}\sum_{\substack{n\\E_p>E_n}} |\vec{\mu}^{pn}|^2k_{pn}^6\left[\frac{1}{k_{pn}^2r^2}+\frac{1}{k_{pn}^4r^4}+\frac{3}{k_{pn}^6r^6}\right]. \tag{2.9.25}$$

At large values of r, this term has an inverse square asymptotic limit, confirming the interpretation of its origin. In a large spherical shell of unit thickness, the energy is

$$\frac{1}{6\pi\varepsilon_0}\sum_{\substack{n\\E_p>E_n}} |\vec{\mu}^{pn}|^2k_{pn}^4, \tag{2.9.26}$$

and is independent of the radius of the shell. The r^{-2} term of (2.9.25) obviously dominates the density in the far zone as the second term of (2.9.17) was shown to produce an r^{-7} dependence in this limit, as in (2.9.22), where the static polarizability is now that for an excited source. Both terms of the energy density (2.9.17), however, exhibit an r^{-6} dependence in the near zone with the contribution of the first term, found from (2.9.25), given by

$$\frac{1}{8\pi^2\varepsilon_0 r^6}\sum_{\substack{n\\E_p>E_n}} |\vec{\mu}^{pn}|^2, \tag{2.9.27}$$

while the contribution from the u-integral term differs in sign for upward relative to downward transitions from $|p\rangle$, and is

$$-\frac{1}{16\pi^2\varepsilon_0 r^6}\sum_{\substack{n\\E_p>E_n}}|\vec{\mu}^{pn}|^2+\frac{1}{16\pi^2\varepsilon_0 r^6}\sum_{\substack{n\\E_p<E_n}}|\vec{\mu}^{pn}|^2. \tag{2.9.28}$$

A direct manifestation of the electromagnetic energy density is the intermolecular interaction energy of a test polarizable body placed in the radiation fields of the source. Hence, the response of a polarizable test molecule in the ground state, with static electric dipole polarizability $\alpha_{\text{test}}(0)$, to the far-zone limit (2.9.22) is

$$-\frac{1}{2\varepsilon_0^2}\alpha_{\text{test}}(0)\vec{d}^{\perp 2}(\vec{r})=-\frac{23\hbar c}{64\pi^3\varepsilon_0^2 r^7}\alpha_{\text{test}}(0)\alpha(0), \tag{2.9.29}$$

which is recognizable as the Casimir–Polder dispersion energy shift at large separation distances (Casimir and Polder, 1948). The far-zone response of a test body to the radiation field giving rise to an energy density due to an excited source is found from (2.9.25) to be

$$-\frac{1}{24\pi^2\varepsilon_0^2 r^2}\sum_{\substack{n\\E_p>E_n}}|\vec{\mu}^{pn}|^2\alpha_{\text{test}}(0)k_{pn}^4, \tag{2.9.30}$$

while from equations (2.9.27) and (2.9.28), the near-zone shift is

$$-\frac{1}{16\pi^2\varepsilon_0^2 r^6}\sum_{\substack{n\\\text{All } E_n}}|\vec{\mu}^{pn}|^2\alpha_{\text{test}}(0) \tag{2.9.31}$$

and has the form of a London-type dispersion potential, also obtainable using electrostatic coupling. In Chapter 5, the pair interaction energy between polarizable molecules in either ground or excited states is calculated from first principles using the Heisenberg fields in a response theory formalism.

The second contribution to the energy density of the electromagnetic field arises from the magnetic field. The calculation is similar to that outlined for the electric dipole-dependent displacement field and only the main results are presented.

Correct to second order in $\vec{\mu}$, the magnetic energy density is

$$\frac{1}{2}\varepsilon_0 c^2[\vec{b}(\vec{r},t)]^2\approx\frac{1}{2}\varepsilon_0 c^2\Big[\vec{b}^{(1)}(\vec{r},t)\cdot\vec{b}^{(1)}(\vec{r},t)+\vec{b}^{(0)}(\vec{r},t)\cdot\vec{b}^{(2)}(\vec{r},t)$$
$$+\vec{b}^{(2)}(\vec{r},t)\cdot\vec{b}^{(0)}(\vec{r},t)\Big]. \tag{2.9.32}$$

Application of expression (2.6.24) gives, for the expectation value for the product of the first-order fields evaluated over the state $|p; 0(\vec{k}, \lambda)\rangle$, the form

$$\frac{1}{2}\varepsilon_0 c^2 \sum_n \langle p|b_i^{(1)}(\vec{r}, t)|n\rangle\langle n|b_i^{(1)}(\vec{r}, t)|p\rangle$$
$$= \frac{1}{32\pi^2\varepsilon_0}\sum_n \mu_j^{pn}\mu_k^{np}k_{pn}^6\bar{G}_{ij}(k_{pn}r)G_{ik}(k_{pn}r), \tag{2.9.33}$$

where

$$G_{ij}(kr) = \frac{i}{k^2}\varepsilon_{ijk}\vec{\nabla}_k\frac{\mathrm{e}^{ikr}}{r} = -\varepsilon_{ijk}\hat{r}_k\left[\frac{1}{kr} + \frac{i}{k^2r^2}\right]\mathrm{e}^{ikr} = g_{ij}(kr)\mathrm{e}^{ikr}. \tag{2.9.34}$$

As for the tensor field $F_{ij}(kr)$, for future use it is convenient to write the form of the $G_{ij}(kr)$ tensor in terms of the complex wavevector $k = iu$. From the definition (2.9.34), one easily finds that

$$G_{ij}(iur) = \frac{1}{iu^2}\varepsilon_{ijk}\vec{\nabla}_k\frac{\mathrm{e}^{-ur}}{r} = i\varepsilon_{ijk}\hat{r}_k\left[\frac{1}{ur} + \frac{1}{u^2r^2}\right]\mathrm{e}^{-ur} = g_{ij}(iur)\mathrm{e}^{-ur}. \tag{2.9.35}$$

It is readily verified from the last two equations that

$$g_{ij}(\pm kr) = -\bar{g}_{ij}(\mp kr) \tag{2.9.36}$$

and

$$g_{ij}(iur) = \bar{g}_{ij}(-iur) = -\bar{g}_{ij}(iur). \tag{2.9.37}$$

Taking the matrix element of the magnetic field operator quadratic in $\vec{\mu}$, given by (2.6.32), over the molecular state $|p\rangle$ results in $\vec{b}^{(2)}(\vec{r}, t)$ operating in the photon space only,

$$\langle p|b_i^{(2)}(\vec{r}, t)|p\rangle = \frac{i}{4\pi\varepsilon_0}\sum_{\vec{k},\lambda}\left(\frac{\hbar k}{2\varepsilon_0 cV}\right)^{1/2}\left[e_k\alpha(0)\mathrm{e}^{-i\omega t}\mathcal{G}_{ki} - \bar{e}_k\alpha^\dagger(0)\mathrm{e}^{i\omega t}\bar{\mathcal{G}}_{ki}\right], \tag{2.9.38}$$

where the tensor field $\mathcal{G}_{ki}$ is defined in terms of the dipole moment and $G_{ik}(kr)$ (2.9.34), analogously to $\mathcal{F}_{ki}$ given by (2.9.6) for the displacement field,

$$\mathcal{G}_{ki} = \sum_n \frac{\mu_j^{pn}\mu_k^{np}}{E_{np}-\hbar\omega} k^3 G_{ij}(kr) - \sum_n \frac{\mu_j^{pn}\mu_k^{np}}{E_{np}-\hbar\omega} k_{np}^3 G_{ij}(k_{np}r)\mathrm{e}^{i(k_{pn}+k)ct}$$
$$+ \sum_n \frac{\mu_k^{pn}\mu_j^{np}}{E_{np}+\hbar\omega} k^3 G_{ij}(kr) - \sum_n \frac{\mu_k^{pn}\mu_j^{np}}{E_{np}+\hbar\omega} k_{pn}^3 G_{ij}(k_{pn}r)\mathrm{e}^{-i(k_{pn}-k)ct}. \tag{2.9.39}$$

Employing the mode expansion for the free magnetic field (2.6.14), along with equation (2.9.38), gives for the last two terms of (2.9.32)

$$\frac{c}{8\pi\varepsilon_0}\sum_{\vec{k},\lambda}\left(\frac{\hbar k}{2V}\right)\left[b_i \mathrm{e}^{i\vec{k}\cdot\vec{r}}\bar{e}_k\bar{\mathcal{G}}_{ki} + e_k\mathcal{G}_{ki}\bar{b}_i\mathrm{e}^{-i\vec{k}\cdot\vec{r}}\right] \tag{2.9.40}$$

for the expectation value taken over the state $|p; 0(\vec{k},\lambda)\rangle$. Carrying out the mode sum in (2.9.40) and adding it to the contribution from the first-order fields (2.9.33) produces for the magnetic energy density of an oriented electric dipole source the expression

$$\frac{1}{16\pi^2\varepsilon_0}\sum_{\substack{n\\E_n<E_p}} \mu_j^{pn}\mu_k^{np}k_{pn}^6\bar{g}_{ij}(k_{pn}r)g_{ik}(k_{pn}r)$$
$$-\frac{1}{16\pi^3\varepsilon_0}\sum_{\substack{n\\\text{All } E_n}} \mu_j^{pn}\mu_k^{np}\int_0^\infty \frac{\mathrm{d}u\, u^6 \mathrm{e}^{-2ur}}{u^2+k_{pn}^2} k_{pn} g_{ij}(iur)g_{ik}(iur). \tag{2.9.41}$$

When the molecule is in the ground state, only the u-integral remains,

$$\frac{\hbar c}{32\pi^3\varepsilon_0}\int_0^\infty \mathrm{d}u\, u^6 \mathrm{e}^{-2ur}\alpha_{jk}(iu)g_{ij}(iur)g_{ik}(iur), \tag{2.9.42}$$

which for an isotropic source reduces to

$$-\frac{\hbar c}{16\pi^3\varepsilon_0}\int_0^\infty \mathrm{d}u\, u^6 \mathrm{e}^{-2ur}\alpha(iu)\left[\frac{1}{u^2r^2}+\frac{2}{u^3r^3}+\frac{1}{u^4r^4}\right], \tag{2.9.43}$$

on using the result (B.4) and which has the far-zone asymptote

$$-\frac{7\hbar c\alpha(0)}{64\pi^3\varepsilon_0 r^7}. \tag{2.9.44}$$

After expanding the tensors in the first term of the magnetic energy density (2.9.41) and rotationally averaging, the additional contribution to the energy density due to downward transitions is

$$\frac{1}{24\pi^2\varepsilon_0}\sum_{\substack{n\\E_p>E_n}} |\vec{\mu}^{pn}|^2 k_{pn}^6 \left[\frac{1}{k_{pn}^2 r^2} + \frac{1}{k_{pn}^4 r^4}\right], \tag{2.9.45}$$

exhibiting r^{-2} and r^{-4} far- and near-zone behavior, respectively. The response of a magnetically susceptible test body to the far-zone limit (2.9.44) then readily gives the dispersion interaction energy between an electric and a magnetic dipole polarizable pair of molecules,

$$-\frac{1}{2\varepsilon_0 c^2}\chi_{\text{test}}(0)\vec{b}^2 = \frac{7\hbar}{64\pi^3\varepsilon_0^2 c r^7}\chi_{\text{test}}(0)\alpha(0), \tag{2.9.46}$$

where $\chi_{\text{test}}(0)$ is the isotropic static magnetic dipole susceptibility,

$$\chi(0) = \frac{2}{3}\sum_n \frac{|\vec{m}^{0n}|^2}{E_{n0}}. \tag{2.9.47}$$

Interestingly, the energy shift (2.9.46) is repulsive. For a source in an electronic excited state, the near- and far-zone interaction energies are found using (2.9.45) to be

$$-\frac{1}{48\pi^2\varepsilon_0^2 c^2 r^4}\sum_{\substack{n\\E_p>E_n}} |\vec{\mu}^{pn}|^2 \chi_{\text{test}}(0) k_{pn}^2 \tag{2.9.48}$$

and

$$-\frac{1}{24\pi^2\varepsilon_0^2 c^2 r^2}\sum_{\substack{n\\E_p>E_n}} |\vec{\mu}^{pn}|^2 \chi_{\text{test}}(0) k_{pn}^4. \tag{2.9.49}$$

The electric and magnetic contributions to the electromagnetic energy density due to an electric dipole source have been evaluated using the quantum electrodynamical Maxwell field operators. Not only were the vacuum and linear fields required, but also the displacement and magnetic fields second order in $\vec{\mu}$ were needed to be employed to correctly account for all terms quadratic in the source moment. Each resulting expression for the energy density is made up of two terms: a u-integral term and the other valid only for downward transitions from an initially excited state. This last type of contribution dominates the density at large field point distances from the source, displaying an inverse square dependence due to emission of a real photon. The energy density arising from the fields is directly observable as an intermolecular energy shift when a test polarizable species is placed in the fields of the source.

2.10 POYNTING'S THEOREM AND POYNTING VECTOR

The law describing the conservation of energy of the electromagnetic field may be formulated in terms of a theorem, due originally to Poynting (Jackson, 1963). Consider the work done by an external radiation field $\vec{e}(\vec{r})$ and $\vec{b}(\vec{r})$ on a single charge e. It is given by $e\dot{\vec{q}}\cdot\vec{e}(\vec{r})$ or $\vec{j}(\vec{r})\cdot\vec{e}(r)$, having the dimensions of power per unit volume, where $\vec{j}(r)$ is the current density. Because the magnetic force is orthogonal to the velocity of the charge $\dot{\vec{q}}$, no work is done by $\vec{b}(\vec{r})$. Recalling from Section 1.4 that the direction of propagation of a plane electromagnetic wave is given by $\vec{e}(\vec{r})\times\vec{b}(\vec{r})$, its divergence, using the vector identity

$$\vec{\nabla}\cdot(\vec{A}\times\vec{B}) = \vec{B}\cdot(\vec{\nabla}\times\vec{A}) - \vec{A}\cdot(\vec{\nabla}\times\vec{B}), \tag{2.10.1}$$

is

$$\vec{\nabla}\cdot(\vec{e}(\vec{r})\times\vec{b}(\vec{r})) = \vec{b}(\vec{r})\cdot(\vec{\nabla}\times\vec{e}(\vec{r})) - \vec{e}(\vec{r})\cdot(\vec{\nabla}\times\vec{b}(\vec{r})). \tag{2.10.2}$$

Substituting for $\vec{\nabla}\times\vec{e}(\vec{r})$ and $\vec{\nabla}\times\vec{b}(\vec{r})$ from the last two microscopic Maxwell field equations (1.3.7) and (1.3.8) produces

$$\begin{aligned}\vec{\nabla}\cdot\left[\vec{e}(\vec{r})\times\vec{b}(\vec{r})\right] &= -\vec{b}(\vec{r})\cdot\frac{\partial\vec{b}(\vec{r})}{\partial t} - \frac{1}{c^2}\vec{e}(\vec{r})\cdot\frac{\partial\vec{e}(\vec{r})}{\partial t} - \frac{1}{\varepsilon_0 c^2}\vec{e}(\vec{r})\cdot\vec{j}(\vec{r}) \\ &= -\frac{1}{2}\frac{\partial}{\partial t}\left[\frac{1}{c^2}\vec{e}^2(\vec{r}) + \vec{b}^2(\vec{r})\right] - \frac{1}{\varepsilon_0 c^2}\vec{e}(\vec{r})\cdot\vec{j}(\vec{r}).\end{aligned} \tag{2.10.3}$$

Defining the electromagnetic energy density by $u(\vec{r}) = (1/2)\varepsilon_0[\vec{e}^2(\vec{r}) + c^2\vec{b}^2(\vec{r})]$ and the Poynting vector by $\vec{S}(\vec{r}) = \varepsilon_0 c^2[\vec{e}(\vec{r})\times\vec{b}(\vec{r})]$, equation (2.10.3) can be written as the conservation law or continuity equation in differential form,

$$\frac{\partial u(\vec{r})}{\partial t} + \vec{\nabla}\cdot\vec{S}(\vec{r}) = -\vec{e}(\vec{r})\cdot\vec{j}(r), \tag{2.10.4}$$

assuming a linear, isotropic, and homogeneous medium. The right-hand side of equation (2.10.4) represents the rate of conversion of electromagnetic energy per unit volume to mechanical or thermal energy, which in turn is balanced by the two terms on the left-hand side, the rate of change of energy per unit volume plus the total outward flow of energy per unit time across a surface bounding the volume. The Poynting vector is commonly interpreted as the flow of energy at a point in the field, that is, the energy crossing per unit area per unit time whose normal is pointing in the direction of $\vec{e}(\vec{r}) \times \vec{b}(\vec{r})$. $\vec{S}(\vec{r})$ is arbitrary to the extent that the curl of any vector field may be added to it without altering any physical consequences, since the divergence of the Poynting vector appears in the conservation law (2.10.4). When integrated over a closed surface, the Poynting vector gives the total outward flow of energy per unit time. Relation (2.10.4) can also be written in integral form by integrating over a volume V bounded by a surface S with element da and normal vector $\vec{n}$ after applying the divergence theorem to the Poynting vector term. Thus,

$$\int_V \frac{\partial u(\vec{r})}{\partial t} \mathrm{d}^3\vec{r} + \int_S \vec{S}(\vec{r}) \cdot \vec{n} \mathrm{d}a = -\int_V \vec{e}(\vec{r}) \cdot \vec{j}(\vec{r}) \mathrm{d}^3\vec{r}. \tag{2.10.5}$$

In quantum mechanics, the Poynting vector is a Hermitian operator given by

$$\vec{S}(\vec{r}, t) = \frac{1}{2}\varepsilon_0 c^2 \left[\vec{e}^{\mathrm{tot}}(\vec{r}, t) \times \vec{b}(\vec{r}, t) - \vec{b}(\vec{r}, t) \times \vec{e}^{\mathrm{tot}}(\vec{r}, t)\right], \tag{2.10.6}$$

in which the time dependence is shown explicitly. The quantum electrodynamical Maxwell field operators due to an electric dipole source calculated earlier in this chapter can be used to evaluate the rate of energy flux in a radiation field (Power and Thirunamachandran, 1992). Equation (2.10.6) may be expressed in terms of multipolar framework variables by remembering that for a neutral molecule the total electric field is proportional to the transverse displacement vector outside the source, so that the ith component of the Poynting vector is

$$S_i(\vec{r}, t) = \frac{1}{2} c^2 \varepsilon_{ijk} \left[d_j^{\perp}(\vec{r}, t) b_k(\vec{r}, t) + b_k(\vec{r}, t) d_j^{\perp}(\vec{r}, t)\right]. \tag{2.10.7}$$

For the molecule in an excited electronic state $|p\rangle$ and the radiation field in the vacuum state, the expectation value of (2.10.7), after expanding the fields in powers of the electric dipole moment and concentrating only on terms second order in $\vec{\mu}$ as before, is calculated from

$$\begin{aligned}\langle 0(\vec{k},\lambda);p|S_i(\vec{r},t)|p;0(\vec{k},\lambda)\rangle \\ &= \frac{1}{2}c^2\varepsilon_{ijk}\langle 0(\vec{k},\lambda);p|(d_j^{(0)}+d_j^{(1)}+d_j^{(2)}+\cdots) \\ &\quad\times(b_k^{(0)}+b_k^{(1)}+b_k^{(2)}+\cdots)|p;0(\vec{k},\lambda)\rangle+\text{c.c.} \\ &\approx\frac{1}{2}c^2\varepsilon_{ijk}\langle 0(\vec{k},\lambda);p|(d_j^{(1)}b_k^{(1)}+d_j^{(2)}b_k^{(0)}+d_j^{(0)}b_k^{(2)}|p;0(\vec{k},\lambda)\rangle+\text{c.c.},\end{aligned} \tag{2.10.8}$$

where c.c. denotes the complex-conjugate term. The contribution arising from the product of the first-order fields is found to be

$$\begin{aligned}&\frac{1}{2}c^2\varepsilon_{ijk}\langle 0;p|d_j^{(1)}(\vec{\mu};\vec{r},t)b_k^{(1)}(\vec{\mu};\vec{r},t)+b_k^{(1)}(\vec{\mu};\vec{r},t)d_j^{(1)}(\vec{\mu};\vec{r},t)|p;0\rangle \\ &=\frac{1}{2}c^2\varepsilon_{ijk}\sum_n[\langle p|d_j^{(1)}|n\rangle\langle n|b_k^{(1)}|p\rangle+\langle p|b_k^{(1)}|n\rangle\langle n|d_j^{(1)}|p\rangle] \\ &=\frac{c}{32\pi^2\varepsilon_0}\varepsilon_{ijk}\sum_n\mu_l^{pn}\mu_m^{np}k_{pn}^6\left[\bar{f}_{jl}(k_{pn}r)g_{km}(k_{pn}r)+\bar{g}_{km}(k_{pn}r)f_{jl}(k_{pn}r)\right].\end{aligned} \tag{2.10.9}$$

For the computation of the last two terms of expression (2.10.8), use is made of the diagonal electron Fock space matrix elements of the quadratic displacement and magnetic fields (2.9.5) and (2.9.38), producing

$$\begin{aligned}&\frac{1}{2}c^2\varepsilon_{ijk}\sum_{\vec{k},\lambda}[\langle 0(\vec{k},\lambda);p|d_j^{(0)}|p;1(\vec{k},\lambda)\rangle\langle 1(\vec{k},\lambda);p|b_k^{(2)}(\vec{\mu}\vec{\mu})|p;0(\vec{k},\lambda)\rangle \\ &\quad+\langle 0(\vec{k},\lambda);p|d_j^{(2)}(\vec{\mu}\vec{\mu})|p;1(\vec{k},\lambda)\rangle\langle 1(\vec{k},\lambda);p|b_k^{(0)}|p;0(\vec{k},\lambda)\rangle]+\text{c.c.} \\ &=\sum_{\vec{k},\lambda}\left(\frac{\hbar c^2k}{16\pi\varepsilon_0V}\right)\varepsilon_{ijk}\left[e_je^{i\vec{k}\cdot\vec{r}}\bar{e}_m\mathcal{G}_{mk}+e_m\mathcal{F}_{mj}\bar{b}_ke^{-i\vec{k}\cdot\vec{r}}\right]+\text{c.c.}\end{aligned} \tag{2.10.10}$$

The mode sum is carried out in the usual way. In contrast to the calculation of the energy density at the analogous step, the wavevector integration occurring in the Poynting vector can be evaluated exactly for the terms independent of time by extending the limits of integration to $(-\infty, \infty)$. Hence, there are no u-integral terms appearing in the results of the present calculation. The contribution from (2.10.10) is found to be

$$\frac{c}{32\pi^2\varepsilon_0}\varepsilon_{ijk}\sum_n \mathrm{sgn}(k_{pn})\mu_l^{pn}\mu_m^{np}k_{pn}^6[\bar{f}_{jl}(k_{pn}r)g_{km}(k_{pn}r)+\bar{g}_{km}(k_{pn}r)f_{jl}(k_{pn}r)], \tag{2.10.11}$$

where sgn(x) is the signum function of x. When this is added to the term obtained from the first-order fields (2.10.9), the Poynting vector of an oriented source is

$$\begin{aligned} S_i(\vec{r},t) &= \frac{c}{16\pi^2\varepsilon_0}\varepsilon_{ijk}\sum_{\substack{n\\E_p>E_n}} \mu_l^{pn}\mu_m^{np}k_{pn}^6[\bar{f}_{jl}(k_{pn}r)g_{km}(k_{pn}r)+\bar{g}_{km}(k_{pn}r)f_{jl}(k_{pn}r)] \\ &= \frac{c}{8\pi^2\varepsilon_0 r^2}\varepsilon_{ijk}\varepsilon_{kmn}\hat{r}_m\sum_{\substack{n\\E_p>E_n}} \mu_l^{pn}\mu_m^{np}k_{pn}^4(\delta_{jl}-\hat{r}_j\hat{r}_l), \end{aligned} \tag{2.10.12}$$

in which only r^{-2}-dependent terms remain after simplifying the geometric tensors. It is interesting that only downward transition terms contribute to the energy flow, the respective terms from upward transitions from $|p\rangle$ arising from contributions (2.10.9) and (2.10.11) have opposite signs and cancel exactly on addition. This aspect also featured in the calculation of the pole terms in the computation of the electromagnetic energy density. After orientational averaging and contracting the tensors, the Poynting vector (2.10.12) becomes

$$\langle S_i(\vec{r},t)\rangle = \frac{c}{12\pi^2\varepsilon_0 r^2}\sum_{\substack{n\\E_p>E_n}} |\vec{\mu}^{pn}|^2 k_{pn}^4\hat{r}_i \tag{2.10.13}$$

from which the rate of energy loss out of a sphere at any radius r is

$$4\pi r^2\hat{r}_i\langle S_i(\vec{r},t)\rangle = \frac{c}{3\pi\varepsilon_0}\sum_{\substack{n\\E_p>E_n}} |\vec{\mu}^{pn}|^2 k_{pn}^4. \tag{2.10.14}$$

The r^{-2} separation distance dependence in (2.10.13) is consistent with the conservation of energy requirement that the energy flow through a spherical surface be independent of the radius, as is evident from the result (2.10.14).

The rate of flow of electromagnetic energy from a radiating electric dipole source (2.10.14) can be calculated from the decay rate of a molecule undergoing spontaneous emission from an excited state (Craig and Thirunamachandran, 1989). This is done by determining the matrix element for the spontaneous emission of a photon from an excited molecule and inserting it into the Fermi golden rule, from which the power loss through a spherical surface by spontaneous emission may be obtained, as shown below.

Consider a molecule initially in an excited state $|p\rangle$ with no photons present. From the form of the quantum electrodynamical Hamiltonian, radiation and matter are perpetually in mutual interaction, even when either one or both components of the system are in their lowest possible energy states. From state $|p\rangle$, the molecule decays via spontaneous emission to a lower lying level $|n\rangle$, in the process emitting a photon with arbitrary mode character $(\vec{k}, \lambda)$. $|n\rangle$ can be the molecular ground state $|0\rangle$. To first order, the matrix element for the transition between initial and final states $|E_n; 1(\vec{k}, \lambda)\rangle \leftarrow |E_p; 0(\vec{k}, \lambda)\rangle$, with use of the electric dipole approximated interaction Hamiltonian $H_{\text{int}} = -\varepsilon_0^{-1}\vec{\mu} \cdot \vec{d}^{\perp}(\vec{r})$, is

$$M_{fi} = i \sum_{\substack{n \\ E_p > E_n}} \left(\frac{\hbar ck}{2\varepsilon_0 V}\right)^{1/2} \bar{e}_i^{(\lambda)}(\vec{k})\mu_i^{np} e^{-i\vec{k}\cdot\vec{r}}, \tag{2.10.15}$$

from which the emission rate into an element of solid angle $d\Omega$ centered around the wavevector of the emitted photon, after rotational averaging and using the Fermi golden rule equation (1.9.33), is

$$\langle d\Gamma(\Omega)\rangle = \sum_{\substack{n \\ E_p > E_n}} \left(\frac{2\pi\rho}{3\hbar}\right)\left(\frac{\hbar ck}{2\varepsilon_0 V}\right)|\vec{\mu}^{pn}|^2, \tag{2.10.16}$$

where ρ is the density of final states. Making use of the fact that the number of modes of wavevector between $\vec{k}$ and $\vec{k} + d\vec{k}$ in a volume V from equation (1.4.13) is

$$\frac{V}{(2\pi)^3} d^3\vec{k} = \frac{V}{(2\pi)^3} k^2\, dk\, d\Omega \tag{2.10.17}$$

in spherical polar coordinates, the number of levels per unit energy interval lying between $\hbar ck$ and $\hbar c(k + dk)$ is then

$$\rho = \frac{k^2 V\, d\Omega}{\hbar c(2\pi)^3}. \tag{2.10.18}$$

Inserting (2.10.18) into equation (2.10.16) and integrating over all possible angles of emission $d\Omega$, along with summation over the two independent polarizations of the emitted photon, gives for the total rate the expression

$$\Gamma_{n \leftarrow p} = \sum_{\substack{n \\ E_p > E_n}} \frac{k_{pn}^3}{3\pi\varepsilon_0\hbar}|\vec{\mu}^{pn}|^2, \tag{2.10.19}$$

since the energy of the emitted photon is conserved subject to $E_{np} + \hbar ck \approx 0$. The rate (2.10.19) is more familiar as the Einstein A coefficient. The power loss through a spherical surface by spontaneous emission is given by $\hbar ck_{pn}$ multiplied by the rate (2.10.19). This is seen to be identical to the Poynting vector (2.10.14). The inclusion of the contribution to the Poynting vector arising from the interference of the quadratic Heisenberg fields with the zeroth-order fields has been shown to be critically important because the energy flow from the product of the first-order fields gives only one half of the spontaneous power rate.

CHAPTER 3

INTERMOLECULAR FORCES

... the theory behind chemistry is quantum electrodynamics.
—R. P. Feynman, *QED: The Strange Theory of Light and Matter*, Princeton University Press, Princeton, NJ, 1985, p. 8.

3.1 CONCEPT OF INTERMOLECULAR POTENTIAL

The forces between atomic and molecular systems are responsible for the overwhelming majority of chemical and physical properties exhibited by matter (Hirschfelder, 1967; Margenau and Kestner, 1969; Maitland et al., 1981). Their evaluation entails the computation of interaction energies between constituent particles comprising the total system, a complex many-body problem in and of itself. At the quantum mechanical level, this is achieved by solving the Schrödinger equation for a total Hamiltonian that is a sum of the individual molecular Hamiltonians and the Coulomb interaction between all of the charged particles within each component of the total system. Such a Hamiltonian is constructed as follows.

Consider a system of N nuclei ξ and n electrons α of masses M_ξ and $m_\alpha = m_e$, where m_e is the mass of an electron, described by position vectors

$\vec{R}_\xi$ and $\vec{r}_\alpha$, respectively. Let the distance between nuclei ξ and ξ' be $R_{\xi\xi'} = |\vec{R}_{\xi\xi'}| = |\vec{R}_\xi - \vec{R}_{\xi'}|$, that between electron α and nucleus ξ be $r_{\alpha\xi} = |\vec{r}_\alpha - \vec{R}_\xi|$, and the separation between electrons α and α' be $r_{\alpha\alpha'} = |\vec{r}_\alpha - \vec{r}_{\alpha'}|$. The nonrelativistic molecular energy E_{mol}, and the wavefunction $\Psi(\vec{R}_\xi, \vec{r}_\alpha)$—a function of nuclear and electronic coordinates, for such a collection of charged particles, is obtained by solving the time-independent Schrödinger equation

$$H_{\text{mol}}\Psi(\vec{R}_\xi, \vec{r}_\alpha) = E_{\text{mol}}\Psi(\vec{R}_\xi, \vec{r}_\alpha), \tag{3.1.1}$$

where H_{mol} is the molecular Hamiltonian operator. For the situation just described, H_{mol} is given explicitly by

$$\begin{aligned} H_{\text{mol}} = &-\sum_{\xi=1}^{N} \frac{\hbar^2}{2M_\xi}\vec{\nabla}_\xi^2 - \sum_{\alpha=1}^{n} \frac{\hbar^2}{2m_e}\vec{\nabla}_\alpha^2 \\ &+ \frac{e^2}{4\pi\varepsilon_0}\left\{ \sum_{\substack{\xi,\xi'=1\\ \xi'>\xi}}^{N} \frac{Z_\xi Z_{\xi'}}{R_{\xi\xi'}} - \sum_{\xi=1}^{N}\sum_{\alpha=1}^{n} \frac{Z_\xi}{r_{\alpha\xi}} + \sum_{\substack{\alpha,\alpha'=1\\ \alpha'>\alpha}}^{n} \frac{1}{r_{\alpha\alpha'}} \right\}, \end{aligned} \tag{3.1.2}$$

where Z_ξ is the charge of nucleus ξ, e is the charge of the proton, and the Laplacians $\vec{\nabla}_\xi^2$ and $\vec{\nabla}_\alpha^2$ operate on nuclear and electronic coordinates, respectively. Each of the five terms of (3.1.2) has a simple physical interpretation. The first term represents the operator for the kinetic energy of the nuclei, while the second term is the operator for the kinetic energy of the electrons. The first and last terms within braces account for the potential energy of repulsions between the nuclei and between the electrons, respectively. Finally, the second term within braces describes the Coulomb attraction between the electrons and the nuclei.

Solution of the general eigenvalue equation (3.1.1) with Hamiltonian (3.1.2) presents a formidable problem. An exact solution is possible only for the simplest element, the hydrogen atom comprising a single proton and electron. In all other cases, approximate methods of solution have to be resorted to. One possible simplification lies in exploiting the considerable difference in mass between nuclei and electrons, for which $M_\xi \gg m_e$. Accordingly the electrons, being significantly less massive, move much faster than the nuclei, which to a first approximation may be taken to be stationary. With the nuclei fixed, the nuclear kinetic energy terms may be neglected in the Hamiltonian (3.1.2). This leaves the Schrödinger equation for electronic motion to be solved,

$$(H_{\text{el}} + V_{\text{nuc}})\psi_{\text{el}} = E\psi_{\text{el}}, \tag{3.1.3}$$

where the purely electronic Hamiltonian describing the motion of n electrons in the field of N nuclear point charges is

$$H_{\mathrm{el}} = -\frac{\hbar^2}{2m_e}\sum_{\alpha=1}^{n}\vec{\nabla}_\alpha^2 - \frac{e^2}{4\pi\varepsilon_0}\sum_{\xi=1}^{N}\sum_{\alpha=1}^{n}\frac{Z_\xi}{r_{\alpha\xi}} + \frac{e^2}{4\pi\varepsilon_0}\sum_{\substack{\alpha,\alpha'=1\\ \alpha'>\alpha}}^{n}\frac{1}{r_{\alpha\alpha'}}, \tag{3.1.4}$$

V_{nuc} is the nuclear repulsion term,

$$V_{\mathrm{nuc}} = \frac{e^2}{4\pi\varepsilon_0}\sum_{\substack{\xi,\xi'=1\\ \xi'>\xi}}^{N}\frac{Z_\xi Z_{\xi'}}{R_{\xi\xi'}}, \tag{3.1.5}$$

and the energy E is the electronic energy including the contribution from internuclear repulsion, namely,

$$E = E_{\mathrm{el}} + V_{\mathrm{nuc}}. \tag{3.1.6}$$

Since V_{nuc} is independent of electronic coordinates, it may be discarded from (3.1.3), leaving the electronic Schrödinger equation,

$$H_{\mathrm{el}}\psi_{\mathrm{el}} = E_{\mathrm{el}}\psi_{\mathrm{el}}. \tag{3.1.7}$$

Its eigenfunction is the electronic wavefunction, ψ_{el},

$$\psi_{\mathrm{el}} = \psi_{\mathrm{el}}[(\vec{r}_\alpha), \{\vec{R}_\xi\}], \tag{3.1.8}$$

which describes the motion of the electron and explicitly depends on the electronic coordinates, but depends only parametrically on the nuclear coordinates. The electronic energy, E_{el}, also has a parametric dependence on $\vec{R}_\xi$, $E_{\mathrm{el}} = E_{\mathrm{el}}[\{\vec{R}_\xi\}]$. Hence, the electronic Schrödinger equation (3.1.7) is solved at differing nuclear configurations yielding E_{el}, with each member of the set $\{\vec{R}_\xi\}$ corresponding to a different molecular electronic state, from which E is then calculated using (3.1.6). After solving the electronic problem, the same assumptions may be used for the nuclear motion. Reliance is made on the notion that as the nuclei move, the electronic energy varies smoothly as a function of the parameters $\{\vec{R}_\xi\}$. Therefore, E now becomes the potential energy for nuclear motion in the average field of the electrons, giving rise to a nuclear Hamiltonian,

$$H_{\mathrm{nuc}}\psi_{\mathrm{nuc}} = E_{\mathrm{mol}}\psi_{\mathrm{nuc}}, \tag{3.1.9}$$

where the nuclear Hamiltonian is given by

$$H_{\mathrm{nuc}} = -\sum_{\xi=1}^{N}\frac{\hbar^2}{2M_\xi}\vec{\nabla}_\xi^2 + E\left[\{\vec{R}_\xi\}\right]. \tag{3.1.10}$$

Recognition of the consequences of the significant differences in mass of electrons and nuclei and their effect on the molecular Hamiltonian was first investigated by Born and Oppenheimer (1927).

Solutions to the nuclear Schrödinger equation (3.1.9) with nuclear wavefunction $\psi_{\text{nuc}}[\{\vec{R}_\xi\}]$ describe vibration, rotation, and translation of a molecule, where the Born–Oppenheimer approximation to the molecular energy, E_{mol}, includes electronic, vibrational, rotational, and translational energy contributions. The corresponding approximation to the molecular wavefunction appearing in (3.1.1) is then

$$\Psi[(\vec{r}_\alpha), \{\vec{R}_\xi\}] = \psi_{\text{el}}[(\vec{r}_\alpha), \{\vec{R}_\xi\}]\psi_{\text{nuc}}[\{\vec{R}_\xi\}]. \tag{3.1.11}$$

Since the electronic energy E_{el} is a function of nuclear coordinates, it may be used to define the concept of an intermolecular potential or an interaction energy.

It is convenient to write the total energy, E, as a sum of one-, two-, and many-body terms

$$\begin{aligned} E &= \sum_{\xi=1}^{N} E_\xi + \sum_{\substack{\xi,\xi'=1 \\ \xi>\xi'}}^{N} E_{\xi\xi'} + \sum_{\substack{\xi,\xi',\xi''=1 \\ \xi>\xi'>\xi''}}^{N} E_{\xi\xi'\xi''} + \cdots \\ &= E_{\text{1-body}} + E_{\text{2-body}} + E_{\text{3-body}} + \cdots, \end{aligned} \tag{3.1.12}$$

where E_ξ is the one-body energy, $E_{\xi\xi'}$ is the two-particle energy, and so on. Frequently, the zero of energy is adjusted so that when the molecules are at an infinite separation from one another, the potential energy is zero. This is given by subtracting the energy of each individual isolated atom or molecule from E. Hence, the intermolecular energy shift is defined by

$$\Delta E = E - \sum_{\xi=1}^{N} E_\xi = \sum_{\substack{\xi,\xi'=1 \\ \xi>\xi'}}^{N} E_{\xi\xi'} + \sum_{\substack{\xi,\xi',\xi''=1 \\ \xi>\xi'>\xi''}}^{N} E_{\xi\xi'\xi''} + \cdots, \tag{3.1.13}$$

a sum of two-, three-, and many-body terms. A common approximation is to assume pairwise additivity and to calculate the leading term of the energy shift, that describing the interaction between two particles and given by $E_{\xi\xi'}$. With this form of partitioning, the three-body contribution $E_{\xi\xi'\xi''}$, the four-body term $E_{\xi\xi'\xi''\xi'''}$, and so on are taken to be nonadditive corrections. Thus, for three interacting molecules A, B, and C, for instance, the energy shift is given by $\Delta E_3 = E_{AB} + E_{BC} + E_{CA} + E_{ABC}$, where the first three terms in the preceding sum represent pairwise contributions between any two of the three entities ignoring the presence of the third body, while the last term is the nonadditive three-body correction.

Forces between molecules are typically separated into short- and long-range contributions. At the former distance regime, the interactions are overall of a repulsive character, the most important terms being due to exchange, repulsion, and charger transfer effects. The last of these is a manifestation of induction forces at short range. In contrast, at long range, attractive forces dominate, arising chiefly from the electrostatic interaction, induction, and dispersion terms. A characteristic plot of the pair potential energy function versus interparticle separation distance, bearing in mind that the potential function between two molecules is not only a function of their relative separation but also of their relative orientation, shows that $\Delta E(R)$ is large and positive at small separations, with $\Delta E(R) \rightarrow \infty$ as $R \rightarrow 0$, while $\Delta E(R)$ has negative value at large R, with $\Delta E(R) \rightarrow 0$ as $R \rightarrow \infty$. These two extremes are connected by a curve containing one negative minimum. In the subsequent sections of this chapter, short-range forces are discussed briefly, followed by an outline of the terms contributing to the long-range part of the interaction energy within the framework of classical electrostatics.

3.2 SHORT-RANGE FORCES

Forces between atoms and molecules at short separation distances occur primarily as a result of overlap of electronic charge clouds associated with each center. Overall, this leads to a net repulsion. Typically, these forces are effective over internuclear separation distances $R < a_0$, where a_0 is the Bohr radius, but also extend to distances $a_0 < R < 10a_0$, termed the intermediate range of separation. The asymptotically correct form at very short range is an exponential function $Ae^{-\kappa R}$, where A and κ are constants. At very short range, electron exclusion effects dominate the interaction. This prevents some electrons from occupying the volume between the two nuclei, thereby reducing the shielding each positive nucleus experiences, and therefore increases the repulsive force in each of them. Like long-range forces, short-range interactions are electromagnetic in character. Unlike interactions occurring at long range, Rayleigh–Schrödinger perturbation theory cannot be used to compute forces at small separation distances. This is due to a number of reasons. When there is significant overlap of charge clouds, for example, the multipole expansion of the electrostatic energy fails to converge. In addition, the wavefunction used in the computation of long-range forces is not required to satisfy the antisymmetry principle, in contrast to evaluation of interactions at short range, for which the wavefunction must obey the Pauli exclusion principle. Moreover, the perturbation approximation for isolated atomic and molecular systems is no longer valid since the strength of the interaction is large.

A common physical interpretation of short-range forces is given in terms of the Pauli exclusion principle and the electrostatic Hellmann–Feynman theorem (Margenau and Kestner, 1969; Maitland et al., 1981). The latter holds for the exact solution of the Schrödinger equation for fixed nuclear positions, from whose solution the total electronic charge density may be evaluated. The theorem itself states that the forces on the nuclei are simply given by the Coulombic forces due to a distribution of charges as calculated by classical electrostatics. When charge overlap is considerable, large distortions in the charge distributions take place because of the exclusion principle, as well as from Coulomb repulsion among the electrons. A net repulsion effect ensues for closed electronic shell species as the respective electronic charge clouds tend to keep out of each others way, decreasing the charge density between the nuclei, leading also to a reduction of nuclear screening by the electrons. For open shell systems, however, increased electron density can occur between the nuclei, resulting in chemical bond formation.

For molecules close together, rigorous use of quantum mechanical results demands that antisymmetrized wavefunctions be employed in the calculation of intermolecular interaction energies. Their use leads to the appearance of terms describing the effects of electron exchange, which is a significant component of short-range forces. This is achieved by writing the total Hamiltonian for the system as a sum of atomic and molecular Hamiltonians, the attraction of an electron associated with one particular atom or molecule to the nucleus of a different center, and interelectron and internuclear repulsion terms. Computing the interaction energy with a wavefunction that satisfies the exclusion principle yields for the total energy the sum of the unperturbed energies of each species, the classical electrostatic interaction as found in long-range theory, and additional contributions arising from the second and third terms of the total Hamiltonian mentioned above that are attributed to exchange and repulsion. For a system comprising two electrons 1 and 2 associated with orbitals a and b of two molecules A and B, respectively, the exchange integral is of the form $\langle a(1)b(2)|r_{12}^{-1}|\, a(2)b(1)\rangle$, where r_{12} is the distance between the two electrons, which gives a negative contribution to the energy. The repulsion term on the other hand, describing overlap of the electrons from the two molecules, is of the form $\langle a|r_{A2}^{-1} + r_{B1}^{-1}|b\rangle$, where r_{A2} and r_{B1} are the distances of electron 2 from nucleus A and of electron 1 from center B, respectively. Together with the remaining terms arising from electron– nuclear attraction and interelectron repulsion, which may be regarded as corrective terms to the electrostatic energy, the total exchange–repulsion energy is overall repulsive.

The approach outlined above has its origins in the method developed by Heitler and London, an early form of valence bond theory, in which the contribution to the intermolecular potential from electrons assigned to specific individual atoms is primary, with the total wavefunction being a properly antisymmetrized product of independent atomic spin orbitals. An alternative approximate method is the now much favored molecular orbital theory, in which the Schrödinger equation is solved for the total system comprising all electrons and nuclei without any prior allocation of electrons, which are in turn described by orbitals possessing the intrinsic symmetry of the molecular framework. The basis functions used in such computations need not be true atomic functions. Any set of localized functions will suffice in principle, although Slater type orbitals and Gaussian functions have been used for quite some time to represent atomic orbitals in the well-known linear combination of atomic orbitals–molecular orbital (LCAO–MO) approach, although the best results are obtained with variationally minimized Hartree–Fock orbitals.

Since nonrelativistic quantum mechanics is used by and large to investigate molecular structure, no effects due to a first principles treatment of electron spin are included in the calculation of intermolecular potentials—both at short and at long range. Instead, for short range, spin orbitals are formed by multiplying spatial orbitals by orthonormal spin eigenfunctions. For orbitals a and b containing electrons with opposite spin, the exchange–repulsion term is zero because the overlap integral, which features as a prefactor in each of the contributions, vanishes. When the electron spins in orbitals a and b are identical, however, the spin functions integrate out.

In cases where antisymmetrization of the wavefunction is necessary, Rayleigh–Schrödinger perturbation theory fails to deal with forces at short range. This has led to the development of exchange–perturbation theories (Claverie, 1978; Szalewicz et al., 2005). These include approaches that employ a set of antisymmetrized unperturbed wavefunctions at the outset, which are nonorthogonal, and so-called symmetry adapted perturbation theories (SAPT), in which a simple nonantisymmetrized product wavefunction is used to represent the unperturbed state with antisymmetrization performed at each order in the perturbation.

3.3 LONG-RANGE FORCES

At large separation distances, typically of the order of $10a_0 < R < 100a_0$, the dominant forces between atomic and molecular systems are attractive. Effects due to exchange are reduced considerably, if not altogether eliminated, now that there is insignificant overlap of molecular electronic clouds. With the integrity of each center remaining largely in tact, the charge distribution produces electrostatic energy of interaction, while the ever-present electronic motion within each species is viewed as giving rise to transient fluctuating electromagnetic fields that polarize the other species, and vice versa. The correlation of these temporary fields leads to net forces of attraction and is the source of induction and dispersion energy shifts. Hence, at long range, the energy may be partitioned according to

$$E_{\text{long range}} = E_{\text{electrostatic}} + E_{\text{induction}} + E_{\text{dispersion}}. \qquad (3.3.1)$$

The long-range interaction energy is normally evaluated by perturbation theory techniques after expanding the electronic charge distribution in the familiar multipolar series. The ensuing pair potentials exhibit a dependence on some inverse power of internuclear separation distance.

Other than the three contributions delineated above in the separation of equation (3.3.1), the remaining dominant interaction between pairs of molecules that is of long-range character is the resonant transfer of excitation energy. This occurs when one of the species is initially excited while the second is in the ground electronic state, with both entities being identical. The other case occurs when at least one of the pairs is in a degenerate state—typically an excited state. The interaction may be either attractive or repulsive, and is not pairwise additive. Resonant migration of energy forms the subject matter of Chapter 4. Meanwhile the electrostatic, induction, and dispersion contributions are elucidated in each of the following three sections after expanding the electrostatic charge distribution in a multipolar series.

3.4 ELECTROSTATIC INTERACTION

Coupling of permanent electric multipole moments in atoms and molecules produces the electrostatic contribution to the intermolecular interaction energy when taken to first order in perturbation theory (Buckingham, 1967). Because permanent magnetic moments in molecules are of considerably lower magnitude than their electric counterparts, interaction of permanent magnetic moments at each center, resulting in a magnetostatic contribution to the energy shift, is usually ignored. The electrostatic interaction energy may be derived and expressed in a number of alternative ways. In the present approach, the potential due to a static distribution of charges is first expanded in a series of electric multipole moments. This is followed by coupling the electric multipole series of a second species to the electrostatic potential of the first, generating the well-known classical interaction energy between two charges, a charge and an electric dipole, a charge and an electric quadrupole, between two electric dipoles, and so forth.

Consider the charge distribution of a molecule A comprising n point charges e_α^A, located at positions $q_{i\alpha}(A)$, $\alpha = 1, 2, \ldots, n$, with respect to an origin within A. The electrostatic potential at a field point $\vec{r}$, due to such an arrangement of charges, is

$$\phi^A(\vec{r}) = \frac{1}{4\pi\varepsilon_0}\sum_{\alpha=1}^{n}\frac{e_\alpha^A}{|\vec{r}-\vec{q}_\alpha(A)|}. \tag{3.4.1}$$

A Taylor series expansion of the potential produces

$$\begin{aligned}4\pi\varepsilon_0\phi^A(\vec{r}) &= \sum_\alpha \frac{e_\alpha^A}{r} - \sum_\alpha e_\alpha^A q_{i\alpha}(A)\vec{\nabla}_i\frac{1}{r} + \frac{1}{2!}\sum_\alpha e_\alpha^A q_{i\alpha}(A)q_{j\alpha}(A)\vec{\nabla}_i\vec{\nabla}_j\frac{1}{r}\\ &\quad - \frac{1}{3!}\sum_\alpha e_\alpha^A q_{i\alpha}(A)q_{j\alpha}(A)q_{k\alpha}(A)\vec{\nabla}_i\vec{\nabla}_j\vec{\nabla}_k\frac{1}{r} + \cdots\\ &= \sum_\alpha \frac{e_\alpha^A}{r} - \mu_i(A)\vec{\nabla}_i\frac{1}{r} + Q_{ij}(A)\vec{\nabla}_i\vec{\nabla}_j\frac{1}{r} - O_{ijk}(A)\vec{\nabla}_i\vec{\nabla}_j\vec{\nabla}_k\frac{1}{r} + \cdots,\end{aligned} \tag{3.4.2}$$

where the sums after the first equality of (3.4.2) are taken over all charged particles α. It is worth pointing out that the electric field, its gradient, the gradient of the field gradient may be evaluated from either of the forms (3.4.1) or (3.4.2). Immediately recognizable after the second equality above is the total charge of the system. Moreover, for the remaining terms after the second equality of (3.4.2), use has been made of the standard definitions of the electric multipole moments in their reducible form. Hence, the second term of (3.4.2) represents the contribution to the electrostatic potential due to an electric dipole, whose ith Cartesian component is

$$\mu_i(A) = \sum_\alpha e_\alpha^A q_{i\alpha}(A), \tag{3.4.3}$$

the third term is the contribution due to an electric quadrupole source,

$$Q_{ij}(A) = \frac{1}{2!}\sum_\alpha e_\alpha^A q_{i\alpha}(A)q_{j\alpha}(A), \tag{3.4.4}$$

and so on.

The electrostatic interaction energy between two molecules may be obtained as follows. Consider a second molecule B, composed of charges e_β^B situated at points $q_{i\beta}(B)$ with respect to an origin within B. Let $\vec{R}$ be the separation distance vector between the origin of the coordinate system of center B and the origin of species A. Assuming that $q_{i\beta}(B) \ll R$ for each charge β, the electrostatic potential between A and B is

$$V(A,B) = \sum_\beta e_\beta^B \phi^A(|\vec{R} + \vec{q}(B)|). \tag{3.4.5}$$

It is convenient for future application to expand the charge distribution of molecule B as a Taylor series and express it in terms of electric multipole moments, giving for the potential

$$V(A,B) = \left\{ \sum_\beta e_\beta^B + \mu_i(B)\vec{\nabla}_i + Q_{ij}(B)\vec{\nabla}_i\vec{\nabla}_j + O_{ijk}(B)\vec{\nabla}_i\vec{\nabla}_j\vec{\nabla}_k + \cdots \right\} \phi^A(R). \tag{3.4.6}$$

Substituting for $\phi^A(R)$ from (3.4.2) into (3.4.6) produces

$$\begin{aligned} V(A,B) &= \frac{1}{4\pi\varepsilon_0}\left\{ \sum_\alpha \frac{e_\alpha^A}{R} - \mu_i(A)\vec{\nabla}_i\frac{1}{R} + Q_{ij}(A)\vec{\nabla}_i\vec{\nabla}_j\frac{1}{R} - O_{ijk}(A)\vec{\nabla}_i\vec{\nabla}_j\vec{\nabla}_k\frac{1}{R} + \cdots \right\} \\ &\quad \times \left\{ \sum_\beta e_\beta^B + \mu_{i'}(B)\vec{\nabla}_{i'} + Q_{i'j'}(B)\vec{\nabla}_{i'}\vec{\nabla}_{j'} + O_{i'j'k'}(B)\vec{\nabla}_{i'}\vec{\nabla}_{j'}\vec{\nabla}_{k'} + \cdots \right\}, \end{aligned} \tag{3.4.7}$$

where the gradient operators now act on R. Multiplying the two terms within braces and grouping terms with a similar physical origin produces for the classical interaction energy the expression

$$\begin{aligned} 4\pi\varepsilon_0 V(A,B) = \sum_{\alpha,\beta} \frac{e_\alpha^A e_\beta^B}{R} &+ \left(\sum_\alpha e_\alpha^A \mu_i(B)\vec{\nabla}_i\frac{1}{R} - \sum_\beta e_\beta^B \mu_i(A)\vec{\nabla}_i\frac{1}{R} \right) \\ &- \mu_i(A)\mu_j(B)\vec{\nabla}_i\vec{\nabla}_j\frac{1}{R} \\ &+ \left(\sum_\alpha e_\alpha^A Q_{ij}(B)\vec{\nabla}_i\vec{\nabla}_j\frac{1}{R} + \sum_\beta e_\beta^B Q_{ij}(A)\vec{\nabla}_i\vec{\nabla}_j\frac{1}{R} \right) \end{aligned}$$

$$-\left(\mu_i(A)Q_{jk}(B)\vec{\nabla}_i\vec{\nabla}_j\vec{\nabla}_k\frac{1}{R}-\mu_i(B)Q_{jk}(A)\vec{\nabla}_i\vec{\nabla}_j\vec{\nabla}_k\frac{1}{R}\right)$$

$$+Q_{ij}(A)Q_{kl}(B)\vec{\nabla}_i\vec{\nabla}_j\vec{\nabla}_k\vec{\nabla}_l\frac{1}{R}$$

$$+\left(\sum_\alpha e_\alpha^A O_{ijk}(B)-\sum_\beta e_\beta^B O_{ijk}(A)\right)\vec{\nabla}_i\vec{\nabla}_j\vec{\nabla}_k\frac{1}{R}+\cdots, \tag{3.4.8}$$

the familiar sum of monopole–monopole, charge–dipole, dipole–dipole, and other contributions. The explicit dependence of the interaction energy (3.4.8) on intermolecular separation distance and relative orientation is obtained on evaluating the gradients. The first few terms are given by

$$\vec{\nabla}_i\frac{1}{R}=-\frac{\hat{R}_i}{R^2}, \tag{3.4.9}$$

$$\vec{\nabla}_i\vec{\nabla}_j\frac{1}{R}=-\frac{1}{R^3}(\delta_{ij}-3\hat{R}_i\hat{R}_j), \tag{3.4.10}$$

$$\vec{\nabla}_i\vec{\nabla}_j\vec{\nabla}_k\frac{1}{R}=\frac{3}{R^4}(\delta_{ij}\hat{R}_k+\delta_{ik}\hat{R}_j+\delta_{jk}\hat{R}_i-5\hat{R}_i\hat{R}_j\hat{R}_k), \tag{3.4.11}$$

and

$$\vec{\nabla}_i\vec{\nabla}_j\vec{\nabla}_k\vec{\nabla}_l\frac{1}{R}=\frac{1}{R^5}\big[3(\delta_{ij}\delta_{kl}+\delta_{ik}\delta_{jl}+\delta_{il}\delta_{jk})-15(\delta_{ij}\hat{R}_k\hat{R}_l+\delta_{ik}\hat{R}_j\hat{R}_l$$
$$+\delta_{il}\hat{R}_j\hat{R}_k+\delta_{jk}\hat{R}_i\hat{R}_l+\delta_{jl}\hat{R}_i\hat{R}_k+\delta_{kl}\hat{R}_i\hat{R}_j)+105\hat{R}_i\hat{R}_j\hat{R}_k\hat{R}_l\big]. \tag{3.4.12}$$

For neutral molecules, the total charge vanishes, leaving for the interaction energy the somewhat simplified form

$$V(A,B)=\frac{1}{4\pi\varepsilon_0}\left\{\begin{aligned}&-\mu_i(A)\mu_j(B)\vec{\nabla}_i\vec{\nabla}_j\frac{1}{R}\\&-\left[\mu_i(A)Q_{jk}(B)-\mu_i(B)Q_{jk}(A)\right]\vec{\nabla}_i\vec{\nabla}_j\vec{\nabla}_k\frac{1}{R}\\&+Q_{ij}(A)Q_{kl}(B)\vec{\nabla}_i\vec{\nabla}_j\vec{\nabla}_k\vec{\nabla}_l\frac{1}{R}+\cdots\end{aligned}\right\}, \tag{3.4.13}$$

a sum of dipole–dipole, dipole–quadrupole, quadrupole–quadrupole, and other contributions. The classical interaction energy may be converted to a quantum mechanical operator by promoting the classical dynamical variables to quantum operators. Perturbation theory may then be used to extract the various contributions to the long-range energy shift.

With the effects of electron exchange contributing negligibly to the intermolecular energy shift at long range, it is safe to consider the electronic charge distribution around molecule A as arising from the electrons assigned to A itself, and similarly for species B. Hence, in the perturbation theory treatment, an appropriate unperturbed Hamiltonian is a sum of the molecular Hamiltonians of the two species,

$$H_0 = H_{\text{mol}}(A) + H_{\text{mol}}(B). \tag{3.4.14}$$

Since H_0 is separable, its eigenstates are product states of the eigenfunctions of $H_{\text{mol}}(A)$ and $H_{\text{mol}}(B)$, designated by $|E_m^A\rangle$ and $|E_n^B\rangle$, to give $|E_m^A\rangle|E_n^B\rangle = |E_m^A, E_n^B\rangle$, which constitutes the zeroth-order wavefunction. Thus,

$$H_0|E_m^A, E_n^B\rangle = E_0|E_m^A, E_n^B\rangle = (E_m^A + E_n^B)|E_m^A, E_n^B\rangle, \tag{3.4.15}$$

where the unperturbed energy $E_0 = E_m^A + E_n^B$, when A and B are described by quantum numbers m and n, respectively. From the Rayleigh–Schrödinger perturbation theory presented in Section 1.9, the zeroth-, first-, second-, and higher order corrections to the energy shift may be evaluated. For the case in which the isolated molecules are in states $|E_r^A\rangle$ and $|E_s^B\rangle$, the energy is given by

$$\begin{aligned} E = {} & E_r^A + E_s^B + \langle E_s^B, E_r^A|V(A,B)|E_r^A, E_s^B\rangle \\ & - \sum_{\substack{m \\ m\neq r}} \sum_{\substack{n \\ n\neq s}} \frac{|\langle E_s^B, E_r^A|V(A,B)|E_r^A, E_s^B\rangle|^2}{(E_m^A - E_r^A) + (E_n^B - E_s^B)} + \cdots, \end{aligned} \tag{3.4.16}$$

where in the last term written above, the sum is executed over all states $|E_m^A\rangle$ of A and $|E_n^B\rangle$ of B except their initial states $|E_r^A\rangle$ and $|E_s^B\rangle$. In formula (3.4.16), the perturbation operator to be used is of the form

$$V(A,B) = \frac{1}{4\pi\varepsilon_0} \left\{ \begin{aligned} & \sum_{\alpha,\beta} \frac{e_\alpha^A e_\beta^B}{R} - \left(\sum_\alpha e_\alpha^A \mu_i(B) - \sum_\beta e_\beta^B \mu_i(A) \right) \hat{R}_i R^{-2} \\ & + \mu_i(A)\mu_j(B)(\delta_{ij} - 3\hat{R}_i\hat{R}_j)R^{-3} \\ & - \left(\sum_\alpha e_\alpha^A Q_{ij}(B) + \sum_\beta e_\beta^B Q_{ij}(A) \right)(\delta_{ij} - 3\hat{R}_i\hat{R}_j)R^{-3} + \cdots \end{aligned} \right\}. \tag{3.4.17}$$

Inserting (3.4.17) into the term corresponding to the first-order correction to the perturbed energy yields, for nondegenerate unperturbed states, the electrostatic energy

$$E_{\text{electrostatic}} = \langle E_s^B, E_r^A | V(A,B) | E_r^A, E_s^B \rangle$$

$$= \frac{1}{4\pi\varepsilon_0} \left\{ \begin{array}{l} \sum\limits_{\alpha,\beta} e_\alpha^A e_\beta^B R^{-1} - \left(\sum\limits_{\alpha} e_\alpha^A \mu_i^{ss}(B) - \sum\limits_{\beta} e_\beta^B \mu_i^{rr}(A) \right) \hat{R}_i R^{-2} \\ + \left(\mu_i^{rr}(A) \mu_j^{ss}(B) - \sum\limits_{\alpha} e_\alpha^A Q_{ij}^{ss}(B) - \sum\limits_{\beta} e_\beta^B Q_{ij}^{rr}(A) \right) \\ \times (\delta_{ij} - 3\hat{R}_i \hat{R}_j) R^{-3} + \cdots \end{array} \right\}, \tag{3.4.18}$$

where $\mu_i^{tt}(\xi), Q_{ij}^{tt}(\xi)$, and so on are the so-called permanent electric moments of species ξ in the unperturbed state $|E_t^\xi\rangle$, comprising electric dipole, quadrupole, and so on, with $\mu_i^{tt}(\xi) = \langle E_t^\xi | \mu_i(\xi) | E_t^\xi \rangle$, and so on. By taking the expectation value of the interaction operator $V(A,B)$ to first order using ground-state unperturbed wavefunctions for A and B, namely, $|E_0^A, E_0^B\rangle$, equation (3.4.18) represents the electrostatic interaction between two ground-state molecules, with ground electronic state permanent moments appearing instead. Electrostatic couplings are strictly pairwise additive and may be of either sign. The contributions to the interaction energy due to the second-order correction term are decomposed in the next two sections.

In the presentation given in this section, no account has been taken of the finite speed of propagation of electromagnetic signals. This is characteristic of semiclassical radiation theory in which the electromagnetic field is viewed as a classical external perturbation, with only the appropriate atomic and molecular dynamical variables subject to quantum conditions. This is in direct contrast to molecular quantum electrodynamics, in which both matter and radiation field are quantized, and the effects of retardation are properly dealt with since all electromagnetic influences travel at the speed of light. It may be recalled that in the multipolar formalism detailed in Section 1.7, no intermolecular electrostatic interaction term, $V_{\text{inter}} = \sum_{\xi < \xi'} V(\xi, \xi')$, appeared in the Hamiltonian. V_{inter} was found to cancel with the intermolecular part of the transverse polarization field, $(1/2\varepsilon_0) \int |\vec{p}^{\perp}(\vec{r})|^2 \, d^3\vec{r}$, as demonstrated explicitly in Section 2.3, leaving an interaction Hamiltonian in which molecules couple directly to the causal electric displacement and magnetic field operators. Finally, it should be remarked that interactions between permanent moments are included in the multipolar formalism and described via transverse photon coupling. This is presented in Section 7.4.

3.5 INDUCTION FORCES

Interaction of a permanent moment in one molecule with a second nonpolar molecule gives rise to an induction force. This is a consequence of the fact that the field due to a static moment distorts the charge distribution in the second species, inducing a multipole moment within it. The induced and inducing moments couple, always resulting in an attractive interaction. Induction effects are, however, nonadditive.

It was pointed out that in the second-order term for the perturbed energy, the summations were to be executed over all individual molecular states of the system except for the initial state. Hence, for both molecules initially in the ground electronic state, this leaves three distinct contributions to be examined individually in what follows. These correspond to a term in which molecule A is excited and B is in the ground state, leading to the induction energy of A; a term in which molecule A is in the ground state and species B is excited, which is the induction energy of B; and the final case in which both entities may be excited. This last situation results in the dispersion energy shift, whose explicit semiclassical expression is presented in the next section. Hence, from formula (3.4.16), the induction energy is obtained from the two terms (Buckingham, 1967),

$$\begin{aligned} E_{\text{induction}} &= -\sum_{m\neq 0} \frac{\langle E_0^B, E_0^A|V(A,B)|E_m^A, E_0^B\rangle\langle E_0^B, E_m^A|V(A,B)|E_0^A, E_0^B\rangle}{E_m^A - E_0^A} \\ &\quad -\sum_{n\neq 0} \frac{\langle E_0^B, E_0^A|V(A,B)|E_0^A, E_n^B\rangle\langle E_n^B, E_0^A|V(A,B)|E_0^A, E_0^B\rangle}{E_n^B - E_0^B} \\ &= E_{\text{induction}}(A) + E_{\text{induction}}(B). \end{aligned} \tag{3.5.1}$$

Substituting for the perturbation operator (3.4.8),

$$V(A,B) = \frac{1}{4\pi\varepsilon_0}\left\{ \begin{aligned} &\sum_{\alpha,\beta}\frac{e_\alpha^A e_\beta^B}{R} + \left(\sum_{\alpha} e_\alpha^A \mu_i(B) - \sum_{\beta} e_\beta^B \mu_i(A)\right)\vec{\nabla}_i\frac{1}{R} \\ &-\left[\mu_i(A)\mu_j(B) - \sum_{\alpha} e_\alpha^A Q_{ij}(B) - \sum_{\beta} e_\beta^B Q_{ij}(A)\right] \\ &\times \vec{\nabla}_i\vec{\nabla}_j\frac{1}{R} + \cdots \end{aligned} \right\}, \tag{3.5.2}$$

into the first term of (3.5.1) gives for the induction energy of molecule A,

$$E_{\text{induction}}(A) = -\sum_{m\neq 0}\frac{(E_m^A - E_0^A)^{-1}}{(4\pi\varepsilon_0)^2}\langle E_0^B, E_0^A|$$

$$\times\left\{\begin{array}{l}\sum_{\alpha,\beta}\frac{e_\alpha^A e_\beta^B}{R}+\left(\sum_\alpha e_\alpha^A\mu_i(B)-\sum_\beta e_\beta^B\mu_i(A)\right)\vec{\nabla}_i\frac{1}{R}\\ -\left[\mu_i(A)\mu_j(B)-\sum_\alpha e_\alpha^A Q_{ij}(B)-\sum_\beta e_\beta^B Q_{ij}(A)\right]\vec{\nabla}_i\vec{\nabla}_j\frac{1}{R}+\cdots\end{array}\right\}|E_m^A, E_0^B\rangle$$

$$\times\langle E_0^B, E_m^A|\left\{\begin{array}{l}\sum_{\alpha,\beta}\frac{e_\alpha^A e_\beta^B}{R}+\left(\sum_\alpha e_\alpha^A\mu_{i'}(B)-\sum_\beta e_\beta^B\mu_{i'}(A)\right)\vec{\nabla}_{i'}\frac{1}{R}\\ -\left[\mu_{i'}(A)\mu_{j'}(B)-\sum_\alpha e_\alpha^A Q_{i'j'}(B)-\sum_\beta e_\beta^B Q_{i'j'}(A)\right]\\ \times\vec{\nabla}_{i'}\vec{\nabla}_{j'}\frac{1}{R}+\cdots\end{array}\right\}|E_0^A, E_0^B\rangle. \tag{3.5.3}$$

Evaluating the matrix elements, it is seen that the expectation value over e_α^A with states $|E_0^A\rangle$ and $|E_m^A\rangle$ vanishes, because the charge is a c-number. Thus,

$$\begin{aligned}E_{\text{induction}}(A) &= -\sum_{m\neq 0}\frac{1}{(4\pi\varepsilon_0)^2}\left(\sum_\beta e_\beta^B\vec{\nabla}_i\frac{1}{R}-\mu_j^{00}(B)\vec{\nabla}_i\vec{\nabla}_j\frac{1}{R}+\cdots\right)\\ &\quad\times\frac{\langle E_0^A|\mu_i(A)|E_m^A\rangle\langle E_m^A|\mu_{i'}(A)|E_0^A\rangle}{E_m^A-E_0^A}\\ &\quad\times\left(\sum_\beta e_\beta^B\vec{\nabla}_{i'}\frac{1}{R}-\mu_{j'}^{00}(B)\vec{\nabla}_{i'}\vec{\nabla}_{j'}\frac{1}{R}+\cdots\right)\\ &= -\frac{1}{2}\frac{1}{(4\pi\varepsilon_0)^2}\left(\sum_\beta e_\beta^B\vec{\nabla}_i\frac{1}{R}-\mu_j^{00}(B)\vec{\nabla}_i\vec{\nabla}_j\frac{1}{R}+\cdots\right)\\ &\quad\times\alpha_{ii'}(A;0)\left(\sum_\beta e_\beta^B\vec{\nabla}_{i'}\frac{1}{R}-\mu_{j'}^{00}(B)\vec{\nabla}_{i'}\vec{\nabla}_{j'}\frac{1}{R}+\cdots\right),\end{aligned} \tag{3.5.4}$$

where $\mu_i^{00}(\xi)$ is the ith Cartesian component of the ground-state permanent electric dipole moment of molecule ξ and $\alpha_{ii'}(\xi;0)$ is the static electric dipole polarizability tensor,

$$\alpha_{ii'}(\xi;0) = 2\sum_{m\neq 0}\frac{\mu_i^{0m}(\xi)\mu_{i'}^{m0}(\xi)}{E_m^{\xi}-E_0^{\xi}}, \tag{3.5.5}$$

in which $\mu_i^{0m}(\xi)$ is the $0m$th matrix element of the transition electric dipole moment operator. Noting that the term within parentheses of (3.5.4) is the static electric field felt by A due to molecule B,

$$\frac{1}{4\pi\varepsilon_0}\left(\sum_{\beta} e_\beta^B \vec{\nabla}_i - \mu_j^{00}(B)\vec{\nabla}_i\vec{\nabla}_j + Q_{jk}^{00}(B)\vec{\nabla}_i\vec{\nabla}_j\vec{\nabla}_k + \cdots\right)\frac{1}{R} = E_i(B;0), \tag{3.5.6}$$

the induction energy of molecule A is then

$$E_{\text{induction}}(A) = -\frac{1}{2}\alpha_{ii'}(A;0)E_i(B;0)E_{i'}(B;0), \tag{3.5.7}$$

correct up to the leading electric dipole approximation. Higher order terms may be obtained in a similar manner, giving

$$E_{\text{induction}}(A) = -\frac{1}{2}\alpha_{ij}(A;0)E_i(B;0)E_j(B;0) - A_{ijk}(A;0)E_i(B;0)\vec{\nabla}_j E_k(B;0)$$
$$-\frac{1}{2}\Theta_{ijkl}(A;0)\vec{\nabla}_i E_j(B;0)\vec{\nabla}_k E_l(B;0) - \cdots, \tag{3.5.8}$$

where $A_{ijk}(\xi;0)$ is the static mixed electric dipole–quadrupole polarizability, $\Theta_{ijkl}(\xi;0)$ is the static pure electric quadrupole polarizability tensor, and so on, analogous to expression (3.5.5) for the static electric dipole polarizability. Terms nonlinear in the electric field may also be included in a systematic manner, giving rise to higher order and higher multipole susceptibilities, such as the electric dipole first and second hyperpolarizability tensors $\beta_{ijk}(\xi;0)$ and $\gamma_{ijkl}(\xi;0)$, and electric dipole–dipole–quadrupole hyperpolarizability $\beta_{ijkl}(\xi;0)$. An expression similar to (3.5.8) holds for the induction energy of molecule B, $E_{\text{induction}}(B)$.

3.6 DISPERSION FORCES

Dispersion forces are ever present between all interacting atomic and molecular systems. They are purely quantum mechanical in origin and arise from the coupling of the fluctuations in charge distribution at each center due to motion of electrons. For the interaction of two neutral nonpolar molecules, the dispersion force is the only force in effect. The dispersion energy shift is also known as the induced multipole-induced multipole interaction, as the coupling is mediated by the temporary distortions in the electronic charge distribution in one species inducing a similar change in the charge density of the second molecule, leading to a transient moment being induced there, with coupling occurring between these induced moments. For a pair of molecules in the ground electronic state, the dispersion energy shift is always attractive.

Like the induction energy, the contribution to the intermolecular interaction energy arising from dispersion forces may be obtained from the second-order correction term to the perturbed energy. But this time, the matrix elements include excited states of both A and B simultaneously. Hence, from the last term written explicitly in (3.4.16), the dispersion energy shift is derived from

$$E_{\text{dispersion}} = -\sum_{m\neq 0}\sum_{n\neq 0} \frac{\langle E_0^B, E_0^A|V(A,B)|E_m^A, E_n^B\rangle\langle E_n^B, E_m^A|V(A,B)|E_0^A, E_0^B\rangle}{(E_m^A - E_0^A) + (E_n^B - E_0^B)}. \tag{3.6.1}$$

On substituting for the perturbation operator $V(A,B)$ from (3.4.8), it is seen that the first nonvanishing term is the electric dipole–electric dipole interaction, followed by the electric dipole–quadrupole, electric quadrupole–quadrupole, electric dipole–octupole terms. For instance, the leading dipole–dipole term is given by

$$\Delta E_{\text{dispersion}}^{\text{d-d}} = -\frac{1}{(4\pi\varepsilon_0 R^3)^2}\sum_{m\neq 0}\sum_{n\neq 0}\frac{\mu_i^{0m}(A)\mu_{i'}^{m0}(A)\mu_j^{0n}(B)\mu_{j'}^{n0}(B)}{(E_m^A - E_0^A) + (E_n^B - E_0^B)} \times (\delta_{ij} - 3\hat{R}_i\hat{R}_j)(\delta_{i'j'} - 3\hat{R}_{i'}\hat{R}_{j'}), \tag{3.6.2}$$

which is the familiar R^{-6} dependent London (1930) dispersion energy shift between a pair of anisotropic electric dipole polarizable molecules. Rotational averaging using the result $\langle \mu_i^{0s}(\xi)\mu_j^{s0}(\xi)\rangle = (1/3)\delta_{ij}|\vec{\mu}^{0s}(\xi)|^2$ yields the recognizable form

$$\Delta E_{\text{dispersion}}^{\text{d-d}} = -\frac{1}{24\pi^2\varepsilon_0^2 R^6}\sum_{m\neq 0}\sum_{n\neq 0}\frac{|\vec{\mu}^{0m}(A)|^2|\vec{\mu}^{n0}(B)|^2}{(E_m^A - E_0^A) + (E_n^B - E_0^B)}. \tag{3.6.3}$$

Since the perturbation operator coupling the two molecules involves an interaction that is instantaneous, no account is taken of the effect of the finite speed of propagation of electromagnetic influences. This deficiency of the semiclassical treatment is remedied in the quantum electrodynamical description of dispersion forces, which is examined fully in Chapter 5. In that chapter, dispersion energy shifts are also calculated between excited molecules, along with contributions arising from higher multipole moments.

CHAPTER 4

RESONANT TRANSFER OF ENERGY

It is neither the point in space, nor the instant in time, at which something happens that has physical reality, but only the event itself.

—A. Einstein, *The Meaning of Relativity*,
Methuen and Co. Ltd., London, 1946, p. 29.

4.1 INTRODUCTION

From the viewpoint of molecular quantum electrodynamics, one of the simplest intermolecular interactions, at least conceptually, is the resonant exchange of energy between a pair of entities A and B, which may be atoms, molecules, chromophores, functional units, and others. This process corresponds to the transfer of energy—typically electronic and/or vibrational energy—resonantly from a species A, located at $\vec{R}_A$, which is initially pre-excited to some quantum state $|n\rangle$ with energy E_n, to a body B, situated at $\vec{R}_B$, which is in the ground electronic state $|0\rangle$ at some initial time $t=0$, but it acquires energy E_n and becomes excited to state $|n\rangle$, with A now decaying to the ground state. Migration of energy between the pair may be represented by the nonchemical equation

$$A^* + B \rightarrow A + B^*, \tag{4.1.1}$$

where the asterisk denotes the localization of excitation energy. In what follows, A and B are taken to be chemically equivalent, although the treatment given is general enough to be applicable to nonidentical A and B so long as they both have overlapping energy spectra.

Because of its fundamental nature, this system has been the topic of considerable study in two distinct but related contexts. One has been its adoption as a prototype to test the foundations of quantum mechanics and measurement theory. A second and equally important feature of the dynamics exemplified by this system is the role it played in helping to elucidate the mechanism underlying the resonant exchange of energy between two particles. This aspect was originally treated by Förster (1948), whose quantum mechanical calculation with dipolar coupling resulted in the transfer rate exhibiting an inverse sixth power dependence on separation distance. This dependence on donor–acceptor separation applied to distances large enough so that there is no overlap of molecular charge distributions associated with each center, but which is short enough so that the coupling may be viewed as occurring instantaneously between the two. At very short separations, a contribution to the rate also arises from a Dexter (1953) type of direct and exchange energy term. This has been treated previously, but will not be considered henceforth, since transfer rates will be computed for pair separation distances $R = |\vec{R}_B - \vec{R}_A|$ outside the region of overlap of molecular wavefunctions.

It is well known that intermolecular interactions are electromagnetic in origin. Therefore, at sufficiently large donor–acceptor separations, the finite speed of propagation of electromagnetic influences must be correctly accounted for. In this regard, molecular quantum electrodynamics, which automatically allows the effects of retardation, has been employed with striking success in the study of intermolecular forces. The first applications of this formalism to resonant energy transfer were carried out by McLone and Power (1964) and Avery (1966). They showed that at separations large relative to characteristic molecular transition wavelengths, the dependence of the transfer rate varied as R^{-2}, as expected from classical considerations. These and subsequent efforts (Andrews and Sherborne, 1987; Andrews, 1989; Craig and Thirunamachandran, 1989; Daniels et al., 2003; Salam, 2005a) have led to a unified theory of resonance energy transfer applicable to all separation distances beyond the region of orbital overlap. In this unified description, the interaction is viewed as being mediated by the exchange of a virtual photon, which carries energy from the donor moiety to the acceptor species. Time–energy uncertainty enables the creation of such a photon—which is undetectable—from the electromagnetic vacuum. From the general expression valid for all R, the short- and long-range limits of the transfer rate follow straightforwardly as asymptotic limits. The near-zone asymptote reproduces the R^{-6} Förster rate, which due to its original derivation using electrostatic dipolar coupling is termed the radiationless transfer mechanism. At the other extreme, the mechanism is described as radiative, yielding an inverse square law. In this case, due to the significant donor–acceptor separation, the propagated photon becomes ever more "real" in character, the exchange of excitation corresponding to the uncorrelated events of emission of light by the donor followed by photon absorption by the acceptor. Indeed, resonance energy transfer is one of the most elementary processes that can involve virtual photon exchange. Another example is the dispersion potential, the interaction of two neutral, nonpolar ground-state molecules, which is interpreted on this basis as arising due to the exchange of two virtual photons, and will be studied in detail in Chapter 5.

Despite its fundamental status as a prototypical system for the study of energy transfer, the donor–acceptor model has been versatile enough to be applied to a variety of chemical and physical systems in which migration of energy occurs (Scholes, 2003). These range from simple bimolecular systems to complexes containing multiple chromophores and other large macromolecular aggregates and include, but are not limited to, phenomena such as Dicke superradiance, the harvesting of light in complexes possessing the photosynthetic unit or other photosensitive centers, the transport of excitation in molecular and ionic crystals via the quasi particle called the exciton, intramolecular resonance energy transfer within dendrimers, and exchange of excitation in nanocomposite materials and photoactive devices such as organic light-emitting diodes.

The present chapter is organized as follows. Section 4.2 details the calculation of the matrix element using standard diagrammatic perturbation theory from which the transfer rate is computed using the Fermi golden rule. Its asymptotic forms at short and large separation distances are examined with a view to understanding the mechanism of energy transfer in play at these extremes of separation. Energy transfer between optically active molecules is then investigated. A number of interesting new features are found to occur, the most important being the discriminatory nature of the exchanged energy. It is then shown how an emitter–absorber model may be used to calculate transfer rates. In Section 4.6, the electric and magnetic displacement fields computed in Chapter 2 are used in a response theory formalism to readily evaluate the matrix element for migration of excitation energy, along with the resonant dipole–dipole coupling tensor. Section 4.7 is devoted to time-dependent energy transfer and questions of causality and Section 4.8 focuses on the proof demonstrating that exchange is causal to all orders in perturbation theory.

4.2 DIAGRAMMATIC PERTURBATION THEORY

The matrix element for the resonant transfer of excitation energy between two molecules is first calculated using diagrammatic perturbation theory methods. Consider a system comprising species A, which at $t=0$ is in excited electronic state $|n\rangle$, with energy E_n^A, and an acceptor B that is initially in the ground electronic state $|0\rangle$, with energy E_0^B, between which energy is exchanged. From (1.7.11), the quantum electrodynamical Hamiltonian operator is given by

$$H = H_{\text{mol}}(A) + H_{\text{mol}}(B) + H_{\text{rad}} + H_{\text{int}}(A) + H_{\text{int}}(B), \tag{4.2.1}$$

which forms the starting point for the calculation. The first two terms denote molecular Hamiltonians $H_{\text{mol}}(\xi)$, $\xi = A, B$, and are Schrödinger equations familiar from the nonrelativistic Born–Oppenheimer approximation of molecular quantum mechanics. Note that the Hamiltonian for the radiation field H_{rad} appears explicitly in the specification of the system, being treated on the same footing as matter in the quantum electrodynamical formalism. The final two terms of equation (4.2.1) describe the coupling of radiation and matter, and they may be viewed as a perturbation on the whole system if the interaction terms are assumed to be small relative to intramolecular Coulomb energies. Hence, the total Hamiltonian (4.2.1) can be separated into a sum of unperturbed and perturbation Hamiltonians H_0 and H_{int}, respectively

$$H = H_0 + H_{\text{int}}, \tag{4.2.2}$$

where

$$H_0 = \sum_{\xi=A,B} H_{\text{mol}}(\xi) + H_{\text{rad}} \tag{4.2.3}$$

and

$$H_{\text{int}} = \sum_{\xi=A,B} H_{\text{int}}(\xi). \tag{4.2.4}$$

From (4.2.3), the base states are seen to be product molecule and field states corresponding to the eigenstates of H_0, namely, the energy of species ξ and the occupation number for the electromagnetic field, the latter quantifying the number of photons. When examining interactions between particles, the preferred choice of quantum electrodynamical formulation to employ is the multipolar version of the theory. Notable among the advantages mentioned in the first two chapters are that molecules couple directly to the radiation field through their multipole moments with no two-, three-, and multicenter terms showing up in the coupling Hamiltonian. All interactions between molecules occur via the exchange of transverse photons that propagate at the speed of light. This description is also appropriate for the present scenario in which there is zero overlap of charge clouds associated with each center and the pair separation distance is large relative to the constituent particle center of mass distances within each body.

For transfer of energy between neutral electric dipole systems, the leading term of the expansion for the electric multipole series is sufficient for the interaction Hamiltonian, namely, the electric dipole approximated form

$$H_{\text{int}} = -\varepsilon_0^{-1}\vec{\mu}(A)\cdot\vec{d}^{\perp}(\vec{R}_A) - \varepsilon_0^{-1}\vec{\mu}(B)\cdot\vec{d}^{\perp}(\vec{R}_B), \tag{4.2.5}$$

where $\vec{\mu}(\xi) = -e(\vec{q}(\xi) - \vec{R}_\xi)$ is the electric dipole moment operator of particle ξ positioned at $\vec{R}_\xi$.

From the viewpoint of quantum electrodynamics, migration of energy may be pictured as arising from the exchange of a single virtual photon between the pair. It may be represented by the two time-ordered diagrams of Fig. 4.1. Time increases vertically upward in these graphs, with a solid line indicating the state of the electron. A wavy, or occasionally dashed line, denotes a photon (real or virtual), which is further specified by its mode,

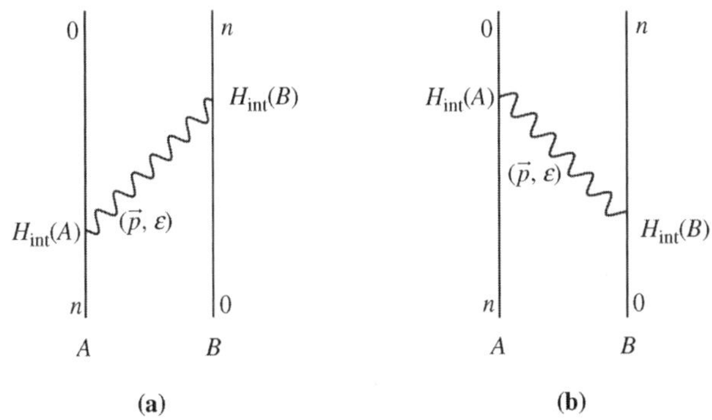

FIGURE 4.1 Time-ordered diagrams for resonant transfer of energy.

with normally only changes in the radiation field being shown. The intersection of a wavy and solid line depicts the coupling of radiation and matter via the relevant term or terms of the interaction Hamiltonian, and is called an interaction vertex. That only two diagrams contribute to the matrix element for energy transfer reflects the two directions in which the virtual photon may propagate between the pair, crossing from $A \leftrightarrows B$. The presence of two electron–photon coupling vertices in each time ordering means that the leading contribution to the matrix element is of second order in H_{int}. The appropriate term from the perturbation theory expansion (1.9.28) is

$$M_{fi} = \sum_I \frac{\langle f|H_{\text{int}}|I\rangle\langle I|H_{\text{int}}|i\rangle}{E_{iI}}, \tag{4.2.6}$$

where $|i\rangle$, $|f\rangle$, and $|I\rangle$ are initial, final, and intermediate states of the total system and are easily read off from the respective Feynman diagram by reading horizontally across that particular graph, with the sum in (4.2.6) executed over all intermediate states that link $|i\rangle$ to $|f\rangle$, with the denominator, $E_{iI} = E_i - E_I$, corresponding to differences in energy between initial and intermediate states. According to Feynman's rules (Feynman 1949a, 1949b), all topologically distinct diagrams that connect the same initial and final states of the system contribute to the matrix element or energy shift, with each time ordering corresponding in a direct one-to-one mapping to a specific term in time-dependent perturbation theory. Hence, the drawing of time-ordered diagrams greatly facilitates the perturbation theory computation of the probability amplitude, and this becomes especially so when higher order processes are tackled in subsequent applications.

For use in formula (4.2.6), the initial and final states describing transfer of energy are given by

$$|i\rangle = \left|E_n^A, E_0^B; 0(\vec{p}, \varepsilon)\right\rangle \tag{4.2.7a}$$

and

$$|f\rangle = \left|E_0^A, E_n^B; 0(\vec{p}, \varepsilon)\right\rangle. \tag{4.2.7b}$$

The former represents excitation energy E_n^A localized on unit A, with B in the electronic ground state with energy E_0^B. In (4.2.7b), A is seen to be in the ground state with species B now excited to state $|n\rangle$, with energy E_n^B. Note that a state of the radiation field without photons characterizes both the initial and final specifications of the system. Two types of intermediate state link $|i\rangle$ to $|f\rangle$, and they are readily written down from Fig. 4.1 as

$$|I_a\rangle = |E_0^A, E_0^B; 1(\vec{p}, \varepsilon)\rangle \tag{4.2.8a}$$

and

$$|I_b\rangle = |E_n^A, E_n^B; 1(\vec{p}, \varepsilon)\rangle. \tag{4.2.8b}$$

In both of these intermediate states, one virtual photon is present, whose mode is designated by $(\vec{p}, \varepsilon)$. In the first intermediate state obtained from the leftmost graph shown in Fig. 4.1a, both moieties are in the ground state, A losing its excitation as a result of virtual emission, while both species are excited in the second intermediate state derived from the second graph drawn in Fig. 4.1b. Evaluating the two contributions to (4.2.6) using (4.2.5), (4.2.7), and (4.2.8) produces

$$M_{fi} = \sum_{\vec{p},\varepsilon} \left(\frac{\hbar cp}{2\varepsilon_0 V}\right) \mu_i^{0n}(A)\mu_j^{n0}(B) \times \left\{ \bar{e}_i^{(\varepsilon)}(\vec{p}) e_j^{(\varepsilon)}(\vec{p}) \frac{e^{i\vec{p}\cdot\vec{R}}}{(E_{n0}^A - \hbar cp)} + e_i^{(\varepsilon)}(\vec{p}) \bar{e}_j^{(\varepsilon)}(\vec{p}) \frac{e^{-i\vec{p}\cdot\vec{R}}}{-(E_{n0}^B + \hbar cp)} \right\}, \tag{4.2.9}$$

where the internuclear separation distance $\vec{R} = \vec{R}_B - \vec{R}_A$. Because the virtual photon traversing the pair is emitted and subsequently absorbed at either center, properties describing its mode behavior must be summed over. These include its polarization and its wavevector. The former is carried out using identity (1.4.56), while the latter is converted to an integral via the prescription

$$\frac{1}{V}\sum_{\vec{p}} \to \frac{1}{(2\pi)^3}\int d^3\vec{p}. \tag{4.2.10}$$

Substituting these relations along with $E_{n0}^{A/B} = \hbar ck$ for the transition energy in either species gives for (4.2.9) the expression

$$\frac{1}{16\pi^3\varepsilon_0}\mu_i^{0n}(A)\mu_j^{n0}(B)\int p(\delta_{ij} - \hat{p}_i\hat{p}_j)\left\{\frac{e^{i\vec{p}\cdot\vec{R}}}{(k-p)} + \frac{e^{-i\vec{p}\cdot\vec{R}}}{-(k+p)}\right\} d^3\vec{p}. \tag{4.2.11}$$

To facilitate evaluation of the angular integral, the volume element is written in terms of spherical polar coordinates as $d^3\vec{p} = p^2 dp d\Omega$. Making use of the result

$$\frac{1}{4\pi}\int(\delta_{ij}-\hat{p}_i\hat{p}_j)e^{\pm i\vec{p}\cdot\vec{R}}d\Omega = \frac{1}{p^3}(-\vec{\nabla}^2\delta_{ij}+\vec{\nabla}_i\vec{\nabla}_j)\frac{\sin pR}{R}, \qquad (4.2.12)$$

(4.2.11) becomes

$$\frac{1}{4\pi^2\varepsilon_0}\mu_i^{0n}(A)\mu_j^{n0}(B)(-\vec{\nabla}^2\delta_{ij}+\vec{\nabla}_i\vec{\nabla}_j)\frac{1}{R}\int\left\{\frac{1}{k-p}+\frac{1}{-(k+p)}\right\}\sin pR\,dp. \qquad (4.2.13)$$

One method of solution of the wavevector integral occurring in (4.2.13) makes use of special functions and circumvents the need for integration in the complex plane (Daniels et al., 2003). After extending the limits of integration from 0 to ∞ to $-\infty$ to ∞, the solution of the Green's function is

$$\frac{1}{2R}\int_{-\infty}^{\infty}\left\{\frac{1}{(k-p)}+\frac{1}{-(k+p)}\right\}\sin pR\,dp = -\frac{\pi}{R}e^{\mp ikR}, \qquad (4.2.14)$$

and it is seen that two equally valid solutions to the wavevetcor integral, and consequently the resonant interaction tensor, emerge. Before going on to write the form of the matrix element in terms of the solution (4.2.14), it is briefly remarked that identical results may be obtained using conventional integration techniques by displacing the pole in (4.2.13) by introducing $\pm i\eta$ to the resonant denominator and employing the identity $(x \pm i\eta)^{-1} = (\mathrm{PV}/x) \mp i\pi\delta(x)$, where *PV* denotes the Cauchy principal value.

Substituting (4.2.14) into (4.2.13) yields for the matrix element the result

$$M_{fi} = -\frac{1}{4\pi\varepsilon_0}\mu_i^{0n}(A)\mu_j^{n0}(B)(-\vec{\nabla}^2\delta_{ij}+\vec{\nabla}_i\vec{\nabla}_j)\frac{e^{\mp ikR}}{R}, \qquad (4.2.15)$$

which can be written succinctly as

$$M_{fi} = \mu_i^{0n}(A)\mu_j^{n0}(B)V_{ij}^{\pm}(k,\vec{R}), \qquad (4.2.16)$$

where $V_{ij}^{\pm}(k,\vec{R})$ is the complex retarded resonant dipole–dipole coupling tensor defined by

$$V_{ij}^{\pm}(k,\vec{R}) = -\frac{1}{4\pi\varepsilon_0}\left(-\vec{\nabla}^2\delta_{ij}+\vec{\nabla}_i\vec{\nabla}_j\right)\frac{e^{\mp ikR}}{R}$$
$$= \frac{1}{4\pi\varepsilon_0 R^3}\left[(\delta_{ij}-3\hat{R}_i\hat{R}_j)(1\pm ikR)-(\delta_{ij}-\hat{R}_i\hat{R}_j)k^2R^2\right]e^{\mp ikR}, \tag{4.2.17}$$

with the second line of the above relation being obtained after performing the tensor calculus. Both choices of sign appearing in the coupling tensor (4.2.17) are permissible, although the lower sign is frequently selected.

The behavior of the transfer matrix element as a function of separation distance is governed by the form of $V_{ij}^{\pm}(k,\vec{R})$. In the near zone, where $kR \ll 1$, the dominant term of (4.2.17) is the first, yielding the matrix element

$$M_{fi}^{\mathrm{NZ}} \approx \frac{\mu_i^{0n}(A)\mu_j^{n0}(B)}{4\pi\varepsilon_0 R^3}(\delta_{ij}-3\hat{R}_i\hat{R}_j), \tag{4.2.18}$$

which is recognizable as the static dipolar coupling interaction with characteristic inverse cube dependence on R. At intermediate separations, the second term of (4.2.17) is important, varying as R^{-2}. The long-range character of $V_{ij}^{\pm}(k,\vec{R})$ is determined by the third term of (4.2.17), exhibiting R^{-1} separation distance dependence, with this term being purely transverse with respect to $\vec{R}$ due to the prefactor $(\delta_{ij}-\hat{R}_i\hat{R}_j)$.

The exchange of excitation energy between the A–B pair is measured through a transfer rate, which may be readily evaluated from the matrix element (4.2.16) by using the Fermi golden rule (1.9.33), being proportional to the modulus square of the matrix element. Thus, the transfer rate is

$$\Gamma = \frac{2\pi\rho_f}{\hbar}\mu_i^{0n}(A)\mu_{i'}^{n0}(A)\mu_j^{0n}(B)\mu_{j'}^{n0}(B)V_{ij}^{\pm}(k,\vec{R})\bar{V}_{i'j'}^{\pm}(k,\vec{R}) \tag{4.2.19}$$

expressed in terms of the transition dipole moment of each species and the resonant interaction tensor coupling the two molecules. The result (4.2.19) holds for dipole moments with specific orientations. Often donor and acceptor species are completely randomly oriented, as in the fluid phase, in which case (4.2.19) is orientationally averaged. After employing result (B.4) of Appendix B, the rate takes the form

$$\langle\Gamma\rangle = \frac{\rho_f}{36\pi\hbar\varepsilon_0^2 R^6}|\vec{\mu}^{0n}(A)|^2|\vec{\mu}^{n0}(B)|^2\left[k^4R^4+k^2R^2+3\right]. \tag{4.2.20}$$

It is noteworthy that the ambiguity in the choice of sign for the resonant coupling tensor appearing in the matrix element has no effect on the rate since the coupling tensor is multiplied by its complex conjugate when evaluating Γ. The asymptotic limits of the transfer rate at short and long separation distances are readily obtained from the result (4.2.20) valid for all R beyond wavefunction overlap. In the near zone, the kR independent term within square brackets dominates, yielding

$$\langle \Gamma^{\mathrm{NZ}} \rangle = \frac{\rho_f}{12\pi\hbar\varepsilon_0^2 R^6} |\vec{\mu}^{0n}(A)|^2 |\vec{\mu}^{n0}(B)|^2, \tag{4.2.21}$$

which has the familiar R^{-6} Förster-type dependence on separation. This asymptote is commonly interpreted as the radiationless exchange mechanism as it arises from static dipolar coupling in which propagation of the electromagnetic signal between the pair is viewed as occurring instantaneously. At the other extreme of separation, $kR \gg 1$, corresponding to the long-range or far-zone limit,

$$\langle \Gamma^{\mathrm{FZ}} \rangle = \frac{k^4 \rho_f}{36\pi\hbar\varepsilon_0^2 R^2} |\vec{\mu}^{0n}(A)|^2 |\vec{\mu}^{n0}(B)|^2, \tag{4.2.22}$$

which exhibits an inverse square dependence on R. At large separations, the transfer of energy is described as radiative since the propagated virtual photon acquires real character, the mechanism for migration of energy in this range being understood to be the result of two separate events of spontaneous emission by the excited donor molecule followed by the absorption of radiation by the unexcited acceptor body. Single virtual photon exchange between an excited and unexcited pair of molecules allows a unified description of resonant transfer of energy to be given within the framework of molecular quantum electrodynamics.

The exchange of energy may be treated with the familiar second-order secular perturbation theory in which the stationary states of the excited A–B pair, $(1/\sqrt{2})(|E_n^A, E_0^B\rangle \pm |E_0^A, E_n^B\rangle)$, are obtained and whose use leads to a zero net transfer rate since these states decay by spontaneous emission by the coupled pair. Migration of excitation can, nevertheless, take place in this case if the two species are identical and transfer is rapid relative to collision or fluorescence-induced decay, which is possible if A and B are close together. For large R, or for reduced transition strength, or for times too short to enable measurement of a stationary state, as well as for a nonidentical A–B pair, excitation energy transfer must be viewed and computed as a time-dependent process. There are situations, however, when even the time-dependent picture is inadequate. This is the case when

A and B are considered as an isolated system and the initial (4.2.7a) and final states (4.2.7b) are taken to be sharp. Then the Fermi golden rule cannot strictly be used to calculate a transfer rate due to the lack of a density of final molecular states. Nonetheless, the method is a useful one in that it allows the basic features of the process to be elucidated and understood.

4.3 STATE SEQUENCE DIAGRAM REPRESENTATION

In this section, it is shown how state sequence diagrams may be used instead of time-ordered graphs for the perturbation theory computation of the matrix element for the resonant transfer of energy. This example serves as an instructive one on which to apply the formal construction scheme presented in Section 1.10, even though the state sequence picture may be drawn directly from the Feynman diagrams for resonance energy transfer. This last fact is also the case for any particular process involving electron–photon coupling.

According to the conventional diagrammatic techniques within time-dependent perturbation theory, the most common being the time-ordered graph inspired by Feynman, the resonance energy transfer is interpreted as arising from the exchange of a single virtual photon between the pair, as illustrated in Fig. 4.1. To leading order, the matrix element is given by the second-order term in perturbation theory. Therefore, the hyperspace number n is equal to two in the present problem, and because the virtual photon creation and annihilation events are distinguishable, the orthonormal set of basis vectors from (1.10.1) is $I = \{1\vec{i}_1, 1\vec{i}_2\}$ and the number of distinct indices j is also two, corresponding to $j=0$ and 1. For these two values of j, the $C_j = 0, 1$. k, on the other hand, has values 0, 1, and 2 and since the base of the hyperspace dimension $B = c_j + 1 = 2$, the coordinate points (k, h) for the construction of the interaction plane network are easily computed, and they along with the pertinent vertex designation r_k^m according to (1.10.7) are given in Table 4.1. By convention, for a fixed value of k, the vertices

TABLE 4.1 Vertex Properties Associated with Two Unique Photonic Events

k	Vertex	Hyperspace Coordinate	Hyperspace Number (Base 2)	h (Base 10)	(k, h)
0	r_0^1	(0, 0)	00	0	(0, 0)
1	r_1^1	(1, 2)	10	2	(1, 2)
	r_1^2	(1, 1)	01	1	(1, 1)
2	r_2^1	(2, 3)	11	3	(2, 3)

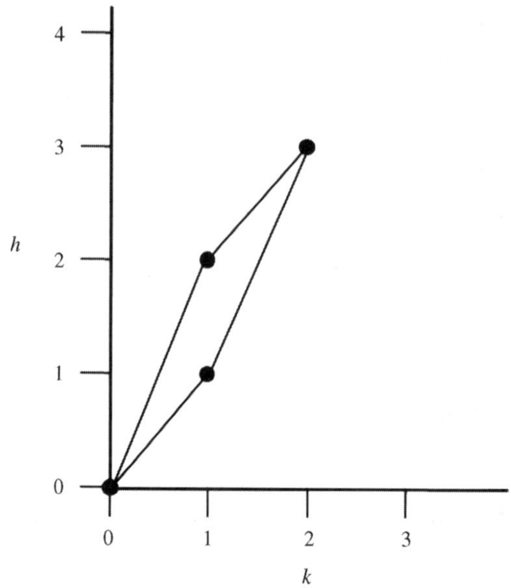

FIGURE 4.2 Interaction plane network for two distinguishable radiation–matter couplings.

are arranged in increasing order of m. Next the appropriate linkage rules given by (1.10.10) are applied, with the total number of paths obtained from (1.10.11) equal to $2!/1!1! = 2$. The structure coefficients ${}^{\{1,1\}}T_k^{2,2}$, $k = 0, 1, 2$ are found from the prescription (1.10.14) to be 1, 2, and 1, which correspond to the second row of Pascal's triangle. The net displayed in Fig. 4.2 is easily seen to follow, with the explicit coordinates given in Table 4.1. It is instructive to point out that the network map shown in Fig. 4.2 forms the basis for the construction of state sequence diagrams for all processes involving two unique radiation–matter interactions, after the appropriate representation of initial and final system states (Jenkins et al., 2002). Therefore, from the stencil shown, the state sequence diagrams for two-photon absorption from two different beams, emission of two photons of differing modes, and all linear forms of light scattering such as Rayleigh and Raman—all of which are unimolecular in origin, as well as for resonant migration of energy—a bimolecular process, may be readily constructed. The salient radiation–matter states and associated energies are then written down straightforwardly. A general state is given by

$$|r_k^m\rangle = \prod_{\xi} |\xi_{r_k^m}\rangle |\mathrm{rad}_{r_k^m}\rangle \equiv |\mathrm{mat}_{r_k^m}\rangle |\mathrm{rad}_{r_k^m}\rangle \equiv |\mathrm{mat}_{r_k^m}; \mathrm{rad}_{r_k^m}\rangle, \qquad (4.3.1)$$

with the relevant energy comprising a sum of radiation and matter energies.

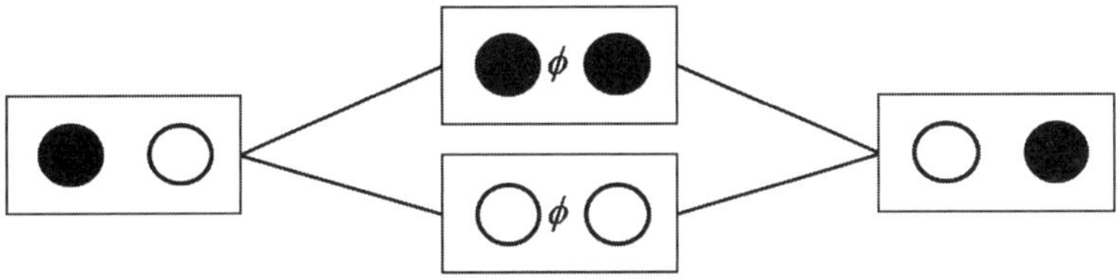

FIGURE 4.3 State sequence diagram for resonant transfer of energy.

The state sequence diagram for resonance excitation transfer is shown in Fig. 4.3. In these pictorial representations, time increases from left to right. The state of the system at a particular time instant at which an event corresponding to an interaction vertex occurs is represented by a box, the left-hand most corresponding to the initial state and the right-hand most to the final state of the system. Between these two extremes in time are drawn the intermediate states that connect $|i\rangle$ to $|f\rangle$. Within each box are depicted the state of the material particle(s) as well as changes in the state of the radiation field. The former are denoted by circles, with one or more arranged horizontally in a line corresponding to species A, B, C, and so on. An open circle designates the species to be in the ground electronic state, while a circle that is filled or contains a letter labels the excited electronic state of the specific unit. The appearance of ϕ in a state box denotes the presence of a photon in the system—either real or virtual, with additional labels being used to differentiate between photons differing in mode character. Hence, in Fig. 4.3, the leftmost box corresponds to state $|i\rangle$ given by (4.2.7a) with the black circle showing that A is initially excited and the open circle alongside corresponds to acceptor species B, which is initially unexcited. No photons—real or virtual—are present, hence an absence of the label ϕ. In the rightmost box of Fig. 4.3, the open circle corresponds to the de-excitation of A to its lowest energy state, and the circle denoting B is filled, indicating transfer of energy from the donor species A. It is the state sequence representation of the state (4.2.7b). The intermediate states (4.2.8a and 4.2.8b) resulting from the two possible time orderings and site of virtual photon creation are illustrated in the center of the figure. The lower box shows that both species are in the ground electronic state, with one virtual photon present after virtual emission by A, the label ϕ denoting the virtual photon is placed in between the two circles, and the latter depicting the two bodies. The upper box illustrates the situation in which both A and B are electronically excited with one virtual photon traversing between the pair, having been created at B. The lower pathway coincides with the time-ordered graph of Fig. 4.1a in which the virtual photon is emitted by A, while the upper path corresponds to creation of a virtual photon by B, which becomes excited in the process, and is illustrated by Feynman graph of Fig. 4.1b. Since the state sequence diagrams represent an alternate pictorial display of time-dependent perturbation theory methodology, the matrix element is calculated using (4.2.6) and yields (4.2.16), with no special computational benefit being gained by the use of state sequences in this case since the order of perturbation theory is low and the number of paths is only two. Additional features and limitations associated with state sequence representations of intermolecular interactions and virtual photon propagation will be detailed in Chapter 5, when retarded dispersion forces are examined.

4.4 ENERGY TRANSFER BETWEEN CHIRAL SYSTEMS

Thus far in this chapter, the leading contribution to excitation energy transfer between a pair of molecules, one of which is excited with the other in the ground electronic state, has been calculated. Only the first term of the multipolar form of interaction Hamiltonian, namely, the electric dipole coupling term, needed to be retained. In many systems, however, especially those of chemical and biological interest, the electric dipole approximation to the transfer rate is no longer sufficient as deviations from the Förster result are found to occur, and the contributions arising from the inclusion of higher multipole moments are required. This is pertinent for species in which the ratio of molecular or chromophore size to interparticle separation distance is sufficiently large. One such studied system is the interaction between the carotenoid S_1 state and chlorophyll (Scholes et al., 1997), where higher order multipole corrections are found to be significant at typical separations occurring in this light-harvesting complex. Another example, of broader applicability, is provided by chiral molecules, between which migration of energy also takes place.

Molecules that lack an improper axis of rotation are termed chiral. These entities exhibit a number of chiroptical properties, such as rotation of the plane of polarization of light and differential absorption, emission, and scattering of circularly polarized radiation. In addition, discriminatory effects occur when optically active species interact with each other, such as through resonance coupling or the dispersion force. Due to the low number of symmetry elements in a chiral body, normally restrictive selection rules applicable to electronic transitions are relaxed. Magnetic dipole, electric quadrupole, and higher order multipole transitions become allowed in addition to the leading electric dipole excitation process. When molecules possess no symmetry, belonging to the C_1 point group, for example, transitions are allowed simultaneously to all multipole orders.

The perturbation theory treatment of bimolecular resonant migration of energy presented in Section 4.2 is now extended by relaxing the electric dipole approximation and including magnetic dipole interaction terms, thereby making the treatment applicable to excitation energy exchange between optically active species (Craig and Thirunamachandran, 1998b). In this case, the total Hamiltonian is again given by equation (4.2.1), but the interaction Hamiltonian (4.2.5) is now modified by the addition of magnetic dipole couplings, and is of the form

$$\begin{aligned} H_{\text{int}}(A)+H_{\text{int}}(B) = & -\varepsilon_0^{-1}\vec{\mu}(A)\cdot\vec{d}^{\perp}(\vec{R}_A)-\vec{m}(A)\cdot\vec{b}(\vec{R}_A) \\ & -\varepsilon_0^{-1}\vec{\mu}(B)\cdot\vec{d}^{\perp}(\vec{R}_B)-\vec{m}(B)\cdot\vec{b}(\vec{R}_B), \end{aligned} \tag{4.4.1}$$

where $\vec{m}(\xi) = -(e/2m)(\vec{q}(\xi)-\vec{R}_\xi)\times\vec{p}$ is the magnetic dipole moment of species ξ located at $\vec{R}_\xi$, whose mass is m and linear momentum is $\vec{p}$, with $\vec{b}(\vec{r})$ the magnetic field operator.

As for pure electric dipole coupling, the matrix element for energy transfer between optically active molecules is evaluated using the second-order perturbation theory formula (4.2.6), in combination with the two time-ordered diagrams shown in Fig. 4.1, but with each interaction Hamiltonian term now a sum of two terms—containing an electric dipole and a magnetic dipole contribution as in (4.4.1). Initial, final, and intermediate states represented by (4.2.7) and (4.2.8) apply to the present case. Making use of the mode expansions for the microscopic displacement and magnetic fields, the matrix element is evaluated in the standard manner to yield

$$M_{fi} = \sum_{\vec{p},\varepsilon}\left(\frac{\hbar cp}{2\varepsilon_0 V}\right)\left\{\begin{aligned} & \left[\bar{e}_i^{(\varepsilon)}(\vec{p})\mu_i^{0n}(A)+\frac{1}{c}\bar{b}_i^{(\varepsilon)}(\vec{p})m_i^{0n}(A)\right] \\ & \quad\times\left[e_j^{(\varepsilon)}(\vec{p})\mu_j^{n0}(B)+\frac{1}{c}b_j^{(\varepsilon)}(\vec{p})m_j^{n0}(B)\right]\frac{e^{i\vec{p}\cdot\vec{R}}}{(E_{n0}-\hbar cp)} \\ & -\left[e_i^{(\varepsilon)}(\vec{p})\mu_i^{0n}(A)+\frac{1}{c}b_i^{(\varepsilon)}(\vec{p})m_i^{0n}(A)\right] \\ & \quad\times\left[\bar{e}_j^{(\varepsilon)}(\vec{p})\mu_j^{n0}(B)+\frac{1}{c}\bar{b}_j^{(\varepsilon)}(\vec{p})m_j^{n0}(B)\right]\frac{e^{-i\vec{p}\cdot\vec{R}}}{(E_{n0}+\hbar cp)} \end{aligned}\right\}. \tag{4.4.2}$$

Unsurprisingly, the leading term above, proportional to $\mu_i^{0n}(A)\mu_j^{n0}(B)$, is identical to matrix element (4.2.9) calculated within the electric dipole approximation, which led to result (4.2.16) in terms of the retarded resonant electric dipole–dipole tensor $V_{ij}^{\mp}(k,\vec{R})$. Recognizing from (1.4.58) that the sum over magnetic polarization vectors is identical to that arising from summation over electric polarization vectors, the term arising from the product of the transition magnetic dipole moments is the same as that for electric dipole coupling (4.2.11), with $\vec{\mu}^{0n}(\xi)$ replaced by $(1/c)\vec{m}^{0n}(\xi)$, $\xi = A,B$. After grouping together pure electric, pure magnetic, and the mixed terms and carrying out the polarization sum, the matrix element (4.4.2) becomes

$$M_{fi} = \sum_{\vec{p}} \left(\frac{\hbar cp}{2\varepsilon_0 V}\right) \left\{ \begin{array}{l} (\delta_{ij} - \hat{p}_i\hat{p}_j)\left[\mu_i^{0n}(A)\mu_j^{n0}(B) + \dfrac{1}{c^2} m_i^{0n}(A) m_j^{n0}(B)\right] \\ + \dfrac{1}{c}\varepsilon_{ijk}\hat{p}_k\left[\mu_i^{0n}(A) m_j^{n0}(B) + m_j^{0n}(A)\mu_i^{n0}(B)\right] \end{array} \right\}$$

$$\times \left\{ \frac{e^{i\vec{p}\cdot\vec{R}}}{(E_{n0} - \hbar cp)} + \frac{e^{-i\vec{p}\cdot\vec{R}}}{-(E_{n0} + \hbar cp)} \right\}. \tag{4.4.3}$$

Hence, the pure magnetic dipole coupling contribution to the matrix element and transfer rate for isotropic systems are written immediately from (4.2.16) and (4.2.20) as

$$M_{fi}^{\text{m-m}} = \frac{1}{c^2} m_i^{0n}(A) m_j^{n0}(B) V_{ij}^{\mp}(k,\vec{R}) \tag{4.4.4}$$

and

$$\langle \Gamma^{\text{m-m}} \rangle = \frac{\rho_f}{36\pi\hbar\varepsilon_0^2 c^2 R^6} |\vec{m}^{0n}(A)|^2 |\vec{m}^{n0}(B)|^2 \left[k^4R^4 + k^2R^2 + 3\right], \tag{4.4.5}$$

where $\hbar ck = E_{n0}$, $V_{ij}^{\mp}(k,\vec{R})$ is defined by (4.2.17), and the superscript m–m denotes the pure magnetic dipole contribution. Like its electric–electric (e–e) counterpart, the transfer rate (4.4.5) does not depend on the handedness of A or B. Since $\vec{m}$ is a factor of the fine structure constant smaller than $\vec{\mu}$, the corrections (4.4.4) and (4.4.5) are usually ignored. The cross-term between the e–e and m–m terms do, however, depend on the chirality of each species and is the source of one of the discriminatory contributions to the rate. It arises from the term proportional to

$\vec{\mu}(A)\vec{m}(A)\vec{\mu}(B)\vec{m}(B)+\vec{m}(A)\vec{\mu}(A)\vec{m}(B)\vec{\mu}(B)$ in (4.4.3). It is explicitly given by

$$\begin{aligned}\langle\Gamma^{\text{disc}}\rangle &= \frac{2\pi\rho_f}{\hbar}(\overline{M}_{fi}^{\text{e-e}}M_{fi}^{\text{m-m}}+M_{fi}^{\text{e-e}}\overline{M}_{fi}^{\text{m-m}}) \\ &= -\frac{\rho_f}{18\pi\hbar\varepsilon_0^2c^2R^6}|\vec{\mu}^{0n}(A)\cdot\vec{m}^{n0}(A)||\vec{\mu}^{0n}(B)\cdot\vec{m}^{n0}(B)|(k^4R^4+k^2R^2+3).\end{aligned} \tag{4.4.6}$$

The second source of discrimination originates from the second term within large braces of (4.4.3). Converting the $\vec{p}$-sum to an integral and performing the angular average using the result

$$\frac{1}{4\pi}\int\hat{p}_k e^{\pm i\vec{p}\cdot\vec{R}}\mathrm{d}\Omega = \mp\frac{i}{p}\vec{\nabla}_k\frac{\sin pR}{pR} = \mp i\left(\frac{\cos pR}{pR}-\frac{\sin pR}{p^2R^2}\right)\hat{R}_k, \tag{4.4.7}$$

the electric–magnetic contribution to the matrix element is

$$\begin{aligned}M_{fi}^{\text{e-m+m-e}} &= -\frac{ik}{2\pi^2\varepsilon_0 c}\varepsilon_{ijk}\hat{R}_k\left[\mu_i^{0n}(A)m_j^{n0}(B)+m_j^{0n}(A)\mu_i^{n0}(B)\right] \\ &\quad\times\int_0^\infty\frac{1}{k^2-p^2}\left(\frac{p^2\cos pR}{R}-\frac{p\sin pR}{R^2}\right)\mathrm{d}p,\end{aligned} \tag{4.4.8}$$

where $k=E_{n0}/\hbar c$. Again, contour integration or use of special functions can be employed to evaluate the integral whose result is

$$\int_0^\infty\frac{1}{k^2-p^2}\left(\frac{p^2\cos pR}{R}-\frac{p\sin pR}{R^2}\right)\mathrm{d}p = \frac{\pi}{2R^2}(1\pm ikR)e^{\mp ikR}, \tag{4.4.9}$$

yielding for (4.4.8),

$$M_{fi}^{\text{e-m+m-e}} = \left[\mu_i^{0n}(A)m_j^{n0}(B)+m_j^{0n}(A)\mu_i^{n0}(B)\right]U_{ij}^{\pm}(k,\vec{R}). \tag{4.4.10}$$

The retarded resonant interaction tensor $U_{ij}^{\pm}(k,\vec{R})$ coupling electric and magnetic dipoles is defined by

$$U_{ij}^{\pm}(k,\vec{R}) = -\frac{ik}{4\pi\varepsilon_0 c}\varepsilon_{ijk}\vec{\nabla}_k\frac{e^{\mp ikR}}{R} = \frac{1}{4\pi\varepsilon_0 cR^3}\varepsilon_{ijk}\hat{R}_k(ikR\mp k^2R^2)e^{\mp ikR}, \tag{4.4.11}$$

with factors that vary as R^{-1} and R^{-2}. From the form of $U_{ij}^{\mp}(k,\vec{R})$, it is apparent that the interaction (4.4.10) occurs through emission and absorption of a real photon of frequency $\omega = \omega_{n0} = E_{n0}/\hbar$, and is therefore completely dynamic and solely transverse in nature. This is true even in the near zone, since in the limit $k \to 0$, $U_{ij}^{\mp}(0,\vec{R}) \to 0$, as expected on physical grounds because static electric and magnetic dipoles do not couple. The coupling (4.4.10) is maximized for the configuration in which $\vec{\mu}(\xi)$, $\vec{m}(\xi')$, and $\hat{R}$ are orthogonal to one another. The transfer rate arising from (4.4.10), which is also discriminatory, is

$$\begin{aligned}\Gamma^{\text{disc}} &= \frac{2\pi\rho_f}{\hbar}|M_{fi}^{\text{e-m}+\text{m-e}}|^2 \\ &= \frac{2\pi\rho_f}{\hbar}\frac{k^2}{(4\pi\varepsilon_0 c)^2}\left[\mu_i^{0n}(A)m_j^{n0}(B) + m_j^{0n}(A)\mu_i^{n0}(B)\right] \\ &\times \left[\bar{\mu}_k^{0n}(A)\bar{m}_l^{n0}(B) + \bar{m}_l^{0n}(A)\bar{\mu}_k^{n0}(B)\right] \\ &\times \left(\varepsilon_{ijm}\vec{\nabla}_m\frac{\mathrm{e}^{\mp ikR}}{R}\right)\left(\varepsilon_{kln}\vec{\nabla}_n\frac{\mathrm{e}^{\pm ikR}}{R}\right),\end{aligned} \tag{4.4.12}$$

which after orientational averaging becomes

$$\langle\Gamma^{\text{disc}}\rangle = -\frac{8\pi\rho_f k^2}{9(4\pi\varepsilon_0 c)^2\hbar R^4}|\vec{\mu}^{0n}(A)\cdot\vec{m}^{n0}(A)||\vec{\mu}^{0n}(B)\cdot\vec{m}^{n0}(B)|(k^2R^2+1). \tag{4.4.13}$$

The total discriminatory transfer rate is given by the sum of (4.4.6) and (4.4.13),

$$\begin{aligned}\langle\Gamma^{\text{disc}}\rangle = &-\frac{\rho_f}{18\pi\hbar\varepsilon_0^2c^2R^6}|\vec{\mu}^{0n}(A)\cdot\vec{m}^{n0}(A)||\vec{\mu}^{0n}(B)\cdot\vec{m}^{n0}(B)| \\ &\times\left[2k^4R^4 + 2k^2R^2 + 3\right].\end{aligned} \tag{4.4.14}$$

It is customary to express the molecular factors appearing in the result (4.4.14) in terms of the optical rotatory strength tensor, defined to be

$$R^{0n}(\xi) = -i\vec{\mu}^{0n}(\xi)\cdot\vec{m}^{n0}(\xi), \tag{4.4.15}$$

so that

$$\langle \Gamma^{\text{disc}} \rangle = \frac{\rho_f}{18\pi\hbar\varepsilon_0^2 c^2 R^6} R^{0n}(A) R^{n0}(B) \left[2k^4R^4 + 2k^2R^2 + 3 \right]. \tag{4.4.16}$$

Because $\vec{\mu}$ is a polar vector and $\vec{m}$ is an axial quantity, their dot product produces a pseudoscalar R^{0n}, which changes sign when one enantiomer is replaced by its mirror image form. The total transfer rate (4.4.16) clearly depends on the handedness of each optical isomer, changing sign when one enantiomer is substituted by its antipode.

The limiting forms of the rate are readily obtainable from expression (4.4.16) after the usual approximations. In the near zone, where $kR \ll 1$, the rate reduces to

$$\left\langle \Gamma^{\text{disc}}_{\text{NZ}} \right\rangle = \frac{\rho_f}{6\pi\hbar\varepsilon_0^2 c^2 R^6} R^{0n}(A) R^{n0}(B), \tag{4.4.17}$$

exhibiting inverse sixth power dependence on intermolecular separation distance. It is of interest to note that this asymptote arises solely from the first contribution to the discriminatory transfer rate (4.4.6) due to the vanishing of (4.4.13) for small k ($k \to 0$). At the other separation extreme, $kR \gg 1$ and the rate is

$$\left\langle \Gamma^{\text{disc}}_{\text{FZ}} \right\rangle = \frac{\rho_f k^4}{9\pi\hbar\varepsilon_0^2 c^2 R^2} R^{0n}(A) R^{n0}(B), \tag{4.4.18}$$

which displays the expected inverse square dependence on R and is characteristic of migration of excitation energy being mediated by real photon emission and absorption.

4.5 EMITTER–ABSORBER MODEL

In the diagrammatic perturbation theory treatment of resonant transfer of energy, coupling between the pair, be they chiral or not, was understood to arise from the exchange of a single virtual photon. For the cases studied, the transfer rate was shown to display an inverse square dependence on interparticle separation distance in the limit of large R. The mechanism at play at this distance extreme was interpreted as occurring as a result of emission of a real photon by the excited donor followed by photon absorption by the unexcited receiver and viewed as two separate events.

The well-known transmitter–receiver or emitter–absorber model (Andrews and Sherborne, 1987; Andrews, 1989; Craig and Thirunamachandran, 1992), therefore, naturally lends itself to the description of radiative energy transfer. It involves picturing the donor as a source of radiation that spontaneously emits a photon, while the second body acts as a receiver entity, absorbing the propagated photon. Intermolecular coupling takes place via their common radiation intensity. This viewpoint is now applied to the long-range migration of energy between optically active molecules (Craig and Thirunamachandran, 1998b) from which the electric dipole-approximated result is easily obtained on letting the transition magnetic dipole moment term vanish.

Consider an excited chiral donor species A spontaneously emitting a circularly polarized photon of mode $(\vec{k}, L/R)$. The initial and final states are given by $|i\rangle = |E_n^A; 0(\vec{k}, L/R)\rangle$ and $|f\rangle = |E_0^A; 1(\vec{k}, L/R)\rangle$, with A initially in excited electronic state $|n\rangle$. The matrix element is easily calculated using the first-order time-dependent perturbation theory and the interaction Hamiltonian

$$H_{\text{int}}(A) = -\varepsilon_0^{-1}\vec{\mu}(A)\cdot\vec{d}^{\perp}(\vec{R}_A) - \vec{m}(A)\cdot\vec{b}(\vec{R}_A). \tag{4.5.1}$$

It is given by

$$M_{fi}^{\text{em}} = i\left(\frac{\hbar ck}{2\varepsilon_0 V}\right)^{1/2} \bar{e}_i^{(L/R)}(\vec{k})\left[\mu_i^{0n}(A) \pm \frac{i}{c}m_i^{0n}(A)\right]\mathrm{e}^{-i\vec{k}\cdot\vec{R}_A}, \tag{4.5.2}$$

where A is positioned at $\vec{R}_A$, the abbreviation em is used to signify emission, and use has been made of the identity

$$b_i^{(L/R)}(\vec{k}) = \mp i e_i^{(L/R)}(\vec{k}), \tag{4.5.3}$$

which may be verified on inserting the definition of the unit circularly polarized electric vector (1.4.12) into the transversality relation $\vec{b}^{(\lambda)}(\vec{k}) = \hat{k}\times\vec{e}^{(\lambda)}(\vec{k})$. Emission occurs with a given polarization in a cone of solid angle $\mathrm{d}\Omega$ centered around the direction of propagation. The rate is calculated from the Fermi golden rule,

$$\mathrm{d}\Gamma_{\text{em}} = \frac{2\pi}{\hbar}|M_{fi}|^2\rho_f, \tag{4.5.4}$$

where the number of levels per unit energy interval of the final state is denoted by ρ_f. Since the number of modes in volume V with wavevector lying between $\vec{k}$ and $\vec{k}+\mathrm{d}\vec{k}$ with energy in the range $\hbar ck$ and $\hbar c(k+\mathrm{d}k)$ within the cone is $V\mathrm{d}^3\vec{k}/(2\pi)^3 = (Vk^2\mathrm{d}k\,\mathrm{d}\Omega/(2\pi)^3)$, the density of final states is this quantity divided by $\hbar c\,\mathrm{d}k$, giving

$$\rho_f = \frac{Vk^2\,\mathrm{d}\Omega}{(2\pi)^3\hbar c}. \tag{4.5.5}$$

Substituting (4.5.2) and (4.5.5) into (4.5.4) produces

$$\mathrm{d}\Gamma_{\mathrm{em}}^{L/R}(\Omega) = \frac{k^3\,\mathrm{d}\Omega}{8\pi^2\varepsilon_0\hbar}\vec{e}^{(L/R)}(\vec{k})\bar{\vec{e}}^{(L/R)}(\vec{k})\left|\vec{\mu}^{0n}(A) \pm \frac{i}{c}\vec{m}^{0n}(A)\right|^2, \tag{4.5.6}$$

which on orientational averaging yields

$$\left\langle \mathrm{d}\Gamma_{\mathrm{em}}^{L/R}(\Omega)\right\rangle = \frac{k^3\mathrm{d}\Omega}{24\pi^2\varepsilon_0\hbar}\left|\vec{\mu}^{0n}(A) \pm \frac{i}{c}\vec{m}^{0n}(A)\right|^2. \tag{4.5.7}$$

Integration over solid angle introduces a factor of 4π and produces an expression for the rate of emission over all directions

$$\begin{aligned}\left\langle \Gamma_{\mathrm{em}}^{L/R}\right\rangle &= \frac{k^3}{6\pi\varepsilon_0\hbar}\left|\vec{\mu}^{0n}(A) \pm \frac{i}{c}\vec{m}^{0n}(A)\right|^2 \\ &= \frac{k^3}{6\pi\varepsilon_0\hbar}\left\{\left|\vec{\mu}^{0n}(A)\right|^2 + \frac{1}{c^2}\left|\vec{m}^{0n}(A)\right|^2 \mp \frac{2i}{c}\vec{\mu}^{0n}(A)\cdot\vec{m}^{n0}(A)\right\},\end{aligned} \tag{4.5.8}$$

where the upper and lower signs refer to *L*/*R*-circular polarization. It should be noted that for pure electric and pure magnetic dipole emissions by a chiral molecule, the rate is independent of the helicity of the emitted photon. Further, the respective contributions apply for a given polarization; an additional factor of two, therefore, arises on summation over the two independent polarizations. It is advantageous to express the emission rate at a point r in terms of the radiant energy flux per unit area per unit frequency interval, $I(\omega;r)$, with

$$I(\omega;r) = \left\langle \Gamma_{\mathrm{em}}^{L/R}\right\rangle\frac{\hbar ck}{4\pi r^2} = \frac{ck^4}{24\pi^2\varepsilon_0 r^2}\left|\vec{\mu}^{0n}(A) \pm \frac{i}{c}\vec{m}^{0n}(A)\right|^2. \tag{4.5.9}$$

Now consider an optically active acceptor molecule B, initially in the ground electronic state $|0\rangle$, undergoing one-photon absorption of circularly polarized light, whose irradiance per unit frequency is given by (4.5.9). The absorption rate is calculated similarly to that given for emission above. Interaction Hamiltonian (4.5.1) is re-employed but with A replaced by B, and with the initial and final states now given by $|i\rangle = |E_0^B; 1(\vec{k}, L/R)\rangle$ and $|f\rangle = |E_n^B; 0(\vec{k}, L/R)\rangle$. The first-order perturbation theory gives for the matrix element

$$M_{fi}^{\text{abs}} = -i\left(\frac{\hbar ck}{2\varepsilon_0 V}\right)^{1/2} e_i^{(L/R)}(\vec{k})\left[\mu_i^{n0}(B) \mp \frac{i}{c} m_i^{n0}(B)\right] e^{i\vec{k}\cdot\vec{R}_B}, \quad (4.5.10)$$

where "abs" designates absorption. Insertion of the above into the Fermi golden rule yields the absorption rate

$$\Gamma_{\text{abs}}^{L/R} = \frac{2\pi\rho_f}{\hbar}\left(\frac{\hbar ck}{2\varepsilon_0 V}\right)\vec{e}^{(L/R)}(\vec{k})\bar{\vec{e}}^{(L/R)}(\vec{k})\left|\vec{\mu}^{n0}(B) \mp \frac{i}{c}\vec{m}^{n0}(B)\right|^2, \quad (4.5.11)$$

which on rotational averaging results in

$$\left\langle \Gamma_{\text{abs}}^{L/R} \right\rangle = \frac{\pi ck\rho_f}{3\varepsilon_0 V}\left|\vec{\mu}^{n0}(B) \mp \frac{i}{c}\vec{m}^{n0}(B)\right|^2. \quad (4.5.12)$$

It is worth pointing out that for absorption from an incident beam containing N photons, the cross-term of equation (4.5.12) leads to the single molecule rate for circular dichroism of species B,

$$\langle \Gamma^{\text{CD}} \rangle = -\frac{4\pi ikN}{3\varepsilon_0 V}\rho_f \vec{\mu}^{0n}(B)\cdot\vec{m}^{n0}(B). \quad (4.5.13)$$

Noting that the intensity of light from a single photon is

$$I = \frac{\hbar c^2 k}{V}, \quad (4.5.14)$$

equation (4.5.12) can be written as

$$\left\langle \Gamma_{\text{abs}}^{L/R} \right\rangle = \frac{\pi I\rho_f}{3\hbar c\varepsilon_0}\left|\vec{\mu}^{n0}(B) \mp \frac{i}{c}\vec{m}^{n0}(B)\right|^2. \quad (4.5.15)$$

Substituting (4.5.9) for the intensity due to radiation emitted by the excited donor into the acceptor absorption rate (4.5.15), with $r = R = |\vec{R}_B - \vec{R}_A|$, gives the radiative contribution to the transfer rate,

$$\langle \Gamma_{\text{rad}} \rangle = \frac{k^4 \rho_f}{72\pi\hbar\varepsilon_0^2 R^2} \left| \vec{\mu}^{0n}(A) \pm \frac{i}{c}\vec{m}^{0n}(A) \right|^2 \left| \vec{\mu}^{n0}(B) \mp \frac{i}{c}\vec{m}^{n0}(B) \right|^2 . \quad (4.5.16)$$

The leading contribution, namely, the pure electric dipole terms of each molecule, is seen to be identical to the far-zone transfer rate (4.2.22) after summing over left- and right-hand circular polarizations. In similar fashion, the discriminatory contribution to the radiative transfer rate may be extracted from (4.5.16), and is

$$\left\langle \Gamma_{\text{rad}}^{\text{disc}} \right\rangle = \frac{k^4 \rho_f}{18\pi\hbar\varepsilon_0^2 c^2 R^2} \left[i\vec{\mu}^{0n}(A) \cdot \vec{m}^{n0}(A) \right] \left[i\vec{\mu}^{0n}(B) \cdot \vec{m}^{n0}(B) \right], \quad (4.5.17)$$

which is equal to (4.4.18) on noting the definition of the optical rotatory strength (4.4.15) and on adding terms arising from each individual helicity.

Toward the end of Section 2.10, it was shown that the spontaneous emission rate multiplied by the photon energy is equal to the Poynting vector. Hence, the multipolar Maxwell field operators in the vicinity of a source presented in Section 2.6 may be used to calculate the net rate of flow of energy from an excited molecule when the former is taken to be the source of radiation incident on the acceptor moiety instead of using the perturbation theory method illustrated above.

4.6 RESPONSE THEORY CALCULATION

Diagrammatic time-dependent perturbation theory was used in Section 4.2 to calculate the matrix element for resonant transfer of excitation energy between an excited and unexcited pair of interacting molecules. An explicit functional form for the retarded resonant electric dipole–dipole coupling tensor valid for all separation distances outside the region of overlap of molecular charge distributions was given. The transition rate, and its asymptotic limits in the near and far zones, was calculated using the Fermi golden rule formula. The treatment was then extended to deal with optically active molecules and the resulting discriminatory transfer rate.

In this section, it is shown how an alternative physical picture and calculational method may be used to evaluate the matrix element for migration of energy resonantly. The approach entails the use of the multipolar Maxwell field operators in the neighborhood of a source computed in Section 2.6 in a response theory calculation (Power and Thirunamachandran, 1983c). The unexcited acceptor molecule B couples via its transition electric dipole moment to the radiation field due to the excited donor molecule A, giving rise to the matrix element directly.

Using this method, the leading contribution to the matrix element for energy transfer is found by treating species B, located at $\vec{R}_B$, and initially in the ground state, as a test dipole in the electric displacement field due to the source dipole of A evaluated at the location of body B. The response of B to the electric dipole-dependent driving electric displacement field of source A, situated at $\vec{R}_A$, and which is undergoing an $0 \leftarrow n$ transition, leads to the interaction energy

$$-\varepsilon_0^{-1}\mu_j^{n0}(B)\mathrm{e}^{-i\omega_{0n}t}d_j^{\perp}(A;\vec{R}_B,t). \tag{4.6.1}$$

In Section 2.6, the electric displacement field was expanded as a power series in the source dipole moment,

$$d_j^{\perp}(A;\vec{R}_B,t) = d_j^{(0)}(A;\vec{R}_B,t) + d_j^{(1)}(\vec{\mu};A;\vec{R}_B,t) + d_j^{(2)}(\vec{\mu}\vec{\mu};A;\vec{R}_B,t) + \cdots, \tag{4.6.2}$$

where the first term on the right-hand side of (4.6.2) is the vacuum field, independent of the source, and the second term is the first-order electric displacement field, linearly dependent on the source moment $\vec{\mu}$. The contribution from the second term of (4.6.2) gives rise to the matrix element for energy transfer that is proportional to the transition electric dipole moment at each center. The first-order displacement field is explicitly given by (2.6.21). Taking its zero–nth matrix element produces

$$\langle 0|d_i^{(1)}(\vec{\mu};A;\vec{r},t)|n\rangle = \frac{1}{4\pi}\mu_j^{0n}(A)\left(-\vec{\nabla}^2\delta_{ij} + \vec{\nabla}_i\vec{\nabla}_j\right)\frac{\mathrm{e}^{ik_{n0}(|\vec{r}-\vec{R}_A|-ct)}}{|\vec{r}-\vec{R}_A|}. \tag{4.6.3}$$

Inserting (4.6.3) into (4.6.1) results in the matrix element

$$-\frac{1}{4\pi\varepsilon_0}\mu_i^{0n}(A)\mu_j^{n0}(B)\left(-\vec{\nabla}^2\delta_{ij} + \vec{\nabla}_i\vec{\nabla}_j\right)\frac{\mathrm{e}^{ik_{n0}|\vec{R}_B-\vec{R}_A|}}{|\vec{R}_B-\vec{R}_A|}, \tag{4.6.4}$$

which on using the definition for the retarded dipole–dipole coupling tensor, $V_{ij}^{\pm}(k,\vec{R})$ given by (4.2.17), is seen to be identical to the result obtained for M_{fi} (4.2.16) using perturbation theory, recalling that $k = k_{n0}$ and $R = |\vec{R}_B - \vec{R}_A|$. The method presented clearly has the advantage of illustrating the role played by radiation fields in the transfer of energy.

It is straightforward to apply the response theory approach to the evaluation of transfer rates between systems containing higher multipole moments, as occurring in the migration of energy between chiral centers (Craig and Thirunamachandran, 1999; Salam, 2005b). To leading order in this latter case, for example, matrix elements are required for the magnetic dipole-dependent electric displacement field and the electric and magnetic dipole-dependent magnetic fields, in addition to (4.6.3). Thus, from (2.7.6) and (2.7.7),

$$\langle 0|d_i^{(1)}(\vec{m};A;\vec{r},t)|n\rangle = -\frac{1}{4\pi c}m_j^{0n}(A)\left(ik_{n0}\varepsilon_{ijk}\vec{\nabla}_k\right)\frac{e^{ik_{n0}(|\vec{r}-\vec{R}_A|-ct)}}{|\vec{r}-\vec{R}_A|} \tag{4.6.5}$$

and

$$\langle 0|b_i^{(1)}(\vec{\mu}+\vec{m};A;\vec{r},t)|n\rangle = \frac{1}{4\pi\varepsilon_0 c}\Big\{\mu_j^{0n}(A)\left(ik_{n0}\varepsilon_{ijk}\vec{\nabla}_k\right) + \frac{1}{c}m_j^{0n}(A)\left(-\vec{\nabla}^2\delta_{ij}+\vec{\nabla}_i\vec{\nabla}_j\right)\Big\}\frac{e^{ik_{n0}(|\vec{r}-\vec{R}_A|-ct)}}{|\vec{r}-\vec{R}_A|}. \tag{4.6.6}$$

For exchange of energy between an optically active pair, the extension of (4.6.1) is

$$\begin{aligned}&-\varepsilon_0^{-1}\mu_j^{n0}(B)e^{-i\omega_{0n}t}d_j^{(1)}\left(\vec{\mu}+\vec{m};A;\vec{R}_B,t\right)\\&-m_j^{n0}(B)e^{-i\omega_{0n}t}b_j^{(1)}\left(\vec{\mu}+\vec{m};A;\vec{R}_B,t\right),\end{aligned} \tag{4.6.7}$$

which directly leads to the matrix element

$$\begin{aligned}M^{\mathrm{e+m}} &= -\frac{1}{4\pi\varepsilon_0}\Bigg\{\Big[\mu_j^{n0}(B)\mu_i^{0n}(A)\left(-\vec{\nabla}^2\delta_{ij}+\vec{\nabla}_i\vec{\nabla}_j\right)\\&\quad+\frac{1}{c}\mu_j^{n0}(B)m_i^{0n}(A)\left(ik_{n0}\varepsilon_{ijk}\vec{\nabla}_k\right)\Big]\\&\quad+\frac{1}{c}\Big[-m_j^{n0}(B)\mu_i^{0n}(A)(ik_{n0}\varepsilon_{ijk}\vec{\nabla}_k)\\&\quad+\frac{1}{c}m_j^{n0}(B)m_i^{0n}(A)\left(-\vec{\nabla}^2\delta_{ij}+\vec{\nabla}_i\vec{\nabla}_j\right)\Big]\Bigg\}\frac{e^{ik_{n0}R}}{R}\\&=\left[\mu_i^{0n}(A)\mu_j^{n0}(B)+\frac{1}{c^2}m_i^{0n}(A)m_j^{n0}(B)\right]V_{ij}^{\pm}(k_{n0},\vec{R})\\&\quad+\left[-\mu_i^{0n}(A)m_j^{n0}(B)+m_i^{0n}(A)\mu_j^{n0}(B)\right]U_{ij}^{\pm}(k_{n0},\vec{R}),\end{aligned} \tag{4.6.8}$$

on using the definition of the retarded dipole–dipole interaction tensors (4.2.17) and (4.4.11). Expression (4.6.8) is a sum of electric, magnetic, and electric–magnetic cross-terms and is identical to the sum of (4.2.16), (4.4.4), and (4.4.10). The discriminatory transfer rate (4.4.16) results from (4.6.8) on extracting the modulus square of the electric– magnetic interference contribution from the first term of the last form of (4.6.8) and adding it to the absolute value squared of the second term of the last equality of (4.6.8).

4.7 TIME-DEPENDENT ENERGY TRANSFER AND CAUSALITY

The importance of the study of the resonant transfer of excitation energy between an excited and an unexcited molecular pair lies not only in the delineation of the underlying mechanism associated with the migration of energy but also in the role played by this system in helping to further understand the nature of causality and signal propagation. These aspects emerge most clearly when exchange of energy takes place between nonidentical systems, for which the Fermi golden rule rate formula does not apply. Instead, the time-dependent probability must be calculated for various state specifications of the system. An early calculation in this context was carried out by Fermi (1932), who used time-dependent perturbation theory and electric dipole coupling to calculate the probability for energy transfer between the pair. Causal behavior resulted only when certain approximations were made in the evaluation of the matrix element, namely, that nonresonant energy denominator terms in the probability amplitude were dropped and that integration limits over frequency were extended from nonnegative values to $(-\infty, \infty)$. Subsequent efforts have led to the accumulation of a large body of literature in which the Fermi problem, and the multitude of scenarios that arise within it corresponding to different possible experimental setups, has been examined in detail. The conditions that ensure strict Einstein causality, that is, the excitation probability vanishes exactly for times $t < R/c$, where R is the distance between the transmitter and receiver objects, are now well understood (Power and Thirunamachandran, 1983c, 1997). The appropriate computational techniques to be adopted and the physical viewpoint to be employed in the solution of the problem are presented below. It is found that the calculation of the transfer probability between two different species is most easily carried out in the Heisenberg picture using the Maxwell fields calculated in Section 2.6.

In Fermi's original formulation of the energy transfer problem, the probability was required to be found for complete specification of the final state of the system, namely, knowledge of the molecular states of donor and acceptor, as well as the state of the radiation field was to be ascertained. A noncausal result was obtained using time-dependent perturbation theory when no mathematical approximations were made. While such a statement of the problem provides a perfectly valid experimental scenario, measurement would prove to be technically challenging. An equally acceptable, alternative statement is to ask the question: Given that initially the system comprises molecule A in an excited state, B in the ground state with no photons present, what is the probability of finding B excited at some time? In this reformulation, possibly experimentally more amenable, explicit mention is made of only the state of the receiver entity, with the state of the source molecule and the radiation field being left unspecified. Since A and B are taken to be nonidentical in the present case, different labels are used to denote excitation. The counterparts to the initial and final states (4.2.7a, 4.2.7b) are

$$|i\rangle = \left|E_m^A, E_0^B; 0(\vec{p}, \varepsilon)\right\rangle \tag{4.7.1}$$

and

$$|f\rangle = \left|A, E_n^B; F\right\rangle. \tag{4.7.2}$$

The initial state is the same as before, but now in the representation of the final state, only the state of the acceptor species B is indicated precisely. The kets $|A\rangle$ and $|F\rangle$, respectively, designate eigenstates of molecule A and number states of the radiation field, which are left arbitrary. The time-dependent probability for the process represented by states (4.7.1) and (4.7.2) is calculated via

$$P(t) = \sum_A \sum_F |\langle f|H_{\text{int}}|i\rangle|^2, \tag{4.7.3}$$

where the sums are executed over a complete set of field states and eigenfunctions of A.

Adopting the multipolar framework, the second quantized Hamiltonian for the system is

$$H = H_{\text{mol}} + H_{\text{rad}} + H_{\text{int}}, \tag{4.7.4}$$

where

$$H_{\text{mol}} = \sum_m b_m^\dagger b_m E_m^A + \sum_n b_n^\dagger b_n E_n^B, \tag{4.7.5}$$

$$H_{\text{rad}} = \sum_{\vec{p},\varepsilon} \left(a^{\dagger(\varepsilon)}(\vec{p}) a^{(\varepsilon)}(\vec{p}) + \frac{1}{2} \right) \hbar\omega, \tag{4.7.6}$$

where the circular frequency $\omega = cp$ and p is the scalar magnitude of the wavevector. In the electric dipole approximation, the interaction Hamiltonian is

$$\begin{aligned} H_{\text{int}} = & -\varepsilon_0^{-1} \sum_{r,s} b_r^\dagger(A,t) b_s(A,t) \vec{\mu}^{rs}(A) \cdot \vec{d}^{\perp}\left(\vec{R}_A\right) \\ & -\varepsilon_0^{-1} \sum_{r,s} b_r^\dagger(B,t) b_s(B,t) \vec{\mu}^{rs}(B) \cdot \vec{d}^{\perp}\left(\vec{R}_B\right), \end{aligned} \tag{4.7.7}$$

where the time dependence of the fermion creation and destruction operators is explicit. Using the method described in Section 2.6, it is straightforward to calculate the electric displacement field for the two-body system. To first and second orders in the source moment, the field is found to be additive and may be written as

$$\vec{d}^{\perp}(\vec{r},t) = \vec{d}^{(0)}(\vec{r},t) + \vec{d}(A;\vec{r},t) + \vec{d}(B;\vec{r},t), \tag{4.7.8}$$

where $\vec{d}^{(0)}(\vec{r},t)$ is the free displacement field and the source-dependent terms are given by

$$\begin{aligned} d_i(\xi;\vec{r},t) = & \frac{1}{4\pi} \sum_{m,n} \beta_m^{\dagger\xi}(t - |\vec{r} - \vec{R}_\xi|/c) \beta_n^{\xi}(t - |\vec{r} - \vec{R}_\xi|/c) \\ & \times \mu_j^{mn}(\xi) \left(-\vec{\nabla}^2 \delta_{ij} + \vec{\nabla}_i \vec{\nabla}_j\right) \frac{e^{i\omega_{mn}^{\xi}(t - |\vec{r} - \vec{R}_\xi|/c)}}{|\vec{r} - \vec{R}_\xi|}, \end{aligned} \tag{4.7.9}$$

for $\xi = A$, B. Note that the interaction picture fermion creation and annihilation operators are evaluated at the retarded time $t - |\vec{r} - \vec{R}|/c$. A consequence of this is that the field corresponding to one particular source is not equal to the value of the field in the absence of the other molecule, and vice versa. It is only at time $t = 0$ do the respective fermion operators act exclusively in the Fock space of the particle concerned. For all other times, $\beta_m^\dagger$ and β_n act in the total system space—in this case the space of atoms A and B and the field.

In terms of the fermion annihilation and creation operators, the probability (4.7.3) can be written as

$$P(t) = \sum_F \sum_{m'} \left| \langle F; 0^A, 0^B | \beta_{m'}^A(t) \beta_n^B(t) | 0^B, m^A; 0(\vec{p},\varepsilon) \rangle \right|^2, \tag{4.7.10}$$

where the molecular state $|0^{\xi}\rangle$ corresponds to a zero-particle fermion state with properties $b_n^{\xi}|0^{\xi}\rangle = 0$ and $b_n^{\dagger\xi}|0^{\xi}\rangle = |n^{\xi}\rangle$. Expression (4.7.10) is computed correct up to the fourth order in the transition moments by extracting contributions proportional to $|\vec{\mu}^A|^2|\vec{\mu}^B|^2$ for times $t > R/c$. For this case, the only field states that contribute are zero-photon $|0(\vec{p},\varepsilon)\rangle$, one-photon $|1(\vec{p},\varepsilon)\rangle$, and two-photon $|1(\vec{p},\varepsilon),1(\vec{p}\,',\varepsilon')\rangle$ states. Hence, (4.7.10) is the sum of three terms

$$\begin{aligned}P(t) = &\left|\langle 0(\vec{p},\varepsilon);0^A,0^B|\beta_0^A(t)\beta_n^B(t)|0^B,m^A;0(\vec{p},\varepsilon)\rangle\right|^2\\ &+\sum_{\text{modes}}\left|\langle 1(\vec{p},\varepsilon);0^A,0^B|\beta_m^A(t)\beta_n^B(t)|0^B,m^A;0(\vec{p},\varepsilon)\rangle\right|^2\\ &+\sum_{\text{modes}}\left|\langle 1(\vec{p},\varepsilon),1(\vec{p}\,',\varepsilon');0^A,0^B|\beta_0^A(t)\beta_n^B(t)|0^B,m^A;0(\vec{p}\,',\varepsilon'),0(\vec{p},\varepsilon)\rangle\right|^2.\end{aligned} \tag{4.7.11}$$

To proceed further, the form of the time-dependent fermion operators are needed, and they are readily obtained from the Heisenberg operator equations of motion together with the Hamiltonian (4.7.4). This procedure was detailed in Section 2.6. Hence,

$$\beta_n^{\xi}(t) = \beta_n^{\xi}(0) + \beta_n^{\xi}(\vec{\mu};t), \tag{4.7.12}$$

where

$$\beta_n^{\xi}(\vec{\mu};t) = \frac{i}{\varepsilon_0\hbar}\sum_m\int_0^t dt'\,\vec{\mu}^{mn}(\xi)\cdot\vec{d}^{\perp}(\vec{R}_{\xi},t')\beta_m^{\xi}(t')e^{-i\omega_{mn}^{\xi}t'}. \tag{4.7.13}$$

Examining in amplitude form the first term of (4.7.11), substituting (4.7.12) produces

$$\begin{aligned}&\left\langle 0(\vec{p},\varepsilon);0^A,0^B\right|\left[\beta_0^A(0)+\beta_0^A(\vec{\mu};t)\right]\left[\beta_n^B(0)+\beta_n^B(\vec{\mu};t)\right]\left|0^B,m^A;0(\vec{p},\varepsilon)\right\rangle\\ &\quad=\left\langle 0(\vec{p},\varepsilon);0^A,0^B\right|\left[\beta_0^A(0)\beta_n^B(\vec{\mu};t)+\beta_0^A(\vec{\mu};t)\beta_n^B(\vec{\mu};t)\right]\left|0^B,m^A;0(\vec{p},\varepsilon)\right\rangle\end{aligned} \tag{4.7.14}$$

on using $\beta_n^{\xi}(0)|0^{\xi}\rangle = 0$. The second term in the second line of (4.7.14) contributes to the probability for $t < R/c$ and is ignored. On inserting (4.7.12), the first term becomes

$$\langle 0(\vec{p},\varepsilon);0^A,0^B|\beta_0^A(0)\frac{i}{\varepsilon_0\hbar}\mu_i(B)\int_0^t dt'\,\beta_0^B(t')d_i^{\perp}(\vec{R}_B,t')e^{i\omega_{n0}^B t'}|0^B,m^A;0(\vec{p},\varepsilon)\rangle, \tag{4.7.15}$$

which can be approximated to

$$\frac{i}{\varepsilon_0\hbar}\mu_i^{0n}(B)\int_0^t dt' e^{i\omega_{n0}^B t'}\langle 0^A|d_i^{(1)}(\vec{\mu};A;\vec{R}_B,t')|m^A\rangle. \tag{4.7.16}$$

Since this term is explicitly proportional to $\mu_i^{0n}(B)$, with the field linear in the moment due to A ensuring that (4.7.16) depends on $\mu_j(A)$ also, the matrix element appearing in this last equation is evaluated solely in the space of the donor species. Incidentally, the contribution (4.7.16) does indeed vanish for $t' < R/c$ since the electric displacement field operator linear in the source moment is causal. From (4.7.9), this field is given by

$$d_i^{(1)}(\vec{\mu};A;\vec{r},t') = \frac{1}{4\pi}\sum_{m,n}\beta_m^{\dagger A}(0)\beta_n^A(0)\mu_j^{mn}(A) \times(-\vec{\nabla}^2\delta_{ij}+\vec{\nabla}_i\vec{\nabla}_j)\frac{e^{-i\omega_{mn}^A(t'-|\vec{r}-\vec{R}_A|/c)}}{|\vec{r}-\vec{R}_A|}, \tag{4.7.17}$$

so that the amplitude to be used in the first term of (4.7.11) is

$$\begin{aligned}&\frac{i}{4\pi\varepsilon_0\hbar}\mu_i^{0n}(B)\mu_j^{m0}(A)\int_{R/c}^t dt' e^{i\omega_{n0}^B t'}(-\vec{\nabla}^2\delta_{ij}+\vec{\nabla}_i\vec{\nabla}_j)\frac{e^{-i\omega_{m0}^A(t'-R/c)}}{R}\\ &=\frac{1}{4\pi\varepsilon_0\hbar}\mu_i^{m0}(A)\mu_j^{n0}(B)e^{i(\omega_{n0}^B-\omega_{m0}^A)R/c}\left[(-\vec{\nabla}^2\delta_{ij}+\vec{\nabla}_i\vec{\nabla}_j)\frac{e^{i\omega_{m0}^A R/c}}{R}\right]\\ &\times\left[\frac{e^{i(\omega_{n0}^B-\omega_{m0}^A)(t-R/c)}-1}{(\omega_{n0}^B-\omega_{m0}^A)}\right].\end{aligned} \tag{4.7.18}$$

Next, the modulus square of (4.7.18) is evaluated to obtain the probability for $R < ct$, recalling that this will be the dominant contribution since the second term of (4.7.14) applies to $R > ct$. Thus, the probability is

$$\begin{aligned}P(t)\approx&\frac{1}{(4\pi\varepsilon_0\hbar)^2}\mu_i^{m0}(A)\mu_j^{n0}(B)\mu_k^{m0}(A)\mu_l^{n0}(B)\left[(-\vec{\nabla}^2\delta_{ij}+\vec{\nabla}_i\vec{\nabla}_j)\frac{e^{i\omega_{m0}^A R/c}}{R}\right]\\ &\times\left[(-\vec{\nabla}^2\delta_{kl}+\vec{\nabla}_k\vec{\nabla}_l)\frac{e^{-i\omega_{m0}^A R/c}}{R}\right]\times\left|\frac{e^{i(\omega_{n0}^B-\omega_{m0}^A)(t-R/c)}-1}{(\omega_{n0}^B-\omega_{m0}^A)}\right|^2.\end{aligned} \tag{4.7.19}$$

In addition, for resonant transfer of energy, $\omega_{n0}^{B} \approx \omega_{m0}^{A}$. This further enhances the probability due to the near-resonant denominator.

Returning to expression (4.7.11) and examining the second term in the form of a probability amplitude, this contribution can be written as a sum of two terms again making use of relation (4.7.12),

$$\begin{aligned}&\langle 1(\vec{p},\varepsilon);0^{A},0^{B}|\beta_{m}^{A}(0)\beta_{n}^{B}(\vec{\mu};t)|0^{B},m^{A};0(\vec{p},\varepsilon)\rangle\\&\quad+\langle 1(\vec{p},\varepsilon);0^{A},0^{B}|\beta_{m}^{A}(\vec{\mu};t)\beta_{n}^{B}(\vec{\mu};t)|0^{B},m^{A};0(\vec{p},\varepsilon)\rangle.\end{aligned} \tag{4.7.20}$$

The first term of (4.7.20) has a nonzero term that is proportional to $\vec{\mu}(B)$, so that the contribution to the third-order amplitude that depends on $\vec{\mu}(B)|\vec{\mu}(A)|^{2}$, namely, quadratically on $\vec{\mu}(A)$ and linearly on $\vec{\mu}(B)$, is required to evaluate the probability correct to $|\vec{\mu}(B)|^{2}|\vec{\mu}(A)|^{2}$. The term linear in $\vec{\mu}(B)$ is given by

$$\begin{aligned}&\frac{i}{\varepsilon_{0}\hbar}\mu_{i}^{0n}(B)\int_{0}^{t}\mathrm{d}t'\mathrm{e}^{i\omega_{n0}^{B}t'}\langle 1(\vec{p},\varepsilon)|d_{i}^{(0)}(\vec{R}_{B},t')|0(\vec{p},\varepsilon)\rangle\\&\quad=\left(\frac{\hbar ck}{2\varepsilon_{0}V}\right)^{1/2}\mu_{i}^{0n}(B)\bar{e}_{i}^{(\varepsilon)}(\vec{p})\mathrm{e}^{-i\vec{p}\cdot\vec{R}_{B}}\left[\frac{\mathrm{e}^{i(\omega+\omega_{n0}^{B})t}-1}{(\omega+\omega_{n0}^{B})}\right].\end{aligned} \tag{4.7.21}$$

The third-order amplitude has contributions arising from both terms of (4.7.20). They are

$$\begin{aligned}&\langle 1(\vec{p},\varepsilon);0^{A},0^{B}|\beta_{m}^{A}(0)\beta_{n}^{(3)B}(\vec{\mu};t)|0^{B},m^{A};0(\vec{p},\varepsilon)\rangle\\&\quad+\langle 1(\vec{p},\varepsilon);0^{A},0^{B}|\beta_{m}^{(2)A}(\vec{\mu};t)\beta_{n}^{(1)B}(\vec{\mu};t)|0^{B},m^{A};0(\vec{p},\varepsilon)\rangle.\end{aligned} \tag{4.7.22}$$

Because the second term of (4.7.22) is found to yield a noncausal result, it is dropped from further consideration. The first matrix element of (4.7.22) is causal, however, and it contributes to the time-dependent probability for energy transfer through interference with the first-order amplitude (4.7.21). The relevant term of the third-order fermion annihilation operator is of the form

$$\beta_{n}^{(3)B}(\vec{\mu};t)=\frac{i}{\varepsilon_{0}\hbar}\mu_{i}^{n0}(B)\int_{0}^{t}\mathrm{d}t'\mathrm{e}^{i\omega_{n0}^{B}t'}\beta_{0}^{B}(0)d_{i}^{(2)}(\vec{\mu}\vec{\mu};A;\vec{R}_{B},t'), \tag{4.7.23}$$

so that the first term of (4.7.22) becomes

$$\frac{i}{\varepsilon_0\hbar}\mu_i^{0n}(B)\int_0^t \mathrm{d}t' \mathrm{e}^{i\omega_{n0}^B t'}\langle 1(\vec{p},\varepsilon)|d_i^{(2)}(\vec{\mu}\vec{\mu};A;\vec{R}_B,t')|0(\vec{p},\varepsilon)\rangle. \tag{4.7.24}$$

Appearing in the last two formulas is the electric dipole-dependent second-order electric displacement field of excited donor molecule A evaluated at the position of particle B, $\vec{R}_B$, whose explicit functional form is given by (2.6.30), which is strictly causal. From (4.7.21) and (4.7.23), the first-order third-order interference term is then

$$\begin{aligned}&\frac{1}{\varepsilon_0^2\hbar^2}\sum_{\vec{p},\,\varepsilon}\mu_i^{0n}(B)\mu_j^{n0}(B)\int_0^t \mathrm{d}t''' \mathrm{e}^{-i\omega_{n0}^B t'''}\int_0^t \mathrm{d}t' \mathrm{e}^{i\omega_{n0}^B t'}\\ &\times\langle 0(\vec{p},\varepsilon)|d_i^{(0)}(\vec{R}_B;t''')|1(\vec{p},\varepsilon)\rangle\langle 1(\vec{p},\varepsilon)|\\ &\times d_j^{(2)}(\vec{\mu}\vec{\mu};A;\vec{R}_B,t)|0(\vec{p},\varepsilon)\rangle+\mathrm{c.c.}\end{aligned} \tag{4.7.25}$$

Alternatively, the quadratic field is readily obtained from (4.7.9). Its diagonal matrix element evaluated over the electronic state $|m\rangle$ is

$$\begin{aligned}\langle m|d_j^{(2)}(\vec{\mu}\vec{\mu};A;\vec{R}_B,t')|m\rangle &= \frac{1}{4\pi}\mu_k^{0n}(B)\left(-\vec{\nabla}^2\delta_{jk}+\vec{\nabla}_j\vec{\nabla}_k\right)\frac{1}{R}\\ &\times\Big[\langle m|\beta_0^{\dagger(1)A}(t'-\vec{R}/c)|0\rangle \mathrm{e}^{-i\omega_{m0}^A(t'-R/c)}\\ &+\langle 0|\beta_0^{(1)A}(t'-\vec{R}/c)|m\rangle \mathrm{e}^{i\omega_{m0}^A(t'-R/c)}\Big].\end{aligned} \tag{4.7.26}$$

Inserting (4.7.26) into (4.7.25) and recognizing from (4.7.13) that

$$\langle 0|\beta_n^{(1)A}(t'-R/c)|m\rangle = \frac{i}{\varepsilon_0\hbar}\mu_l^{0m}(A)\int_0^{t'-R/c} \mathrm{d}t'' \mathrm{e}^{-i\omega_{m0}^A t''}\,\mathrm{d}_l^{(0)}(\vec{R}_A,t''), \tag{4.7.27}$$

produces for the interference term,

$$\frac{i}{4\pi\varepsilon_0^2\hbar^3}\sum_{\vec{p},\,\varepsilon}\mu_i^{0n}(B)\mu_j^{n0}(B)\mu_k^{0m}(A)\mu_l^{m0}(A)$$
$$\times\int_0^t \mathrm{d}t'''\mathrm{e}^{-i\omega_{n0}^B t'''}\int_{R/c}^t \mathrm{d}t'\mathrm{e}^{i\omega_{n0}^B t'}\left(-\vec{\nabla}^2\delta_{jk}+\vec{\nabla}_j\vec{\nabla}_k\right)\frac{1}{R}$$
$$\times\int_0^{t'-R/c}\mathrm{d}t''\langle 0(\vec{p},\varepsilon)|d_i^{(0)}(\vec{R}_B,t''')|1(\vec{p},\varepsilon)\rangle\langle 1(\vec{p},\varepsilon)|d_l^{(0)}(\vec{R}_A,t'')|0(\vec{p},\varepsilon)\rangle$$
$$\times\left[\mathrm{e}^{-i\omega_{m0}^A t''}\mathrm{e}^{i\omega_{m0}^A(t'-R/c)}-\mathrm{e}^{i\omega_{m0}^A t''}\mathrm{e}^{-i\omega_{m0}^A(t'-R/c)}\right]+\text{c.c.}\tag{4.7.28}$$

When $\omega_{n0}^B \approx \omega_{m0}^A$, the leading term of (4.7.28) after carrying out the mode sum and time integrals is found to be

$$\frac{1}{(4\pi\varepsilon_0)^2}\frac{1}{9\pi\hbar^2}\sum_{m,n}|\vec{\mu}^{0m}(A)|^2|\vec{\mu}^{n0}(B)|^2$$
$$\times\left[\left(-\vec{\nabla}^2\delta_{ij}+\vec{\nabla}_i\vec{\nabla}_j\right)\frac{\mathrm{e}^{i\omega_{m0}^A R/c}}{R}\right]\frac{1}{\left(\omega_{n0}^B-\omega_{m0}^A\right)}\int_0^\infty\frac{\mathrm{d}\omega}{\left(\omega_{n0}^B+\omega\right)\left(\omega_{m0}^A+\omega\right)}$$
$$\times\left[\left(-\vec{\nabla}^2\delta_{ij}+\vec{\nabla}_i\vec{\nabla}_j\right)\frac{\sin(\omega R/c)}{R}\right]$$
$$\times\left(\mathrm{e}^{-i(\omega_{n0}^B+\omega)t}-1\right)\left(\mathrm{e}^{i(\omega_{n0}^B-\omega_{m0}^A)t}-\mathrm{e}^{i(\omega_{n0}^B-\omega_{m0}^A)R/c}\right)+\text{c.c.},\tag{4.7.29}$$

for isotropically averaged species A and B.

Finally, returning to (4.7.11) and evaluating the third term, which involves two-photon states and is proportional to $\vec{\mu}(A)\vec{\mu}(B)$. Its amplitude is computed from

$$\langle 1(\vec{p},\varepsilon),1(\vec{p}',\varepsilon');0^A,0^B|\beta_0^A(t)\beta_n^B(t)|0^B,m^A;0(\vec{p}',\varepsilon'),0(\vec{p},\varepsilon)\rangle$$
$$=-\frac{1}{\varepsilon_0^2\hbar^2}\mu_i^{0n}(B)\mu_j^{m0}(A)\int_0^t\mathrm{d}t'\mathrm{e}^{-i\omega_{m0}^A t'}\int_0^t\mathrm{d}t''\mathrm{e}^{i\omega_{n0}^B t''}$$
$$\times\langle 1(\vec{p},\varepsilon),1(\vec{p}',\varepsilon')|d_j^{(0)}(\vec{R}_A,t')d_i^{(0)}(\vec{R}_B,t'')|0(\vec{p},\varepsilon),0(\vec{p}',\varepsilon')\rangle.\tag{4.7.30}$$

Expression (4.7.30) contains the off-diagonal matrix element of the spatiotemporal vacuum field correlation function, which is noncausal and therefore does not contribute to the time-dependent probability for $t > R/c$. The transfer probability is, therefore, given by the sum of (4.7.19) (after orientational averaging) and (4.7.29). The dominant contribution to $P(t)$ is given by (4.7.19) since it is proportional to $(\omega_{n0}^B - \omega_{m0}^A)^{-2}$, in contrast to (4.7.29) that has a $(\omega_{n0}^B - \omega_{m0}^A)^{-1}$ dependence. After random averaging, (4.7.19) becomes

$$
\begin{aligned}
P(t) &= \frac{1}{(4\pi\varepsilon_0)^2}\frac{1}{9\hbar^2}\sum_{m,n}|\vec{\mu}^{0m}(A)|^2|\vec{\mu}^{n0}(B)|^2 \\
&\times\left[\left(-\vec{\nabla}^2\delta_{ij}+\vec{\nabla}_i\vec{\nabla}_j\right)\frac{e^{i\omega_{m0}^A R/c}}{R}\right]\left[\left(-\vec{\nabla}^2\delta_{ij}+\vec{\nabla}_i\vec{\nabla}_j\right)\frac{e^{-i\omega_{m0}^A R/c}}{R}\right] \\
&\times\frac{\sin^2\left\{\left[(\omega_{n0}^B-\omega_{m0}^A)/2\right](t-R/c)\right\}}{\left[(\omega_{n0}^B-\omega_{m0}^A)/2\right]^2} \\
&= \frac{2}{9\hbar^2}\frac{1}{(4\pi\varepsilon_0 R^3)^2}\sum_{m,n}|\vec{\mu}^{0m}(A)|^2|\vec{\mu}^{n0}(B)|^2\left[3+(pR)^2+(pR)^4\right] \\
&\times\frac{\sin^2\left\{\left[(\omega_{n0}^B-\omega_{m0}^A)/2\right](t-R/c)\right\}}{\left[(\omega_{n0}^B-\omega_{m0}^A)/2\right]^2}, \quad t > R/c, \qquad (4.7.31)
\end{aligned}
$$

where $\omega_{n0}^B \approx \omega_{m0}^A = cp$.

4.8 PROOF OF CAUSALITY OF ENERGY TRANSFER TO ALL ORDERS IN PERTURBATION THEORY

The time-dependent probability for resonant transfer of energy has been calculated correct up to terms proportional to the product of the modulus squares of the transition dipole moments of A and B, and the result was shown to hold for $R < ct$. When $t < R/c$, $P(t)$ vanishes at this order of approximation. In this section, it is demonstrated that the probability is strictly causal to all orders in perturbation theory (Power and Thirunamachandran, 1997). Again, the possible experimental scenario posited in the previous section is considered, namely, the situation in which species B is excited at some time t for $t < R/c$, ignoring the state of donor A—which was initially excited—and the state of the radiation field. For this purpose, it is convenient to define a projection operator for entity B in terms of the electron Fock space operators,

$$\mathcal{P}_{mn}(B;t) = \beta_m^{\dagger}(B;t)\beta_n(B;t). \tag{4.8.1}$$

The time-dependent probability (4.7.11) can be written as the expectation value of the projection operator $\mathcal{P}_{nn}(B;t)$ for B to be found in state $|n\rangle$ at time t,

$$P(t) = \langle 0(\vec{p},\varepsilon); m^A, 0^B|\mathcal{P}_{nn}(B;t)|0^B, m^A; 0(\vec{p},\varepsilon)\rangle. \tag{4.8.2}$$

Differentiating (4.8.1) with respect to time and inserting

$$\dot{\beta}_n(B;t) = \frac{i}{\varepsilon_0\hbar}\sum_m \vec{\mu}^{mn}(B)\cdot\vec{d}^{\perp}(\vec{R}_B,t)\beta_m(B;t)\mathrm{e}^{-i\omega_{mn}^B t}, \tag{4.8.3}$$

with the last relation obtained from (4.7.13), an equation of motion may be written for $\mathcal{P}_{mn}(B;t)$. It is

$$\begin{aligned}\dot{\mathcal{P}}_{mn}(B;t) &= \dot{\beta}_m^{\dagger}(B;t)\beta_n(B;t) + \beta_m^{\dagger}(B;t)\dot{\beta}_n(B;t)\\ &= -\frac{i}{\varepsilon_0\hbar}\vec{\mu}^{uv}(B)\cdot\vec{d}^{\perp}(\vec{R}_B,t)\mathcal{P}_{rs}(B;t)T_{mn;uv;rs}(t),\end{aligned} \tag{4.8.4}$$

where

$$T_{mn;uv;rs}(t) = \delta_{ur}\delta_{ns}\delta_{mv}\,\mathrm{e}^{-i\omega_{uv}t} - \delta_{us}\delta_{mr}\delta_{nv}\,\mathrm{e}^{i\omega_{uv}t}. \tag{4.8.5}$$

Note that in the classical quantity defined in equation (4.8.5), the indices appearing as subscripts in the Kronecker deltas refer to electronic states of molecule B and do not designate Cartesian tensor components. As usual, the Kronecker delta is nonvanishing only when the two suffixes are equal to each other, in which case they give unity. From (4.8.4) is obtained an integral relation for the projection operator,

$$\mathcal{P}_{mn}(B;t) = \mathcal{P}_{mn}(B;0) - \frac{i}{\varepsilon_0\hbar}\int_0^t \mathrm{d}t'\,\vec{\mu}^{uv}(B)\cdot\vec{d}^{\perp}(\vec{R}_B,t')T_{mn;uv;rs}(t')\mathcal{P}_{rs}(B;t'). \tag{4.8.6}$$

Iterating generates the following series correct to Nth order,

$$\begin{aligned}
\mathcal{P}_{mn}(B;t) &= \mathcal{P}_{mn}(B;0) - \frac{i}{\varepsilon_0\hbar}\int_0^t \mathrm{d}t'\left[\vec{\mu}^{uv}(B)\cdot\vec{d}^{\perp}(\vec{R}_B,t')\right]T_{mn;uv;rs}(t')\mathcal{P}_{rs}(B;0)\\
&+\left(-\frac{i}{\varepsilon_0\hbar}\right)^2\int_0^t \mathrm{d}t'\int_0^{t'}\mathrm{d}t''\left[\vec{\mu}^{uv}(B)\cdot\vec{d}^{\perp}(\vec{R}_B,t')\right]\\
&\times\left[\vec{\mu}^{u'v'}(B)\cdot\vec{d}^{\perp}(\vec{R}_B,t'')\right]T_{mn;uv;rs}(t')T_{rs;u'v';r's'}(t'')\mathcal{P}_{r's'}(B;0)\\
&+\cdots+\left(-\frac{i}{\varepsilon_0\hbar}\right)^N\int_0^t \mathrm{d}t_1\int_0^{t_1}\mathrm{d}t_2\ldots\int_0^{t_{N-1}}\mathrm{d}t_N\left[\vec{\mu}(B)\cdot\vec{d}^{\perp}(\vec{R}_B,t_1)\right]\\
&\times\left[\vec{\mu}(B)\cdot\vec{d}^{\perp}(\vec{R}_B,t_2)\right]\ldots\left[\vec{\mu}(B)\cdot\vec{d}^{\perp}(\vec{R}_B,t_N)\right]\\
&\times T(t_1)T(t_2)\ldots T(t_N)\mathcal{P}_{r_Ns_N}(B;0),
\end{aligned} \tag{4.8.7}$$

where in the last term of (4.8.7), the molecular labels have been omitted. The displacement field appearing in the formula for the projection operator is the total field evaluated at B. This is of the form

$$\vec{d}^{\perp}(\vec{R}_B,t_i) = \vec{d}^{(0)}(\vec{R}_B,t_i) + \vec{d}(B;\vec{R}_B,t_i), \tag{4.8.8}$$

since for $0 \le t \le R/c$, t_i lies within this range and $\vec{d}(A;\vec{R}_B,t_i) \equiv 0$. Thus, $\vec{d}(B;\vec{R}_B,t)$ is independent of $\vec{\mu}(A)$ and hence of the intermolecular separation distance R. From this it can be concluded that after N iterations, with N arbitrary, the projection operator $\mathcal{P}_{nn}(B;t)$ is independent of R for $0 \le t < R/c$, thereby proving the causal nature of $P(t)$ to all orders of $\vec{\mu}(B)$.

CHAPTER 5

RETARDED DISPERSION FORCES

That is obviously a question of zero point energy.
—Remark by Niels Bohr to H. B. G. Casimir, from H. B. G. Casimir, *Niels Bohr, A Centenary Volume*, A. P. French and P.J. Kennedy (Eds), Harvard University Press, Cambridge, MA, 1985, p. 180.

5.1 INTRODUCTION

A key triumph of the theory of quantum electrodynamics is its application to the study of molecular interactions. Problems of this type may be examined using the formalism developed and successfully applied to a single atomic or molecular center coupled to one or more sources of external radiation, as exemplified by numerous and wide-ranging spectroscopic processes and quantum optical phenomena. This transferability of the theory is due to the fact that ultimately all intermolecular couplings are electromagnetic in origin and can be rationalized at one level as resulting from the emission and absorption of virtual photons, as most clearly personified by the use of the multipolar version of molecular quantum electrodynamics. As demonstrated in the previous chapter, the perturbation theory calculation of the resonant exchange of excitation energy is most easily understood as arising from the transfer of a single such particle between the pair. Creation and destruction of virtual photons are described by the same theoretical techniques that are used to treat emission and absorption of real quantized particles of light, with the added requirement that all possible modes of the one or more virtual photons must be summed over. This is due to the rapid (subject to time–energy uncertainty) appearance and disappearance of this virtual particle. Hence, energy is conserved only between initial and final states of the interacting system, but may be violated for intermediate states.

Another fundamental intermolecular process that proves amenable to study by the methods of molecular quantum electrodynamics, which forms the subject of the present chapter, is the van der Waals dispersion force. For two neutral nonpolar molecules in their ground electronic states, this potential was first calculated using the methods of a quantized field theory by Casimir and Polder (1948). Employing the minimal-coupling Hamiltonian and invoking the long-wavelength approximation, in which the spatial variations of the vector potential were neglected, and leading to the electric dipole approximation, they computed the interaction energy as a function of the internuclear separation distance R and found the remarkable result that at separations large relative to reduced characteristic transition wavelengths occurring within the molecular species, the energy shift varied as R^{-7}. This was in direct contrast to the accepted inverse sixth power dependence on R of the dispersion interaction first found by London (1930), whose perturbative calculation used a static dipolar coupling potential. The diminution in the strength of the interaction at long range was attributed by Casimir and Polder to proper allowance being made for the finite speed of propagation of electromagnetic signals in the fully quantum mechanical formulation of the theory. Fluctuations in the charge distribution taking place at one center, which induce a similar distortion in the electron density of the other particle, are therefore felt by the second molecule after a time delay R/c, by which further time the charge cloud in the first species will have subsequently redistributed, no longer coinciding with its original configuration. London's familiar result was found to follow as the short-range limit of the general form of the potential valid for all R, and was understood to arise from instantaneous mediation of electromagnetic influences, a singularly unphysical feature of the semiclassical method. While there has not yet been explicit experimental verification of the Casimir–Polder potential in the microscopic regime since the derivation of their result, there have been macroscopic measurements as well as a number of experiments concerned in general with Casimir effects. The final section of this chapter is devoted to this topic.

In this chapter, three different physical viewpoints and calculational schemes within the multipolar framework are adopted to evaluate the van der Waals dispersion force. Each comes with its own set of merits and drawbacks. In the first method to be presented, diagrammatic time-dependent perturbation theory is employed to calculate the energy shift. This is the often used method of attack and has the advantage of providing a visual representation of the interpretation of the interaction as due to the exchange of two virtual photons. One disadvantage of this technique is the difficulty associated with its application to the computation of the interaction energy when one of the pairs is electronically excited. Nonetheless, the calculation is detailed in Section 5.6. Afterward, a second approach is introduced. It relies on the Maxwell field operators derived in Chapter 2. The method takes the form of a response theory in which one molecule is viewed as a test polarizable body in the electromagnetic field of a second source object, and vice versa. A distinct benefit of this approach is that the energy shift when both entities are in electronically excited states is easily calculated from the outset, with potentials when one or none of the molecules are excited reducing as special cases of the more general result. Finally, a third variant is presented, which like the second method enables both ground- and excited-state dispersion energies to be readily calculated and is commonly known as the induced multipole moment method. This approach is based on the fact that a moment is induced in a polarizable molecule by the action of a radiation field. The moments induced at each center are coupled to the retarded resonant multipole–multipole interaction tensor, whose dipole (both electric and magnetic terms) form was calculated and appeared in the treatment of the resonant transfer of energy to low order in the previous chapter. By taking the expectation value of this product of moments and the coupling tensor over appropriate molecular and field states, the energy shift is obtained in a facile manner. This particular approach has a number of advantages over the first two mentioned above when other intermolecular process are evaluated, as will be shown in subsequent chapters.

Each of the three alternative physical and computational approaches is then used to calculate higher multipole moment contributions to the retarded dispersion interaction. Chief among them is the term proportional to the product of mixed electric–magnetic dipole polarizability of each molecule that characterizes the energy shift between optically active species, which is found to be discriminatory. Other contributions include those involving a magnetic dipole susceptible molecule, as well as the effect of the diamagnetic coupling term, and electric quadrupole and octupole and diamagnetic interaction terms, which have recently acquired significance due to the computation of highly accurate dispersion potentials for homo atom dimers comprising hydrogen, helium, and alkali metals. Of the four mentioned interaction terms, only the magnetic dipole and diamagnetic contribution to the dispersion energy shift will be examined in detail. Using the methods to be described in this chapter, higher order terms involving electric quadrupole and octupole interaction terms may be evaluated in similar fashion to that used for electric and magnetic dipole coupling terms.

5.2 CASIMIR–POLDER POTENTIAL: PERTURBATION THEORY

The dispersion energy shift is most commonly calculated by means of diagrammatic time-dependent perturbation theory (Craig and Thirunamachandran, 1998a). Within the multipolar formalism, this interaction has a simple interpretation. It is viewed as arising from the exchange of two virtual photons of modes $(\vec{p},\varepsilon)$ and $(\vec{p}',\varepsilon')$ between the pair. For two neutral nonpolar molecules A and B, both in their electronic ground states, the total Hamiltonian for the system is given by

$$H = H_0 + H_{\mathrm{int}}, \tag{5.2.1}$$

where

$$H_0 = H_{\mathrm{mol}}(A) + H_{\mathrm{mol}}(B) + H_{\mathrm{rad}} \tag{5.2.2}$$

and

$$H_{\mathrm{int}} = H_{\mathrm{int}}(A) + H_{\mathrm{int}}(B). \tag{5.2.3}$$

To leading order in perturbation theory, the dispersion potential is computed using the fourth-order formula

$$\Delta E = -\sum_{I,\,II,\,III} \frac{\langle 0|H_{\mathrm{int}}|III\rangle\langle III|H_{\mathrm{int}}|II\rangle\langle II|H_{\mathrm{int}}|I\rangle\langle I|H_{\mathrm{int}}|0\rangle}{(E_{III}-E_0)(E_{II}-E_0)(E_I-E_0)}, \tag{5.2.4}$$

when the interaction Hamiltonian is linear in the electromagnetic field. On making the electric dipole approximation, (5.2.3) assumes the form

$$H_{\mathrm{int}} = -\varepsilon_0^{-1}\vec{\mu}(A)\cdot\vec{d}^{\perp}(\vec{R}_A) - \varepsilon_0^{-1}\vec{\mu}(B)\cdot\vec{d}^{\perp}(\vec{R}_B), \tag{5.2.5}$$

where A and B are positioned at $\vec{R}_A$ and $\vec{R}_B$, respectively. An additional term in the perturbation theory expression for the total energy shift (5.2.4) due to normalization of the wavefunction has been omitted since it applies only when the molecules are polar, a property that is not being considered in the present treatment. The initial and final states appearing in the equation for the interaction energy are the same in this problem and are denoted by the ket $|0\rangle = |E_0^A, E_0^B; 0(\vec{p}, \varepsilon), 0(\vec{p}', \varepsilon')\rangle$ corresponding to both molecules in the their lowest energy state, E_0^ξ, $\xi = A, B$, and the field without photons. That the dispersion force is a manifestation of fluctuations associated with the electromagnetic vacuum is evident from the state representation of the system. In expression (5.2.4), the sums are taken over all possible intermediate states that link $|0\rangle$ via the coupling Hamiltonian (5.2.3). To facilitate evaluation of the energy shift and the writing of intermediate states, as well as calculating differences in energy between intermediate and ground states, as occurring in the denominator of (5.2.4), time-ordered diagrams may be drawn. For the dispersion interaction, which involves traversal of two virtual photons between A and B, there are 4! distinct permutations of electron–photon coupling vertices of the type exemplified by the interaction Hamiltonian (5.2.5). Because virtual photons are ultimately indistinguishable, with the primed and unprimed labels characterizing the two modes being introduced merely as a device to aid calculation, with each mode being individually summed over, the overall number of graphs depicting this process is reduced by a factor of two, thereby avoiding any double counting of virtual photons. The relevant graphs are illustrated in Fig. 5.1 in which the two virtual photons and their respective modes are designated by ϕ and ϕ'. The excited energy levels of molecules A and B are labeled by $|r\rangle$ and $|s\rangle$, respectively. Virtual photon emission from a molecular ground state may be understood by recourse to the time–energy uncertainty principle. For time intervals short subject to $\Delta E \Delta t \geq (\hbar/2)$, sufficient energy is acquired from the vacuum field to permit creation of such a virtual photon. From the various time-ordered sequences, four types of intermediate states are clearly evident. They comprise (i) both species in the ground state with two virtual photons in transit, (ii) both molecules excited with simultaneous propagation of two virtual photons, (iii) both molecules excited with no photons present, and (iv) one species excited with one virtual photon being exchanged. The intermediate states along with the ket $|0\rangle$ are used to form and compute the numerator in (5.2.4), with the energy denominators read off from the individual graph and the last of these are listed explicitly in Table 5.1. The interaction energy is obtained from the

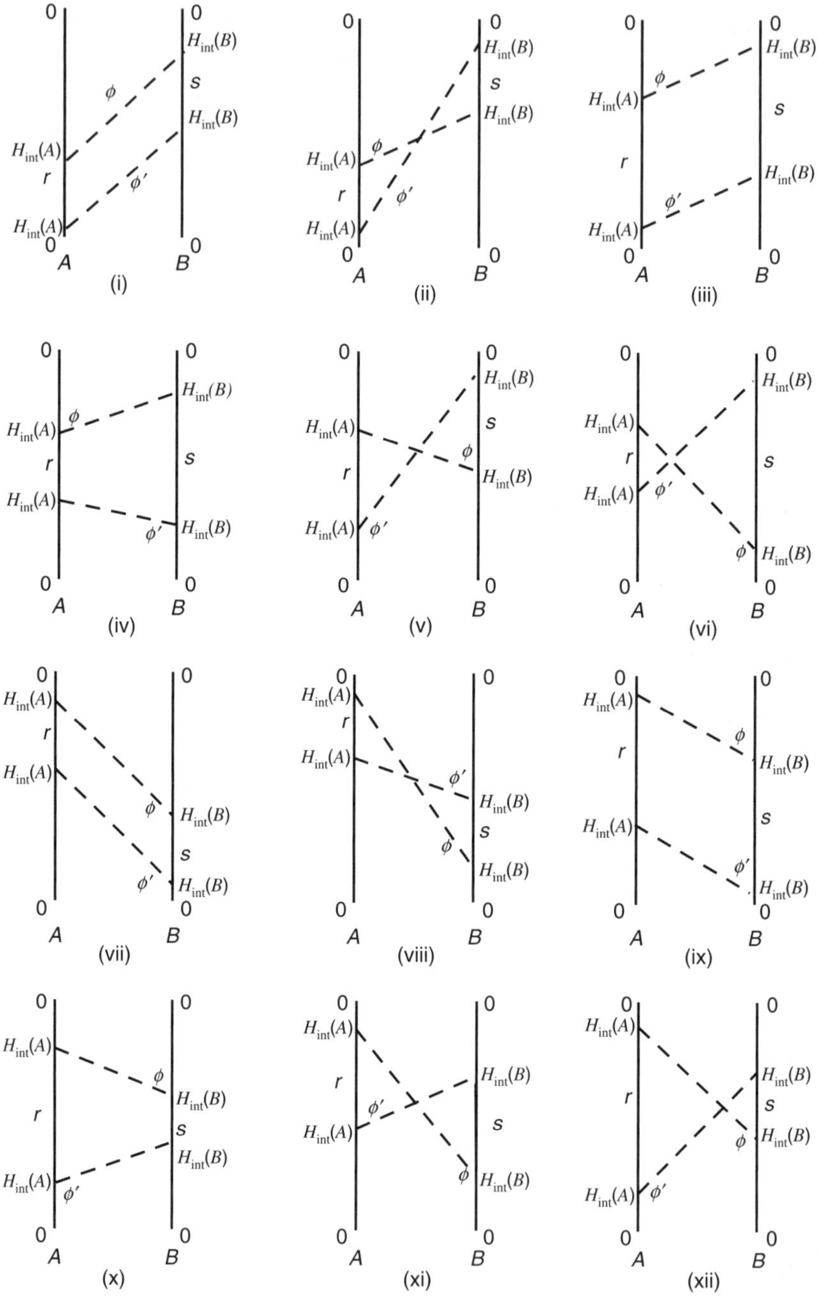

FIGURE 5.1 Twelve time-ordered graphs used for the calculation of ground-state dispersion potential.

TABLE 5.1 Energy Denominators Used in the Calculation of Casimir–Polder Potential

Graph	Denominator
(i)	$(E_{s0}+\hbar cp)(\hbar cp+\hbar cp')(E_{r0}+\hbar cp')$
(ii)	$(E_{s0}+\hbar cp')(\hbar cp+\hbar cp')(E_{r0}+\hbar cp')$
(iii)	$(E_{s0}+\hbar cp)(E_{r0}+E_{s0})(E_{r0}+\hbar cp')$
(iv)	$(E_{s0}+\hbar cp)(E_{r0}+E_{s0})(E_{s0}+\hbar cp')$
(v)	$(E_{s0}+\hbar cp')(E_{r0}+E_{s0}+\hbar cp+\hbar cp')(E_{r0}+\hbar cp')$
(vi)	$(E_{s0}+\hbar cp')(E_{r0}+E_{s0}+\hbar cp+\hbar cp')(E_{s0}+\hbar cp)$
(vii)	$(E_{s0}+\hbar cp')(\hbar cp+\hbar cp')(E_{r0}+\hbar cp)$
(viii)	$(E_{s0}+\hbar cp)(\hbar cp+\hbar cp')(E_{r0}+\hbar cp)$
(ix)	$(E_{s0}+\hbar cp')(E_{r0}+E_{s0})(E_{r0}+\hbar cp)$
(x)	$(E_{r0}+\hbar cp)(E_{r0}+E_{s0})(E_{r0}+\hbar cp')$
(xi)	$(E_{s0}+\hbar cp)(E_{r0}+E_{s0}+\hbar cp+\hbar cp')(E_{r0}+\hbar cp)$
(xii)	$(E_{r0}+\hbar cp')(E_{r0}+E_{s0}+\hbar cp+\hbar cp')(E_{r0}+\hbar cp)$

expression (5.2.4) by adding the contribution from each of the 12 graphs. It is found to be

$$\Delta E=-\sum_{\vec{p},\vec{p}'}\sum_{\varepsilon,\varepsilon'}\sum_{r,s}\left(\frac{\hbar cp}{2\varepsilon_0 V}\right)\left(\frac{\hbar cp'}{2\varepsilon_0 V}\right)e_i^{(\varepsilon)}(\vec{p})\bar{e}_k^{(\varepsilon)}(\vec{p})e_j^{(\varepsilon')}(\vec{p}')\bar{e}_l^{(\varepsilon')}(\vec{p}')$$

$$\times\mu_i^{0r}(A)\mu_j^{r0}(A)\mu_k^{0s}(B)\mu_l^{s0}(B)\,\mathrm{e}^{i(\vec{p}+\vec{p}')\cdot\vec{R}}\sum_{a=i}^{xii}D_a^{-1}. \tag{5.2.6}$$

On deriving (5.2.6), use has been made of the fact that the factor preceding the exponential function is invariant to the sign of $\vec{p}$ and/or $\vec{p}'$, so that the sign or signs of these vectors may be changed in the summands to give the form of the exponent written. Appearing in the expression for the energy shift are the transition electric dipole moment matrix elements of each molecule taken over ground and excited electronic states, the internuclear separation distance vector $\vec{R}=\vec{R}_B-\vec{R}_A$, with the energy denominator of graph (a) signified by D_a^{-1}.

The energy denominators from the 12 graphs may be summed to give

$$\sum_{a=i}^{xii} D_a^{-1} = \frac{4(k_{r0}+k_{s0}+p)}{\hbar^3 c^3 (k_{r0}+k_{s0})(k_{r0}+p)(k_{s0}+p)} \left(\frac{1}{(p+p')} - \frac{1}{(p-p')} \right), \tag{5.2.7}$$

so that the energy shift becomes

$$\Delta E = -\sum_{\vec{p},\vec{p}'} \sum_{r,s} \left(\frac{pp'}{\hbar c \varepsilon_0^2 V^2} \right) (\delta_{ik} - \hat{p}_i \hat{p}_k)(\delta_{jl} - \hat{p}'_j \hat{p}'_l) \mu_i^{0r}(A) \mu_j^{r0}(A) \mu_k^{0s}(B)$$
$$\times \mu_l^{s0}(B)\, \mathrm{e}^{i(\vec{p}+\vec{p}')\cdot\vec{R}} \frac{(k_{r0}+k_{s0}+p)}{(k_{r0}+k_{s0})(k_{r0}+p)(k_{s0}+p)}$$
$$\times \left(\frac{1}{(p+p')} - \frac{1}{(p-p')} \right), \tag{5.2.8}$$

after carrying out the polarization sum using the identity (1.4.56). Converting the wavevector sums to integrals and carrying out the angular averages using

$$\frac{1}{4\pi} \int (\delta_{ij} - \hat{p}_i \hat{p}_j) \mathrm{e}^{i\vec{p}\cdot\vec{R}} \mathrm{d}\Omega = \mathrm{Im}\, F_{ij}(pR), \tag{5.2.9}$$

where

$$F_{ij}(pR) = \left\{ (\delta_{ij} - \hat{R}_i \hat{R}_j) \frac{1}{pR} + (\delta_{ij} - 3\hat{R}_i \hat{R}_j) \left(\frac{i}{p^2R^2} - \frac{1}{p^3R^3} \right) \right\} \mathrm{e}^{ipR}, \tag{5.2.10}$$

produces

$$-\frac{1}{4\pi^4 \hbar c \varepsilon_0^2} \sum_{r,s} \mu_i^{0r}(A) \mu_j^{r0}(A) \mu_k^{0s}(B) \mu_l^{s0}(B) \frac{1}{(k_{r0}+k_{s0})}$$
$$\times \int_0^\infty \int_0^\infty p^3 p'^3 \frac{(k_{r0}+k_{s0}+p)}{(k_{r0}+p)(k_{s0}+p)} \left(\frac{1}{(p+p')} - \frac{1}{(p-p')} \right)$$
$$\times \mathrm{Im}[F_{ik}(pR)] \mathrm{Im}[F_{jl}(p'R)] \mathrm{d}p\, \mathrm{d}p'. \tag{5.2.11}$$

Because $\mathrm{Im}[F_{jl}(p'R)]$ is an even function of p', the limits of the p'-integral in (5.2.11) can be extended to $-\infty$ to ∞ and the principal value taken at the pole $p' = -p$, so that

$$\int_{-\infty}^{\infty} \frac{p'^3}{(p+p')} \mathrm{Im}\left[F_{jl}(p'R)\right] \mathrm{d}p' = \pi p^3 \,\mathrm{Re}\left[F_{jl}(pR)\right], \tag{5.2.12}$$

giving for ΔE the expression

$$\Delta E = -\frac{1}{4\pi^3 \hbar c \varepsilon_0^2} \sum_{r,s} \mu_i^{0r}(A)\mu_j^{r0}(A)\mu_k^{0s}(B)\mu_l^{s0}(B) \frac{1}{(k_{r0}+k_{s0})}$$
$$\times \int_0^{\infty} p^6 \frac{(k_{r0}+k_{s0}+p)}{(k_{r0}+p)(k_{s0}+p)} \mathrm{Re}\left[F_{jl}(pR)\right] \mathrm{Im}[F_{ik}(pR)] \mathrm{d}p. \tag{5.2.13}$$

On multiplying the real and imaginary parts of the tensor field $F_{ij}(pR)$ appearing in (5.2.13), the integral in the equation for the energy shift can be written as a difference of two integrals,

$$\frac{1}{4i} \int_0^{\infty} \mathrm{d}p \frac{(k_{r0}+k_{s0}+p)}{(k_{r0}+p)(k_{s0}+p)} \mathrm{e}^{2ipR} p^6$$
$$\times \left[\frac{\alpha_{ik}\alpha_{jl}}{p^2R^2} + \frac{i(\alpha_{ik}\beta_{jl}+\beta_{ik}\alpha_{jl})}{p^3R^3} - \frac{(\alpha_{ik}\beta_{jl}+\beta_{ik}\alpha_{jl}+\beta_{ik}\beta_{jl})}{p^4R^4} - \frac{2i\beta_{ik}\beta_{jl}}{p^5R^5} + \frac{\beta_{ik}\beta_{jl}}{p^6R^6}\right]$$
$$-\frac{1}{4i} \int_0^{\infty} \mathrm{d}p \frac{(k_{r0}+k_{s0}+p)}{(k_{r0}+p)(k_{s0}+p)} \mathrm{e}^{-2ipR} p^6$$
$$\times \left[\frac{\alpha_{ik}\alpha_{jl}}{p^2R^2} - \frac{i(\alpha_{ik}\beta_{jl}+\beta_{ik}\alpha_{jl})}{p^3R^3} - \frac{(\alpha_{ik}\beta_{jl}+\beta_{ik}\alpha_{jl}+\beta_{ik}\beta_{jl})}{p^4R^4} + \frac{2i\beta_{ik}\beta_{jl}}{p^5R^5} + \frac{\beta_{ik}\beta_{jl}}{p^6R^6}\right], \tag{5.2.14}$$

where $\alpha_{ij} = \delta_{ij} - \hat{R}_i\hat{R}_j$ and $\beta_{ij} = \delta_{ij} - 3\hat{R}_i\hat{R}_j$. Inserting $p = iu$ in the first integral and $p = -iu$ in the second one transforms the integral to a complex variable, after which their sum gives

$$\frac{1}{4} \int_0^{\infty} \left[\frac{(k_{r0}+k_{s0}+iu)}{(k_{r0}+iu)(k_{s0}+iu)} + \frac{(k_{r0}+k_{s0}-iu)}{(k_{r0}-iu)(k_{s0}-iu)}\right] \left[\frac{\alpha_{ik}\alpha_{jl}}{u^2R^2} + \frac{(\alpha_{ik}\beta_{jl}+\beta_{ik}\alpha_{jl})}{u^3R^3}\right.$$
$$\left. + \frac{(\alpha_{ik}\beta_{jl}+\beta_{ik}\alpha_{jl}+\beta_{ik}\beta_{jl})}{u^4R^4} + \frac{2\beta_{ik}\beta_{jl}}{u^5R^5} + \frac{\beta_{ik}\beta_{jl}}{u^6R^6}\right] \mathrm{e}^{-2uR} u^6 \mathrm{d}u. \tag{5.2.15}$$

Substituting (5.2.15) into the integral occurring in equation (5.2.13) gives the Casimir–Polder energy shift for a pair of molecules in fixed orientation relative to each other,

$$\Delta E = -\frac{1}{8\pi^3\varepsilon_0^2\hbar c}\sum_{r,s}\mu_i^{0r}(A)\mu_j^{r0}(A)\mu_k^{0s}(B)\mu_l^{s0}(B)\int_0^\infty \frac{k_{r0}k_{s0}}{(k_{r0}^2+u^2)(k_{s0}^2+u^2)}$$
$$\times\left[\frac{\alpha_{ik}\alpha_{jl}}{u^2R^2}+\frac{(\alpha_{ik}\beta_{jl}+\beta_{ik}\alpha_{jl})}{u^3R^3}+\frac{(\alpha_{ik}\beta_{jl}+\beta_{ik}\alpha_{jl}+\beta_{ik}\beta_{jl})}{u^4R^4}\right.$$
$$\left.+\frac{2\beta_{ik}\beta_{jl}}{u^5R^5}+\frac{\beta_{ik}\beta_{jl}}{u^6R^6}\right]e^{-2uR}u^6 du. \quad (5.2.16)$$

Orientationally averaging the molecular multipole moments using the result

$$\langle\mu_i^{0r}(A)\mu_j^{r0}(A)\rangle\langle\mu_k^{0s}(B)\mu_l^{s0}(B)\rangle = \frac{1}{9}\delta_{ij}\delta_{kl}|\vec{\mu}^{0r}(A)|^2|\vec{\mu}^{0s}(B)|^2 \quad (5.2.17)$$

and contracting the geometrical tensors yields the familiar Casimir–Polder potential for isotropic systems,

$$\Delta E = -\frac{1}{36\pi^3\varepsilon_0^2\hbar c}\sum_{r,s}|\vec{\mu}^{0r}(A)|^2|\vec{\mu}^{0s}(B)|^2\int_0^\infty \frac{k_{r0}k_{s0}}{(k_{r0}^2+u^2)(k_{s0}^2+u^2)}$$
$$\times\left[\frac{1}{u^2R^2}+\frac{2}{u^3R^3}+\frac{5}{u^4R^4}+\frac{6}{u^5R^5}+\frac{3}{u^6R^6}\right]e^{-2uR}u^6 du, \quad (5.2.18)$$

which holds for all R beyond wavefunction overlap. An expression equivalent to (5.2.18) may be written in terms of the isotropic polarizability $\alpha(\xi;iu)$ of species $\xi = A,B$ at the imaginary frequency $\omega = icu$, where

$$\alpha(\xi;iu) = \frac{2}{3}\sum_t E_{t0}\frac{|\vec{\mu}^{t0}(\xi)|^2}{E_{t0}^2+(\hbar cu)^2}. \quad (5.2.19)$$

It is given by

$$\Delta E = -\frac{\hbar c}{16\pi^3\varepsilon_0^2}\int_0^\infty du\, u^6 e^{-2uR}\alpha(A;iu)\alpha(B;iu)$$
$$\times\left[\frac{1}{u^2R^2}+\frac{2}{u^3R^3}+\frac{5}{u^4R^4}+\frac{6}{u^5R^5}+\frac{3}{u^6R^6}\right]. \quad (5.2.20)$$

It is instructive to examine the asymptotic behavior of the energy shift (5.2.18) in the limits of large and small intermolecular separation. At short range, R is much smaller than characteristic reduced transition wavelengths, so that $kR \ll 1$. With this condition $e^{-2uR} \approx 1$, and the dominant term within square brackets is that proportional to R^{-6}. Employing the integral relation

$$\frac{\pi}{(a+b)} = 2\int_0^\infty \frac{ab\,du}{(a^2+u^2)(b^2+u^2)}, \quad a,b>0, \tag{5.2.21}$$

the energy shift in the near zone is

$$\Delta E_{\mathrm{NZ}} = -\frac{1}{24\pi^2\varepsilon_0^2 R^6}\sum_{r,s}\frac{|\vec{\mu}^{0r}(A)|^2|\vec{\mu}^{0s}(B)|^2}{(E_{r0}+E_{s0})}, \tag{5.2.22}$$

which is instantly recognizable as the London dispersion energy. At the other extreme of separation, $kR \gg 1$ and u^2 may be ignored in relation to k_{r0}^2 and k_{s0}^2 in the denominators of the general result. After performing the u-integral using the standard integral result

$$\int_0^\infty x^n e^{-\alpha x}dx = n!\alpha^{-n-1}, \quad \mathrm{Re}\,\alpha > 0, \tag{5.2.23}$$

the far zone asymptote is

$$\Delta E_{\mathrm{FZ}} = -\frac{23\hbar c}{64\pi^3\varepsilon_0^2 R^7}\alpha(A;0)\alpha(B;0), \tag{5.2.24}$$

in which the static polarizabilities appear, which may be obtained from (5.2.19) on letting $u=0$. The potential at large separations exhibits an inverse seventh power dependence on R and is attributed to the effects of retardation.

As indicated earlier, the calculation of the dispersion potential between a pair of neutral, nonpolar molecules in their ground electronic states constituted one of the foremost successes of the theory of quantum electrodynamics. This pioneering calculation was first carried out by Casimir and Polder in 1948. These researchers employed time-dependent perturbation theory on the minimal-coupling form of Hamiltonian for the interaction of radiation with matter, instead of the multipolar framework treatment presented herein. Nevertheless, in their computations, all contributory terms correct up to the fourth order in the electronic charge still had to be retained. These comprise single-photon interaction vertices of the form $(e/m)\vec{p}\cdot\vec{a}$ that arise in the fourth order of perturbation theory, a third-order term involving coupling of the $(e^2/2m)\vec{a}^2$ two-photon interaction vertex at one site with the $(e/m)\vec{p}\cdot\vec{a}$ interaction at the other center, and a second-order contribution that arises from purely two-photon coupling vertices at each molecule of the $(e^2/2m)\vec{a}^2$ type. Knowing that the static intermolecular Coulomb interaction appears explicitly in the minimal-coupling scheme necessitates the inclusion of the effects of this term. This is composed of a term in the second order of perturbation theory that involves only the static Coulomb interaction and a third-order term arising from the coupling of the mixed single-photon $(e/m)\vec{p}\cdot\vec{a}$ contribution and the Coulomb interaction. An intricate calculation accounting for each of these contributions yields the energy shift (5.2.20).

5.3 NEAR-ZONE POTENTIAL: LONDON DISPERSION ENERGY

In the diagrammatic perturbation theory calculation of the dispersion potential applicable to all intermolecular separation distances outside the region of charge overlap, the interaction was viewed as being mediated by the exchange of two virtual photons. The near- and far-zone limiting energy shifts were obtained from the full expression after making the appropriate physical and mathematical approximations. In this and the following sections, it is shown how the short- and long-range asymptotic interaction energies may be evaluated directly by retaining and summing over a subset of the time-ordered graphs shown in Fig. 5.1, with the appropriate graphs to be summed over being justified on physical grounds.

To be effective at short separation distances, virtual photons with high wavevector values will dominate the interaction. This is most easily understood by appeal to the time–energy uncertainty principle. In the near zone, photons will be emitted and absorbed after a very short time, during which the energy borrowed by them from the vacuum will be large. Table 5.1 reveals that denominators arising from graphs (iii), (iv), (ix), and (x) are the smallest, and consequently, make the biggest contribution to the energy shift. These four energy denominators can, therefore, all be approximated to $\hbar cp\hbar cp'(E_{r0}+E_{s0})$. It is interesting to note that in each of the four graphs singled out, only one virtual photon propagates at any given time instant. Summing the contributions from the four graphs mentioned above in the usual way produces

$$\Delta E = -4\sum_{\vec{p},\vec{p}'}\sum_{r,s}\left(\frac{\hbar cp}{2\varepsilon_0 V}\right)\left(\frac{\hbar cp'}{2\varepsilon_0 V}\right)(\delta_{ik}-\hat{p}_i\hat{p}_k)(\delta_{jl}-\hat{p}'_j\hat{p}'_l)$$
$$\times\mu_i^{0r}(A)\mu_j^{r0}(A)\mu_k^{0s}(B)\mu_l^{s0}(B)\,\mathrm{e}^{i(\vec{p}+\vec{p}')\cdot\vec{R}}\frac{1}{(\hbar c)^2 pp'(E_{r0}+E_{s0})}, \qquad (5.3.1)$$

after performing the polarization sums. Converting the $\vec{p},\vec{p}'$-sums to integrals gives

$$\Delta E = -\frac{1}{\varepsilon_0^2}\sum_{r,s}\frac{1}{(E_{r0}+E_{s0})}\mu_i^{0r}(A)\mu_j^{r0}(A)\mu_k^{0s}(B)\mu_l^{s0}(B)$$
$$\times\frac{1}{(2\pi)^6}\int\!\!\int(\delta_{ik}-\hat{p}_i\hat{p}_k)(\delta_{jl}-\hat{p}'_j\hat{p}'_l)\mathrm{e}^{i(\vec{p}+\vec{p}')\cdot\vec{R}}\mathrm{d}^3\vec{p}\,\mathrm{d}^3\vec{p}'. \qquad (5.3.2)$$

The integrals featuring in (5.3.2) may be evaluated using the relation

$$\frac{1}{(2\pi)^3}\int(\delta_{ij}-\hat{p}_i\hat{p}_j)\mathrm{e}^{i\vec{p}\cdot\vec{R}}\,\mathrm{d}^3\vec{p} = -\frac{1}{4\pi R^3}(\delta_{ij}-3\hat{R}_i\hat{R}_j), \qquad (5.3.3)$$

which holds for positive R only, resulting in the near-zone energy shift for an oriented A–B pair,

$$\Delta E = -\frac{1}{16\pi^2\varepsilon_0^2 R^6}\sum_{r,s}\frac{1}{(E_{r0}+E_{s0})}\mu_i^{0r}(A)\mu_j^{r0}(A)\mu_k^{0s}(B)\mu_l^{s0}(B)$$
$$\times(\delta_{ik}-3\hat{R}_i\hat{R}_k)(\delta_{jl}-3\hat{R}_j\hat{R}_l). \qquad (5.3.4)$$

The energy shift (5.2.22) follows immediately from (5.3.4) after random averaging.

In the near zone, the energy shift may be understood as arising from instantaneous coupling between the two molecules. It is commonly calculated using the second-order perturbation theory when the static dipolar coupling potential,

$$V_{AB} = \frac{1}{4\pi\varepsilon_0 R^3}\mu_i(A)\mu_j(B)(\delta_{ij}-3\hat{R}_i\hat{R}_j), \qquad (5.3.5)$$

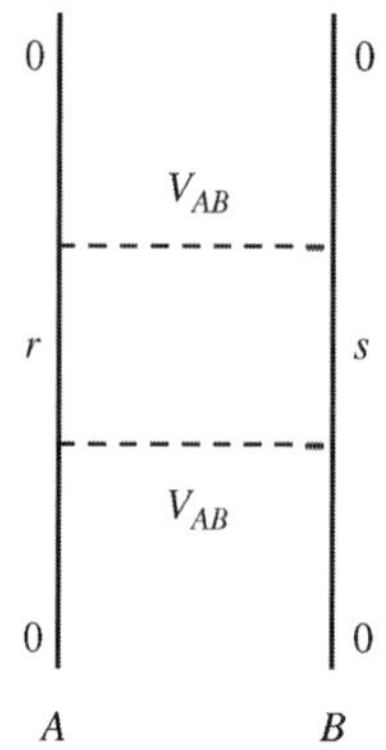

FIGURE 5.2 Diagram used in the calculation of London dispersion energy depicting static dipolar coupling.

is used as the perturbation operator. Diagrammatically, the interaction may be visualized as illustrated in Fig. 5.2. Because the interaction is not delayed, A and B are coupled by V_{AB} at the same time. Applying (5.3.5) in the formula for the second-order energy shift,

$$\Delta E = -\sum_I \frac{\langle 0|V_{AB}|I\rangle\langle I|V_{AB}|0\rangle}{(E_I - E_0)}, \tag{5.3.6}$$

on using the ground-state representation of the total system $|0\rangle$ given in the Section 5.2, the result (5.3.4) is easily obtained.

To express the London dispersion energy in terms of polarizabilities, it is common to make the average energy approximation or to invoke Unsold's theorem. An identical result can be achieved on making a two-level approximation. Denoting the lowest energy transition in molecule ξ as E^{ξ}, the static isotropic electric dipole polarizability is

$$\alpha(\xi;0) = \frac{2}{3}\frac{|\vec{\mu}(\xi)|^2}{E^{\xi}}, \tag{5.3.7}$$

where $\vec{\mu}(\xi)$ is the transition dipole between the two levels. This enables the London dispersion energy (5.2.22) to be written as

$$\Delta E = -\left(\frac{3}{64\pi^2\varepsilon_0^2 R^6}\right)\alpha(A;0)\alpha(B;0)\tilde{E}, \tag{5.3.8}$$

if the scaled energy is

$$\tilde{E} = \frac{2E^A E^B}{E^A + E^B}. \tag{5.3.9}$$

Finally, it is interesting to note that the averaged near-zone energy shift (5.2.22) can be written in closed form as an integral over the product of the polarizabilities at imaginary frequency of A and B in a form that resembles the Casimir–Polder potential (5.2.20). This is accomplished by using the integral representation (5.2.21) after extending the domain to cover both positive and negative imaginary frequencies to give

$$\Delta E = -\frac{1}{24\pi^3\varepsilon_0^2R^6}\sum_{r,\,s}\int_{-\infty}^{\infty} \mathrm{d}u \frac{E_{r0}E_{s0}|\vec{\mu}^{0r}(A)|^2|\vec{\mu}^{s0}(B)|^2}{(E_{r0}^2+u^2)(E_{s0}^2+u^2)}. \tag{5.3.10}$$

Substituting for the isotropic dynamic polarizabilities from (5.2.19) results in

$$\Delta E = -\left(\frac{3}{32\pi^3\varepsilon_0^2R^6}\right)\int_{-\infty}^{\infty} \alpha(A;iu)\alpha(B;iu)\mathrm{d}u. \tag{5.3.11}$$

5.4 FAR-ZONE DISPERSION POTENTIAL

In the wave zone, a pair of interacting molecules is separated by a distance considerably greater than the reduced wavelengths of molecular transitions. Hence, at this separation regime, the most important contribution to the energy shift is made by virtual photons with low values of wavevector p and p'. This means that energy may be borrowed from the vacuum for longer durations and can consequently be effective over larger distances. From Table 5.1, it is seen that denominators that satisfy this condition each contain the factor $\hbar c(p+p')$. Once again four graphs of Fig. 5.1 contribute to the potential in the far zone. They are diagrams (i), (ii), (vii), and (viii). Since $E_{r0}, E_{s0} \gg \hbar cp, \hbar cp'$, each energy denominator product may be approximated to $E_{r0}E_{s0}(\hbar cp+\hbar cp')$. Adding the contribution from the four graphs leads to

$$\Delta E = -4\sum_{\vec{p},\,\vec{p}'}\sum_{\varepsilon,\,\varepsilon'}\sum_{r,\,s}\left(\frac{\hbar cp}{2\varepsilon_0V}\right)\left(\frac{\hbar cp'}{2\varepsilon_0V}\right)e_i^{(\varepsilon)}(\vec{p})\bar{e}_k^{(\varepsilon)}(\vec{p})e_j^{(\varepsilon')}(\vec{p}')\bar{e}_l^{(\varepsilon')}(\vec{p}')$$
$$\times\mu_i^{0r}(A)\mu_j^{r0}(A)\mu_k^{0s}(B)\mu_l^{s0}(B)\,\mathrm{e}^{i(\vec{p}+\vec{p}')\cdot\vec{R}}\frac{1}{E_{r0}E_{s0}(\hbar cp+\hbar cp')}. \tag{5.4.1}$$

After performing the polarization sums, converting the wavevector sums to integrals, orientationally averaging the transition dipole moments, and expressing the molecular part in terms of the isotropic static polarizability

$$\alpha(\xi;0) = \frac{2}{3}\sum_t \frac{|\vec{\mu}^{t0}(\xi)|^2}{E_{t0}}, \tag{5.4.2}$$

expression (5.4.1) becomes

$$\Delta E = -\frac{\hbar c}{4\varepsilon_0^2}\alpha(A;0)\alpha(B;0)\iint \frac{pp'}{(p+p')}(\delta_{ij}-\hat{p}_i\hat{p}_j)(\delta_{ij}-\hat{p}'_i\hat{p}'_j)$$
$$\times e^{i(\vec{p}+\vec{p}')\cdot\vec{R}}\frac{d^3\vec{p}}{(2\pi)^3}\frac{d^3\vec{p}'}{(2\pi)^3}. \tag{5.4.3}$$

To separate the variables p and p', the following integral representation is applied:

$$\frac{1}{(p+p')} = R\int_0^\infty d\eta\, e^{-(p+p')R\eta}, \tag{5.4.4}$$

enabling the energy shift to be written as

$$\Delta E = -\frac{\hbar c R}{4\varepsilon_0^2}\alpha(A;0)\alpha(B;0)\int_0^\infty d\eta\left\{\frac{1}{(2\pi)^3}\int p^3(\delta_{ij}-\hat{p}_i\hat{p}_j)e^{i\vec{p}\cdot\vec{R}}e^{-pR\eta}dp\,d\Omega\right\}^2. \tag{5.4.5}$$

Concentrating on the factor occurring within braces, the angular integration may be carried out straightforwardly to leave

$$\frac{1}{2\pi^2}\int_0^\infty dp\,p^3\left[(\delta_{ij}-\hat{R}_i\hat{R}_j)\frac{\sin pR}{pR}+(\delta_{ij}-3\hat{R}_i\hat{R}_j)\left(\frac{\cos pR}{p^2R^2}-\frac{\sin pR}{p^3R^3}\right)\right]e^{-pR\eta}. \tag{5.4.6}$$

The p-integrals are carried out using the standard integrals

$$\int_0^\infty e^{-ax}\sin mx\,dx = \frac{m}{a^2+m^2},\quad a>0, \tag{5.4.7a}$$

$$\int_0^\infty x e^{-ax}\cos mx\,dx = \frac{a^2-m^2}{(a^2+m^2)^2},\quad a>0, \tag{5.4.7b}$$

and

$$\int_0^\infty x^2 e^{-ax} \sin mx \, dx = \frac{2m(3a^2 - m^2)}{(a^2 + m^2)^3}, \quad a > 0, \tag{5.4.7c}$$

to give for (5.4.6),

$$\frac{1}{\pi^2 R^4}\left[(\delta_{ij} - \hat{R}_i\hat{R}_j)\frac{(3\eta^2 - 1)}{(\eta^2 + 1)^3} - (\delta_{ij} - 3\hat{R}_i\hat{R}_j)\frac{1}{(\eta^2 + 1)^2}\right]. \tag{5.4.8}$$

Squaring (5.4.8) and inserting into (5.4.5) produces

$$\Delta E = -\frac{\hbar c}{\pi^4 \varepsilon_0^2 R^7}\alpha(A;0)\alpha(B;0)\int_0^\infty d\eta \frac{(3 - 2\eta^2 + 3\eta^4)}{(\eta^2 + 1)^6}. \tag{5.4.9}$$

Performing the η-integral using

$$\int_0^\infty \frac{x^{2m}}{(ax^2 + c)^n} dx = \frac{(2m-1)!!(2n-2m-3)!!\pi}{2\cdot(2n-2)!!a^m c^{n-m-1}\sqrt{ac}}, \tag{5.4.10}$$

with $(2r + 1)!! = 1\cdot 3\cdot 5 \ldots (2r + 1)$ and $(2r)!! = 2\cdot 4\cdot 6 \ldots (2r)$, results in

$$\Delta E = -\frac{23\hbar c}{64\pi^3 \varepsilon_0^2 R^7}\alpha(A;0)\alpha(B;0), \tag{5.4.11}$$

which is the far-zone Casimir–Polder potential (5.2.24).

Finally, it is shown how second-order time-dependent perturbation theory may be used to obtain the far-zone energy shift (5.4.11). This is a viable option due to the observation that in each of the four time-ordered diagrams used to calculate the wave-zone result, intermediate state $|II\rangle$ represents a state in which two virtual photons are simultaneously in transit, in one case emitted by species A and in the other by molecule B. These four, linear in the interaction vertex graphs, may be reduced to two diagrams by collapsing the two one-photon coupling vertices at each center to produce a two-photon vertex at each site, as shown in Fig. 5.3. Each vertex is represented by an effective interaction Hamiltonian of the form

$$H_{\text{eff, int}} = -\frac{1}{2\varepsilon_0^2}\sum_{\text{modes}}\sum_{\xi}\alpha_{ij}(\xi;0)d_i^\perp(\vec{R}_\xi)d_j^\perp(\vec{R}_\xi), \quad \xi = A, B, \tag{5.4.12}$$

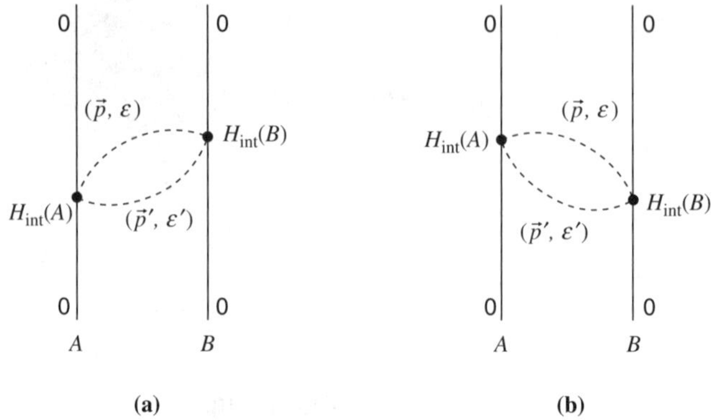

FIGURE 5.3 Collapsed two-photon graphs for the far-zone Casimir–Polder potential.

which is able to describe all combinations of annihilation/destruction events involving two-photons since the coupling Hamiltonian is now quadratic in the electric displacement field and in the process reducing the required order of perturbation theory by two. The new interaction Hamiltonian is obtained from the multipolar Hamiltonian (5.2.1) by performing a canonical transformation on the latter, with the generator chosen specifically to cancel the coupling terms linear in the electric dipole moment and displacement field (5.2.5). Interaction Hamiltonian (5.4.12) is also known as the Craig–Power Hamiltonian (Craig and Power, 1969).

From the two Feynman graphs of Fig. 5.3, the states to be used in the calculation of ΔE are easily written down. As before, the initial and final states are the same and are identical to the ket $|0\rangle$ used previously, namely, $|0\rangle = |E_0^A, E_0^B; 0(\vec{p}, \varepsilon), 0(\vec{p}', \varepsilon')\rangle$. Only one ket is required to represent the intermediate state of both graphs (Fig. 5.3a) and (Fig. 5.3b) in the second-order of perturbation theory, as the two states are identical. They are given by $|I_a\rangle = |I_b\rangle = |E_0^A, E_0^B; 1(\vec{p}, \varepsilon), 1(\vec{p}', \varepsilon')\rangle$. The energy shift is computed from

$$\Delta E = -\sum_I \frac{\langle 0|H_{\text{eff, int}}|I\rangle\langle I|H_{\text{eff, int}}|0\rangle}{(E_I - E_0)}. \tag{5.4.13}$$

To facilitate calculation, it is convenient to expand the effective interaction Hamiltonian in terms of annihilation and creation operators for the two exchanged virtual photons using the mode expansion for $\vec{d}^{\perp}(\vec{r})$. This results in

$$H_{\text{eff, int}}(\xi) = \frac{1}{2\varepsilon_0^2}\sum_{\text{modes}} \alpha_{ij}(\xi;0)\left(\frac{\hbar cp\varepsilon_0}{2V}\right)^{1/2}\left(\frac{\hbar cp'\varepsilon_0}{2V}\right)^{1/2}$$

$$\times\Big[e_i^{(\varepsilon)}(\vec{p})e_j^{(\varepsilon')}(\vec{p}')a^{(\varepsilon)}(\vec{p})a^{(\varepsilon')}(\vec{p}')\mathrm{e}^{i(\vec{p}+\vec{p}')\cdot\vec{R}_\xi}$$

$$-e_i^{(\varepsilon)}(\vec{p})\bar{e}_j^{(\varepsilon')}(\vec{p}')a^{(\varepsilon)}(\vec{p})a^{\dagger(\varepsilon')}(\vec{p}')\mathrm{e}^{i(\vec{p}-\vec{p}')\cdot\vec{R}_\xi}$$

$$-\bar{e}_i^{(\varepsilon)}(\vec{p})e_j^{(\varepsilon')}(\vec{p}')a^{\dagger(\varepsilon)}(\vec{p})a^{(\varepsilon')}(\vec{p}')\mathrm{e}^{-i(\vec{p}-\vec{p}')\cdot\vec{R}_\xi}$$

$$+\bar{e}_i^{(\varepsilon)}(\vec{p})\bar{e}_j^{(\varepsilon')}(\vec{p}')a^{\dagger(\varepsilon)}(\vec{p})a^{\dagger(\varepsilon')}(\vec{p}')\mathrm{e}^{-i(\vec{p}+\vec{p}')\cdot\vec{R}_\xi}\Big]. \quad (5.4.14)$$

Only the first and last terms of (5.4.14), namely, those that destroy and create two photons, respectively, are required in the calculation. Summing over the contributions from the two graphs gives

$$\Delta E = -\frac{\hbar c}{2\varepsilon_0^2}\sum_{\vec{p},\,\vec{p}'}\sum_{\varepsilon,\,\varepsilon'}\frac{pp'}{(p+p')}\alpha_{ij}(A;0)\alpha_{kl}(B;0)\bar{e}_i^{(\varepsilon)}(\vec{p})e_k^{(\varepsilon)}(\vec{p})$$

$$\times\bar{e}_j^{(\varepsilon')}(\vec{p}')e_l^{(\varepsilon')}(\vec{p}')\mathrm{e}^{i(\vec{p}+\vec{p}')\cdot\vec{R}}. \quad (5.4.15)$$

Proceeding in the usual way by carrying out the ε, ε'-sums, performing a molecular average and absorbing a factor of $\left(\frac{1}{3}\right)^2$ into the product of the isotropic polarizabilities, transforming the $\vec{p}$, $\vec{p}'$-sums to integrals, and dividing by two to avoid double counting of the virtual photons, which are indistinguishable, yield expression (5.4.3) for the interaction energy shift. Hence, the remainder of the computation leading to the far-zone limiting form (5.4.11) is identical to that presented earlier in this section.

5.5 STATE SEQUENCE DIAGRAMS FOR DISPERSION FORCE

Formal rules were given in Section 1.10 for the generation of interaction plane networks that depict the allowed sequence of photon absorption and emission events for any nth order process occurring in a hyperspace of dimension n. After designation of the relevant process-specific initial and final states, the intermediate levels connecting $|i\rangle$ to $|f\rangle$ in correct time-ordered sequences may be written systematically and employed in the construction of the appropriate state sequence diagram (Jenkins et al., 2002). Perturbation theory may then be used in the usual manner to arrive at the probability amplitude or energy shift for the process. This *modus operandi* was carried out in Section 4.3 for the problem of resonant exchange of excitation energy. Since migration of energy was understood to arise from single virtual photon exchange in the perturbative treatment, its representation in terms of state sequences was especially facile and could quite easily and legitimately been sketched directly from the two Feynman diagrams of Fig. 4.1, which are usually used in the calculation of the transfer rate.

As has been demonstrated in Section 5.2, the molecular QED perturbation theory computation of the van der Waals dispersion energy shift involves summation over $4!/2! = 12$ two-photon exchange diagrams. To further extend the range of application of the alternative diagrammatic approach, in particular to higher order intermolecular processes, and to elicit pros and cons of the method, the state sequence representation of the dispersion potential is obtained in this section (Alligood and Salam, 2007).

Since there are four distinguishable photon–matter interactions in the representation of the dispersion force, due to virtual emission and absorption occurring twice at each center, the hyperspace dimension n corresponds to four in this problem. With each vertex labeled by an index and each index in turn denoted by a vector whose multiplicity is one, the set of four orthogonal basis vectors for the problem are given by $I = \{1\vec{i}_1, 1\vec{i}_2, 1\vec{i}_3, 1\vec{i}_4\}$. As perturbation theory formulas demand summation over all intermediate states linking initial to final, which is achieved in the graphical method by the drawing of all topologically distinct diagrams, the analogous procedure in the state sequence formulation is accomplished via index manipulation. Using the prescription detailed in Section 1.10, hyperspace coordinates of the form (C_1, C_2, C_3, C_4) with $C_j = 0, 1$, only for $j = 1$–4, are generated from the set of vectors I. These coordinates are subsequently converted from a binary base $B = 2$ (since $c_j = 1$) to a decimal base, thereby allowing the (k, h) coordinates to be obtained for the plotting of the general 4-space interaction plane, and the writing of system states $|r_k^m\rangle$ for each vertex. This is summarized in Table 5.2. From the entries in the right-hand most column, the network plane for $n = 4$ is plotted, as illustrated in Fig. 5.4. As for the network map Fig. 4.2, which involved all processes described by two distinguishable photon emissions and/or absorptions, the interaction plane network shown in Fig. 5.4 serves as a blueprint for the construction of state sequence diagrams for any process that contains four unique photonic events. As to be expected and calculated from (1.10.11), there are 24 paths from the i-terminus to the f-terminus, $4!/1!1!1!1! = 24$. Since the two exchanged virtual photons are differentiated for calculational purposes by attaching different labels to them, a factor of one-half is introduced in the computation to account for the fact that mode properties associated with both

TABLE 5.2 Hyperspace and Interaction Plane Coordinates for Four Distinct Radiation–Matter Couplings

k	Vertex	Hyperspace Coordinate	Hyperspace Number (Base 2)	h (Base 10)	(k, h)
0	r_0^1	(0, 0, 0, 0)	0000	0	(0, 0)
1	r_1^1	(0, 0, 0, 1)	0001	1	(1, 1)
	r_1^2	(0, 0, 1, 0)	0010	2	(1, 2)
	r_1^3	(0, 1, 0, 0)	0100	4	(1, 4)
	r_1^4	(1, 0, 0, 0)	1000	8	(1, 8)
2	r_2^1	(0, 0, 1, 1)	0011	3	(2, 3)
	r_2^2	(0, 1, 0, 1)	0101	5	(2, 5)
	r_2^3	(0, 1, 1, 0)	0110	6	(2, 6)
	r_2^4	(1, 0, 0, 1)	1001	9	(2, 9)
	r_2^5	(1, 0, 1, 0)	1010	10	(2, 10)
	r_2^6	(1, 1, 0, 0)	1100	12	(2, 12)
3	r_3^1	(0, 1, 1, 1)	0111	7	(3, 7)
	r_3^2	(1, 0, 1, 1)	1011	11	(3, 11)
	r_3^3	(1, 1, 0, 1)	1101	13	(3, 13)
	r_3^4	(1, 1, 1, 0)	1110	14	(3, 14)
4	r_4^1	(1, 1, 1, 1)	1111	15	(4, 15)

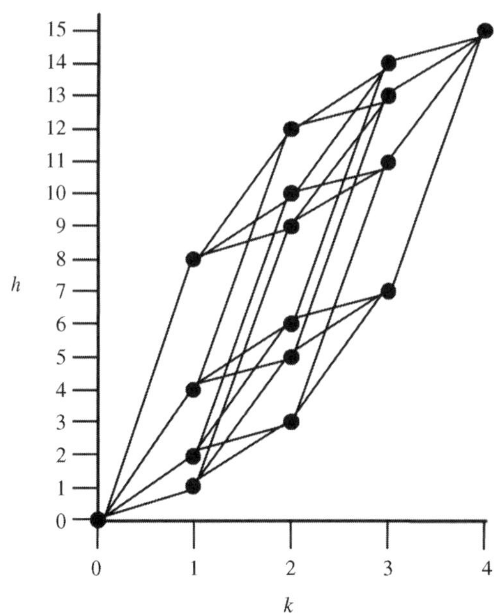

FIGURE 5.4 Network map for four unique photonic events.

virtual photons are summed over and ultimately all virtual photons are indistinguishable, being emitted and subsequently reabsorbed. Furthermore, the structure coefficients are given by the fourth row of Pascal's triangle: 1 4 6 4 1, which may be obtained from formula (1.10.14), ${}^{\{1,1,1,1\}}T_k^{4,4}$ for $k=0,1,2,3,4$. As a consequence, these structural features will appear in any ensuing state sequence diagram when degeneracy is absent.

Returning to the Casimir–Polder potential, associating indices $\vec{i}_1$ with virtual emission of photon ϕ', $\vec{i}_2$ with emission of virtual photon ϕ, $\vec{i}_3$ with virtual absorption of ϕ', and $\vec{i}_4$ with absorption of ϕ leads to the state sequence diagram shown in Fig. 5.5. Entities *A* and *B* are represented by circles on the left and right of each box, respectively, with an open circle denoting that a species is in the ground state. Easily identifiable from Fig. 5.5 are the initial and final state sequences, depicted by the left- and right-hand most boxes in the time evolution, corresponding to the states $|i\rangle = |f\rangle = |0\rangle$, denoting the vacuum state in which no real or virtual photons are present and both molecules are in their lowest energy states. After virtual excitation, entity *A* jumps to a higher level $|r\rangle$, which is represented in a cell by a circle with label *r* enclosed, while the encircled *s*

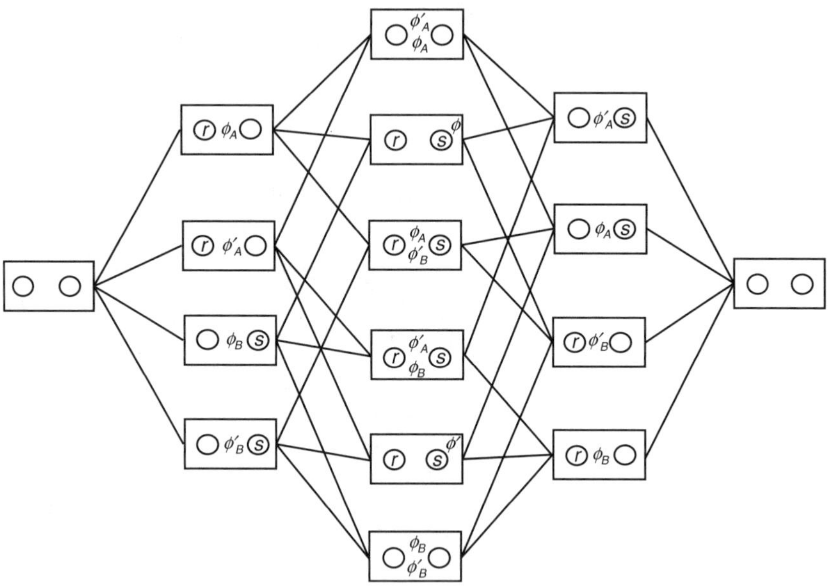

FIGURE 5.5 State sequence diagram for the Casimir–Polder potential. The pathway corresponding to graph (i) of Fig. 5.1 is

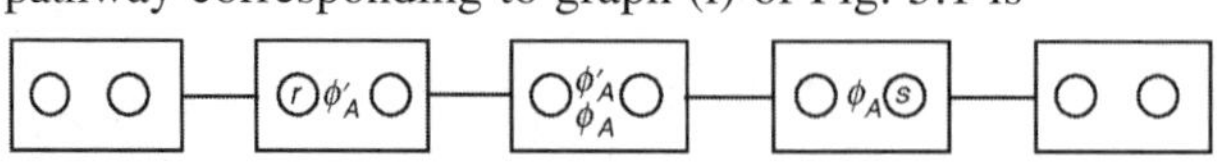.

denotes a state $|s\rangle$ to which molecule B has been electronically excited. The modes corresponding to the two exchanged virtual photons ϕ and ϕ' are designated by $(\vec{p}, \varepsilon)$ and $(\vec{p}', \varepsilon')$, respectively. A symbol ϕ or ϕ' in the upper right-hand corner of a box representing intermediate states $|II\rangle$ denotes which one of the two virtual photons has completed its propagation between the two centers. Moreover, the subscripts appearing on ϕ or ϕ' indicate the site of virtual emission. Although strictly unnecessary, these last two additional labels prevent possible ambiguity, particularly for interactions to and from the second intermediate state in which A is in excited state $|r\rangle$, B is in state $|s\rangle$, and no virtual photons are present, which occurs when only one of the virtual photons is traversing between the pair. Each of the 24 paths that can be traced out in Fig. 5.5 corresponds to one of the 4! possible time orderings of a two-virtual photon exchange Feynman diagram, half of whose number are shown explicitly in Fig. 5.1.

All of the relevant matter–field states to be employed in the fourth-order perturbation theory formula for the energy shift (5.2.4) are readily obtained from (4.3.1) and the state sequence diagram of Fig. 5.5, and the matrix elements are computed in the usual manner to yield the Casimir–Polder potential (5.2.18) along the same lines detailed in Section 5.2. This aspect is not entirely unexpected since the state sequencing approach provides an alternate pictorial representation of conventional time-dependent perturbation theory methodology. Nevertheless, a number of distinct advantages result from the use of state sequence diagrams. One obvious benefit, already mentioned, is the capturing of all time orderings associated with photon creation–destruction events for a specific process in one picture and the systematic generation of all of the states required in the calculation. Another advantage is the construction of the precursor to the state sequence diagram—the n-interaction plane. This two-dimensional network map displays the permitted connectivities between initial, intermediate, and final states for any process comprised of n distinguishable radiation–matter interactions. From this most general of situations, interaction planes and state sequence diagrams can be easily generated for the special case when two or more interaction vertices are indistinguishable, as is the case when degenerate photons are emitted or absorbed. In each of these two scenarios, the underlying isomorphism between seemingly disparate and unrelated processes is manifest. A further benefit of using state sequencing techniques, though one that did not feature in the calculation of the dispersion potential, is the expediting of computation through the potential to exploit any inherent symmetry present in a problem and the ability to group together contributory terms to the probability amplitude that are similar in structural form, aspects that are more easily identifiable relative to the standard method of calculation, which greatly aids in facilitating overall computational analysis. This becomes especially apparent when processes involving the emission and absorption of a large number of real and/or virtual photons from one or more centers are tackled.

Despite these positive attributes and the successful formulation of the state sequence methodology, there is one serious limitation of the approach presented in Section 1.10. It concerns construction of diagrams used to visualize intermolecular interactions occurring between molecules at extremes of separation. From the applications of the state sequence technique to intermolecular processes considered thus far, namely, resonance energy transfer and van der Waals dispersion, all state sequence pathways are generated, resulting in the diagram being applicable to the full range of internuclear separation distances beyond wavefunction overlap. While the limiting functional forms of the interaction may be obtained from the result valid for all R after making the appropriate physical and mathematical approximations, additional valuable insight is obtained if near- and far-zone asymptotic energy shifts or transfer rates can be arrived at directly. This was the case in Sections 5.3 and 5.4, where arguments based upon the time–energy uncertainty principle were made to ascertain which of the 12 Feynman diagrams needed to be retained and whose contributions when evaluated to yield the London dispersion formula and the Casimir–Polder limit. Unlike the procedure detailed in Section 1.10, which generates all time orderings, no formal procedure is currently available for independently constructing state sequence diagrams applicable to short- and long-range asymptotic limits. This does not, however, prevent the drawing of appropriate state sequence pictures for these two limiting separations. Construction is accomplished by converting Figs. 5.2 and 5.3 to state sequence notation. For the near zone, the pertinent state sequence diagram is illustrated in Fig. 5.6. The four pathways correspond to the time orderings of graphs (iii), (iv), (ix), and (x) of Fig. 5.1, which were shown in Section 5.3

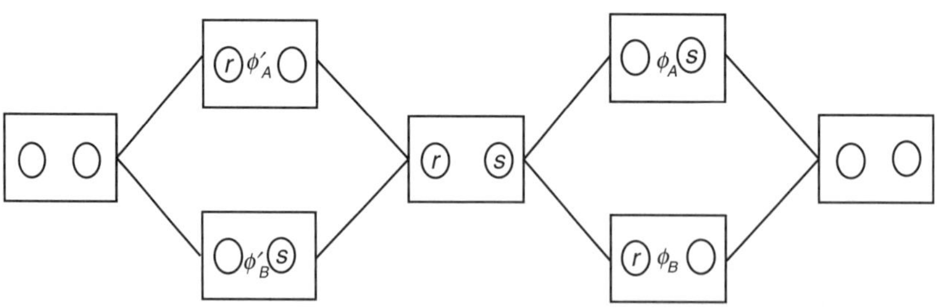

FIGURE 5.6 State sequence diagram representing near-zone dispersion interaction.

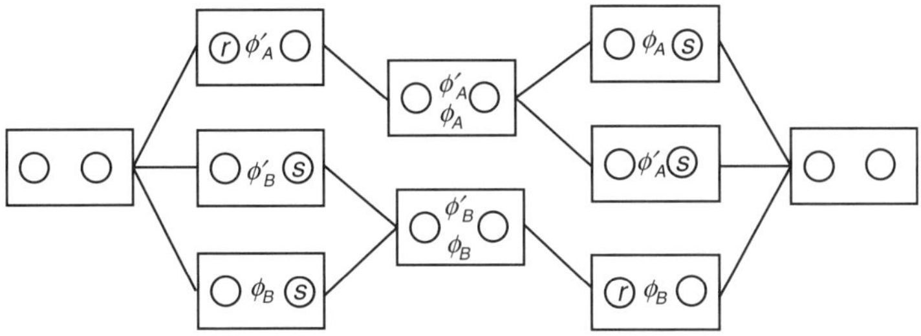

FIGURE 5.7 Single-photon interaction vertex state sequence diagram for far-zone dispersion energy shift.

to yield the London dispersion energy shift (5.3.4). When the pair separation distance is large relative to characteristic reduced transition wavelengths, it was demonstrated in Section 5.4 that summation over the four time-ordered graphs (i), (ii), (vii) and (viii) of Fig. 5.1 produced the far-zone energy shift (5.4.11). From these four diagrams, the state sequence representation of the far-zone limit may be drawn. This is shown in Fig. 5.7. Recalling that the long-range limit to the dispersion potential may be derived by employing the effective two-photon interaction Hamiltonian (5.4.12), which corresponds to collapsing the two one-photon coupling vertices occurring at each center to a single two-photon vertex at either site, as displayed in Fig. 5.3, the pictorial representation in the state sequence scheme has the most simple form, as shown in Fig. 5.8. Since in perturbation theory the far-zone limiting behavior is interpreted as arising from the simultaneous transit of two virtual photons, the intermediate states $|II\rangle$ of Fig. 5.7 and intermediate states $|I\rangle$ of Fig. 5.8 are identical, as expected.

5.6 DISPERSION INTERACTION BETWEEN ONE GROUND AND ONE EXCITED MOLECULE: PERTURBATION THEORY

Dispersion energy shifts are not limited only to interactions occurring between ground-state species. The coupling arising from the fluctuations in

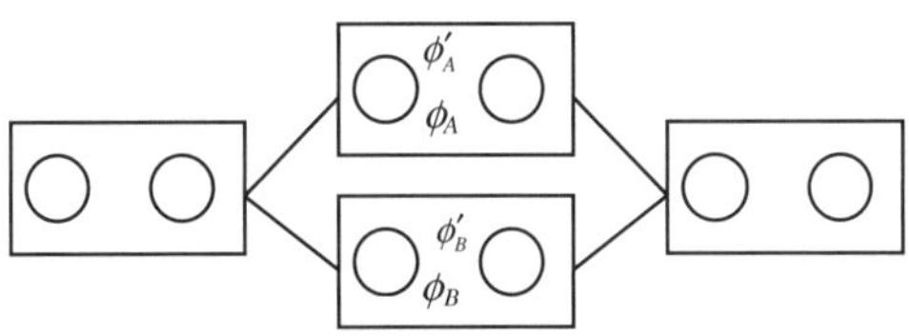

FIGURE 5.8 State sequence pathways for wave-zone asymptote of dispersion potential involving two-photon coupling vertices.

electron density at each center is manifest when one or both of the pairs are in electronically excited states. The concept of an intermolecular interaction energy for the situation in which the initial state of A or B corresponds to an excited state holds as long as the excited state or states in question are sufficiently long lived relative to the time taken for the photon to propagate between the two sites. Now both upward and downward transitions have to be taken into account in contrast to the calculation of the ground-state dispersion potential in which transitions occur only from the lowest energy level to a higher lying state. As a result, it is found that the energy shift contains a term arising from the emission of a real photon due to possible downward transitions, in addition to a contribution due solely to the exchange of virtual photons, the last of which is similar in structure to the Casimir–Polder interaction energy. A direct consequence of the appearance of this new contribution is the change in the form of the potential in the limit where the intermolecular separation distance is much larger than the characteristic reduced transition wavelengths. Instead of an inverse seventh power dependence on R, the potential exhibits an R^{-2} behavior in the far zone because the emitted photon has all of the properties associated with a real quantized particle of light. While the near-zone interaction energy still obeys an R^{-6} power law, it is repulsive when both molecules are initially excited, but when one of the pairs is excited, the overall sign of ΔE is determined by the comparative magnitudes of the relevant transition energies of both molecules. This aspect is in contradistinction to the London dispersion energy shift between two ground-state molecules, which is always attractive.

In this section, the time-dependent perturbation theory calculation of the dispersion interaction between one ground- and one excited-state molecules is presented (Power and Thirunamachandran, 1995a). The more general case in which both species are electronically excited will be expounded in Section 5.7 using response theory.

Consider two different molecules A and B, the former initially in excited electronic state $|q\rangle$ and the latter in the ground state $|0\rangle$. The total Hamiltonian comprising these two particles and the radiation field, as well as their mutual interaction, is given by

$$H = H_{\text{mol}}(A) + H_{\text{mol}}(B) + H_{\text{rad}} + H_{\text{int}}. \tag{5.6.1}$$

Continuing to work within the leading electric dipole approximation, the interaction Hamiltonian H_{int} is given by equation (5.2.5) as for the calculation of the ground-state dispersion energy shift. Again, the fourth-order perturbation theory formula for the interaction energy (5.2.4) is to be used for the calculation involving one ground and one excited molecule, since the force is still mediated by the exchange of two virtual photons between the pair. As in the calculation of the Casimir–Polder potential, 12 time-ordered diagrams aid in the evaluation of the result. The relevant graphs are identical to those shown in Fig. 5.1, but with the initial and final state labels of molecule A changed to q. As before, the intermediate states of A and B are labeled $|r\rangle$ and $|s\rangle$, respectively, subject to $E_r^A < E_q^A$ and $E_s^B > E_0^B$. The former restriction means that only downward transitions from the excited state of A are to be considered. Although transitions from $|q\rangle$ to higher lying intermediate states $|r\rangle$ in A are of course possible, the resulting contribution is identical in form to the energy shift between two ground-state species, for which only upward transitions are allowed, and is given by the Casimir–Polder result (5.2.20), but with the excited-state dynamic electric dipole polarizability of A appearing in the expression for ΔE instead of the ground-state one and replacement of k_{r0} by k_{rq}.

For this application, the initial and final molecule-field states are written as

$$|0\rangle = |E_q^A, E_0^B; 0(\vec{p}, \varepsilon), 0(\vec{p}\,', \varepsilon')\rangle, \tag{5.6.2}$$

with the modes of the two virtual photons being denoted by $(\vec{p}, \varepsilon)$ and $(\vec{p}\,', \varepsilon')$. The contributing intermediate states to be summed over are read off from each graph, along with the corresponding energy denominator product. For convenience and due to important differences compared to the analogous quantity occurring in the ground-state calculation, the energy denominators for the present case are listed in Table 5.3 for each of the suitably modified time-ordered diagrams of Fig. 5.1.

Unlike the case when both molecules are in the ground electronic state, one or more factors in the denominators of Table 5.3 can be zero due to the absorption or emission of a real photon. To deal with this when carrying out integrations over virtual photon momenta p and p', damping factors $\pm i\gamma$ are introduced. Adding the contributions from each of the 12 graphs, after evaluating the matrix elements in the usual way, produces for the energy shift between an excited- and a ground-state molecule the real part of the expression

$$\begin{aligned}\Delta E^{\mathrm{E-G}} = -\sum_{\vec{p}, \vec{p}\,'} \sum_{\varepsilon, \varepsilon'} \sum_{r, s} \left(\frac{\hbar c p}{2\varepsilon_0 V}\right)\left(\frac{\hbar c p'}{2\varepsilon_0 V}\right) \bar{e}_i^{(\varepsilon)}(\vec{p}) e_k^{(\varepsilon)}(\vec{p}) \bar{e}_j^{(\varepsilon')}(\vec{p}\,') \\ \times e_l^{(\varepsilon')}(\vec{p}\,') \mu_i^{qr}(A) \mu_j^{rq}(A) \mu_k^{0s}(B) \mu_l^{s0}(B)\, \mathrm{e}^{i(\vec{p}+\vec{p}\,')\cdot\vec{R}} \sum_{\alpha=i}^{\mathrm{xii}} E_\alpha^{-1},\end{aligned} \tag{5.6.3}$$

TABLE 5.3 Energy Denominator Products E_α^{-1}, α = i–xii for Dispersion Interaction Between One Excited- and One Ground-State Molecules

Graph	$(\hbar c)^{-3}E_\alpha$
(i)	$(p+k_{s0})(p+p')(p'-k_{rq}+i\gamma)$
(ii)	$(p'+k_{s0})(p+p')(p'-k_{rq}+i\gamma)$
(iii)	$(p+k_{s0})(k_{s0}-k_{rq})(p'-k_{rq}+i\gamma)$
(iv)	$(p+k_{s0})(k_{s0}-k_{rq})(p'+k_{s0})$
(v)	$(p'+k_{s0})(p+p'+k_{s0}-k_{rq})(p'-k_{rq}+i\gamma)$
(vi)	$(p'+k_{s0})(p+p'+k_{s0}-k_{rq})(p+k_{s0})$
(vii)	$(p-k_{rq}-i\gamma)(p+p')(p'+k_{s0})$
(viii)	$(p-k_{rq}-i\gamma)(p+p')(p+k_{s0})$
(ix)	$(p-k_{rq}-i\gamma)(k_{s0}-k_{rq})(p'+k_{s0})$
(x)	$(p-k_{rq}-i\gamma)(k_{s0}-k_{rq})(p'-k_{rq}+i\gamma)$
(xi)	$(p-k_{rq}-i\gamma)(p+p'+k_{s0}-k_{rq})(p+k_{s0})$
(xii)	$(p-k_{rq}-i\gamma)(p+p'+k_{s0}-k_{rq})(p'-k_{rq}+i\gamma)$

where $\vec{R}=\vec{R}_B-\vec{R}_A$. Carrying out the polarization sums and converting the wavevector sums to integrals and performing the angular averages yields

$$\begin{aligned}\Delta E^{\mathrm{E-G}} &= -\frac{(\hbar c)^3}{144\pi^4\varepsilon_0^2}\sum_{r,\,s}|\vec{\mu}^{rq}(A)|^2|\vec{\mu}^{s0}(B)|^2\left(-\vec{\nabla}^2\delta_{ij}+\vec{\nabla}_i\vec{\nabla}_j\right)^R\frac{1}{R}\\ &\quad\times\left(-\vec{\nabla}^2\delta_{ij}+\vec{\nabla}_i\vec{\nabla}_j\right)^{\bar{R}}\frac{1}{\bar{R}}\int_0^\infty\!\!\int_0^\infty\frac{1}{2}(\sin pR\sin p'\bar{R}+\sin p'R\sin p\bar{R})\\ &\quad\times\sum_{\alpha=i}^{\mathrm{xii}}E_\alpha^{-1}\,\mathrm{d}p\,\mathrm{d}p'.\end{aligned}\tag{5.6.4}$$

In going from expression (5.6.3) to (5.6.4), the position vectors present in the exponentials of the former equation have been formally distinguished by using R and $\bar{R}$; at the end of the calculation they will be put equal. Further, formula (5.6.4) applies to isotropic A and B as a rotational average has been carried out. Table 5.3 reveals that the contribution to the energy shift arising from graphs (x) and (xii) has, in addition to the principal value of the integral—which arises from all 12 graphs, a real part. The integrals over p and p' are performed with the aid of the identity

$$\frac{1}{x \pm i\gamma} = \frac{\mathrm{PV}}{x} \mp i\pi\delta(x), \tag{5.6.5}$$

where PV denotes the principal value, from which it readily follows that

$$\frac{1}{x - i\gamma}\frac{1}{y + i\gamma} = \frac{\mathrm{PV}}{x}\frac{\mathrm{PV}}{y} + i\pi\left(\frac{\mathrm{PV}}{y}\delta(x) - \frac{\mathrm{PV}}{x}\delta(y)\right) + \pi^2\delta(x)\delta(y). \tag{5.6.6}$$

The first term on the right-hand side of (5.6.6) can be further written as

$$\frac{\mathrm{PV}}{x}\frac{\mathrm{PV}}{y} = \frac{\mathrm{PV}}{x-y}\left(\frac{\mathrm{PV}}{y} - \frac{\mathrm{PV}}{x}\right) + \pi^2\delta(x)\delta(y). \tag{5.6.7}$$

Thus, the additional real term from denominators E_{x}^{-1} and E_{xii}^{-1} due to the product of delta functions in (5.6.7) is

$$-\frac{1}{144\pi^2\varepsilon_0^2\hbar c}\sum_{r,s}|\vec{\mu}^{rq}(A)|^2|\vec{\mu}^{s0}(B)|^2(-\vec{\nabla}^2\delta_{ij} + \vec{\nabla}_i\vec{\nabla}_j)^R\frac{1}{R}(-\vec{\nabla}^2\delta_{ij} + \vec{\nabla}_i\vec{\nabla}_j)^{\bar{R}}\frac{1}{\bar{R}}$$

$$\times\int_0^\infty\int_0^\infty\frac{1}{2}(\sin pR\sin p'\bar{R} + \sin p'R\sin p\bar{R})\left(\frac{1}{k_{s0} - k_{rq}} + \frac{1}{k_{s0} + k_{rq}}\right)$$

$$\times\pi^2\delta(p - k_{rq})\delta(p' - k_{rq})\mathrm{d}p\,\mathrm{d}p' = -\frac{1}{144\pi^2\varepsilon_0^2\hbar c}\sum_{r,s}|\vec{\mu}^{rq}(A)|^2|\vec{\mu}^{s0}(B)|^2$$

$$\times(-\vec{\nabla}^2\delta_{ij} + \vec{\nabla}_i\vec{\nabla}_j)^R\frac{1}{R}(-\vec{\nabla}^2\delta_{ij} + \vec{\nabla}_i\vec{\nabla}_j)^{\bar{R}}\frac{1}{\bar{R}}\times\frac{2k_{s0}}{k_{s0}^2 - k_{rq}^2}\sin(k_{rq}R)\sin(k_{rq}\bar{R}), \tag{5.6.8}$$

which in fact appears twice, since an identical contribution arises from the third term on the right-hand side of (5.6.6). The last step in the calculation of the energy shift involves the evaluation of the contribution independent of δ-functions. This is done using the first term of identity (5.6.7). The sums over energy denominators are nearly identical to that carried out for the calculation of the ground-state dispersion energy shift, and equation

(5.2.7) may be used to give

$$-\frac{1}{144\pi^4\varepsilon_0^2\hbar c}\sum_{r,s}|\vec{\mu}^{rq}(A)|^2|\vec{\mu}^{s0}(B)|^2(-\vec{\nabla}^2\delta_{ij}+\vec{\nabla}_i\vec{\nabla}_j)^R\frac{1}{R}(-\vec{\nabla}^2\delta_{ij}+\vec{\nabla}_i\vec{\nabla}_j)^{\bar{R}}\frac{1}{\bar{R}}$$

$$\times\int_0^\infty\int_0^\infty\frac{1}{2}(\sin pR\sin p'\bar{R}+\sin p'R\sin p\bar{R})$$

$$\times\frac{2(-k_{rq}+k_{s0}+p)}{(p-k_{rq})(p+k_{s0})(k_{s0}-k_{rq})}\left(\frac{1}{(p+p')}-\frac{1}{(p-p')}\right)\mathrm{d}p\,\mathrm{d}p', \tag{5.6.9}$$

which on performing the p' integral yields

$$-\frac{\pi}{144\pi^4\varepsilon_0^2\hbar c}\sum_{r,s}|\vec{\mu}^{rq}(A)|^2|\vec{\mu}^{s0}(B)|^2(-\vec{\nabla}^2\delta_{ij}+\vec{\nabla}_i\vec{\nabla}_j)^R\frac{1}{R}$$

$$\times(-\vec{\nabla}^2\delta_{ij}+\vec{\nabla}_i\vec{\nabla}_j)^{\bar{R}}\frac{1}{\bar{R}}\frac{1}{k_{s0}-k_{rq}}\int_0^\infty\int_0^\infty\frac{1}{2}(\sin pR\sin p'\bar{R}+\sin p'R\sin p\bar{R})$$

$$\times\frac{2(-k_{rq}+k_{s0}+p)}{(p-k_{rq})(p+k_{s0})(k_{s0}-k_{rq})}\mathrm{d}p. \tag{5.6.10}$$

Using the integral result

$$\mathrm{PV}\int_0^\infty\frac{\sin ax}{x-b}\mathrm{d}x=-f(ab)+\pi\cos(ab),\quad a,b>0, \tag{5.6.11}$$

where

$$f(x)=\int_0^\infty\frac{\mathrm{e}^{-ux}}{1+u^2}\mathrm{d}u=ci(x)\sin(x)-si(x)\cos(x), \tag{5.6.12}$$

equation (5.6.10) becomes

$$-\frac{1}{72\pi^3\varepsilon_0^2\hbar c}\sum_{r,s}\frac{|\vec{\mu}^{rq}(A)|^2|\vec{\mu}^{s0}(B)|^2}{(k_{rq}^2-k_{s0}^2)}(-\vec{\nabla}^2\delta_{ij}+\vec{\nabla}_i\vec{\nabla}_j)^R\frac{1}{R}$$

$$\times(-\vec{\nabla}^2\delta_{ij}+\vec{\nabla}_i\vec{\nabla}_j)^{\bar{R}}\frac{1}{\bar{R}}[-k_{rq}f(k_{s0}(R+\bar{R}))+k_{s0}f(k_{rq}(R+\bar{R}))$$

$$-\pi k_{s0}\cos(k_{rq}(R+\bar{R}))]. \tag{5.6.13}$$

It is worth noting that for $|q\rangle = |0\rangle$, the f-dependent terms of (5.6.13) are equal and opposite to the Casimir–Polder potential and provides an alternative expression for the energy shift between a pair of ground-state molecules. Explicitly,

$$\Delta E^{\mathrm{G-G}} = -\frac{1}{72\pi^3\varepsilon_0^2\hbar c}\sum_{r,s}\frac{|\vec{\mu}^{0r}(A)|^2|\vec{\mu}^{0s}(B)|^2}{(k_{r0}^2-k_{s0}^2)}\left(-\vec{\nabla}^2\delta_{ij}+\vec{\nabla}_i\vec{\nabla}_j\right)^R\frac{1}{R}$$

$$\times\left(-\vec{\nabla}^2\delta_{ij}+\vec{\nabla}_i\vec{\nabla}_j\right)^{\bar{R}}\frac{1}{\bar{R}}[k_{r0}f(k_{s0}(R+\bar{R}))-k_{s0}f(k_{r0}(R+\bar{R}))]|_{\bar{R}=R}$$

$$= -\frac{1}{72\pi^3\varepsilon_0^2\hbar cR^2}\sum_{r,s}|\vec{\mu}^{0r}(A)|^2|\vec{\mu}^{0s}(B)|^2\frac{k_{r0}k_{s0}}{(k_{r0}^2-k_{s0}^2)}$$

$$\times\left\{\begin{aligned}&k_{r0}^3\left[\frac{1}{k_{r0}R}-f(2k_{r0}R)\left(2-\frac{10}{k_{r0}^2R^2}+\frac{6}{k_{r0}^4R^4}\right)+g(2k_{r0}R)\left(\frac{4}{k_{r0}R}-\frac{12}{k_{r0}^3R^3}\right)\right]\\&-k_{s0}^3\left[\frac{1}{k_{s0}R}-f(2k_{s0}R)\left(2-\frac{10}{k_{s0}^2R^2}+\frac{6}{k_{s0}^4R^4}\right)+g(2k_{s0}R)\left(\frac{4}{k_{s0}R}-\frac{12}{k_{s0}^3R^3}\right)\right]\end{aligned}\right\}, \tag{5.6.14}$$

after differentiating and setting $\bar{R}=R$, with

$$g(x) = \int_0^\infty \frac{u\mathrm{e}^{-ux}}{1+u^2}\mathrm{d}u = -ci(x)\cos(x) - si(x)\sin(x). \tag{5.6.15}$$

Since downward transitions from the excited state $|q\rangle$ are being considered, the first two terms of (5.6.13) can be identified as the contribution from upward transitions and designated as

$$\Delta E^{\mathrm{up}} = -\frac{1}{72\pi^3\varepsilon_0^2\hbar c}\sum_{r,s}\frac{|\vec{\mu}^{rq}(A)|^2|\vec{\mu}^{s0}(B)|^2}{(k_{rq}^2-k_{s0}^2)}\left(-\vec{\nabla}^2\delta_{ij}+\vec{\nabla}_i\vec{\nabla}_j\right)^R\frac{1}{R}$$

$$\times\left(-\vec{\nabla}^2\delta_{ij}+\vec{\nabla}_i\vec{\nabla}_j\right)^{\bar{R}}\frac{1}{\bar{R}}[k_{rq}f(k_{s0}(R+\bar{R}))-k_{s0}f(k_{rq}(R+\bar{R}))]|_{\bar{R}=R}. \tag{5.6.16}$$

Hence, the energy shift between an excited molecule A and a ground-state one B is given by the sum of (5.6.13) and twice (5.6.8),

$$\Delta E^{\mathrm{E-G}} = -\Delta E^{\mathrm{up}} + \frac{2}{144\pi^2\varepsilon_0^2\hbar c}\sum_{r,s}\frac{k_{s0}}{(k_{rq}^2 - k_{s0}^2)}|\vec{\mu}^{rq}(A)|^2|\vec{\mu}^{s0}(B)|^2$$
$$\times(-\vec{\nabla}^2\delta_{ij}+\vec{\nabla}_i\vec{\nabla}_j)^R\frac{1}{R}(-\vec{\nabla}^2\delta_{ij}+\vec{\nabla}_i\vec{\nabla}_j)^{\bar{R}}\frac{1}{\bar{R}}$$
$$\times\left[\cos(k_{rq}(R+\bar{R}))+2\sin k_{rq}R\sin k_{rq}\bar{R}\right]|_{\bar{R}=R}. \tag{5.6.17}$$

Inserting (5.6.16) into (5.6.17), evaluating the gradients and simplifying produces

$$\Delta E^{\mathrm{E-G}} = -\frac{1}{24\pi^3\varepsilon_0^2}\sum_r\int_0^\infty du\frac{k_{rq}}{(k_{rq}^2+u^2)}|\vec{\mu}^{qr}(A)|^2\alpha(B;icu)u^6\mathrm{e}^{-2uR}$$
$$\times\left[\frac{1}{u^2R^2}+\frac{2}{u^3R^3}+\frac{5}{u^4R^4}+\frac{6}{u^5R^5}+\frac{3}{u^6R^6}\right]-\frac{1}{24\pi^2\varepsilon_0^2}\sum_{\substack{r\\E_q>E_r}}|\vec{\mu}^{qr}(A)|^2$$
$$\times\alpha(B;k_{qr})k_{qr}^6\left[\frac{1}{k_{qr}^2R^2}+\frac{1}{k_{qr}^4R^4}+\frac{3}{k_{qr}^6R^6}\right]. \tag{5.6.18}$$

The first term of the energy shift is similar in structure to the Casimir–Polder result (5.2.20), but with excited state $|q\rangle$ replacing ground state $|0\rangle$ of A, and includes contributions from both upward and downward transitions from $|q\rangle$, since the sum over r is unrestricted. The second term of (5.6.18), however, applies only for downward transitions and arises from real photon emission. When both molecules are in the ground state, (5.6.18) reduces to the Casimir–Polder potential as expected. Discussion of the asymptotic behavior of the energy shift $\Delta E^{\mathrm{E-G}}$ is deferred until Section 5.7 when the interaction energy between two excited molecules is computed for all R using response theory.

If A and B are identical or have resonant energy levels, the formula given for the ground-state interaction energy (5.6.14) is invalid since not only does the energy denominator vanish but also the term within braces. An expression for the energy shift may be obtained in this case by carrying out the following limiting procedure on (5.6.14),

$$\lim_{k_{s0}\to k_{r0}(\equiv k)} \Delta E^{\mathrm{G-G}} = -\frac{1}{72\pi^3\varepsilon_0^2\hbar c}\sum_r |\vec{\mu}^{0r}|^4 (-\vec{\nabla}^2\delta_{ij}+\vec{\nabla}_i\vec{\nabla}_j)^R \frac{1}{R}$$

$$\times(-\vec{\nabla}^2\delta_{ij}+\vec{\nabla}_i\vec{\nabla}_j)^{\bar{R}}\frac{1}{\bar{R}} \lim_{k_{s0}\to k_{r0}(\equiv k)} \left.\frac{[k_{r0}f(k_{s0}(R+\bar{R}))-k_{s0}f(k_{r0}(R+\bar{R}))]}{k_{r0}^2-k_{s0}^2}\right|_{\bar{R}=R}$$

$$= -\frac{1}{72\pi^3\varepsilon_0^2\hbar c}\sum_r |\vec{\mu}^{0r}|^4 (-\vec{\nabla}^2\delta_{ij}+\vec{\nabla}_i\vec{\nabla}_j)^R \frac{1}{R}(-\vec{\nabla}^2\delta_{ij}+\vec{\nabla}_i\vec{\nabla}_j)^{\bar{R}}\frac{1}{\bar{R}}$$

$$\times\left.\frac{[f(k(R+\bar{R}))-k^2(R+\bar{R})g(k(R+\bar{R}))]}{2k}\right|_{\bar{R}=R}, \tag{5.6.19}$$

where the common energy spacing is denoted by $\hbar ck$. Carrying out the differentiations produces

$$-\frac{1}{72\pi^3\varepsilon_0^2\hbar cR^2}\sum_r |\vec{\mu}^{0r}|^4 k^3\left[-\frac{1}{kR}+\frac{6}{k^3R^3}\right.$$

$$\left.-f(2kR)\left(-1+\frac{7}{k^2R^2}-\frac{3}{k^4R^4}\right)+g(2kR)\left(2kR-\frac{6}{kR}+\frac{6}{k^3R^3}\right)\right]. \tag{5.6.20}$$

The total energy shift is then given by the addition of (5.6.20) to (5.6.14), with summation in this last equation excluding the term for which $k_{r0}=k_{s0}$.

It should also be remarked that the limiting process cannot be used to obtain the potential between two identical molecules when one of them is excited and the other is in the ground state. The limit $k_{s0}\to k_{rq}$ does not exist, as is easily seen from the energy denominators associated with graphs (iii), (iv), (ix), and (x) in Table 5.3, in which intermediate state $|II\rangle$ is degenerate with the initial state. Before perturbation theory can be used, the degeneracy must be removed in lower order.

5.7 RESPONSE THEORY CALCULATION OF DISPERSION FORCES

In Chapter 4, it was shown how a form of response theory could be used instead of the routinely applied diagrammatic perturbation theory method for the calculation of the matrix element for the resonant transfer of energy between an excited and a ground-state pair of molecules. Adoption of the former approach proved to be conceptually simpler and computationally more direct. It relied on the coupling of the transition electric dipole moment of the acceptor species to the electric dipole-dependent driving electric displacement field linear in the source molecule. This viewpoint is now extended to the treatment of retarded dispersion interactions (Power and Thirunamachandran, 1993a; Salam, 2008). The physical picture is one in which a molecule responds, via its dynamic polarizability, to the Maxwell fields of a second body. Symmetry is maintained in this model as both species are permitted to simultaneously take on the roles of source and test molecules. In addition to the advantages already mentioned in using this formulation to calculate interactions between molecules, it is particularly advantageous for evaluating dispersion energy shifts when one or both of the molecules are electronically excited. In fact, response theory enables the Casimir–Polder potential to be extracted as a special case of the general result valid when the two entities are initially in higher lying energy levels rather than in the ground state. This is because intermediate-state resonances due to possible downward transitions from an excited state, in which a real photon is emitted and subsequently absorbed, are easily located, characterized, and handled by utilizing the electromagnetic field operators.

Consider two neutral, polarizable molecules A and B, positioned at $\vec{R}_A$ and $\vec{R}_B$, respectively, with interparticle separation distance vector $\vec{R} = \vec{R}_B - \vec{R}_A$. Further, take A and B initially to be in excited electronic states $|p\rangle$ and $|r\rangle$. To leading order of approximation in the multipolar coupling scheme, let electric dipole allowed transitions occur of the form $|q^A\rangle \leftarrow |p\rangle$ and $|s^B\rangle \leftarrow |r\rangle$ to intermediate states $|q^A\rangle$ and $|s^B\rangle$ from the initial states. The intermediate levels may lie above or below the initial states, thereby enabling the contribution from both upward and downward transitions to be properly accounted for. Another advantage of working with Maxwell field operators in the Heisenberg formalism is the formal equivalence between quantum mechanical observable quantities and their expression in terms of dynamical variables in classical theory. Hence, the first term in the expansion of the interaction energy is given by the familiar formula,

$$\Delta E = -\frac{1}{2\varepsilon_0^2}\alpha_{ij}(A;k)d_i^{\perp}(B;k;\vec{R}_A)d_j^{\perp}(B;k;\vec{R}_A)$$
$$-\frac{1}{2\varepsilon_0^2}\alpha_{kl}(B;k)d_k^{\perp}(A;k;\vec{R}_B)d_l^{\perp}(A;k;\vec{R}_B). \qquad (5.7.1)$$

The energy shift is interpreted as arising from the response of molecule ξ, $\xi = A, B$, through its frequency-dependent polarizability $\alpha_{ij}(\xi;k)$ at frequency $\omega = ck$ to the electric displacement field of the other molecule

ξ' at the position of the first body, $d_i^{\perp}(\xi'; k; \vec{R}_\xi)$. In the electric dipole approximation, the excited-state dynamic electric dipole polarizability of molecule A is

$$\alpha_{ij}(A;k) = \sum_q \left\{ \frac{\mu_i^{pq}(A)\mu_j^{qp}(A)}{E_{qp} - \hbar ck} + \frac{\mu_j^{pq}(A)\mu_i^{qp}(A)}{E_{qp} + \hbar ck} \right\}. \tag{5.7.2}$$

A similar expression holds for the polarizability of B, with states $|p\rangle$ and $|q\rangle$ replaced by $|r\rangle$ and $|s\rangle$, respectively. Recalling from Section 2.6 that the electric displacement field in the proximity of a source molecule may be expanded in series of powers of molecular multipole moments, inserting the first three terms of the expansion (2.6.11) in equation 5.7.1 and collecting together all terms proportional to the second power of the transition electric dipole moment at each center produces for the two excited molecules the explicit expression

$$\begin{aligned}
\Delta E = &-\frac{1}{2\varepsilon_0^2}\sum_s \alpha_{ij}(A;k_{rs}) d_i^{(1)}(B;\vec{\mu};k_{rs};\vec{R}_A) d_j^{(1)}(B;\vec{\mu};k_{rs};\vec{R}_A) \\
&-\frac{1}{2\varepsilon_0^2}\sum_q \alpha_{kl}(B;k_{pq}) d_k^{(1)}(A;\vec{\mu};k_{pq};\vec{R}_B) d_l^{(1)}(A;\vec{\mu};k_{pq};\vec{R}_B) \\
&-\frac{1}{2\varepsilon_0^2}\sum_{\text{modes}} \alpha_{ij}(A;k)\Big[d_i^{(0)}(k;\vec{R}_A) d_j^{(2)}(B;\vec{\mu}\vec{\mu};k;\vec{R}_A) \\
&+ d_i^{(2)}(B;\vec{\mu}\vec{\mu};k;\vec{R}_A) d_j^{(0)}(k;\vec{R}_A)\Big] \\
&-\frac{1}{2\varepsilon_0^2}\sum_{\text{modes}} \alpha_{kl}(B;k)\Big[d_k^{(0)}(k;\vec{R}_B) d_l^{(2)}(A;\vec{\mu}\vec{\mu};k;\vec{R}_B) \\
&+ d_k^{(2)}(A;\vec{\mu}\vec{\mu};k;\vec{R}_B) d_l^{(0)}(k;\vec{R}_B)\Big].
\end{aligned} \tag{5.7.3}$$

Interestingly, the first two terms arising from the product of the displacement fields do not contribute to the energy shift. These two terms comprise (i) product of the free radiation field, which is independent of $\vec{\mu}$ and, therefore, does not enter into the formula for ΔE, being simply a corrective zero-point energy term, and (ii) interference of the vacuum field with the first-order electric displacement field. This product is also noncontributory as its expectation value over the ground state of the electromagnetic field results in a change in the number of photons. The first contributing term is that arising from the product of the fields linear in the transition dipole moments, as evidenced by the first two lines of (5.7.3). In the first term of this equation, for instance, molecule A responds through its dynamic polarizability to the first-order electric dipole-dependent driving fields of species B at frequency $\omega_{rs} = ck_{rs}$. Surprisingly, the zeroth-order field is used in the calculation of the potential. As seen from the third and fourth terms of (5.7.3), there is a contribution from the interference of the vacuum field with the displacement field that is quadratic in the electric dipole moment, giving a contribution that overall is second order in the source moment and that must be added to the first two terms of (5.7.3) for consistency. The various contributions to the energy shift are now evaluated by utilizing the source-dependent displacement fields derived in Chapter 2.

Concentrating for the moment on the very first term of (5.7.3), its contribution to the energy shift is calculated by taking its expectation value over the molecular state $|r\rangle$ and the vacuum state of the radiation field. Since the first-order displacement field operates only in the fermion space, integrals over radiation field states are unity. Thus,

$$-\frac{1}{2\varepsilon_0^2}\sum_s \langle r|\alpha_{ij}(A;k_{rs})d_i^{(1)}(B;\vec{\mu};k_{rs};\vec{R}_A,t)|s\rangle\langle s|d_j^{(1)}(B;\vec{\mu};k_{rs};\vec{R}_A,t)|r\rangle$$
$$= -\frac{1}{32\pi^2\varepsilon_0^2}\sum_s \alpha_{ij}(A;k_{rs})\mu_k^{rs}\mu_l^{sr}k_{rs}^6\bar{f}_{ik}(k_{rs}R)f_{jl}(k_{rs}R), \tag{5.7.4}$$

on using the displacement field operator linear in the electric dipole moment (2.6.21), with the tensor field $f_{ij}(kr)$ given by (2.9.4). For the third term of (5.7.3) involving the response of A to the product of the free and quadratic fields of B, only the diagonal matrix element over the electronic state $|r\rangle$ is required of the second-order field because the vacuum field operates exclusively in the boson space. This quantity has been worked out in Section 2.9 and is given by equation (2.9.5). Also, making use of the mode expansion for the free displacement field, the third term of (5.7.3) can be written as

$$-\frac{1}{2\varepsilon_0^2}\sum_{\vec{k},\lambda}\alpha_{ij}(A;k)\Big[\langle 0(\vec{k},\lambda);r|d_i^{(0)}(k;\vec{R}_A)|r;1(\vec{k},\lambda)\rangle\langle 1(\vec{k},\lambda);r|$$
$$\times d_j^{(2)}(B;\vec{\mu}\vec{\mu};k;\vec{R}_A)|r;0(\vec{k},\lambda)\rangle + \langle 0(\vec{k},\lambda);r|d_i^{(2)}(B;\vec{\mu}\vec{\mu};k;\vec{R}_A)$$
$$\times|r;1(\vec{k},\lambda)\rangle\langle 1(\vec{k},\lambda);r|d_i^{(0)}(k;\vec{R}_A)|r;0(\vec{k},\lambda)\rangle\Big]$$
$$= -\frac{1}{8\pi^2\varepsilon_0}\sum_{\vec{k},\lambda}\left(\frac{\hbar ck}{2V}\right)\alpha_{ij}(A;k)\Big[e_i^{(\lambda)}(\vec{k})e^{i\vec{k}\cdot\vec{R}}\bar{e}_k^{(\lambda)}(\vec{k})\bar{\mathcal{F}}_{kj}$$
$$+e_k^{(\lambda)}(\vec{k})\mathcal{F}_{kj}\bar{e}_i^{(\lambda)}(\vec{k})e^{-i\vec{k}\cdot\vec{R}}\Big], \tag{5.7.5}$$

with $\mathcal{F}_{kj}$ given by (2.9.6). After carrying out the polarization sum and angular integration, the first term of (5.7.5) becomes

$$-\frac{1}{8\pi^2\varepsilon_0^2}\sum_{\vec{k},\lambda}\left(\frac{\hbar ck}{2V}\right)\alpha_{ij}(A;k)e_i^{(\lambda)}(\vec{k})\bar{e}_l^{(\lambda)}(\vec{k})e^{i\vec{k}\cdot\vec{R}}\bar{\mathcal{F}}_{lj}$$
$$=-\frac{\hbar c}{32\pi^2\varepsilon_0^2}\frac{1}{2\pi i}\int_0^\infty dk k^3\alpha_{ij}(A;k)[F_{il}(kR)-\bar{F}_{il}(kR)]\bar{\mathcal{F}}_{lj}$$
$$=-\frac{1}{32\pi^2\varepsilon_0^2}\sum_s \mu_k^{rs}(B)\mu_l^{sr}(B)\frac{\mathrm{PV}}{2\pi i}\int_0^\infty dk k^3\alpha_{ij}(A;k)$$
$$\times\left\{\begin{array}{l}\dfrac{[F_{il}(kR)-\bar{F}_{il}(kR)][k^3\bar{F}_{jk}(kR)-k_{sr}^3\bar{F}_{jk}(k_{sr}R)e^{-i(k_{rs}+k)ct}]}{(k_{sr}-k)}\\ +\dfrac{[F_{il}(kR)-\bar{F}_{il}(kR)][k^3\bar{F}_{jk}(kR)-k_{rs}^3\bar{F}_{jk}(k_{rs}R)e^{i(k_{rs}-k)ct}]}{(k_{sr}+k)}\end{array}\right\}. \tag{5.7.6}$$

To evaluate (5.7.6), the Cauchy principal value is taken for the integral, which is appropriate since exact resonances are excluded in the k-integral when making the continuum approximation to the mode sum and transforming the integral from one along the real axis to one along the imaginary axis in the complex plane on inserting $k=-iu$. On substituting the explicit form of the tensor field $F_{ij}(kR)$ and expanding, (5.7.6) is seen to contain both time-dependent and time-independent terms. The former is given by

$$\frac{1}{64\pi^3\varepsilon_0^2}\sum_s \mu_k^{rs}(B)\mu_l^{sr}(B)(-k_{rs})^3\int_0^\infty du u^3\alpha_{ij}(A;icu)\left[\bar{f}_{il}(k_{sr}R)e^{ik_{rs}(R-ct)}\right.$$
$$\times f_{jk}(-iuR)\frac{e^{-uc(t-R/c)}}{u+ik_{rs}}-\bar{f}_{il}(k_{sr}R)e^{ik_{rs}(R-ct)}\bar{f}_{jk}(-iuR)\frac{e^{-uc(t+R/c)}}{u+ik_{rs}}$$
$$+\bar{f}_{il}(k_{rs}R)e^{ik_{sr}(R-ct)}f_{jk}(-iuR)\frac{e^{-uc(t-R/c)}}{u-ik_{rs}}-\bar{f}_{il}(k_{rs}R)e^{ik_{sr}(R-ct)}$$
$$\left.\times\bar{f}_{jk}(-iuR)\frac{e^{-uc(t+R/c)}}{u-ik_{rs}}\right]. \tag{5.7.7}$$

For times $t \gg R/c$, the contribution (5.7.7) tends to zero as the integrals have exponentially decreasing values. Furthermore, the average of (5.7.7) over a finite duration approaches zero due to the modulating factors $e^{\pm i k_{rs} ct}$, resulting in the neglect of these oscillatory terms henceforth. Returning to (5.7.6) and evaluating the k-integral for the two cases $k_{rs} > 0$ and $k_{sr} > 0$ produces for the time-independent part

$$-\frac{1}{64\pi^2\varepsilon_0^2}\sum_s \operatorname{sgn}(k_{rs})\alpha_{ij}(A;k_{rs})\mu_k^{rs}(B)\mu_l^{sr}(B)k_{rs}^6\bar{f}_{ik}(k_{rs}R)f_{jl}(k_{rs}R)$$
$$+\frac{\hbar c}{64\pi^3\varepsilon_0^2}\int_0^\infty du\,u^6 e^{-2uR}\alpha_{ij}(A;icu)\alpha_{kl}(B;icu)f_{ik}(iuR)f_{jl}(iuR), \tag{5.7.8}$$

where sgn(x) is the signum function. An identical contribution to (5.7.8) is obtained on evaluating the second term of (5.7.5), so that the contribution to the energy shift arising from the response of A to the interference of the free and second-order fields of B is twice (5.7.8). It is important to note that for states for which $E_r < E_s$, the pole term of (5.7.8) is equal and opposite to the contribution arising from the product of the first-order fields (5.7.4). When $E_r > E_s$, however, twice the first term of (5.7.8) is identical to (5.7.4), doubling this contribution overall. This addition and cancellation of pole contributions from the zeroth and quadratic fields, with terms from the product of the fields linear in the source moment, also occurred in the computation of the Poynting vector and electromagnetic energy density, as detailed in Chapter 2.

After evaluating the second and fourth terms of (5.7.3), namely, those in which species B is viewed as a test polarizable body that responds to the source fields of entity A, and adding to twice the value of the contribution from (5.7.8), the energy shift between two oriented and excited molecules is found to be

$$\Delta E = -\frac{1}{16\pi^2\varepsilon_0^2}\sum_{\substack{s\\E_r>E_s}} \alpha_{ij}(A;k_{rs})\mu_k^{rs}(B)\mu_l^{sr}(B)k_{rs}^6\bar{f}_{ik}(k_{rs}R)f_{jl}(k_{rs}R)$$
$$-\frac{1}{16\pi^2\varepsilon_0^2}\sum_{\substack{q\\E_p>E_q}} \alpha_{kl}(B;k_{pq})\mu_i^{pq}(A)\mu_j^{qp}(A)k_{pq}^6\bar{f}_{ik}(k_{pq}R)f_{jl}(k_{pq}R)$$
$$+\frac{\hbar c}{32\pi^3\varepsilon_0^2}\int_0^\infty du\,u^6 e^{-2uR}\alpha_{ij}(A;icu)\alpha_{kl}(B;icu)f_{ik}(iuR)f_{jl}(iuR), \tag{5.7.9}$$

where the u-integral term has been counted only once. For isotropic A and B, rotational averaging of (5.7.9) produces

$$\Delta E = -\frac{1}{24\pi^2\varepsilon_0^2}\sum_{\substack{s\\E_r>E_s}} \alpha(A;k_{rs})|\vec{\mu}^{rs}(B)|^2 k_{rs}^6\left[\frac{1}{k_{rs}^2R^2}+\frac{1}{k_{rs}^4R^4}+\frac{3}{k_{rs}^6R^6}\right]$$
$$-\frac{1}{24\pi^2\varepsilon_0^2}\sum_{\substack{q\\E_p>E_q}} \alpha(B;k_{pq})|\vec{\mu}^{pq}(A)|^2 k_{pq}^6\left[\frac{1}{k_{pq}^2R^2}+\frac{1}{k_{pq}^4R^4}+\frac{3}{k_{pq}^6R^6}\right]$$
$$-\frac{1}{36\pi^3\varepsilon_0^2\hbar c}\sum_{q,\,s}|\vec{\mu}^{pq}(A)|^2|\vec{\mu}^{rs}(B)|^2\int_0^\infty du\,u^6 e^{-2uR}\frac{k_{qp}k_{sr}}{(k_{qp}^2+u^2)(k_{sr}^2+u^2)}$$
$$\times\left[\frac{1}{u^2R^2}+\frac{2}{u^3R^3}+\frac{5}{u^4R^4}+\frac{6}{u^5R^5}+\frac{3}{u^6R^6}\right]. \tag{5.7.10}$$

It is interesting to note that the first two terms of the energy shift (5.7.10) apply only to downward transitions from the initial state and correspond to real photon emission. The third term, on the other hand, contains both upward and downward transition terms since the sums over intermediate states $|q\rangle$ and $|s\rangle$ are unrestricted. From the result (5.7.10), which applies when both of the molecules are excited, it is a simple matter to obtain expressions for the energy shift when one or none of the molecular pairs is excited. The last of these cases is examined first.

When both species are in the ground electronic state, the first two terms of (5.7.10) vanish since only upward transitions from the initial state are possible. Left behind from (5.7.10) is the u-integral term with $|p\rangle = |r\rangle = |0\rangle$,

$$-\frac{1}{36\pi^3\varepsilon_0^2\hbar c}\sum_{q,\,s}|\vec{\mu}^{0q}(A)|^2|\vec{\mu}^{0s}(B)|^2\int_0^\infty du\,u^6 e^{-2uR}\frac{k_{q0}k_{s0}}{(k_{q0}^2+u^2)(k_{s0}^2+u^2)}$$
$$\times\left[\frac{1}{u^2R^2}+\frac{2}{u^3R^3}+\frac{5}{u^4R^4}+\frac{6}{u^5R^5}+\frac{3}{u^6R^6}\right], \tag{5.7.11}$$

whose composition is made up solely from contributions due to virtual transitions and is recognized straight away as the Casimir–Polder potential (5.2.18). This potential and the form of its asymptotic limits were discussed in Section 5.2.

If molecule A is taken to be excited while B remains in its ground state, for example, only the second and third terms of (5.7.10) survive, with $|r\rangle = |0\rangle$, and the result (5.6.18) is recovered. It is instructive to examine the asymptotic limits of these two terms. In the far zone, the dominant contribution arises from the second term of (5.7.10) and has an R^{-2} dependence of the form

$$-\frac{1}{24\pi^2\varepsilon_0^2 R^2}\sum_{\substack{q\\ E_p>E_q}}\alpha(B;k_{pq})|\vec{\mu}^{pq}(A)|^2 k_{pq}^4, \tag{5.7.12}$$

characteristic of emission of a real photon from the excited state $|p\rangle$ of A. This contrasts with the inverse seventh power dependence at long range of the contribution from the third term of (5.7.10),

$$-\frac{1}{36\pi^3\varepsilon_0^2\hbar c}\sum_{q,\,s}|\vec{\mu}^{pq}(A)|^2|\vec{\mu}^{0s}(B)|^2\int_0^\infty du\,u^6 e^{-2uR}\frac{k_{qp}k_{s0}}{(k_{qp}^2+u^2)(k_{s0}^2+u^2)}$$
$$\times\left[\frac{1}{u^2R^2}+\frac{2}{u^3R^3}+\frac{5}{u^4R^4}+\frac{6}{u^5R^5}+\frac{3}{u^6R^6}\right], \tag{5.7.13}$$

where the summation over q includes both upward and downward transitions, which has the familiar form

$$-\frac{23\hbar c}{64\pi^3\varepsilon_0^2 R^7}\alpha(A;0)\alpha(B;0), \tag{5.7.14}$$

with $\alpha(A;0)$ the static excited polarizability of molecule A. For small R, the leading term of (5.7.13) is

$$-\frac{1}{24\pi^2\varepsilon_0^2 R^6}\sum_{q,\,s}\mathrm{sgn}(E_{qp})\frac{|\vec{\mu}^{qp}(A)|^2|\vec{\mu}^{0s}(B)|^2}{(|E_{qp}|+E_{s0})}, \tag{5.7.15}$$

while that from the second term of (5.7.10) is

$$-\frac{1}{8\pi^2\varepsilon_0^2 R^6}\sum_{\substack{q\\ E_p>E_q}}|\vec{\mu}^{pq}(A)|^2\alpha(B;k_{pq}). \tag{5.7.16}$$

Adding the last two equations results in the total small R limit

$$-\frac{1}{24\pi^2\varepsilon_0^2 R^6}\sum_{q,\,s}\frac{|\vec{\mu}^{qp}(A)|^2|\vec{\mu}^{0s}(B)|^2}{(E_{qp}+E_{s0})}, \tag{5.7.17}$$

in which both real and virtual photon terms contribute to the energy shift, which exhibits R^{-6} dependence on separation distance. The limiting behavior coincides with results obtained by calculating the response of a polarizable test body to the fields of the source, the latter giving rise to an electromagnetic energy density, as presented in Section 2.9.

As already pointed out, when both molecules are excited, all three terms of (5.7.10) contribute to the interaction energy. To simplify the analysis, it is convenient to decompose the contributions into three types of terms arising from transitions $|q\rangle \leftarrow |p\rangle$ and $|s\rangle \leftarrow |r\rangle$ that are both upward, one upward and one downward, or both downward. Details concerning the first two scenarios have already been examined, as the functional forms of the contributions are similar to that found for the ground-state dispersion interaction and the energy shift when one of the two molecules is excited. The pertinent formulas are identical except for replacement by the appropriate ground- or excited-state polarizability tensor. Downward transitions from the initial states of both species can occur only when A and B are both excited and the contribution in this case is new relative to the other two.

When characteristic molecular transition wavelengths are larger than the internuclear separation, R^{-6}-dependent limiting terms arise from all three parts of (5.7.10),

$$\begin{aligned}&-\frac{1}{12\pi^2\varepsilon_0^2\hbar c R^6}\sum_{\substack{q\\E_p>E_q}} k_{sr}\frac{|\vec{\mu}^{pq}(A)|^2|\vec{\mu}^{rs}(B)|^2}{k_{sr}^2-k_{pq}^2}\\&-\frac{1}{12\pi^2\varepsilon_0^2\hbar c R^6}\sum_{\substack{s\\E_r>E_s}} k_{qp}\frac{|\vec{\mu}^{pq}(A)|^2|\vec{\mu}^{rs}(B)|^2}{k_{qp}^2-k_{sr}^2}\\&-\frac{1}{24\pi^2\varepsilon_0^2\hbar c R^6}\sum_{q,\,s}\mathrm{sgn}(k_{qp})\mathrm{sgn}(k_{sr})\frac{|\vec{\mu}^{pq}(A)|^2|\vec{\mu}^{rs}(B)|^2}{(|k_{qp}|+|k_{sr}|)},\end{aligned} \tag{5.7.18}$$

whose addition simplifies to

$$-\frac{1}{24\pi^2\varepsilon_0^2 R^6}\sum_{\substack{q,\,s\\E_p>E_q\\E_r>E_s}}\frac{|\vec{\mu}^{pq}(A)|^2|\vec{\mu}^{rs}(B)|^2}{(E_{qp}+E_{sr})}. \tag{5.7.19}$$

At large separation distances, the dominant contribution is again proportional to R^{-2} and originates from the first two terms of (5.7.10) due to real photon exchange. Their summation may be simplified to

$$-\frac{1}{36\pi^2\varepsilon_0^2(\hbar c)^4R^2}\sum_{\substack{q,s\\ E_p>E_q\\ E_r>E_s}}\frac{|\vec{\mu}^{pq}(A)|^2|\vec{\mu}^{rs}(B)|^2}{(E_{pq}+E_{rs})}E_{pq}E_{rs}(E_{pq}^2+E_{pq}E_{rs}+E_{rs}^2). \tag{5.7.20}$$

In addition to the advantages mentioned in the introduction to this section regarding employing response theory for the calculation of energy shifts, the method also clearly shows the role played by both vacuum and source fields as well as the necessity of including terms correct to second order in the dipole moment so as to achieve correct results, in particular the reinforcing or canceling of relevant terms from upward and downward transition contributions.

5.8 DISPERSION POTENTIAL VIA THE METHOD OF INDUCED MULTIPOLE MOMENTS

An alternative approach to perturbation and response theories for the calculation of the retarded van der Waals dispersion energy, including the additional contributions arising when molecules are excited, is the induced multipole moment method (Power and Thirunamachandran, 1993b). This particular technique provides an intuitive physical picture of dispersion forces and a simplified computational procedure. It relies on the fact that action of an electric displacement field on a pair of neutral, electrically polarizable molecules induces an electric dipole moment to leading order at each center. Even though the expectation value of the radiation field over the electromagnetic vacuum state vanishes, fluctuations of the field operators persist for this state and are nonzero. Hence, vacuum fluctuations of the electromagnetic field can momentarily distort the molecular charge distribution and induce a temporary dipole moment. The moments induced at each molecular center are coupled via the retarded resonant interaction tensor, a quantity that features in the transfer of energy between an excited and a ground-state species, giving rise to a dispersion energy shift on taking the expectation value of the ground state of the total molecule plus field system. In this regard, the viewpoint is similar to that originally adopted by London in his semiclassical treatment of the dispersion interaction in which virtual dipole transition moments were coupled by a static dipolar coupling potential, giving rise to an R^{-6} dependence on intermolecular separation distance in second order of perturbation theory. In this section, the causal quantized displacement field operator is used instead of a classical external electric field, enabling the dispersion potential to be obtained for all R beyond wavefunction overlap and correctly incorporating the effects of retardation.

Let A and B be neutral polarizable molecules situated at $\vec{R}_A$ and $\vec{R}_B$, respectively, with relative displacement vector $\vec{R} = \vec{R}_B - \vec{R}_A$. In the presence of an electric displacement field of specific mode character at the field point $\vec{r}$, $d_i^{\perp}(\vec{k}, \lambda; \vec{r})$, the ith component of the induced moment at site ξ is given by

$$\mu_i^{\mathrm{ind}}(\xi; \vec{k}) = \varepsilon_0^{-1} \alpha_{ij}(\xi; k) d_j^{\perp}(\vec{k}, \lambda; \vec{R}_\xi), \tag{5.8.1}$$

where $\alpha_{ij}(\xi; k)$ is the dynamic electric dipole polarizability tensor of species ξ,

$$\alpha_{ij}(\xi; k) = \sum_n \left\{ \frac{\mu_i^{0n}(\xi)\mu_j^{n0}(\xi)}{E_{n0} - \hbar ck} + \frac{\mu_j^{0n}(\xi)\mu_i^{n0}(\xi)}{E_{n0} + \hbar ck} \right\}. \tag{5.8.2}$$

Coupling of the electric moments induced at A and B occurs through the resonant interaction tensor at the single frequency $\omega = ck$,

$$\begin{aligned} V_{ij}^{\pm}(k, \vec{R}) &= -\frac{1}{4\pi\varepsilon_0}\left(-\vec{\nabla}^2 \delta_{ij} + \vec{\nabla}_i \vec{\nabla}_j\right)\frac{\mathrm{e}^{\mp ikR}}{R} \\ &= \frac{1}{4\pi\varepsilon_0 R^3}\left[(\delta_{ij} - 3\hat{R}_i\hat{R}_j)(1 \pm ikR) - (\delta_{ij} - \hat{R}_i\hat{R}_j)k^2R^2\right]\mathrm{e}^{\mp ikR}. \end{aligned} \tag{5.8.3}$$

An expression for the energy shift is obtained on summing over all modes of the radiation field,

$$\Delta E = \sum_{\vec{k}, \lambda} \mu_i^{\mathrm{ind}}(A; \vec{k}) \mu_j^{\mathrm{ind}}(B; \vec{k}) \mathrm{Re}\, V_{ij}(k, \vec{R}). \tag{5.8.4}$$

Since it is the real part of the resonant interaction tensor (5.8.3) that appears in the energy shift formula (5.8.4), which is independent of the signs occurring in the former, the $\pm$ superscript is left unwritten in the last relation. Inserting equation (5.8.1) into (5.8.4) yields an expression for ΔE that explicitly depends upon the polarizabilities of each molecule and the electric displacement field. Thus,

$$\Delta E = \sum_{\vec{k},\,\lambda} \varepsilon_0^{-2} \alpha_{ik}(A;k)\alpha_{jl}(B;k) d_k^{\perp}(\vec{k},\,\lambda;\vec{R}_A) d_l^{\perp}(\vec{k},\,\lambda;\vec{R}_B) \mathrm{Re}\, V_{ij}(k,\vec{R}). \tag{5.8.5}$$

Clearly evident from formula (5.8.5) is the presence of the field–field spatial correlation function, namely, the product of the electric displacement fields at two different points in space. To compute the ground-state dispersion potential, the expectation value of equation (5.8.5) is taken over the state $|E_0^A, E_0^B; 0(\vec{k},\,\lambda)\rangle$ corresponding to both entities in the ground electronic state and the radiation field in the vacuum state without photons. The expectation value over the molecular factor is elementary, yielding ground-state molecular polarizabilities of each species. For the radiation field part, use is made of the mode expansion for the transverse displacement field equation (1.7.17), the required quantity easily shown to be

$$\begin{aligned} &\langle 0(\vec{k},\,\lambda)|d_i^{\perp}(\vec{k},\,\lambda;\vec{R}_A) d_j^{\perp}(\vec{k},\,\lambda;\vec{R}_B)|0(\vec{k},\,\lambda)\rangle \\ &\quad = \left(\frac{\hbar c k \varepsilon_0}{2V}\right) e_i^{(\lambda)}(\vec{k}) \bar{e}_j^{(\lambda)}(\vec{k}) \mathrm{e}^{-i\vec{k}\cdot\vec{R}}. \end{aligned} \tag{5.8.6}$$

Substituting (5.8.6) into (5.8.5) produces

$$\Delta E = \sum_{\vec{k},\,\lambda} \left(\frac{\hbar c k}{2\varepsilon_0 V}\right) \alpha_{ik}(A;k)\alpha_{jl}(B;k) e_k^{(\lambda)}(\vec{k}) \bar{e}_l^{(\lambda)}(\vec{k}) \mathrm{e}^{-i\vec{k}\cdot\vec{R}} \mathrm{Re}\, V_{ij}(k,\vec{R}). \tag{5.8.7}$$

Next the mode sum is carried out. First, by performing the sum over polarizations using identity (1.4.56) and, second, by converting the wavevector sum to an integral via the replacement $(1/V)\sum_{\vec{k}} \Rightarrow \frac{1}{(2\pi)^3}\int \mathrm{d}^3\vec{k}$ and expressing the volume element in spherical polar coordinates, $\mathrm{d}^3\vec{k} = k^2 \mathrm{d}k\mathrm{d}\Omega$, where $\mathrm{d}\Omega$ is an infinitesimal element of solid angle. Angular integration is then executed using the result

$$\begin{aligned} &\frac{1}{4\pi}\int (\delta_{ij} - \hat{k}_i\hat{k}_j) \mathrm{e}^{\pm i\vec{k}\cdot\vec{R}} \mathrm{d}\Omega \\ &\quad = \left[(\delta_{ij} - \hat{R}_i\hat{R}_j)\frac{\sin kR}{kR} + (\delta_{ij} - 3\hat{R}_i\hat{R}_j)\left(\frac{\cos kR}{k^2R^2} - \frac{\sin kR}{k^3R^3}\right)\right]. \end{aligned} \tag{5.8.8}$$

Following this procedure and inserting the real part of $V_{ij}(k,\vec{R})$ from equation (5.8.3), the energy shift (5.8.7) becomes

$$\Delta E = \frac{\hbar c}{16\pi^3\varepsilon_0^2 R^3}\int_0^\infty dk k^3 \alpha_{ik}(A;k)\alpha_{jl}(B;k)$$
$$\times [(\delta_{ij}-3\hat{R}_i\hat{R}_j)(\cos kR + kR\sin kR) - (\delta_{ij}-\hat{R}_i\hat{R}_j)k^2R^2\cos kR]$$
$$\times \left[(\delta_{kl}-\hat{R}_k\hat{R}_l)\frac{\sin kR}{kR} + (\delta_{kl}-3\hat{R}_k\hat{R}_l)\left(\frac{\cos kR}{k^2R^2} - \frac{\sin kR}{k^3R^3}\right)\right]. \tag{5.8.9}$$

It is convenient to reexpress the energy shift more compactly in terms of the tensor field $F_{ij}(kR)$ defined by equation (2.9.4) on noting that the angular integral (5.8.8) is simply the imaginary part of $F_{ij}(kR)$, while the real part of the coupling tensor $V_{ij}(k,\vec{R})$ is $(-k^3/4\pi\varepsilon_0)\mathrm{Re}\,F_{ij}(kR)$. This leads to

$$\Delta E = -\frac{\hbar c}{16\pi^3\varepsilon_0^2}\int_0^\infty dk k^6 \alpha_{ik}(A;k)\alpha_{jl}(B;k)\mathrm{Re}[F_{ij}(kR)]\mathrm{Im}[F_{kl}(kR)] \tag{5.8.10}$$

or alternatively as

$$\Delta E = -\frac{\hbar c}{16\pi^3\varepsilon_0^2}\int_0^\infty dk \alpha_{ik}(A;k)\alpha_{jl}(B;k)\left[\left(-\vec{\nabla}^2\delta_{ij} + \vec{\nabla}_i\vec{\nabla}_j\right)\frac{\cos kR}{R}\right]$$
$$\times\left[\left(-\vec{\nabla}^2\delta_{kl} + \vec{\nabla}_k\vec{\nabla}_l\right)\frac{\sin kR}{R}\right]. \tag{5.8.11}$$

To obtain the result for isotropic A and B, use is made of the average over the product of the molecular polarizabilities, $\langle\alpha_{ik}(A;k)\alpha_{jl}(B;k)\rangle = \delta_{ik}\delta_{jl}\alpha(A;k)\alpha(B;k)$, where a factor of $1/3$ has been absorbed into each of the orientationally averaged polarizabilities. Contracting with the tensors occurring in equation (5.8.10) or (5.8.11) yields

$$\Delta E = -\frac{\hbar c}{16\pi^3\varepsilon_0^2 R^2}\int_0^\infty dk \alpha(A;k)\alpha(B;k)k^4\left[\sin kR\cos kR\left(2 - \frac{10}{k^2R^2} + \frac{6}{k^4R^4}\right)\right.$$
$$\left. + (\cos^2 kR - \sin^2 kR)\left(\frac{2}{kR} - \frac{6}{k^3R^3}\right)\right]. \tag{5.8.12}$$

Recognizing that the term within the square brackets above can be written as

$$\sin 2kR\left(1-\frac{5}{k^2R^2}+\frac{3}{k^4R^4}\right)+\cos 2kR\left(\frac{2}{kR}-\frac{6}{k^3R^3}\right)$$
$$=\mathrm{Im}\left[1+\frac{2i}{kR}-\frac{5}{k^2R^2}-\frac{6i}{k^3R^3}+\frac{3}{k^4R^4}\right]e^{2ikR} \tag{5.8.13}$$

and transforming to an imaginary wavevector variable $k = iu$ results in the Casimir–Polder potential

$$\Delta E = -\frac{\hbar c}{16\pi^3\varepsilon_0^2 R^2}\int_0^\infty du\, u^4 e^{-2uR}\alpha(A;icu)\alpha(B;icu)$$
$$\times\left[1+\frac{2}{uR}+\frac{5}{u^2R^2}+\frac{6}{u^3R^3}+\frac{3}{u^4R^4}\right], \tag{5.8.14}$$

after a 90° rotation in the line of integration in the complex plane, where the polarizabilities have been taken to be real at both real and imaginary frequencies.

The versatility of the induced moment approach is illustrated by its application to the calculation of the dispersion potential between electronically excited molecules, reproducing in a facile way the results obtained via diagrammatic perturbation theory in Section 5.6 and response theory in Section 5.7. To bring to the fore the essential physics and to simplify the treatment, molecule A is taken to be excited, initially in electronic state $|p\rangle$, with species B in the ground electronic level. In addition to upward transitions from $|p\rangle$ to higher lying intermediate states $|q\rangle$, molecule A can now make downward transitions from $|p\rangle$, a scenario that was previously forbidden when A was in its lowest electronic state. The contribution from upward transitions when A is excited and B is in the ground state can be obtained using the induced moment method in a manner similar to that leading to the result (5.8.14) on taking the expectation value of formula (5.8.5) over the state $|E_p^A, E_0^B; 0(\vec{k}, \lambda)\rangle$, where now the excited-state polarizability of A, equation (5.7.2), appears in the expression for the energy shift. The explicit form is given by the first term of equation (5.6.18) or equivalently by (5.7.13).

To evaluate the additional contribution to the interaction energy due to downward transitions occurring in species A, in which a real photon is emitted from the excited state, the starting point in the calculation is the formula for the energy shift (5.8.5), modified to

$$\Delta E^{\mathrm{RES}} = \sum_{\vec{k},\,\lambda} \varepsilon_0^{-2} \alpha_{ik}(A;k)\alpha_{jl}(B;k) d_k^{\perp}(\vec{R}_A) d_l^{\perp}(\vec{R}_B) V_{ij}^{\mathrm{RES}}(k_{pq},\vec{R}), \tag{5.8.15}$$

where the superscript "RES" denotes the contribution of the resonant term, corresponding to a downward transition $|q\rangle \leftarrow |p\rangle$ in A of circular frequency $\omega_{pq} = ck_{pq}$, and $V_{ij}^{\mathrm{RES}}(k_{pq},\vec{R})$ is the retarded coupling tensor (5.8.3) evaluated at the resonant frequency of the downward transition. Employing the vacuum field spatial correlation function (5.8.6), carrying out the polarization sum, converting the $\vec{k}$-sum to an integral, and applying the integral for the angular average in the form

$$\frac{1}{4\pi}\int(\delta_{ij}-\hat{k}_i\hat{k}_j)\mathrm{e}^{\pm i\vec{k}\cdot\vec{R}}\mathrm{d}\Omega = \frac{1}{2ik^3}\left(-\vec{\nabla}^2\delta_{ij}+\vec{\nabla}_i\vec{\nabla}_j\right)\frac{1}{R}(\mathrm{e}^{ikR}-\mathrm{e}^{-ikR}), \tag{5.8.16}$$

produces

$$\Delta E^{\mathrm{RES}} = \frac{1}{4\pi^2\varepsilon_0 i}\left(-\vec{\nabla}^2\delta_{kl}+\vec{\nabla}_k\vec{\nabla}_l\right)\frac{1}{R}\int_0^\infty \mathrm{d}k\,\alpha_{jl}(B;k)$$
$$\times \sum_q \frac{k_{qp}}{k_{qp}^2-k^2}\mu_i^{pq}(A)\mu_k^{qp}(A)V_{ij}^{\mathrm{RES}}(k_{pq},\vec{R})(\mathrm{e}^{ikR}-\mathrm{e}^{-ikR}). \tag{5.8.17}$$

Integration yields

$$\Delta E^{\mathrm{RES}} = -\frac{1}{16\pi^2\varepsilon_0^2}\sum_{\substack{q\\E_p>E_q}} \mu_i^{pq}(A)\mu_k^{qp}(A)\alpha_{jl}(B;k_{pq})$$
$$\times\left[\left(-\vec{\nabla}^2\delta_{ij}+\vec{\nabla}_i\vec{\nabla}_j\right)\frac{\mathrm{e}^{ik_{qp}R}}{R}\right]\left[\left(-\vec{\nabla}^2\delta_{kl}+\vec{\nabla}_k\vec{\nabla}_l\right)\frac{\mathrm{e}^{-ik_{qp}R}}{R}\right] \tag{5.8.18}$$

after substituting for $V_{ij}^{\mathrm{RES}}(k_{pq},\vec{R})$. In terms of the tensor $F_{ij}(kR)$, (5.8.18) becomes

$$\Delta E^{\mathrm{RES}} = -\frac{1}{16\pi^2\varepsilon_0^2}\sum_{\substack{q\\E_p>E_q}} \mu_i^{pq}(A)\mu_k^{qp}(A)\alpha_{jl}(B;k_{pq})k_{qp}^6 F_{ij}(k_{qp}R)\bar{F}_{kl}(k_{qp}R), \tag{5.8.19}$$

where the overbar designates the complex conjugate. It is interesting to note that the resonant contribution holds only for downward transitions from $|p\rangle$, with B responding through its polarizability to the frequency $\omega_{pq} = (E_p - E_q)/\hbar$. Evaluating the gradient operators in (5.8.18) and performing a rotational average results in the following contribution due to exchanged photons being on the energy shell in agreement with the second term of (5.6.18):

$$\Delta E^{\mathrm{RES}} = -\frac{1}{24\pi^2\varepsilon_0^2}\sum_{\substack{q \\ E_p>E_q}} |\vec{\mu}^{pq}(A)|^2 \alpha(B; k_{pq}) k_{pq}^6 \left[\frac{1}{k_{pq}^2 R^2} + \frac{1}{k_{pq}^4 R^4} + \frac{3}{k_{pq}^6 R^6}\right] \tag{5.8.20}$$

Many of the difficulties associated with the use of diagrammatic perturbation theory techniques in the calculation of dispersion energy shifts between ground and/or excited molecules are avoided in the induced moment approach. For instance, integration over wavevector, if at all, is tackled more easily in the latter method and summation over a large number of graphs is unnecessary. Further, the viewpoint is conceptually simple and physically intuitive. It is easy to extend to include the effects of higher multipole moments and may also be applied to the computation of other intermolecular interactions such as the modification of the dispersion force by external radiation or laser-induced resonance energy transfer. Advantages also occur relative to response theory. In the response technique, *a priori* knowledge of the functional forms of the Maxwell field operators in the proximity of a source molecule are needed correct to at least second order in the electronic charge.

5.9 DISCRIMINATORY DISPERSION INTERACTIONS

It was shown in Section 4.4 that the resonant transfer of energy between two chiral molecules is discriminatory, being proportional to the optical rotatory strength tensor of each unit and changing sign when one enantiomer is replaced by its mirror image form. Migration of energy is not the only intermolecular process that depends on molecular handedness—a characteristic of interactions occurring between optically active species in general, but is also manifest in processes such as dispersion energy shifts, radiation-induced chiral discrimination, and molecule-induced circular dichroism and luminescence, for example. Once again the discriminatory aspect is attributed to the low symmetry of chiral entities, with transitions in these species including contributions from higher multipole moments, with the leading term dependent on the handedness of each molecule being proportional to the interference of electric and magnetic dipole coupling terms. In this section, the retarded dispersion interaction between a pair of optically active molecules is computed. As for the calculation of the Casimir–Polder potential, three different physical viewpoints and computational procedures are employed. First, diagrammatic time-dependent perturbation theory is used to evaluate the energy shift between two chiral molecules in the ground electronic state. Second, response theory is employed to evaluate the interaction potential for two excited optically active species. In this approach, a test chiral molecule responds through its mixed electric–magnetic dipole polarizability to the electric–magnetic dipole-dependent electric displacement and magnetic field operators of a second source molecule. Third, the induced moment method introduced in Section 5.8 is extended to deal with coupling between chiral chromophores.

5.9.1 Perturbation Theory

In the time-dependent perturbation theory calculation of the van der Waals dispersion potential between a pair of neutral electric dipole polarizable molecules, the interaction was interpreted as arising from the exchange of two virtual photons between the two centers. Employing an interaction Hamiltonian that is linear in the electric displacement field necessitated the use of fourth-order perturbation theory for the calculation of the energy shift. A similar viewpoint may be adopted for the evaluation of the dispersion interaction between two chiral molecules (Jenkins et al., 1994a,1994b). Let these two bodies be labeled A and B, both be in the ground electronic state, and be located at $\vec{R}_A$ and $\vec{R}_B$, respectively. The radiation– molecule Hamiltonian for this two-particle system is

$$H = H_{\text{mol}}(A) + H_{\text{mol}}(B) + H_{\text{rad}} + H_{\text{int}}(A) + H_{\text{int}}(B). \tag{5.9.1}$$

To correctly describe optically active molecules, the electric dipole approximated form of the perturbation operator (5.2.5) is now no longer sufficient. It is modified by adding the first term of the magnetic multipole series, namely, the $-\vec{m}(\xi)\cdot\vec{b}(\vec{R}_\xi)$ interaction term, where $\vec{m}(\xi)$ is the magnetic dipole moment operator of species ξ and $\vec{b}(\vec{R}_\xi)$ is the magnetic field operator, since selection rules now permit magnetic dipole allowed transitions to take place in addition to electric dipole allowed ones to leading order, as well as contributions from higher multipole moment terms. Therefore, the interaction Hamiltonian can be written as

$$H_{\rm int}(A)+H_{\rm int}(B) = -\varepsilon_0^{-1}\vec{\mu}(A)\cdot\vec{d}^{\perp}(\vec{R}_A)-\vec{m}(A)\cdot\vec{b}(\vec{R}_A)$$

$$-\varepsilon_0^{-1}\vec{\mu}(B)\cdot\vec{d}^{\perp}(\vec{R}_B)-\vec{m}(B)\cdot\vec{b}(\vec{R}_B), \tag{5.9.2}$$

which is identical to the coupling operator (4.4.1) used in the computation of the discriminatory transfer rate. Although the electric quadrupole interaction term $-\varepsilon_0^{-1}Q_{ij}(\xi)\vec{\nabla}_j d_i^{\perp}(\vec{R}_\xi)$, where Q_{ij} is the electric quadrupole moment tensor, is of a comparable order of magnitude to the magnetic dipole moment, the electric dipole–quadrupole contribution to the dispersion potential vanishes for isotropic molecules and is, therefore, excluded from further consideration.

It may be recalled from Section 5.2 that in the perturbation theory calculation of the dispersion energy shift, 12 two-photon exchange diagrams involving electric dipole interaction vertices needed to be summed. If $H_{\rm int}(\xi)$, $\xi = A, B$, given by equation (5.9.2) is used as the interaction Hamiltonian at each respective center, the resulting energy shift will comprise three different types of contribution. The leading electric dipole interaction term will again yield the Casimir–Polder expression, while the pure magnetic dipole coupling will give rise to the dispersion potential between two paramagnetically susceptible molecules, in essence the magnetic dipole analogue of the Casimir–Polder potential. It is considerably smaller in magnitude than the electric dipole–dipole interaction and has the functional form

$$\Delta E = -\frac{\hbar}{16\pi^3\varepsilon_0^2c^3R^2}\int_0^\infty du\,u^4 e^{-2uR}\chi(A;iu)\chi(B;iu)$$

$$\times\left[1+\frac{2}{uR}+\frac{5}{u^2R^2}+\frac{6}{u^3R^3}+\frac{3}{u^4R^4}\right], \tag{5.9.3}$$

where the isotropic magnetic dipole susceptibility at imaginary frequency is defined as

$$\chi(\xi;iu) = \frac{2}{3}\sum_t \frac{E_{t0}|\vec{m}^{0t}(\xi)|^2}{E_{t0}^2+(\hbar cu)^2}. \tag{5.9.4}$$

A contribution of similar order of magnitude to the chiral discrimination dispersion potential, which survives orientational averaging, is the energy shift between an electrically polarizable molecule and a magnetically susceptible one, and details of this calculation are presented in the following section.

To obtain the energy shift between two chiral molecules, the contribution proportional to the product of the electric–magnetic dipole moments at each center is extracted. This means replacing one electric dipole interaction vertex in A and B in the time-ordered graphs of Fig. 5.1 by a magnetic dipole coupling term. Instead of adding the contributions from 12 diagrams, now 48 Feynman graphs have to be summed. The four graphs ensuing from diagram (i) of Fig. 5.1 are illustrated in Fig. 5.9. For the

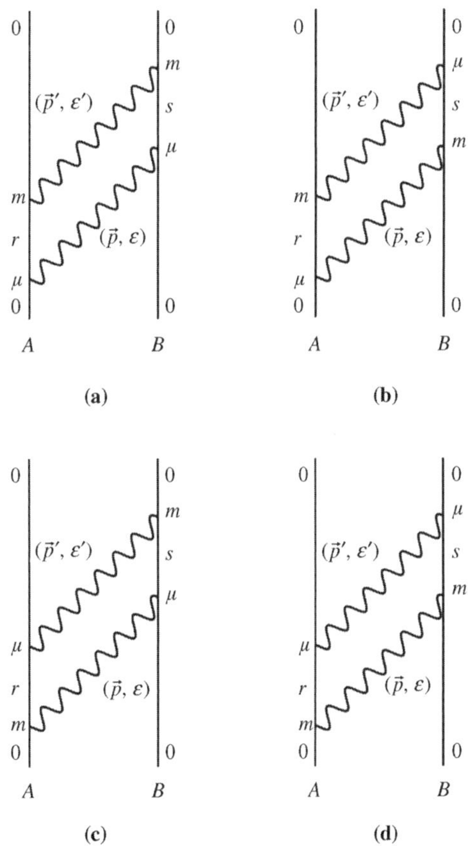

FIGURE 5.9 One set of four time-ordered graphs used in the calculation of the chiral discrimination dispersion interaction. The labels μ and m are shorthand for electric and magnetic dipole coupling vertices.

current problem, the perturbation theory solution involves techniques and formulas previously employed in Sections 4.4 and 5.2. The energy shift is calculated using the fourth-order perturbation theory expression (5.2.4), but with coupling Hamiltonian (5.9.2). As before, the initial and final states are represented by the ket $|0\rangle = |E_0^A, E_0^B; 0(\vec{p}, \varepsilon), 0(\vec{p}', \varepsilon')\rangle$, corresponding to both molecules in their lowest electronic level, with no photons present. The 48 possible time orderings may be grouped into 12 sets of 4 diagrams, with the product of the energy denominators occurring in each set given in Table 5.1. Evaluating the four graphs in diagram (i) of Fig. 5.1, which have the common energy denominator product $[(E_{r0} + \hbar cp')(\hbar cp + \hbar cp')\ (E_{s0} + \hbar cp)]^{-1}$, where r and s denote excited states of A and B, respectively, corresponding to the denominator of graph (i) of Fig. 5.1 as displayed in Table 5.1, the contribution is

$$
\begin{aligned}
\Delta E_{i(a-d)} = &-\sum_{\vec{p},\vec{p}'}\sum_{\varepsilon,\varepsilon'}\sum_{r,s}\left(\frac{\hbar p}{2\varepsilon_0 V}\right)\left(\frac{\hbar p'}{2\varepsilon_0 V}\right) \\
&\times [m_j^{0r}(A)\mu_i^{r0}(A)m_l^{0s}(B)\mu_k^{s0}(B)\bar{b}_j^{(\varepsilon)}(\vec{p})b_l^{(\varepsilon)}(\vec{p})e_k^{(\varepsilon')}(\vec{p}')\bar{e}_i^{(\varepsilon')}(\vec{p}') \\
&+\mu_i^{0r}(A)m_j^{r0}(A)m_l^{0s}(B)\mu_k^{s0}(B)\bar{e}_i^{(\varepsilon)}(\vec{p})b_l^{(\varepsilon)}(\vec{p})e_k^{(\varepsilon')}(\vec{p}')\bar{b}_j^{(\varepsilon')}(\vec{p}') \\
&+m_j^{0r}(A)\mu_i^{r0}(A)\mu_k^{0s}(B)m_l^{s0}(B)\bar{b}_j^{(\varepsilon)}(\vec{p})e_k^{(\varepsilon)}(\vec{p})\bar{e}_i^{(\varepsilon')}(\vec{p}')b_l^{(\varepsilon')}(\vec{p}') \\
&+\mu_i^{0r}(A)m_j^{r0}(A)\mu_k^{0s}(B)m_l^{s0}(B)\bar{e}_i^{(\varepsilon)}(\vec{p})e_k^{(\varepsilon)}(\vec{p})\bar{b}_j^{(\varepsilon')}(\vec{p}')b_l^{(\varepsilon')}(\vec{p}')] \\
&\times e^{i(\vec{p}+\vec{p}')\cdot\vec{R}}[(E_{r0}+\hbar cp')(\hbar cp+\hbar cp')(E_{s0}+\hbar cp)]^{-1}.
\end{aligned}
\tag{5.9.5}
$$

Using the fact that $\mu_i^{0r}(A)m_j^{r0}(A) = -m_j^{0r}(A)\mu_i^{r0}(A)$ and $\mu_k^{0s}(B)m_l^{s0}(B) = -m_l^{0s}(B)\mu_k^{s0}(B)$ for transition dipole moment matrix elements taken over real wavefunctions and carrying out the sums over polarization vectors, equation (5.9.5) becomes

$$
\begin{aligned}
&-\sum_{\vec{p},\vec{p}'}\sum_{r,s}\left(\frac{\hbar p}{2\varepsilon_0 V}\right)\left(\frac{\hbar p'}{2\varepsilon_0 V}\right)\mu_i^{0r}(A)m_j^{r0}(A)\mu_k^{0s}(B)m_l^{s0}(B) \\
&\times\left[(\delta_{ik}-\hat{p}_i\hat{p}_k)(\delta_{jl}-\hat{p}'_j\hat{p}'_l)+(\delta_{ik}-\hat{p}'_i\hat{p}'_k)(\delta_{jl}-\hat{p}_j\hat{p}_l)+\varepsilon_{ils}\varepsilon_{jkt}(\hat{p}_s\hat{p}'_t+\hat{p}'_s\hat{p}_t)\right] \\
&\times e^{i(\vec{p}+\vec{p}')\cdot\vec{R}}[(E_{r0}+\hbar cp')(\hbar cp+\hbar cp')(E_{s0}+\hbar cp)]^{-1}.
\end{aligned}
\tag{5.9.6}
$$

Evaluating the contribution from the remaining 44 diagrams and adding to (5.9.6) enables the energy shift to be written as

$$\Delta E = -\sum_{\vec{p},\vec{p}'}\sum_{r,s}\left(\frac{\hbar p}{2\varepsilon_0 V}\right)\left(\frac{\hbar p'}{2\varepsilon_0 V}\right)\mu_i^{0r}(A)m_j^{r0}(A)\mu_k^{0s}(B)m_l^{s0}(B)$$
$$\times\Big[(\delta_{ik}-\hat{p}_i\hat{p}_k)(\delta_{jl}-\hat{p}'_j\hat{p}'_l)+(\delta_{ik}-\hat{p}'_i\hat{p}'_k)(\delta_{jl}-\hat{p}_j\hat{p}_l)$$
$$+\varepsilon_{ils}\varepsilon_{jkt}(\hat{p}_s\hat{p}'_t+\hat{p}'_s\hat{p}_t)\Big]\mathrm{e}^{i(\vec{p}+\vec{p}')\cdot\vec{R}}\sum_{a=i}^{xii}D_a^{-1}, \quad (5.9.7)$$

the 12 energy denominator products being listed in Table 5.1. Their summation may be performed in a manner similar to that carried out in the calculation of the Casimir–Polder potential, since the energy denominators are the same in both cases. Converting the wavevector sums in (5.9.7) to integrals and performing the angular averages using relations (4.2.12) and (4.4.7) produces

$$\Delta E = -\frac{1}{4\pi^4\varepsilon_0^2\hbar c^3}\sum_{r,s}\mu_i^{0r}(A)m_j^{r0}(A)\mu_k^{0s}(B)m_l^{s0}(B)\frac{1}{(k_{r0}+k_{s0})}$$
$$\times\int_0^\infty\int_0^\infty \mathrm{d}p\,\mathrm{d}p'\,p^3p'^3\left(\frac{1}{p+p'}-\frac{1}{p-p'}\right)$$
$$\times\Bigg[\Bigg\{\Bigg[(\delta_{ik}-\hat{R}_i\hat{R}_k)\frac{\sin pR}{pR}+(\delta_{ik}-3\hat{R}_i\hat{R}_k)\left(\frac{\cos pR}{p^2R^2}-\frac{\sin pR}{p^3R^3}\right)\Bigg]$$
$$\times\Bigg[(\delta_{jl}-\hat{R}_j\hat{R}_l)\frac{\sin p'R}{p'R}+(\delta_{jl}-3\hat{R}_j\hat{R}_l)\left(\frac{\cos p'R}{p'^2R^2}-\frac{\sin p'R}{p'^3R^3}\right)\Bigg]$$
$$+\Bigg[(\delta_{ik}-\hat{R}_i\hat{R}_k)\frac{\sin p'R}{p'R}+(\delta_{ik}-3\hat{R}_i\hat{R}_k)\left(\frac{\cos p'R}{p'^2R^2}-\frac{\sin p'R}{p'^3R^3}\right)\Bigg]$$
$$\times\Bigg[(\delta_{jl}-\hat{R}_j\hat{R}_l)\frac{\sin pR}{pR}+(\delta_{jl}-3\hat{R}_j\hat{R}_l)\left(\frac{\cos pR}{p^2R^2}-\frac{\sin pR}{p^3R^3}\right)\Bigg]\Bigg\}$$
$$\times\left(\frac{p}{(k_{r0}+p)(k_{s0}+p)}\right)+2\varepsilon_{ils}\varepsilon_{kjt}\left(\frac{\cos pR}{pR}-\frac{\sin pR}{p^2R^2}\right)$$
$$\times\left(\frac{\cos p'R}{p'R}-\frac{\sin p'R}{p'^2R^2}\right)\left(\frac{p'}{(k_{r0}+p')(k_{s0}+p')}\right)\Bigg], \quad (5.9.8)$$

which has been separated into two parts, one symmetric and one antisymmetric in p and p'. Performing the p' integral and transforming p to an imaginary wavevector results in an expression for the dispersion energy shift between two chiral molecules in fixed relative orientation

$$\Delta E = -\frac{1}{4\pi^3\varepsilon_0^2\hbar c^3}\sum_{r,s}\mu_i^{0r}(A)m_j^{r0}(A)\mu_k^{0s}(B)m_l^{s0}(B)\int_0^\infty \frac{du\,u^8\,\mathrm{e}^{-2uR}}{(k_{r0}^2+u^2)(k_{s0}^2+u^2)}$$

$$\times\Big[(\alpha_{ik}\alpha_{jl}-\varepsilon_{ils}\varepsilon_{jkt}\hat{R}_s\hat{R}_t)(uR)^{-2}+(\alpha_{ik}\beta_{jl}+\beta_{ik}\alpha_{jl}-2\varepsilon_{ils}\varepsilon_{jkt}\hat{R}_s\hat{R}_t)(uR)^{-3}$$

$$+(\alpha_{ik}\beta_{jl}+\beta_{ik}\alpha_{jl}+\beta_{ik}\beta_{jl}-\varepsilon_{ils}\varepsilon_{jkt}\hat{R}_s\hat{R}_t)(uR)^{-4}+2\beta_{ik}\beta_{jl}(uR)^{-5}$$

$$+\beta_{ik}\beta_{jl}(uR)^{-6}\Big], \tag{5.9.9}$$

where the dyadics α_{ij} and β_{ij} are defined as $\alpha_{ij}=\delta_{ij}-\hat{R}_i\hat{R}_j$ and $\beta_{ij}=\delta_{ij}-3\hat{R}_i\hat{R}_j$. To obtain the energy shift for a pair of freely rotating optically active molecules, an orientational average is carried out using the result

$$\left\langle \mu_i^{0t}(\xi)m_j^{0t}(\xi)\right\rangle=\frac{1}{3}\delta_{ij}|\vec{\mu}^{0t}(\xi)\cdot\vec{m}^{t0}(\xi)|, \tag{5.9.10}$$

yielding the interaction energy

$$\Delta E = -\frac{1}{18\pi^3\varepsilon_0^2\hbar c^3R^4}\sum_{r,s}|\vec{\mu}^{0r}(A)\cdot\vec{m}^{r0}(A)||\vec{\mu}^{0s}(B)\cdot\vec{m}^{s0}(B)|$$

$$\times\int_0^\infty\frac{du\,u^4\,\mathrm{e}^{-2uR}}{(k_{r0}^2+u^2)(k_{s0}^2+u^2)}\left[4+\frac{6}{uR}+\frac{3}{u^2R^2}\right], \tag{5.9.11}$$

which holds for all separation distances outside the charge overlap region. In contrast to the Casimir–Polder potential, the interaction energy (5.9.11) depends on the chirality of each molecule and is discriminatory. This is due to the pseudoscalar nature of the dot product of the transition electric and magnetic dipole moments that features in the energy shift. When one optical isomer is changed to its antipodal form, $\vec{\mu}$, a polar vector, changes sign on inversion while $\vec{m}$, an axial vector, is symmetric to $\vec{r}$ being transformed to $-\vec{r}$. The energy shift (5.9.11) may also be expressed in terms of the rotatory strength tensor defined in equation (4.4.15). For chemically distinct species, it is not possible to determine the absolute sign of the interaction because the rotatory strength may be of either sign. When the two molecules are chemically identical, however, the energy shift for opposite isomers is attractive, while that for like isomers it is positive in sign.

From the general result (5.9.11) valid for all R, it is a simple matter to obtain the limiting forms of the potential at long and short separation distances. The physical and mathematical approximations are identical to those made in calculating the asymptotic behavior of the Casimir–Polder potential in Section 5.2. In the far zone, after dropping the u^2 factor relative to k_{r0} and k_{s0} in the wavevector denominator product and doing the u-integral,

$$\Delta E_{\mathrm{FZ}} = -\frac{\hbar^3 c}{3\pi^3\varepsilon_0^2 R^9}\sum_{r,\,s}\frac{R^{0r}(A)R^{s0}(B)}{E_{r0}^2 E_{s0}^2}, \tag{5.9.12}$$

which is expressed in terms of the rotatory strength tensor and is seen to exhibit an R^{-9} power law dependence. To obtain the near-zone limit in which $kR \ll 1$, e^{-2uR} is approximated to unity, the term $3(uR)^{-2}$ in square brackets is retained, and the integral is evaluated using relation (5.2.21) to give

$$\Delta E_{\mathrm{NZ}} = -\frac{1}{12\pi^2\varepsilon_0^2 c^2 R^6}\sum_{r,\,s}\frac{R^{0r}(A)R^{s0}(B)}{E_{r0}+E_{s0}}. \tag{5.9.13}$$

It should be mentioned that just as the R^{-6} London dispersion formula (5.2.22) can be derived from the static electric dipolar coupling potential (5.3.5) and second-order perturbation theory as shown in Section 5.3, the near-zone discriminatory shift (5.9.13) can likewise be obtained by adding the static magnetic dipolar coupling to the electric part to give

$$V_{AB} = \frac{1}{4\pi\varepsilon_0 R^3}\left[\mu_i(A)\mu_j(B) + \frac{1}{c^2}m_i(A)m_j(B)\right](\delta_{ij} - 3\hat{R}_i\hat{R}_j), \tag{5.9.14}$$

and using the expression for the second-order energy shift (5.3.6) and extracting the electric–magnetic cross-term.

5.9.2 Response Theory

Having demonstrated in Section 5.7 that response theory is advantageous for the computation of dispersion potentials between electric dipole polarizable molecules relative to diagrammatic time-dependent perturbation theory, the interaction between a polarizable species and the source Maxwell fields of a second body is used to evaluate the discriminatory dispersion force between two excited chiral molecules in this section (Jenkins et al., 1994a, 1994b; Salam,1996), extending previous results for such systems that were limited to ground-state interactions. Labeling the initial excited electronic states of A and B as $|p^A\rangle$ and $|r^B\rangle$ as before, with electric and magnetic dipole allowed upward and downward transitions to intermediate states $|q^A\rangle$ and $|s^B\rangle$, respectively, the extension of the energy shift formula (5.7.1) applicable to the dispersive coupling between two optically active molecules is

$$\begin{aligned}\Delta E = {} & \mathrm{Im}\frac{i}{\varepsilon_0}G_{ij}(A,\omega)d_i^{\perp}(B;\vec{R}_A;t)b_j(B;\vec{R}_A;t)\\ & + \mathrm{Im}\frac{i}{\varepsilon_0}G_{kl}(B,\omega)d_k^{\perp}(A;\vec{R}_B;t)b_l(A;\vec{R}_B;t). \qquad (5.9.15)\end{aligned}$$

A consequence of less restrictive selection rules for electronic transitions in chiral molecules is that such substances are characterized by a mixed electric–magnetic dipole dynamic polarizability tensor $G_{ij}(\xi;\omega)$, defined analogously to the pure electric dipole polarizability as

$$G_{ij}(\xi;\omega) = \sum_n \left\{ \frac{\mu_i^{mn}(\xi)m_j^{nm}(\xi)}{E_{nm}-\hbar\omega} + \frac{m_j^{mn}(\xi)\mu_i^{nm}(\xi)}{E_{nm}+\hbar\omega} \right\}. \qquad (5.9.16)$$

In contrast to the electric dipole polarizability tensor $\alpha_{ij}(\xi;\omega)$, the mixed tensor $G_{ij}(\xi;\omega)$ changes sign when one enantiomer is replaced by another of opposite handedness, thereby permitting differentiation of species with differing chirality to occur. From expression (5.9.15), it is seen that each optically active molecule responds via $G_{ij}(\xi;\omega)$, $\xi = A, B$, to the electric displacement and magnetic field operators due to a second source molecule at the position at which the first species is located. Since the electric–magnetic dipole response tensor (5.9.16) is imaginary for real wavefunctions, the imaginary part is taken in the formula for the energy shift. To find the interaction energy between two chiral molecules using the response theory approach, the electric dipole-dependent Maxwell field operators in the proximity of a molecule are no longer sufficient. Contributions from higher multipole moments are required. This necessitates using the electric and magnetic dipole-dependent first-order Maxwell fields and the second-order electric displacement and magnetic fields dependent bilinearly on $\vec{\mu}$ and $\vec{m}$. The first of these were found in Section 2.7 and are given by equations (2.7.6) and (2.7.7). Appendix A contains the second-order fields correct up to and including the electric quadrupole coupling, with the relevant fields to be employed in the present application given by expressions (A.2) and (A.8). Inserting the expansion of the Maxwell fields in series of powers of the first two multipole moments and retaining terms proportional to the product of $\vec{\mu}$ and $\vec{m}$ at each center, the energy shift (5.9.15) becomes

$$\begin{aligned}\Delta E = {} & \mathrm{Im}\frac{i}{\varepsilon_0}\sum_s G_{ij}(A;\omega_{rs}) \\ & \times\Big[d_i^{(1)}(B;\vec{\mu};\omega_{rs};\vec{R}_A)b_j^{(1)}(B;\vec{m};\omega_{rs};\vec{R}_A)+d_i^{(1)}(B;\vec{m};\omega_{rs};\vec{R}_A) \\ & \times b_j^{(1)}(B;\vec{\mu};\omega_{rs};\vec{R}_A)\Big]+\mathrm{Im}\frac{i}{\varepsilon_0}\sum_q G_{kl}(B;\omega_{pq}) \\ & \times\Big[d_k^{(1)}(A;\vec{\mu};\omega_{pq};\vec{R}_B)b_l^{(1)}(A;\vec{m};\omega_{pq};\vec{R}_B)+d_k^{(1)}(A;\vec{m};\omega_{pq};\vec{R}_B) \\ & \times b_l^{(1)}(A;\vec{\mu};\omega_{pq};\vec{R}_B)\Big]-\mathrm{Im}\frac{i}{\varepsilon_0}\sum_{\text{modes}} G_{ij}(A;\omega) \\ & \times\Big[b_j^{(0)}(\omega;\vec{R}_A)d_i^{(2)}(B;\vec{\mu}\vec{m};\omega;\vec{R}_A)+b_j^{(2)}(B;\vec{\mu}\vec{m};\omega;\vec{R}_A)d_i^{(0)}(\omega;\vec{R}_A)\Big] \\ & -\mathrm{Im}\frac{i}{\varepsilon_0}\sum_{\text{modes}} G_{kl}(B;\omega) \\ & \times\Big[b_l^{(0)}(\omega;\vec{R}_B)d_k^{(2)}(A;\vec{\mu}\vec{m};\omega;\vec{R}_B)+b_l^{(2)}(A;\vec{\mu}\vec{m};\omega;\vec{R}_B)d_k^{(0)}(\omega;\vec{R}_B)\Big].\end{aligned} \tag{5.9.17}$$

As in the corresponding calculation of the Casimir–Polder potential in Section 5.7, A responds to the source fields of B, the latter species undergoing transitions with energy $E_{rs} = \hbar ck_{rs}$. Meanwhile, B reacts to the fields of molecule A, for which transitions take place between states $|q\rangle \leftarrow |p\rangle$ with energy $E_{pq} = \hbar ck_{pq}$. Concentrating for the moment on the first two terms of (5.9.17) arising from the product of the radiation fields linear in the moments, use of (2.7.6) and (2.7.7) leads to

$$\mathrm{Im}\left\{\begin{aligned}&-\frac{i}{16\pi^2\varepsilon_0^2c^2}\sum_s G_{ij}(A;k_{rs})k_{rs}^6\left[\mu_k^{rs}(B)m_l^{sr}(B)\bar{f}_{ik}(k_{rs}R)f_{jl}(k_{rs}R)\right.\\&\left.-m_k^{rs}(B)\mu_l^{sr}(B)\bar{g}_{ik}(k_{rs}R)g_{jl}(k_{rs}R)\right]\end{aligned}\right\}$$
$$+\mathrm{Im}\left\{\begin{aligned}&-\frac{i}{16\pi^2\varepsilon_0^2c^2}\sum_q G_{kl}(B;k_{pq})k_{pq}^6\left[\mu_i^{pq}(A)m_j^{qp}(A)\bar{f}_{ki}(k_{pq}R)f_{lj}(k_{pq}R)\right.\\&\left.-m_j^{pq}(A)\mu_i^{qp}(A)\bar{g}_{ki}(k_{pq}R)g_{lj}(k_{pq}R)\right]\end{aligned}\right\}. \tag{5.9.18}$$

For the evaluation of the last two terms of (5.9.17), a procedure identical to that carried out for the calculation of the Casimir–Polder potential using response theory, detailed in Section 5.7, is followed. Employing the appropriate second-order Maxwell fields in addition to the free displacement and magnetic fields, equations (2.6.13) and (2.6.14), the contribution arising from the interference of the vacuum and bilinear fields is

$$\mathrm{Im}\left\{\begin{aligned}&-\frac{i}{16\pi^2\varepsilon_0^2c^2}\sum_s \mathrm{sgn}(k_{rs})G_{ij}(A;k_{rs})k_{rs}^6\left[\mu_k^{rs}(B)m_l^{sr}(B)\bar{f}_{ik}(k_{rs}R)f_{jl}(k_{rs}R)\right.\\&\left.-m_k^{rs}(B)\mu_l^{sr}(B)\bar{g}_{ik}(k_{rs}R)g_{jl}(k_{rs}R)\right]\end{aligned}\right\}$$
$$+\mathrm{Im}\left\{\begin{aligned}&-\frac{i}{16\pi^2\varepsilon_0^2c^2}\sum_q \mathrm{sgn}(k_{pq})G_{kl}(B;k_{pq})k_{pq}^6\left[\mu_i^{pq}(A)m_j^{qp}(A)\bar{f}_{ki}(k_{pq}R)\right.\\&\left.\times f_{lj}(k_{pq}R)-m_j^{pq}(A)\mu_i^{qp}(A)\bar{g}_{ki}(k_{pq}R)g_{lj}(k_{pq}R)\right]\end{aligned}\right\}$$
$$-\frac{1}{32\pi^3\varepsilon_0^2c^2}\sum_s \mu_k^{rs}(B)m_l^{sr}(B)\int_0^\infty du\frac{2iu^7e^{-2uR}}{(k_{rs}^2+u^2)}G_{ij}(A;icu)$$
$$\times\left[f_{ik}(iuR)f_{jl}(iuR)-g_{ik}(iuR)g_{jl}(iuR)\right]$$
$$-\frac{1}{32\pi^3\varepsilon_0^2c^2}\sum_q \mu_i^{pq}(A)m_j^{qp}(A)\int_0^\infty du\frac{2iu^7e^{-2uR}}{(k_{pq}^2+u^2)}G_{kl}(B;icu)$$
$$\times\left[f_{ki}(iuR)f_{lj}(iuR)-g_{ki}(iuR)g_{lj}(iuR)\right]. \tag{5.9.19}$$

Appearing in the u-integral terms of the energy shift is the mixed electric–magnetic dipole polarizability tensor at imaginary wavevector $k=iu$,

$$G_{ij}(\xi;icu)=\sum_n\left\{\frac{\mu_i^{mn}(\xi)m_j^{nm}(\xi)}{E_{nm}-i\hbar cu}+\frac{m_j^{mn}(\xi)\mu_i^{nm}(\xi)}{E_{nm}+i\hbar cu}\right\}$$

$$=2i\sum_n\frac{\mu_i^{mn}(\xi)m_j^{nm}(\xi)\hbar cu}{E_{nm}^2+(\hbar cu)^2}, \tag{5.9.20}$$

and the geometric tensors $f_{ij}(iuR)$ and $g_{ij}(iuR)$ are given by equations (2.9.15) and (2.9.35), respectively. The total energy shift is obtained by adding equation (5.9.19) to the contribution (5.9.18). As in the field theoretic computation of the Poynting vector, the electromagnetic energy density due to an excited source, and the dispersion force between two excited molecules, the contribution arising from fields linear in the moments exactly cancels the term from upward transitions from the initial state originating from the product of the vacuum and second-order Maxwell fields, producing

$$\Delta E=-\frac{1}{8\pi^2\varepsilon_0^2c^2}\sum_{\substack{s\\E_r>E_s}}G_{ij}(A;k_{rs})k_{rs}^6\left[\mu_k^{rs}(B)m_l^{sr}(B)\bar{f}_{ik}(k_{rs}R)f_{jl}(k_{rs}R)\right.$$
$$\left.-m_k^{rs}(B)\mu_l^{sr}(B)\bar{g}_{ik}(k_{rs}R)g_{jl}(k_{rs}R)\right]$$
$$-\frac{1}{8\pi^2\varepsilon_0^2c^2}\sum_{\substack{q\\E_p>E_q}}G_{kl}(B;k_{pq})k_{pq}^6\left[\mu_i^{pq}(A)m_j^{qp}(A)\bar{f}_{ki}(k_{pq}R)f_{lj}(k_{pq}R)\right.$$
$$\left.-m_j^{pq}(A)\mu_i^{qp}(A)\bar{g}_{ki}(k_{pq}R)g_{lj}(k_{pq}R)\right]$$
$$-\frac{1}{4\pi^3\varepsilon_0^2\hbar c^3}\sum_{q,s}\mu_i^{pq}(A)m_j^{qp}(A)\mu_k^{rs}(B)m_l^{sr}(B)$$
$$\times\int_0^\infty du\frac{u^8e^{-2uR}}{(k_{pq}^2+u^2)(k_{rs}^2+u^2)}\left[f_{ik}(iuR)f_{jl}(iuR)-g_{ik}(iuR)g_{jl}(iuR)\right]. \tag{5.9.21}$$

As before, the first two terms of the energy shift apply only for downward transitions from the initial state, corresponding to real photon emission, while the u-integral term contains contributions from both upward and downward transitions. Inserting the geometric tensors and simplifying produces the following result for the dispersion interaction energy between a pair of excited chiral molecules in fixed relative orientation,

$$\Delta E = -\frac{1}{8\pi^2\varepsilon_0^2c^2}\sum_{\substack{s\\E_r>E_s}} G_{ij}(A;k_{rs})\mu_k^{rs}(B)m_l^{sr}(B)k_{sr}^6 \times \left[\frac{(\alpha_{ik}\alpha_{jl}+\varepsilon_{ils}\varepsilon_{jkt}\hat{R}_s\hat{R}_t)}{k_{sr}^2R^2} + \frac{(\beta_{ik}\beta_{jl}-\alpha_{ik}\beta_{jl}-\beta_{ik}\alpha_{jl}+\varepsilon_{ils}\varepsilon_{jkt}\hat{R}_s\hat{R}_t)}{k_{sr}^4R^4} + \frac{\beta_{ik}\beta_{jl}}{k_{sr}^6R^6}\right]$$
$$-\frac{1}{8\pi^2\varepsilon_0^2c^2}\sum_{\substack{q\\E_p>E_q}} G_{kl}(B;k_{pq})\mu_i^{pq}(A)m_j^{qp}(A)k_{qp}^6 \times \left[\frac{(\alpha_{ik}\alpha_{jl}+\varepsilon_{ils}\varepsilon_{jkt}\hat{R}_s\hat{R}_t)}{k_{qp}^2R^2} + \frac{(\beta_{ik}\beta_{jl}-\alpha_{ik}\beta_{jl}-\beta_{ik}\alpha_{jl}+\varepsilon_{ils}\varepsilon_{jkt}\hat{R}_s\hat{R}_t)}{k_{qp}^4R^4} + \frac{\beta_{ik}\beta_{jl}}{k_{qp}^6R^6}\right]$$
$$+\frac{\hbar}{16\pi^3\varepsilon_0^2c}\int_0^\infty du\, u^6 e^{-2uR} G_{ij}(A;icu)G_{kl}(B;icu) \times \left[\frac{(\alpha_{ik}\alpha_{jl}-\varepsilon_{ils}\varepsilon_{jkt}\hat{R}_s\hat{R}_t)}{u^2R^2} + \frac{(\alpha_{ik}\beta_{jl}+\beta_{ik}\alpha_{jl}-2\varepsilon_{ils}\varepsilon_{jkt}\hat{R}_s\hat{R}_t)}{u^3R^3} + \frac{(\alpha_{ik}\beta_{jl}+\beta_{ik}\alpha_{jl}+\beta_{ik}\beta_{jl}-\varepsilon_{ils}\varepsilon_{jkt}\hat{R}_s\hat{R}_t)}{u^4R^4} + \frac{2\beta_{ik}\beta_{jl}}{u^5R^5} + \frac{\beta_{ik}\beta_{jl}}{u^6R^6}\right]. \tag{5.9.22}$$

For freely tumbling A and B, an orientational average of equation (5.9.22) results in the energy shift

$$\Delta E = -\frac{1}{12\pi^2\varepsilon_0^2c^2}\sum_{\substack{s\\E_r>E_s}} G(A;k_{rs})|\vec{\mu}^{rs}(B)\cdot\vec{m}^{sr}(B)|k_{sr}^6\left[\frac{2}{k_{sr}^2R^2}+\frac{2}{k_{sr}^4R^4}+\frac{3}{k_{sr}^6R^6}\right]$$
$$-\frac{1}{12\pi^2\varepsilon_0^2c^2}\sum_{\substack{q\\E_p>E_q}} G(B;k_{pq})|\vec{\mu}^{pq}(A)\cdot\vec{m}^{qp}(A)|k_{qp}^6\left[\frac{2}{k_{qp}^2R^2}+\frac{2}{k_{qp}^4R^4}+\frac{3}{k_{qp}^6R^6}\right]$$
$$+\frac{\hbar}{72\pi^3\varepsilon_0^2cR^4}\int_0^\infty du\, G(A;icu)G(B;icu)u^2e^{-2uR}\left[4+\frac{6}{uR}+\frac{3}{u^2R^2}\right]. \tag{5.9.23}$$

Of the three terms of equation (5.9.23), only the u-integral term is present when both species are in the ground electronic state, as the first two terms hold only for downward transitions from the initial state. The u-integral is seen to be identical to the result (5.9.11) obtained via diagrammatic time-dependent perturbation theory on inserting ground-state mixed electric–magnetic dipole polarizability tensors $G_{ij}(\xi;\omega)$, $\xi = A,\ B$. Examining the case in which molecule B is in the ground state and A is excited, it is seen that only the second and third terms of equation (5.9.23) contribute. Once again the u-integral term has the same functional form, with the expectation value of $G(B;\omega)$ taken over the ground state of B, $|0^B\rangle$. Since the far-zone limit of the u-integral was shown to exhibit inverse ninth power separation distance dependence in equation (5.9.12), the dominant contribution at large separations arises from the pole term, having R^{-2} dependence,

$$-\frac{1}{6\pi^2\varepsilon_0^2c^2R^2}\sum_{\substack{q\\E_p>E_q}} G(B;k_{pq})|\vec{\mu}^{pq}(A)\cdot\vec{m}^{qp}(A)|k_{qp}^4. \tag{5.9.24}$$

For small R, the near-zone limit of the second term of (5.9.23) is

$$-\frac{1}{4\pi^2\varepsilon_0^2c^2R^6}\sum_{\substack{q\\E_p>E_q}} G(B;k_{pq})|\vec{\mu}^{pq}(A)\cdot\vec{m}^{qp}(A)|, \tag{5.9.25}$$

while the asymptotic form in the near zone from the u-integral is

$$-\frac{1}{12\pi^2\varepsilon_0^2c^2R^6}\sum_{q,s}\mathrm{sgn}(E_{qp})\frac{|\vec{\mu}^{pq}(A)\cdot\vec{m}^{qp}(A)||\vec{\mu}^{0s}(B)\cdot\vec{m}^{s0}(B)|}{(|E_{qp}|+E_{s0})}, \tag{5.9.26}$$

both terms displaying R^{-6} behavior. Their sum gives for the total small R limit

$$-\frac{1}{12\pi^2\varepsilon_0^2c^2R^6}\sum_{\substack{q,s\\\text{All }E_q}}\frac{|\vec{\mu}^{pq}(A)\cdot\vec{m}^{qp}(A)||\vec{\mu}^{0s}(B)\cdot\vec{m}^{s0}(B)|}{(E_{qp}+E_{s0})}, \tag{5.9.27}$$

and is composed of both real and virtual photon terms.

When both chiral molecules are excited, all three terms of the energy shift (5.9.23) remain. The far-zone limit of the first term is

$$-\frac{1}{6\pi^2\varepsilon_0^2c^2R^2}\sum_{\substack{s\\E_r>E_s}} G(A;k_{rs})|\vec{\mu}^{rs}(B)\cdot\vec{m}^{sr}(B)|k_{sr}^4, \tag{5.9.28}$$

which when added to (5.9.24) results in the total large R limiting form of the potential. In the near zone, the contribution to the interaction energy arising solely from downward transitions contains terms from all three parts of the result (5.9.23). Simplifying

$$-\frac{1}{4\pi^2\varepsilon_0^2c^2R^6}\sum_{\substack{s\\E_r>E_s}} G(A;k_{rs})|\vec{\mu}^{rs}(B)\cdot\vec{m}^{sr}(B)|$$
$$-\frac{1}{4\pi^2\varepsilon_0^2c^2R^6}\sum_{\substack{q\\E_p>E_q}} G(B;k_{pq})|\vec{\mu}^{pq}(A)\cdot\vec{m}^{qp}(A)|$$
$$-\frac{1}{12\pi^2\varepsilon_0^2c^2R^6}\sum_{q,s}\mathrm{sgn}(E_{qp})\mathrm{sgn}(E_{sr})\frac{|\vec{\mu}^{pq}(A)\cdot\vec{m}^{qp}(A)||\vec{\mu}^{rs}(B)\cdot\vec{m}^{sr}(B)|}{(|E_{qp}|+|E_{sr}|)} \tag{5.9.29}$$

to

$$\frac{1}{12\pi^2\varepsilon_0^2c^2R^6}\sum_{\substack{q,s\\E_p>E_q\\E_r>E_s}}\frac{|\vec{\mu}^{pq}(A)\cdot\vec{m}^{qp}(A)||\vec{\mu}^{rs}(B)\cdot\vec{m}^{rs}(B)|}{(E_{pq}+E_{rs})}, \tag{5.9.30}$$

results in a repulsive force.

5.9.3 Induced Moment Approach

In the previous section, it was shown how the method of induced multipole moments allowed the dispersion potential between ground or excited electric dipole polarizable molecules to be obtained directly in a physically transparent and calculationally simple way. It is now shown how the approach outlined earlier may be extended to treat interactions between optically active molecules (Craig and Thirunamachandran, 1999). As an application, the dispersion energy shift between a pair of chiral molecules in the ground state is recalculated. The techniques presented will serve as a basis for evaluating the radiation-induced chiral discrimination interaction energy in Chapter 7.

To a first approximation, electric and magnetic dipole transitions are allowed simultaneously in optically active units. A measure of chiral response is provided by the mixed electric–magnetic dipole dynamic polarizability tensor $G_{ij}(\xi;\omega)$. In the induced moment method for evaluating dispersion forces, in which fluctuations of the vacuum electromagnetic field induce molecular multipole moments, both electric and magnetic dipole moments are induced to leading order. In a magnetically polarizable system, for example, application of a magnetic field $\vec{b}(\vec{r})$ induces a magnetic dipole moment $\vec{m}^{\mathrm{ind}}$ as the first moment,

$$m_i^{\mathrm{ind}}(\xi;\vec{k}) = \chi_{ij}(\xi;k)b_j(\vec{k};\vec{R}_\xi), \tag{5.9.31}$$

where $\chi_{ij}(\xi;\vec{k})$ is the frequency-dependent magnetic dipole susceptibility tensor. Action of both an electric and a magnetic field on a chiral molecule characterized by the $G_{ij}(\xi;k)$ tensor causes both an electric and a magnetic dipole moment to be induced. With the ground-state mixed electric–magnetic dipole polarizability tensor at real wavevector defined by (5.9.16), the two induced multipole moments are

$$\mu_i^{\mathrm{ind}}(\xi) = G_{ij}(\xi;k)b_j(\vec{k};\vec{R}_\xi) \tag{5.9.32}$$

and

$$m_j^{\mathrm{ind}}(\xi) = \varepsilon_0^{-1}G_{ij}(\xi;k)d_i^{\perp}(\vec{k};\vec{R}_\xi). \tag{5.9.33}$$

Like the interaction of induced electric dipole moments at each center, the induced magnetic dipoles of each molecule couple via the resonant dipole–dipole coupling tensor $V_{ij}^{\pm}(k,\vec{R})$ given by equation (5.8.3). Hence, the contribution to the energy shift from this term is of the form

$$m_i^{\mathrm{ind}}(A;\vec{k})m_j^{\mathrm{ind}}(B;\vec{k})\mathrm{Re}\, V_{ij}(k,\vec{R}). \tag{5.9.34}$$

For the coupling of an induced electric dipole moment at one center ξ with the induced magnetic dipole at a second site ξ', coupling no longer occurs via $V_{ij}^{\pm}(k,\vec{R})$, but now takes place through the interaction tensor $U_{ij}^{\pm}(k,\vec{R})$, a quantity first encountered in the resonant transfer of excitation energy between an electric and a magnetic dipole, defined by equation (4.4.11). The contribution to the energy shift from such coupling is of the form

$$\mu_i^{\mathrm{ind}}(A;\vec{k})m_j^{\mathrm{ind}}(B;\vec{k})\mathrm{Im}U_{ij}(k,\vec{R}), \tag{5.9.35}$$

in addition to a similar term arising from interchange of A and B. Along with the induced electric dipole–electric dipole coupling term (5.8.4), it is seen that the expression for the energy shift proportional to the product of the electric and magnetic dipole moments at each center comprises four terms,

$$\Delta E = \sum_{\vec{k},\,\lambda} \left\{ \left[\mu_i^{\text{ind}}(A)\mu_j^{\text{ind}}(B) + \frac{1}{c^2} m_i^{\text{ind}}(A) m_j^{\text{ind}}(B) \right] \text{Re}\, V_{ij}(k, \vec{R}) \right.$$
$$\left. + \left[\mu_i^{\text{ind}}(A) m_j^{\text{ind}}(B) + m_i^{\text{ind}}(A)\mu_j^{\text{ind}}(B) \right] \text{Im} U_{ij}(k, \vec{R}) \right\}. \qquad (5.9.36)$$

As for $V_{ij}(k, \vec{R})$, the $\pm$ superscript formerly appearing on $U_{ij}(k, \vec{R})$ has been dropped, since $\text{Im}U_{ij}^{\pm}(k, \vec{R})$ is invariant to the signs explicitly appearing in the functional form of the tensor. Inserting the induced dipole moments from equations (5.9.32) and (5.9.33) into (5.9.36) produces an expression for the energy shift that highlights the dependence of ΔE on the chiral response tensors of each molecule and on the radiation field operators. Therefore,

$$\Delta E = \sum_{\vec{k},\,\lambda} \left\{ \left[G_{ik}(A;k) G_{jl}(B;k) b_k(\vec{R}_A) b_l(\vec{R}_B) \right. \right.$$
$$\left. + \frac{1}{\varepsilon_0^2 c^2} G_{ki}(A;k) G_{lj}(B;k) d_k^{\perp}(\vec{R}_A) d_l^{\perp}(\vec{R}_B) \right] \text{Re}\, V_{ij}(k, \vec{R})$$
$$+ \varepsilon_0^{-1} \left[G_{ik}(A;k) G_{lj}(B;k) b_k(\vec{R}_A) d_l^{\perp}(\vec{R}_B) \right.$$
$$\left. \left. + G_{ki}(A;k) G_{jl}(B;k) d_k^{\perp}(\vec{R}_A) b_l(\vec{R}_B) \right] \text{Im} U_{ij}(k, \vec{R}) \right\}. \qquad (5.9.37)$$

To evaluate the ground-state dispersion potential between both A and B chirals, the expectation value of formula (5.9.37) is taken over the familiar radiation–matter state $|0\rangle = |E_0^A, E_0^B; 0(\vec{k}, \lambda), 0(\vec{k}', \lambda')\rangle$ and the sum over all modes of the electromagnetic field is carried out. As in the calculation of the Casimir–Polder energy using this approach, the expectation value of the molecular factors leads to the ground-state polarizability tensors of the form (5.9.16) with $|m\rangle = 0$. The radiation field part again involves the expectation value of the spatial field–field correlation function, this time featuring products of the magnetic field at each center, the electric displacement–magnetic field correlation function, as well as $d_k^{\perp}(\vec{R}_A) d_l^{\perp}(\vec{R}_B)$, with the last of these contained in the second term of (5.9.37), which was the sole contribution to the electric–electric dispersion energy shift, whose expectation value over the vacuum field state was given by expression (5.8.6). For future convenience, the expectation value over the number state of the radiation field $|N(\vec{k}, \lambda)\rangle$, corresponding to an occupation number N of photons of mode $(\vec{k}, \lambda)$, for the combination of four electromagnetic field correlation functions is given below (Salam, 2006a). They are obtained from the mode expansions for $\vec{d}^{\perp}(\vec{r})$ and $\vec{b}(\vec{r})$ from equations (1.7.17) and (1.4.53), respectively. Thus,

$$\langle N(\vec{k}, \lambda)|d_i^{\perp}(\vec{k}, \lambda; \vec{R}_A)d_j^{\perp}(\vec{k}, \lambda; \vec{R}_B)|N(\vec{k}, \lambda)\rangle = \left(\frac{\hbar c k \varepsilon_0}{2V}\right)\left[(N+1)e_i^{(\lambda)}(\vec{k})\bar{e}_j^{(\lambda)}(\vec{k})\mathrm{e}^{-i\vec{k}\cdot\vec{R}} + N\bar{e}_i^{(\lambda)}(\vec{k})e_j^{(\lambda)}(\vec{k})\mathrm{e}^{i\vec{k}\cdot\vec{R}}\right], \tag{5.9.38}$$

$$\langle N(\vec{k}, \lambda)|b_i(\vec{k}, \lambda; \vec{R}_A)b_j(\vec{k}, \lambda; \vec{R}_B)|N(\vec{k}, \lambda)\rangle = \left(\frac{\hbar k}{2\varepsilon_0 c V}\right)\left[(N+1)b_i^{(\lambda)}(\vec{k})\bar{b}_j^{(\lambda)}(\vec{k})\mathrm{e}^{-i\vec{k}\cdot\vec{R}} + N\bar{b}_i^{(\lambda)}(\vec{k})b_j^{(\lambda)}(\vec{k})\mathrm{e}^{i\vec{k}\cdot\vec{R}}\right], \tag{5.9.39}$$

$$\langle N(\vec{k}, \lambda)|d_i^{\perp}(\vec{k}, \lambda; \vec{R}_A)b_l(\vec{k}, \lambda; \vec{R}_B)|N(\vec{k}, \lambda)\rangle = \left(\frac{\hbar k}{2V}\right)\left[(N+1)e_i^{(\lambda)}(\vec{k})\bar{b}_j^{(\lambda)}(\vec{k})\mathrm{e}^{-i\vec{k}\cdot\vec{R}} + N\bar{e}_i^{(\lambda)}(\vec{k})b_j^{(\lambda)}(\vec{k})\mathrm{e}^{i\vec{k}\cdot\vec{R}}\right], \tag{5.9.40}$$

$$\langle N(\vec{k}, \lambda)|b_i(\vec{k}, \lambda; \vec{R}_A)d_j^{\perp}(\vec{k}, \lambda; \vec{R}_B)|N(\vec{k}, \lambda)\rangle = \left(\frac{\hbar k}{2V}\right)\left[(N+1)b_i^{(\lambda)}(\vec{k})\bar{e}_j^{(\lambda)}(\vec{k})\mathrm{e}^{-i\vec{k}\cdot\vec{R}} + N\bar{b}_i^{(\lambda)}(\vec{k})e_j^{(\lambda)}(\vec{k})\mathrm{e}^{i\vec{k}\cdot\vec{R}}\right]. \tag{5.9.41}$$

Inserting $N = 0$ into each of these four relations results in the expectation values for the vacuum field correlation functions. Each of the four terms of the energy shift (5.9.37) are now examined and evaluated in turn, taking the expectation value for the ground state of the system (molecules plus field). From (5.9.39), the vacuum field correlation function of the magnetic field is

$$\langle 0(\vec{k}, \lambda)|b_i(\vec{k}, \lambda; \vec{R}_A)b_j(\vec{k}, \lambda; \vec{R}_B)|0(\vec{k}, \lambda)\rangle = \left(\frac{\hbar k}{2\varepsilon_0 c V}\right)b_i^{(\lambda)}(\vec{k})\bar{b}_j^{(\lambda)}(\vec{k})\mathrm{e}^{-i\vec{k}\cdot\vec{R}}. \tag{5.9.42}$$

Substituting this relation and carrying out the polarization sum gives for the first term of (5.9.37),

$$\sum_{\vec{k}} \left(\frac{\hbar k}{2\varepsilon_0 cV} \right) G_{ik}(A;k) G_{jl}(B;k) (\delta_{kl} - \hat{k}_k \hat{k}_l) \mathrm{e}^{-i\vec{k}\cdot\vec{R}} \mathrm{Re}\, V_{ij}(k,\vec{R}). \qquad (5.9.43)$$

After converting the $\vec{k}$-sum to an integral and performing the angular integral using the result (5.8.8) and substituting for $\mathrm{Re}\, V_{ij}(k,\vec{R})$ from (5.8.3), the above becomes

$$\frac{\hbar}{16\pi^3 \varepsilon_0^2 c R^3} \int_0^\infty \mathrm{d}k k^3 G_{ik}(A;k) G_{jl}(B;k)$$
$$\times [(\delta_{ij} - 3\hat{R}_i \hat{R}_j)(\cos kR + kR \sin kR) - (\delta_{ij} - \hat{R}_i \hat{R}_j) k^2 R^2 \cos kR]$$
$$\times \left[(\delta_{kl} - \hat{R}_k \hat{R}_l) \frac{\sin kR}{kR} + (\delta_{kl} - 3\hat{R}_k \hat{R}_l) \left(\frac{\cos kR}{k^2 R^2} - \frac{\sin kR}{k^3 R^3} \right) \right]. \qquad (5.9.44)$$

Comparing expression (5.9.44) with equation (5.8.9), the k-dependent part is seen to be identical in both formulas, with only the prefactor differing. Following steps identical to that carried out on (5.8.9), which led to the result (5.8.14), produces the functional form

$$-\frac{\hbar}{16\pi^3 \varepsilon_0^2 c R^2} \int_0^\infty du u^4 \mathrm{e}^{-2uR} G(A;icu) G(B;icu) \left[1 + \frac{2}{uR} + \frac{5}{u^2 R^2} + \frac{6}{u^3 R^3} + \frac{3}{u^4 R^4} \right]. \qquad (5.9.45)$$

Returning to equation (5.9.37) and examining the second term, inserting the vacuum electric displacement field correlation function (5.8.6) yields

$$\sum_{\vec{k},\lambda} \left(\frac{\hbar k}{2\varepsilon_0 cV} \right) G_{ki}(A;k) G_{lj}(B;k) e_k^{(\lambda)}(\vec{k}) \bar{e}_l^{(\lambda)}(\vec{k}) \mathrm{e}^{-i\vec{k}\cdot\vec{R}} \mathrm{Re}\, V_{ij}(k,\vec{R}). \qquad (5.9.46)$$

Performing the polarization sum gives rise to a term identical to equation (5.9.43), whose subsequent evaluation yields a contribution equal to (5.9.45).

To calculate the third term of the energy shift equation (5.9.37) requires the $N=0$ value of the magnetic field–displacement field correlation function (5.9.41),

$$\langle 0(\vec{k},\,\lambda)|b_i(\vec{k},\,\lambda;\vec{R}_A)d_j^{\perp}(\vec{k},\,\lambda;\vec{R}_B)|0(\vec{k},\,\lambda)\rangle = \left(\frac{\hbar k}{2V}\right)b_i^{(\lambda)}(\vec{k})\bar{e}_j^{(\lambda)}(\vec{k})\mathrm{e}^{-i\vec{k}\cdot\vec{R}}. \tag{5.9.47}$$

Inserting this into the third term of ΔE produces

$$\sum_{\vec{k},\,\lambda}\left(\frac{\hbar k}{2\varepsilon_0 V}\right)G_{ik}(A;k)G_{lj}(B;k)b_k^{(\lambda)}(\vec{k})\bar{e}_l^{(\lambda)}(\vec{k})\mathrm{e}^{-i\vec{k}\cdot\vec{R}}\mathrm{Im}U_{ij}(k,\vec{R}). \tag{5.9.48}$$

Performing the λ-sum using identity (1.4.57) and converting the $\vec{k}$-sum to an integral gives for (5.9.48),

$$-\frac{\hbar}{16\pi^3\varepsilon_0}\int \mathrm{d}k\mathrm{d}\Omega k^3 G_{ik}(A;k)G_{lj}(B;k)\varepsilon_{klm}\hat{k}_m\mathrm{e}^{-i\vec{k}\cdot\vec{R}}\mathrm{Im}\,U_{ij}(k,\vec{R}). \tag{5.9.49}$$

Carrying out the angular average using relation (4.4.7) and substituting for $\mathrm{Im}\,U_{ij}(k,\vec{R})$ given below,

$$\mathrm{Im}\,U_{ij}(k,\vec{R}) = \frac{1}{4\pi\varepsilon_0 cR^3}\varepsilon_{ijm}\hat{R}_m(kR\cos kR + k^2R^2\sin kR), \tag{5.9.50}$$

expression (5.9.49) becomes

$$\mathrm{Im}-\frac{i\hbar}{16\pi^3\varepsilon_0^2 cR^3}\int_0^\infty \mathrm{d}kk^3 G_{ik}(A;k)G_{lj}(B;k)\varepsilon_{ijm}\varepsilon_{kln}\hat{R}_m\hat{R}_n$$
$$\times\left(\frac{\cos kR}{kR} - \frac{\sin kR}{k^2R^2}\right)(kR\cos kR + k^2R^2\sin kR). \tag{5.9.51}$$

Performing an orientational average over the molecular polarizabilities to obtain the contribution valid for a randomly oriented A–B pair, using the result $\langle G_{ik}(A;k)G_{lj}(B;k)\rangle = \delta_{ik}\delta_{jl}G(A;k)G(B;k)$, where a factor of 1/3 has been absorbed into the definition of each of the isotropic mixed electric–magnetic dipole polarizability tensors, defined by

$$G(\xi;k) = \frac{2}{3}\sum_{t} \frac{|\vec{\mu}^{0t}(\xi)\cdot\vec{m}^{t0}(\xi)|\hbar ck}{E_{t0}^2-(\hbar ck)^2}, \tag{5.9.52}$$

and contracting the tensors, noting that $\delta_{ik}\delta_{jl}\varepsilon_{ijm}\varepsilon_{kln}\hat{R}_m\hat{R}_n = 2$, enables (5.9.51) to be written as

$$\mathrm{Im} - \frac{i\hbar}{16\pi^3\varepsilon_0^2 cR^2}\int_0^\infty \mathrm{d}k k^4 G(A;k)G(B;k) \times \left[\sin 2kR + \frac{2}{kR}\cos 2kR - \frac{1}{k^2R^2}\sin 2kR\right]. \tag{5.9.53}$$

Recognizing that

$$\mathrm{Im}\left[1 + \frac{2i}{kR} - \frac{1}{k^2R^2}\right]\mathrm{e}^{2ikR} = \left[\sin 2kR + \frac{2}{kR}\cos 2kR - \frac{1}{k^2R^2}\sin 2kR\right], \tag{5.9.54}$$

expression (5.9.53) becomes

$$\mathrm{Im} - \frac{i\hbar}{16\pi^3\varepsilon_0^2 cR^2}\int_0^\infty \mathrm{d}k k^4 G(A;k)G(B;k)\left[1 + \frac{2i}{kR} - \frac{1}{k^2R^2}\right]\mathrm{e}^{2ikR}. \tag{5.9.55}$$

Transforming k to the imaginary variable $k = iu$ finally results in the above becoming

$$\frac{\hbar}{16\pi^3\varepsilon_0^2 cR^2}\int_0^\infty \mathrm{d}u u^4 \mathrm{e}^{-2uR} G(A;icu)G(B;icu)\left[1 + \frac{2}{uR} + \frac{1}{u^2R^2}\right]. \tag{5.9.56}$$

For the fourth term of (5.9.37), use is made of the $N = 0$ value of the electric field–magnetic field spatial correlation function (5.9.40). The calculation follows the same lines as for the third term of (5.9.37), and a contribution identical to (5.9.56) is obtained. Hence, the energy shift is given by twice the sum of equations (5.9.45) and (5.9.56),

$$\Delta E = -\frac{\hbar}{8\pi^3\varepsilon_0^2 cR^2}\int_0^\infty du u^4 e^{-2uR} G(A;icu)G(B;icu)$$
$$\times\left\{\left[1+\frac{2}{uR}+\frac{5}{u^2R^2}+\frac{6}{u^3R^3}+\frac{3}{u^4R^4}\right]-\left[1+\frac{2}{uR}+\frac{1}{u^2R^2}\right]\right\}, \quad (5.9.57)$$

which simplifies to

$$\Delta E = -\frac{\hbar}{8\pi^3\varepsilon_0^2 cR^2}\int_0^\infty du u^4 e^{-2uR} G(A;icu)G(B;icu)\left[\frac{4}{u^2R^2}+\frac{6}{u^3R^3}+\frac{3}{u^4R^4}\right], \quad (5.9.58)$$

which is seen to be identical to the chiral discrimination dispersion potential between ground-state molecules obtained in the two previous sub-sections using alternative physical viewpoints.

5.10 INTERACTIONS INVOLVING MAGNETICALLY SUSCEPTIBLE MOLECULES

By relaxing the electric dipole approximation and allowing each molecule to also interact with the radiation field through magnetic dipole coupling, it was shown in Section 5.9 that the interaction energy between two molecules each possessing mixed electric–magnetic dipole polarizability was discriminatory, depending upon the chirality of each species. Including coupling of the magnetic dipole moment to the magnetic field permits dispersion forces between molecules with polarizability characteristics different to $G_{ij}(\xi;\omega)$, $\xi = A, B$, that are of a similar order of magnitude to the discriminatory dispersion potential between two optically active molecules to be examined. One such interaction on which the present section focuses is the dispersion potential between an electric dipole polarizable molecule and a magnetic dipole susceptible molecule (Thirunamachandran, 1988; Jenkins et al., 1994b; Salam, 1996). The energy shift between the pair is computed using response theory, since this method enables contributions to be easily calculated when one or both of the species are electronically excited. This potential is important when molecules with small electric dipole polarizability interact with species having a large magnetic dipole polarizability.

Correct to second order in the molecular moments, with A being electric dipole polarizable and B responding to the electric dipole-dependent magnetic field of A through its magnetic dipole susceptibility tensor $\chi_{ij}(B;k)$, defined by

$$\chi_{ij}(B;k) = \sum_s \left\{ \frac{m_i^{rs}(B)m_j^{sr}(B)}{E_{sr}-\hbar ck} + \frac{m_j^{rs}(B)m_i^{sr}(B)}{E_{sr}+\hbar ck} \right\} = \sum_s \frac{2E_{sr}m_i^{rs}(B)m_j^{sr}(B)}{E_{sr}^2-(\hbar ck)^2}, \quad (5.10.1)$$

the interaction energy is evaluated from the formula

$$\begin{aligned}\Delta E = &-\frac{1}{2\varepsilon_0^2}\sum_s \alpha_{ij}(A;k_{rs})d_i^{(1)}(B;\vec{m};k_{rs};\vec{R}_A)d_j^{(1)}(B;\vec{m};k_{rs};\vec{R}_A) \\ &-\frac{1}{2}\sum_q \chi_{kl}(B;k_{pq})b_k^{(1)}(A;\vec{\mu};k_{pq};\vec{R}_B)b_l^{(1)}(A;\vec{\mu};k_{pq};\vec{R}_B) \\ &-\frac{1}{2\varepsilon_0^2}\sum_{\vec{k},\lambda}\alpha_{ij}(A;k)\Big[d_i^{(0)}(\vec{k};\vec{R}_A)d_j^{(2)}(B;\vec{m}\vec{m};\vec{k};\vec{R}_A) \\ &+d_i^{(2)}(B;\vec{m}\vec{m};\vec{k};\vec{R}_A)d_j^{(0)}(\vec{k};\vec{R}_A)\Big] \\ &-\frac{1}{2}\sum_{\vec{k},\lambda}\chi_{kl}(B;k)\Big[b_k^{(0)}(\vec{k};\vec{R}_B)b_l^{(2)}(A;\vec{\mu}\vec{\mu};\vec{k};\vec{R}_B) \\ &+b_k^{(2)}(A;\vec{\mu}\vec{\mu};\vec{k};\vec{R}_B)b_l^{(0)}(\vec{k};\vec{R}_B)\Big],\end{aligned} \quad (5.10.2)$$

with body A initially in excited state $|p\rangle$ and able to undergo electric dipole allowed transitions to higher or lower lying intermediate states $|q\rangle$ and with magnetic dipole allowed transitions from initial to intermediate state of type $|s\rangle \leftarrow |r\rangle$, with $E_{rs} > 0$ or $E_{rs} < 0$, similarly possible in species B.

Employing the first-order electric displacement field due to a magnetic dipole, equation (2.7.6), and the magnetic field linear in the electric dipole source, equation (2.7.7), the first two terms of (5.10.2) are readily found to be

$$\begin{aligned}&-\frac{1}{32\pi^2\varepsilon_0^2c^2}\sum_s \alpha_{ij}(A;k_{rs})m_k^{rs}(B)m_l^{sr}(B)k_{rs}^6\bar{g}_{ik}(k_{rs}R)g_{jl}(k_{rs}R) \\ &-\frac{1}{32\pi^2\varepsilon_0^2c^2}\sum_q \chi_{kl}(B;k_{pq})\mu_i^{pq}(A)\mu_j^{qp}(A)k_{pq}^6\bar{g}_{ik}(k_{pq}R)g_{jl}(k_{pq}R).\end{aligned} \quad (5.10.3)$$

For the evaluation of the third term of (5.10.2), proportional to the second-order magnetic dipole-dependent electric displacement field, use is made of operator (A.3) along with the vacuum displacement field (2.6.13), for the fourth term of the energy shift, the quadratic, electric dipole-dependent magnetic field is required that is given by equation (2.6.32), as well as the zeroth-order magnetic field (2.6.14). As in earlier calculations using the response formalism, upward and downward transitions from both $|p\rangle$ and $|r\rangle$ arising from the interference of zeroth- and second-order fields, respectively, cancel and reinforce with corresponding terms from equation (5.10.3). The energy shift for oriented A and B, after following the familiar computational procedure, is

$$\begin{aligned}\Delta E = &-\frac{1}{16\pi^2\varepsilon_0^2c^2}\sum_{\substack{s\\E_r>E_s}}\alpha_{ij}(A;k_{rs})m_k^{rs}(B)m_l^{sr}(B)k_{rs}^6\bar{g}_{ik}(k_{rs}R)g_{jl}(k_{rs}R)\\ &-\frac{1}{16\pi^2\varepsilon_0^2c^2}\sum_{\substack{q\\E_p>E_q}}\chi_{kl}(B;k_{pq})\mu_i^{pq}(A)\mu_j^{qp}(A)k_{pq}^6\bar{g}_{ik}(k_{pq}R)g_{jl}(k_{pq}R)\\ &-\frac{\hbar}{32\pi^3\varepsilon_0^2c}\int_0^\infty du\,u^6e^{-2uR}\alpha_{ij}(A;icu)\chi_{kl}(B;icu)g_{ik}(iuR)g_{jl}(iuR).\end{aligned} \tag{5.10.4}$$

After expanding the geometrical tensors using the definition of $G_{ij}(kr)$ and $G_{ij}(iur)$ given by equations (2.9.34) and (2.9.35), expression (5.10.4) can be written as

$$\begin{aligned}\Delta E = &-\frac{1}{16\pi^2\varepsilon_0^2c^2}\sum_{\substack{s\\E_r>E_s}}\alpha_{ij}(A;k_{rs})m_k^{rs}(B)m_l^{sr}(B)k_{rs}^6\varepsilon_{iks}\varepsilon_{jlt}\hat{R}_s\hat{R}_t\left[\frac{1}{k_{rs}^2R^2}+\frac{1}{k_{rs}^4R^4}\right]\\ &-\frac{1}{16\pi^2\varepsilon_0^2c^2}\sum_{\substack{q\\E_p>E_q}}\chi_{kl}(B;k_{pq})\mu_i^{pq}(A)\mu_j^{qp}(A)k_{pq}^6\varepsilon_{iks}\varepsilon_{jlt}\hat{R}_s\hat{R}_t\left[\frac{1}{k_{pq}^2R^2}+\frac{1}{k_{pq}^4R^4}\right]\\ &+\frac{1}{8\pi^3\varepsilon_0^2\hbar c^3}\varepsilon_{iks}\varepsilon_{jlt}\hat{R}_s\hat{R}_t\sum_{q,s}\mu_i^{pq}(A)\mu_j^{qp}(A)m_k^{rs}(B)m_l^{sr}(B)\\ &\times\int_0^\infty du\,u^6e^{-2uR}\frac{k_{qp}k_{sr}}{(k_{qp}^2+u^2)(k_{sr}^2+u^2)}\left[\frac{1}{u^2R^2}+\frac{2}{u^3R^3}+\frac{1}{u^4R^4}\right].\end{aligned} \tag{5.10.5}$$

On rotational averaging and contracting the tensors, the result for freely tumbling A and B is

$$\Delta E = -\frac{1}{24\pi^2\varepsilon_0^2 c^2}\sum_{\substack{s\\E_r>E_s}} \alpha(A;k_{rs})|\vec{m}^{rs}(B)|^2 k_{rs}^6\left[\frac{1}{k_{rs}^2R^2}+\frac{1}{k_{rs}^4R^4}\right]$$
$$-\frac{1}{24\pi^2\varepsilon_0^2 c^2}\sum_{\substack{q\\E_p>E_q}} \chi(B;k_{pq})|\vec{\mu}^{pq}(A)|^2 k_{pq}^6\left[\frac{1}{k_{pq}^2R^2}+\frac{1}{k_{pq}^4R^4}\right]$$
$$+\frac{1}{36\pi^3\varepsilon_0^2\hbar c^3}\sum_{q,s}|\vec{\mu}^{pq}(A)|^2|\vec{m}^{rs}(B)|^2\int_0^\infty du\, u^6 e^{-2uR}\frac{k_{qp}k_{sr}}{(k_{qp}^2+u^2)(k_{sr}^2+u^2)}$$
$$\times\left[\frac{1}{u^2R^2}+\frac{2}{u^3R^3}+\frac{1}{u^4R^4}\right]. \quad (5.10.6)$$

When A and B are both electronically excited, all three terms of (5.10.6) contribute to the energy shift. The far-zone behavior is governed by the limiting forms of the first two terms of the result (5.10.6), each exhibiting an inverse square separation distance dependence, whose summation produces

$$-\frac{1}{36\pi^2\varepsilon_0^2 c^2(\hbar c)^4 R^2}\sum_{\substack{p,q\\E_p>E_q\\E_r>E_s}}\frac{|\vec{\mu}^{pq}(A)|^2|\vec{m}^{rs}(B)|^2}{(E_{qp}^2-E_{rs}^2)}(E_{rs}^4E_{qp}+E_{qp}^4E_{sr}), \quad (5.10.7)$$

while in the near zone, both of these terms have R^{-4} character

$$-\frac{1}{24\pi^2\varepsilon_0^2 c^2 R^4}\sum_{\substack{s\\E_r>E_s}}\alpha(A;k_{sr})|\vec{m}^{rs}(B)|^2 k_{sr}^2, \quad (5.10.8)$$

from the first term of (5.10.6) and

$$-\frac{1}{24\pi^2\varepsilon_0^2 c^2 R^4}\sum_{\substack{q\\E_p>E_q}}\chi(B;k_{pq})|\vec{\mu}^{pq}(A)|^2 k_{pq}^2, \quad (5.10.9)$$

from the second term of (5.10.6). The following asymptotic limits follow from the u-integral term of the energy shift

$$\Delta E_{\mathrm{NZ}} = \frac{1}{72\pi^2\varepsilon_0^2\hbar^2c^4R^4}\sum_{q,s}\mathrm{sgn}(E_{qp})\mathrm{sgn}(E_{sr})|E_{qp}||E_{sr}|\frac{|\vec{\mu}^{pq}(A)|^2|\vec{m}^{rs}(B)|^2}{(|E_{qp}|+|E_{sr}|)} \quad (5.10.10)$$

and

$$\Delta E_{\mathrm{FZ}} = \frac{7\hbar}{64\pi^3\varepsilon_0^2 cR^7}\alpha(A;0)\chi(B;0), \tag{5.10.11}$$

where in the last expression, $\alpha(0)$ and $\chi(0)$ represent isotropic static excited molecular susceptibilities. The energy shift in the far zone displays an inverse seventh power law. The near-zone limit is not a true static limit, since there is no static coupling between an electric dipole and a magnetic dipole, but is retarded, being codependent on the transition wavevectors within each species. The last five results agree with those obtained from the response of one body to the fields due to the electromagnetic energy density of the other found in Section 2.9.

When both molecules are in the ground state, the u-integral of equation (5.10.6) alone contributes to the energy shift, as in

$$\frac{1}{36\pi^3\varepsilon_0^2\hbar c^3}\sum_{q,s}|\vec{\mu}^{0q}(A)|^2|\vec{m}^{s0}(B)|^2\int_0^\infty du\,u^6 e^{-2uR} \times \frac{k_{q0}k_{s0}}{(k_{q0}^2+u^2)(k_{s0}^2+u^2)}\left[\frac{1}{u^2R^2}+\frac{2}{u^3R^3}+\frac{1}{u^4R^4}\right], \tag{5.10.12}$$

with limiting results

$$\Delta E_{\mathrm{NZ}} = \frac{1}{72\pi^2\varepsilon_0^2\hbar^2 c^4 R^4}\sum_{q,s}E_{q0}E_{s0}\frac{|\vec{\mu}^{q0}(A)|^2|\vec{m}^{s0}(B)|^2}{(E_{q0}+E_{s0})} \tag{5.10.13}$$

and

$$\Delta E_{\mathrm{FZ}} = \frac{7\hbar}{64\pi^3\varepsilon_0^2 cR^7}\alpha(A;0)\chi(B;0), \tag{5.10.14}$$

where the polarizability tensors are those for molecules in the ground electronic state. An important aspect of the dispersion potential (5.10.12) is that it is repulsive.

As it stands, the results obtained above are incomplete in the sense that no account has been taken of the diamagnetic coupling term, which like the term $(-1/2)\chi(B;k)\vec{b}^2(A;\vec{R}_B)$ is also proportional to the square of the magnetic field (Salam, 2000a, 2000b).

Consider the ground-state dispersion interaction between an electric dipole polarizable molecule A and a second species B that is diamagnetic. From equation (1.7.16), the diamagnetic coupling term is $(e^2/8m)$ $\{\vec{q}(B) \times \vec{b}(\vec{R}_B)\}^2$, which for a freely rotating source is

$$\frac{e^2}{12m} q^2 \vec{b}^2(\vec{R}_B). \tag{5.10.15}$$

The ensuing interaction energy is given by the expectation value

$$\frac{e^2}{12m} \langle 0(\vec{k}, \lambda); 0^B, 0^A | q^2(B) \vec{b}^2(\vec{R}_B) | 0^A, 0^B; 0(\vec{k}, \lambda) \rangle, \tag{5.10.16}$$

in which diamagnetic molecule B responds to electric dipole-dependent source fields of A. The evaluation of the radiation field part of (5.10.16) is identical to the calculation of the magnetic energy density presented in Section 2.9, but instead of B being excited, here B is taken to be in its lowest electronic level. The form of the contribution is similar to the second and fourth terms of (5.10.2). Computing the expectation value using the expansion of the magnetic field due to an electric dipole source correct up to second order in $\vec{\mu}(A)$ yields the result

$$\Delta E^{\alpha-\mathrm{dia}} = -\frac{e^2}{144\pi^3 \varepsilon_0^2 c^2 m} \sum_n |\vec{\mu}^{0n}(A)|^2 \langle q^2(B) \rangle^{00} \times \int_0^\infty du\, u^6 e^{-2uR} \frac{k_{n0}}{(k_{n0}^2 + u^2)} \left[\frac{1}{u^2 R^2} + \frac{2}{u^3 R^3} + \frac{1}{u^4 R^4} \right], \tag{5.10.17}$$

where the complete set of intermediate state of A is given the new label n,

$$\langle q^2(B) \rangle^{00} = \langle 0^B | q^2(B) | 0^B \rangle, \tag{5.10.18}$$

with the energy shift (5.10.17) holding for all A–B separation distances R beyond charge overlaps. After making the usual approximations, the near-zone asymptote is found to be

$$\Delta E_{\mathrm{NZ}}^{\alpha-\mathrm{dia}} = -\frac{e^2}{288\pi^3 \varepsilon_0^2 c^2 m R^5} \sum_n |\vec{\mu}^{0n}(A)|^2 \langle q^2(B) \rangle^{00} k_{n0}. \tag{5.10.19}$$

Comparing (5.10.19) with the near-zone limit between electric and magnetic dipole polarizable molecules (5.10.13), the ratio of the two is

$$\frac{\Delta E_{\mathrm{NZ}}^{\alpha-\chi}}{\Delta E_{\mathrm{NZ}}^{\alpha-\mathrm{dia}}} \approx kR. \tag{5.10.20}$$

Since in the near zone $kR \ll 1$, the contribution from the diamagnetic coupling term can dominate the interaction.

Returning to equation (5.10.17) and neglecting u^2 relative to k_{n0}^2 in the energy denominator and using the standard integral (5.2.23) to evaluate the integral over imaginary wavevector,

$$\int_0^\infty du e^{-2uR}\left[\frac{u^4}{R^2}+\frac{2u^3}{R^3}+\frac{u^2}{R^4}\right]=\frac{7}{4R^7}, \tag{5.10.21}$$

the far-zone energy shift is found to be

$$\Delta E_{\mathrm{FZ}}^{\alpha-\mathrm{dia}} = -\frac{7e^2}{384\pi^3\varepsilon_0^2 cmR^7}\alpha(A;0)\langle q^2(B)\rangle^{00} \tag{5.10.22}$$

on using the definition for the static isotropic electric dipole polarizability of A,

$$\alpha(A;0) = \frac{2}{3}\sum_n \frac{|\vec{\mu}^{0n}(A)|^2}{E_{n0}}. \tag{5.10.23}$$

Like the wave-zone limiting form of the potential between an electric dipole polarizable molecule and a magnetic dipole susceptible species (5.10.14), the far-zone energy shift when one of the pairs is diamagnetic exhibits an R^{-7} separation distance dependence. The two far-zone asymptotes may be combined to yield a long-range limiting energy shift between an electric dipole polarizable molecule and a magnetically susceptible one, as in

$$\Delta E_{\mathrm{FZ}} = \frac{7\hbar}{64\pi^3\varepsilon_0^2 cR^7}\alpha(A;0)\chi'(B;0), \tag{5.10.24}$$

where the modified magnetic susceptibility tensor of species ξ takes the form

$$\chi'(\xi;0) = \chi(\xi;0) - \frac{e^2}{6m}\langle q^2(\xi)\rangle^{00}. \tag{5.10.25}$$

The first term of (5.10.25), $\chi(\xi;0)$, is termed the static paramagnetic susceptibility tensor, while the second is the diamagnetic contribution. Although both $\alpha(\xi;0)$ and $\chi(\xi;0)$ are positive quantities for the ground-state molecules, $\chi'(\xi;0)$ may be of either sign depending on the relative magnitudes of the two components of (5.10.25). If $\chi'(\xi;0)$ is negative, the molecule is said to be diamagnetic.

Finally, it is of interest to discuss the expression for the ground-state dispersion potential between two magnetic dipole polarizable molecules, equation (5.9.3), even though it is much smaller than the leading Casimir–Polder term and is also nondiscriminatory. It may be obtained straightforwardly using any of the three methods detailed in this chapter or written immediately from the Casimir–Polder potential. Making the usual approximations, the asymptotic limits readily follow from (5.9.3). The pure magnetic dipole correction to the London dispersion formula is therefore

$$\Delta E_{\text{NZ}}^{\chi-\chi} = -\frac{1}{24\pi^2\varepsilon_0^2c^2R^6}\sum_{n,\,s}\frac{|\vec{m}^{0n}(A)|^2|\vec{m}^{0s}(B)|^2}{(E_{n0}+E_{s0})}. \tag{5.10.26}$$

At the other separation extreme,

$$\Delta E_{\text{FZ}}^{\chi-\chi} = -\frac{23\hbar}{64\pi^3\varepsilon_0^2c^3R^7}\chi(A;0)\chi(B;0). \tag{5.10.27}$$

Moreover, noting that the dispersion interaction energy between two diamagnetic molecules is given by (Salam, 2000a)

$$\Delta E^{\text{dia-dia}} = -\frac{23e^4\hbar}{2304\pi^3\varepsilon_0^2m^2c^3R^7}\langle q^2(A)\rangle^{00}\langle q^2(B)\rangle^{00}, \tag{5.10.28}$$

which has inverse seventh power law for *all R*, the last two results may be combined and reexpressed in terms of the modified magnetic susceptibility tensor (5.10.25) to read

$$\Delta E_{\text{FZ}}^{\text{mag-mag}} = -\frac{23\hbar}{64\pi^3\varepsilon_0^2c^3R^7}\chi'(A;0)\chi'(B;0). \tag{5.10.29}$$

It is worth pointing out that the result (5.10.29) also contains the far-zone limit of the dispersion interaction between a magnetic dipole susceptible molecule and a diamagnetic molecule.

5.11 MEASUREMENTS OF CASIMIR EFFECT

The change in the mode structure of the electromagnetic field due to the presence of a pair of bodies relative to their absence is commonly termed the "Casimir effect" and is frequently attributed to the zero-point energy $(1/2)\hbar\omega$, associated with each mode of the radiation field. The label is also often applied to a number of other interactions occurring between bodies at large separation distances, each having a common physical origin interpretable in terms of fluctuations of the vacuum electromagnetic field. Arguably, the best known among these various manifestations of the change in the zero-point energy is the so-called Casimir force—the force of attraction felt by a pair of neutral conducting flat plates separated by a distance d. Casimir (1948) derived a remarkable formula for the force F per unit area A between the plates,

$$\frac{F(d)}{A} = \frac{\pi^2}{240}\frac{\hbar c}{d^4}, \qquad (5.11.1)$$

which is independent of the material properties of the plate and contains the fundamental constants $\hbar$ and c.

Other well-known phenomena described as Casimir effects include the interaction between an atom and a surface—also known as the Casimir–Polder interaction (Casimir and Polder, 1948)—and the energy shift between a pair of uncharged atoms or molecules—often referred to as the van der Waals interaction, whose dispersion component has been extensively dealt with in this chapter. It is an interesting historical note that Casimir's original motivation lay in exploring the nature of long-range van der Waals forces in colloidal suspensions (Verwey and Overbeek, 1948), a work carried out with Polder. In each of the examples mentioned, an important consequence of the application of a quantized field treatment to their study is the retarded aspect of the coupling between the pair of particles or bodies in question, which is due to the explicit and automatic appearance of the photon in the formalism, propagating with speed c *in vacuo*.

The upsurge of interest in the Casimir effect (Bordag et al., 2009) that has occurred in the last decade and a half has largely been due to advances in experimentation, thus enabling the highly accurate measurements to be made of the Casimir force (Bordag et al., 2001; Lamoreaux, 2005). This in turn has stimulated a rapid growth in the number of theoretical papers devoted to the subject. These range from alternative derivations of Casimir's result to more transparent presentations of the theory of Lifshitz

(1956), which applied to a material body characterized by a frequency-dependent complex electromagnetic permittivity, and implications resulting from his theory. These include Casimir effects taking place in metals, nonconductors, and dielectrics, which has led to the study of a multitude of different geometrical scenarios and boundary conditions. Additional corrections examined include the consequences of imperfect conductivity and surfaces being nonsmooth to the effects of finite temperature and thermal Casimir effects.

The Casimir effects and concepts and ideas inherent to it are now playing an important role in a number of subdisciplines—both within and outside its traditional domain of atomic, molecular, optical and condensed matter physics, and quantum field theory. Topics include Hawking radiation, the Unruh effect, and the dynamical Casimir effect—whose origins are electromagnetic, manifestations in space–times with nontrivial topology, the influence of vacuum polarization on inflationary models of the universe in gravitation, cosmology, and astrophysics, the development of advanced regularization and renormalization methods in mathematical physics, and applications of technological value such as the manufacture of nanoelectromechanical devices in which it is now possible to observe repulsive as well as attractive Casimir forces.

Prior to recent experiments, very few attempts were made to measure various Casimir effects, while theoretical work was confined primarily to alternative derivations of the results of Casimir and Polder, rationalizing the physical basis of the phenomenon along with probing the finer details of the Lifshitz theory. Part of the imbalance seen in the number of articles reporting experimental versus theoretical results is the sheer technical difficulty associated with the measurements concerned. This aspect is evident from studies carried out by Abrikosova and Derjaguin (1957), who employed dielectric surfaces and later first made use of curved surfaces such as a lens, sphere, or a cylinder, thereby avoiding the need for the flat plates to be parallel to one another, and by Sparnaay (1958), whose work included the earliest use of metal plates. While Sparnaay's research is often cited as providing the first experimental verification of the Casimir effect, the exponent of d featuring in the expression (5.11.1) for the force per unit area between the plates was uncertain to ± 1.

One major limitation of any potential experiment is that the force of attraction between two parallel plates of infinite conductivity is measurable only for separation distances of a micrometer or less and is minute. For instance, a pair of flat surfaces of area 1 cm^2 separated from each other by a distance of 1 μm produces a force of the order of 100 nN. In addition to requiring extremely sensitive force measurements, since the force is a function of separation distance, d must also be measured to a high level of accuracy and be reliably reproduced. Additional requirements spelled out by Sparnaay to further improve agreement between experiment and theory include ensuring plate surfaces are as clean as possible with low electrostatic charge on them and keeping the voltage between the surfaces to a minimum. These problems persist yet and prevent to make accurate measurements even with modern experimental apparatus.

Improvements in experiments involving metallic surfaces were made by van Blokland and Overbeek (1978), who reported a relative uncertainty in the measured force of around 25% at separation close to 150 nm, but grew substantially at half a micron. Meanwhile, force measurements were also made on nonconductive surfaces. Notable among these were the experiments of Tabor and Winterton (1969) on muscovite mica. This enabled smooth atomic surfaces to be produced, which in turn allowed the two surfaces to approach very closely, from which it was then possible to measure the crossover between the retarded and the nonretarded van der Waals force regimes. This transition was found to occur at approximately 12 nm.

Another manifestation of the Casimir effect, this time in the microscopic regime, which offers the potential of measurement, is found to occur in the excited-state fine structure of the helium atom (Lundeen, 1993), first predicted by Spruch and Kelsey (1978). At the very long range, the potential between an electron and a polarizable atom is of the form

$$V(R) = \frac{11e^2\hbar\alpha(0)}{16\pi^2\varepsilon_0 mcR^5}, \tag{5.11.2}$$

and is repulsive. Since such a force like the Casimir–Polder potential between two neutral species is not easily amenable to measurement, it was suggested that the coupling between an electron and an ion would yield experimental results more readily as the electron may be bound in stable Rydberg orbits at long range. Microwave spectroscopy of the Rydberg fine structure could then be used to compare the binding energies of different Rydberg states. Unlike the dispersion force, which is the only contribution to the interaction energy between two neutral nonpolar bodies, the potential (5.11.2) is the smallest of three terms contributing to the overall force in the Rydberg states of helium. The other two contributions are due to the dipole polarization, which varies as R^{-4}, and the dominant Coulomb potential between the electron and the He^+ ion, which has an inverse separation power law. At long range, where $R \gg 137a_0$, the relative ratios of the three potentials, in the order in which they were mentioned, is $1:10^4:10^{11}$. Working within the confines of nonrelativistic theory, however, the

Rydberg fine structure is effectively independent of the Coulomb force because states with equal principal quantum number are degenerate in a pure Coulomb field. Even then the contribution to the long-range interaction is very small and comparable to relativistic and possibly radiative corrections to the fine structure.

In addition to these theoretical difficulties, there are challenges that have to be met experimentally, the most serious being the level of precision with which the radiative widths of the states can be measured. For the pertinent Rydberg states, this width is usually two orders of magnitude greater than the expectation value of the potential (5.11.2). Nevertheless, microwave spectroscopy of the fine structure of helium has been carried out by Palfrey and Lundeen (1984) and by Hessels et al. (1990) using fast beam techniques. This permitted transitions between high orbital angular momentum states L to be studied, such as between $10I$, K, L, and M. The difference between the measured mean-averaged fine structure interval and the largest contribution, namely, that arising from the nonrelativistic energies of the states, theoretically calculated from the Coulomb Hamiltonian leaves a sum of contributions attributable to relativistic, radiative, and retardation terms. The last of these contains the long-range Casimir force as given by the retarded two-photon exchange potential (Au et al., 1984; Babb and Spruch, 1988). For the H–I, I–K, and K–L fine-structure levels corresponding to $n = 10$, comparison with measurements demonstrates an agreement to approximately 1 kHz. It should be noted that this is precise to only about 10% of the retardation contribution and applies to separations in the range $10a_0$–$50a_0$ and not in the asymptotically large region.

It was only in 1993 that the first known definitive measurement of the force experienced by an isolated atom—also known as the Casimir–Polder force, was made by Sukenik et al. (1993). Their experiment consisted of passing a beam of sodium atoms in the 3s ground state between a pair of parallel plates that formed a cavity. Due to the variation of the vacuum field with position, the atoms are subject to a Casimir–Polder force that pushes them toward the walls of the cavity. The cavity itself comprises two gold plates of height 3 cm and length 8 mm, which forms a wedge that varies in width from 0.5 to 8 μm. The transmitted intensity of the atomic beam is observed as a function of cavity width. Atoms that pass through the cavity and do not stick to the walls are then excited resonantly to the 12s state by two superimposed laser beams of wavelengths 589 and 425 nm and detected by a channel electron multiplier. Comparison of the transmission function is made with curves predicted theoretically based on a near-zone van der Waals potential and the retarded long-range Casimir–Polder potential. Agreement between theory and experiment was found to be excellent. Treating the strength of the interaction as a variable parameter, a least-squares fit of the data recorded gave

$$\frac{\Delta E(\text{expt})}{\Delta E(\text{theory})} = 1.02 \pm 0.13, \tag{5.11.3}$$

providing unambiguous verification of the inverse fourth-power distance dependence of the Casimir–Polder force.

Two experiments that prompted the recent renewed interest in Casimir effects and constitute the first in a succession of modern measurements are those by Lamoreaux (1997), who used a torsion pendulum balance, and Mohideen and Roy (1998), who employed atomic force microscopy (AFM) to measure the Casimir force.

In the first of these experiments, the force between a spherical lens coated with gold and a flat metal plate was measured. The former was mounted on a piezoelectric stack, and the latter on one arm of the torsion balance while the second arm formed the middle electrode of dual parallel plate capacitors, whose position and orientation could be controlled by application of a potential difference to the capacitor plates. A torque is produced by the Casimir force causing a change in the angle of the torsion balance. The ensuing change in the capacitances of the two capacitors was then detected via a phase-sensitive circuit. To counteract angular changes of the torsion balance, compensating voltages were applied to the capacitors by employing a feedback circuit, thereby providing a direct measurement of the Casimir force. This procedure was repeated for separations of 10 μm down to touching distances of the two surfaces in 16 steps. The electrostatic force and surface separation were then determined by curve fitting part of the data to the expected Casimir force. Measurements of the force were accurate to about 10%.

In the second of the modern experiments, AFM techniques were employed to measure the Casimir force between a 0.3 mm polystyrene sphere coated with a 300 nm layer of Al, further covered with a 20 nm layer of Au/Pd, attached to an AFM cantilever, and a similarly coated optically polished sapphire plate. The latter was fixed to a piezoelectric transducer and moved toward the sphere with the attractive force measured by reflecting a laser beam from the tip of the cantilever. The movement of the beam on a pair of photodiodes yields a difference signal in direct proportion to the bending angle of the cantilever. AFM measurements have the advantage of being reproducible and reliable. Agreement between theory and experiment was excellent. With only second-order conductivity and surface roughness corrections included in the comparison, the root mean square deviation of the measured force from the predicted value was found to be 1.6 pN in the total range of measurement from contact distance (approximately 30 nm) to 1 μm separation. This corresponds to a statistical precision of 1% at the smallest distances of separation. Subsequent improvements in experimental procedure as well as in the treatment of surface roughness and finite conductivity of the metal by accounting for fourth-order corrections enabled the same level of precision to be achieved over the complete measurement range.

Experimental efforts continue to be made, especially via AFM techniques, to extend the measurement both to smaller separation distances and larger distances above the current 1 μm limit. The former will enable further light to be shed on possible deviations from Newton's universal law of gravitation as predicted by string theory in which extra compactified dimensions appear, giving rise to the so-called hypothetical fifth force that is effective at distances large relative to nuclear dimensions, whose characteristic length scale in some models is 10^{-6} m. Similar reasons apply to examining the force at distances beyond 1000 nm. In this case, limits could be imposed on the coupling constants of the aforementioned conjectured forces to test predictions of supersymmetry and string theory and to search for new elementary particles. In addition, at large separations, the effects of temperature come into play and significantly modify the nature of Casimir forces, especially between real metals.

CHAPTER 6

MANY-BODY FORCES

I just wanted to remind you that the effects that we see on a large scale and the strange phenomena we see on a small scale are both produced by the interaction of electrons and photons, and all are described, ultimately, by the theory of quantum electrodynamics.

—R. P. Feynman, *QED: The Strange Theory of Light and Matter*, Princeton University Press, Princeton, NJ, 1985, p. 123.

6.1 INTRODUCTION

The total interaction energy for a collection of more than two atoms or molecules is not simply equal to the sum of all of the pairwise interactions. As detailed in Section 3.1, there are terms involving coupling of three-, four-, and many-bodies that contribute to the energy shift, and which are nonadditive in nature. While interactions between molecules are dominated by terms involving pairs of particles, with the approximation of pairwise additivity proving to be a highly useful device for evaluating contributions to the interaction energy even when the force itself is inherently nonpairwise additive, the inclusion of leading nonadditive contributions is found to be essential for a number of chemical systems, ensuring even better agreement with experiment for a variety of chemical and physical properties (Maitland et al., 1981). For example, retaining only the pairwise additive terms in the two-body potential energy functions for Ar and Xe results in the computation of the energy of the crystal differing from experimental values by 10% and 13%, respectively. This is attributed to the neglect of nonadditive contributions. Accounting for the leading triple-dipole dispersion energy term, however, improves agreement to a few percent for solid-state properties. This is to be expected since nuclei of atoms and molecules are in proximity to one another in a crystal lattice. In condensed phases, the contribution from the triple-dipole dispersion potential correction to the interaction energy is, in general, positive for the majority of molecular configurations. In a related context, the addition of the three-body contribution to the dispersion potential was found to be vital in explaining the unusually large differences occurring between experimental third virial coefficients, $C(T)$, and those computed using the pairwise additive assumption, in particular for rare gas atoms, with $C(T)$ being more sensitive to the precise form of the intermolecular potential function than its lower order counterpart, the second virial coefficient, $B(T)$. Many-body forces are also expected to play a significant role in the modification of intermolecular interactions taking place in a medium, for instance in a solvent.

The most common nonadditive contribution to the total interaction energy is the dispersion potential between three atoms or molecules. Hence, this term along with the four- and many-body dispersion potential forms the main focus of this chapter. Section 6.2 contains the derivation of the triple-dipole energy shift between three neutral ground-state atoms or molecules within the framework of semiclassical theory. This was calculated for the first time independently by Axilrod and Teller (1943) and by Muto (1943). Their computation involves use of static dipolar coupling potentials and third-order perturbation theory. Results for different geometrical arrangements of the three atoms are presented. In the following section, a diagrammatic perturbation theory calculation is carried out within the multipolar formalism of molecular quantum electrodynamics. In this viewpoint, the dispersion potential between the three bodies is understood as arising from the exchange of three virtual photons, each of differing mode, one between A and B, a second between B and C, and a third between C and A. After allowing for all possible time orderings and summing over contributions, this yields the retarded correction to the static triple-dipole dispersion potential, with the latter being shown to hold only in the near-zone regime. The computation of the general result is involved, and requires use of the sixth-order formula for the perturbed energy shift. A simplification is made possible by employing an effective 2-photon interaction Hamiltonian, which results in a 60-fold reduction in the number of Feynman diagrams that have to be summed over with third-order perturbation theory again being employed. The calculation is given in Section 6.4. In Section 6.5, it is shown how the correlations of dipole moments induced in two atoms by a source dressed vacuum field, leads directly to the retarded triple-dipole dispersion energy shift in a method that is a variation of the induced multipole moment approach introduced in Section 5.8 to evaluate two-body dispersion forces. For the computation of a general formula for the N-body dispersion energy shift presented in Section 6.6, a different approach to perturbation theory is adopted. It is based on a response theory formalism, used successfully in the previous chapter to calculate two-body dispersion potentials between molecules in either ground or excited states. In the present problem, the Maxwell field operators due to an assembly of N atoms or molecules considered as sources of charge are first evaluated. This is followed by determining the response of each particle, taken one at a time, to the fields due to the remaining $N - 1$ entities. The limiting forms of the N-body potential in the radiation- and near-zones are also given. From the general result applicable to N bodies, the retarded four-body dispersion potential is evaluated in Section 6.7. Special attention is given to a tetrahedral configuration of molecules. In Section 6.8, two contrasting methods are detailed for the calculation of the retarded three-body dispersion energy shift when one of the species is initially in an electronically excited state. One approach involves time-dependent perturbation theory, and is similar to the calculation presented in Section 5.6 for a pair of molecules when one of them is excited. For the case of three-bodies, it is found that two terms contribute to the energy shift. A term arising solely from real photon emission due to downward transitions in the excited molecule and an imaginary wavevector integral term of similar functional form to the ground-state interaction energy between three molecules, which describes contributions from both upward and downward transitions in all three bodies and downward transitions in the excited source. In the second technique, the electric displacement field due to the excited source molecule induces dipole moments in each of the two ground-state bodies. The two induced dipoles couple to the retarded dipole-dipole interaction tensor at the resonant frequency of the downward transition in the excited molecule. Taking an expectation value over this quantity for a state of the system in which one molecule is excited, two are in the ground state and no photons are present—real or virtual, results in the additional contribution due to downward transitions occurring in the excited species. Finally, in

Section 6.9, the effect of a third body in mediating the resonant transfer of excitation energy between two molecules is investigated. In the near zone, direct transfer between the pair is compared with the rate modified by the presence of a third species, along with the interference term occurring due to both of these mechanisms, to gain insight into microscopic and macroscopic limits of migration of energy taking place in a medium.

6.2 AXILROD-TELLER-MUTO DISPERSION ENERGY SHIFT

It was shown in Section 3.1 that the first nonadditive contribution to the interaction energy occurs between three bodies, arising from the simultaneous interaction of three species A, B, and C, ΔE_{ABC}. This term provides a correction to the energy shift arising from the sum of the pairwise interactions between any two of the three particles. Since the dispersion potential is nonpairwise additive, evaluation of the three-body energy shift is crucial as it is the leading nonadditive contribution to the interaction potential. Its calculation was first carried out independently by Axilrod and Teller (1943) and by Muto (1943). Each of them employed third-order perturbation theory and static dipolar coupling potentials to compute the dispersion energy shift. Consequently, the effects of retardation were not accounted for. Details of their calculation are presented below.

Consider three atoms A, B, and C in the ground electronic state with energies E_0^{ξ}, $\xi = A, B, C$. Let them be separated by distances $R^{\xi\xi'} = |\vec{R}^{\xi} - \vec{R}^{\xi'}|$, $\xi \neq \xi' = A, B, C$. Excluding the effects of the radiation field, the total Hamiltonian for the system can be written as

$$H = H_{\text{mol}}^A + H_{\text{mol}}^B + H_{\text{mol}}^C + H_{\text{int}}. \tag{6.2.1}$$

In the absence of interaction, the first three terms of (6.2.1) constitute the unperturbed Hamiltonian,

$$H_0 = H_{\text{mol}}^A + H_{\text{mol}}^B + H_{\text{mol}}^C, \tag{6.2.2}$$

a sum of Hamiltonians for the three isolated bodies. The unperturbed state of the system is then simply a product state of eigenfunctions of each atom, $|E_{n_\xi}^{\xi}\rangle$, $\xi = A, B, C$, described by quantum number n_ξ of species ξ and whose ground state is represented by

$$|0\rangle = |E_0^A, E_0^B, E_0^C\rangle. \tag{6.2.3}$$

Since the unperturbed Hamiltonian (6.2.2) is separable, the unperturbed energy of the system is given by the sum of unperturbed energies of each body, $E^{\xi}_{n_\xi}$, $\xi = A, B, C$. Hence, the energy of the unperturbed ground state is

$$E_0 = E_0^A + E_0^B + E_0^C. \tag{6.2.4}$$

Interaction of the atoms with each other perturbs the system, causing a shift in energy relative to the energy associated with each individual entity. Within the framework of semiclassical theory, the perturbation operator describing the coupling of the atoms at long range, in which exchange effects can be safely ignored is given, to leading order, by static dipole-dipole couplings between pairs of particles. Thus,

$$H_{\text{int}} = H_{\text{int}}^{AB} + H_{\text{int}}^{BC} + H_{\text{int}}^{CA}, \tag{6.2.5}$$

where

$$H_{\text{int}}^{\xi\xi'} = \frac{\mu_i(\xi)\mu_j(\xi')}{4\pi\varepsilon_0 R^3_{\xi\xi'}}\left(\delta_{ij} - 3\hat{R}_i^{\xi\xi'}\hat{R}_j^{\xi\xi'}\right), \quad \xi \neq \xi' = A, B, C. \tag{6.2.6}$$

With this form of coupling Hamiltonian, the interaction energy between the three atoms is given by the third-order perturbation theory formula for the energy shift,

$$\Delta E = \sum_{I,II} \frac{\langle 0|H_{\text{int}}|II\rangle\langle II|H_{\text{int}}|I\rangle\langle I|H_{\text{int}}|0\rangle}{E_{II0}E_{I0}}, \tag{6.2.7}$$

where the sum is carried out over all intermediate states excluding the initial and final states, with the latter being identical to the initial state in the case of the dispersion interaction. In expression (6.2.7), $E_{I0} = E_I - E_0$ and $E_{II0} = E_{II} - E_0$, denote differences in energy between intermediate and initial states. In the present formulation, the three-body dispersion potential is viewed as arising from the instantaneous exchange of a virtual photon between each pair of particles, that is, between A and B, between B and C, and between A and C. Since this can occur in 3! possible time orderings, 6 diagrams may be drawn depicting the coupling between A, B, and C and consequently, 6 terms contribute to the energy shift in (6.2.7). Each species is therefore involved in the exchange of two virtual photons and can subsequently return to the electronic ground state after virtual excitation to an excited intermediate state. Let the set of excited virtual states in the three bodies A, B, and C be denoted by $|p\rangle$, $|q\rangle$, and $|r\rangle$, respectively, with energy E_p^A, E_q^B, and E_r^C, which may also be used as state labels. Electric dipole allowed transitions are assumed to occur between the electronic states of each atom, which are nonvanishing only when the states have differing parity.

Consider the sequence in which the coupling between A and B occurs first, followed by the interaction of B and C, with the coupling between C and A taking place last. With the initial and final states given by (6.2.3), the two types of intermediate state linking the states of the system before and after interaction are easily seen to be of the form $|I\rangle = |E_p^A, E_q^B, E_0^C\rangle$ and $|II\rangle = |E_p^A, E_0^B, E_r^C\rangle$, with energy difference $E_{I0} = E_{p0} + E_{q0}$ and $E_{II0} = E_{p0} + E_{r0}$, where the superscripts labeling atoms have been omitted from these two terms. The first intermediate state reflects interaction between A and B having taken place, with these two particles undergoing virtual excitation to higher lying states $|p\rangle$ and $|q\rangle$, respectively. The second intermediate state represents the result of coupling between B and C, with B returning to the ground state from $|q\rangle$ and C excited to virtual state $|r\rangle$ from the ground state. This contribution to the energy shift is calculated from

$$\sum_{I,II} \frac{\langle E_0^C, E_0^B, E_0^A | H_{\text{int}}^{AC} | E_p^A, E_0^B, E_r^C\rangle \langle E_r^C, E_0^B, E_p^A | H_{\text{int}}^{BC} | E_p^A, E_q^B, E_0^C\rangle \langle E_0^C, E_q^B, E_p^A | H_{\text{int}}^{AB} | E_0^A, E_0^B, E_0^C\rangle}{(E_{p0}+E_{q0})(E_{p0}+E_{r0})}. \tag{6.2.8}$$

Substituting the respective ground and intermediate states, inserting the appropriate form of the perturbation operator (6.2.6), and evaluating the matrix elements produces

$$\begin{aligned} &\frac{1}{(4\pi\varepsilon_0)^3} \frac{1}{R_{AB}^3 R_{BC}^3 R_{CA}^3} \sum_{p,q,r} \mu_i^{p0}(A)\mu_j^{0p}(A)\mu_k^{q0}(B)\mu_l^{0q}(B)\mu_m^{r0}(C)\mu_n^{0r}(C) \\ &\times \left(\delta_{ik} - 3\hat{R}_i^{AB}\hat{R}_k^{AB}\right)\left(\delta_{lm} - 3\hat{R}_l^{BC}\hat{R}_m^{BC}\right)\left(\delta_{jn} - 3\hat{R}_j^{CA}\hat{R}_n^{CA}\right) \\ &\times (E_{p0}+E_{q0})^{-1}(E_{p0}+E_{r0})^{-1}. \end{aligned} \tag{6.2.9}$$

The contributions to the energy shift from the five remaining sequences of excitation events gives terms identical to (6.2.9) except for the energy denominators. There are in fact three different energy denominator products. Hence, the total interaction energy is

$$\Delta E = \frac{2}{(4\pi\varepsilon_0)^3}\frac{1}{R_{AB}^3 R_{BC}^3 R_{CA}^3}\sum_{p,q,r}\mu_i^{p0}(A)\mu_j^{0p}(A)\mu_k^{q0}(B)\mu_l^{0q}(B)\mu_m^{r0}(C)\mu_n^{0r}(C)$$

$$\times\left(\delta_{ik}-3\hat{R}_i^{AB}\hat{R}_k^{AB}\right)\left(\delta_{lm}-3\hat{R}_l^{BC}\hat{R}_m^{BC}\right)\left(\delta_{jn}-3\hat{R}_j^{CA}\hat{R}_n^{CA}\right)$$

$$\times\left[\frac{1}{(E_{p0}+E_{q0})(E_{p0}+E_{r0})}+\frac{1}{(E_{p0}+E_{q0})(E_{q0}+E_{r0})}\right.$$

$$\left.+\frac{1}{(E_{p0}+E_{r0})(E_{q0}+E_{r0})}\right]. \tag{6.2.10}$$

Summing the energy denominators yields

$$\left[\frac{1}{(E_{p0}+E_{q0})(E_{p0}+E_{r0})}+\frac{1}{(E_{p0}+E_{q0})(E_{q0}+E_{r0})}+\frac{1}{(E_{p0}+E_{r0})(E_{q0}+E_{r0})}\right]$$

$$=\frac{2(E_{p0}+E_{q0}+E_{r0})}{(E_{p0}+E_{q0})(E_{p0}+E_{r0})(E_{q0}+E_{r0})}. \tag{6.2.11}$$

Replacing the sum of energy denominators in (6.2.10) by the right-hand form of (6.2.11) produces the recognizable expression for the triple-dipole interaction energy between oriented systems

$$\Delta E = \frac{1}{16\pi^3\varepsilon_0^3}\frac{1}{R_{AB}^3 R_{BC}^3 R_{CA}^3}\sum_{p,q,r}\mu_i^{p0}(A)\mu_j^{0p}(A)\mu_k^{q0}(B)\mu_l^{0q}(B)\mu_m^{r0}(C)\mu_n^{0r}(C)$$

$$\times\left(\delta_{ik}-3\hat{R}_i^{AB}\hat{R}_k^{AB}\right)\left(\delta_{lm}-3\hat{R}_l^{BC}\hat{R}_m^{BC}\right)\left(\delta_{jn}-3\hat{R}_j^{CA}\hat{R}_n^{CA}\right)$$

$$\times\frac{(E_{p0}+E_{q0}+E_{r0})}{(E_{p0}+E_{q0})(E_{p0}+E_{r0})(E_{q0}+E_{r0})}. \tag{6.2.12}$$

For a collection of three isotropic species, a rotational average must be performed on the result (6.2.12). This may be carried out as three independent orientational averages using result (B.4) from Appendix B, with the molecular part factored according to

$$
\begin{aligned}
&\langle \mu_i^{p0}(A)\mu_j^{0p}(A)\mu_k^{q0}(B)\mu_l^{0q}(B)\mu_m^{r0}(C)\mu_n^{0r}(C)\rangle \\
&\quad = \langle \mu_i^{p0}(A)\mu_j^{0p}(A)\rangle\langle \mu_k^{q0}(B)\mu_l^{0q}(B)\rangle\langle \mu_m^{r0}(C)\mu_n^{0r}(C)\rangle \\
&\quad = \left(\frac{1}{3}\right)^3 |\vec{\mu}^{0p}(A)|^2|\vec{\mu}^{0q}(B)|^2|\vec{\mu}^{0r}(C)|^2\delta_{ij}\delta_{kl}\delta_{mn}.
\end{aligned}
\tag{6.2.13}
$$

Contracting the tensors of (6.2.13) with the orientational factors appearing in (6.2.12) results in the averaged energy shift becoming

$$
\begin{aligned}
\Delta E = {} & \frac{1}{16(3\pi\varepsilon_0)^3}\frac{1}{R_{AB}^3R_{BC}^3R_{CA}^3}\sum_{p,q,r}|\vec{\mu}^{0p}(A)|^2|\vec{\mu}^{0q}(B)|^2|\vec{\mu}^{0r}(C)|^2 \\
& \times \frac{(E_{p0}+E_{q0}+E_{r0})}{(E_{p0}+E_{q0})(E_{p0}+E_{r0})(E_{q0}+E_{r0})} \\
& \times \left\{-6+9\left[\left(\hat{R}^{BC}\cdot\hat{R}^{CA}\right)^2+\left(\hat{R}^{CA}\cdot\hat{R}^{AB}\right)^2+\left(\hat{R}^{AB}\cdot\hat{R}^{BC}\right)^2\right]\right. \\
& \left.\quad -27\left(\hat{R}^{AB}\cdot\hat{R}^{BC}\right)\left(\hat{R}^{BC}\cdot\hat{R}^{CA}\right)\left(\hat{R}^{CA}\cdot\hat{R}^{AB}\right)\right\}.
\end{aligned}
\tag{6.2.14}
$$

A much-studied configuration of three bodies is when they form a triangular geometry. To simplify the notation, it is convenient to introduce the unit vectors $\hat{a} = \hat{R}^B - \hat{R}^C$, $\hat{b} = \hat{R}^C - \hat{R}^A$, and $\hat{c} = \hat{R}^A - \hat{R}^B$ and the internal angles θ_A, θ_B, and θ_C opposite sides BC, CA, and AB, respectively. The direction cosines appearing in braces in expression (6.2.14) can be written as

$$
\left\{-6+9\left[(\hat{a}\cdot\hat{b})^2+(\hat{b}\cdot\hat{c})^2+(\hat{c}\cdot\hat{a})^2\right]-27(\hat{a}\cdot\hat{b})(\hat{b}\cdot\hat{c})(\hat{c}\cdot\hat{a})\right\},
\tag{6.2.15}
$$

with $\cos\theta_A = \hat{b}\cdot\hat{c}$, and so on. Using the identity for internal angles of a triangle,

$$
\cos^2\theta_A + \cos^2\theta_B + \cos^2\theta_C = 1 - 2\cos\theta_A\cos\theta_B\cos\theta_C,
\tag{6.2.16}
$$

(6.2.15) simplifies to

$$
3[1 + 3\cos\theta_A\cos\theta_B\cos\theta_C].
\tag{6.2.17}
$$

Hence, the energy shift for a triangular arrangement of three interacting atoms is

$$\Delta E = \frac{1}{144\pi^3\varepsilon_0^3}\frac{1}{a^3b^3c^3}\sum_{p,q,r}|\vec{\mu}^{0p}(A)|^2|\vec{\mu}^{0q}(B)|^2|\vec{\mu}^{0r}(C)|^2$$
$$\times\frac{(E_{p0}+E_{q0}+E_{r0})}{(E_{p0}+E_{q0})(E_{p0}+E_{r0})(E_{q0}+E_{r0})}[1+3\cos\theta_A\cos\theta_B\cos\theta_C]. \tag{6.2.18}$$

It is instructive to examine the geometrical factor

$$\frac{[1+3\cos\theta_A\cos\theta_B\cos\theta_C]}{a^3b^3c^3}, \tag{6.2.19}$$

appearing in the result (6.2.18) for particular configurations of the atoms. For an equilateral triangular arrangement, the lengths are all equal, $a=b$ $c=R$, say, and $\theta_A=\theta_B=\theta_C=60°$; the term within square brackets simplifies to 11/8 and (6.2.19) to $11/8R^9$ and the energy shift is positive and therefore repulsive. For a right triangle in which $\theta_A=90°$, say, the trigonometric factor in square brackets of the geometrical term (6.2.19) vanishes leaving unity and the energy shift is again repulsive. When the three atoms are collinear, $\theta_A=\theta_B=0°$, and $\theta_C=180°$, the term within square brackets above simplifies to -2, giving rise to an attractive interaction energy.

It is interesting to note that the energy shift (6.2.18) can be expressed in terms of the dynamic electric dipole polarizability at imaginary frequency for each atom analogously to expression (5.3.11) for the London dispersion energy. Thus,

$$\Delta E = \frac{3\hbar c}{256\pi^4\varepsilon_0^3}\int_{-\infty}^{\infty} du\,\alpha(A;iu)\alpha(B;iu)\alpha(C;iu)\frac{[1+3\cos\theta_A\cos\theta_B\cos\theta_C]}{a^3b^3c^3}. \tag{6.2.20}$$

Like formula (5.3.8), it is common to express the near-zone dispersion energy shift in terms of static polarizabilities. This may be achieved by carrying out an average energy or Unsöld approximation. Alternatively, a two-level coupling scheme may be chosen in which the static polarizabilites are taken to be dependent only on the single, lowest energy transition within particle ξ at an energy, E_ξ, with transition electric dipole moment $\vec{\mu}(\xi)$. In such a situation, the static electric dipole polarizability is given by

$$\alpha(\xi;0) = \frac{2}{3}\frac{|\vec{\mu}(\xi)|^2}{E_\xi}. \tag{6.2.21}$$

Redefining the energy as

$$\tilde{E} = \frac{2E_A E_B E_C(E_A+E_B+E_C)}{(E_A+E_B)(E_B+E_C)(E_A+E_C)}, \tag{6.2.22}$$

and using (6.2.21), the triple-dipole dispersion energy (6.2.18) can be approximated as

$$\Delta E \approx \frac{3}{256\pi^3\varepsilon_0^3}\alpha(A;0)\alpha(B;0)\alpha(C;0)\frac{[1+3\cos\theta_A\cos\theta_B\cos\theta_C]}{a^3b^3c^3}\tilde{E}. \tag{6.2.23}$$

6.3 RETARDED TRIPLE-DIPOLE DISPERSION POTENTIAL: PERTURBATION THEORY

In the previous section, the Axilrod–Teller–Muto triple-dipole dispersion energy shift was computed. Since the potentials coupling each of the species were taken to be of the static dipolar variety, the effect of the finite speed of propagation of light signals was not properly accounted for. Including retardation effects will modify the form of the nonadditive interaction energy, as was seen to occur in the comparison of the Casimir–Polder energy shift with the London dispersion formula. In this section, the computation of the retarded triple-dipole dispersion potential is carried out using time-dependent perturbation theory. The calculation is similar to that given by Axilrod and Teller, but instead of employing a static electric dipole form of coupling as the perturbation operator, the electrodynamic multipolar coupling operator in electric dipole approximation is used as the interaction Hamiltonian. Instead of employing third-order perturbation theory, now the energy shift is calculated via the sixth-order term when the interaction Hamiltonian is linear in the electric displacement field. Considerable simplification of the calculation is found to occur if an effective two-photon interaction Hamiltonian is employed instead of the usual $-\varepsilon_0^{-1}\vec{\mu}(\xi)\cdot\vec{d}^{\perp}(\vec{R}_\xi)$ form. This was also the case in the calculation of the Casimir–Polder shift as shown in Section 5.4 and details applicable to the present situation are given in the next section.

As for the Axilrod–Teller calculation, the triple-dipole dispersion potential is viewed as arising from the exchange of a single virtual photon between A and B, between B and C, and between C and A. Again each species undergoes two virtual photon emission/absorption events so that each body can return by a downward transition to the ground electronic state. The total Hamiltonian for the three-particle system, including the effect of a radiation field, is given by

$$H = H_{\text{mol}}(A) + H_{\text{mol}}(B) + H_{\text{mol}}(C) + H_{\text{rad}} + H_{\text{int}}(A) + H_{\text{int}}(B) + H_{\text{int}}(C). \tag{6.3.1}$$

In the electric dipole approximation, the interaction Hamiltonian is of the form

$$H_{\text{int}}(\xi) = -\varepsilon_0^{-1}\vec{\mu}(\xi)\cdot\vec{d}^{\perp}(\vec{R}_\xi), \tag{6.3.2}$$

where $\vec{\mu}(\xi)$ is the electric dipole moment operator of species $\xi = A, B, C$ and $\vec{d}^{\perp}(\vec{r})$ is the transverse electric displacement field operator evaluated at the field point $\vec{r}$. The leading term of the energy shift is of sixth order and is calculated from

$$\Delta E = -\sum_{I-V} \frac{\langle 0|H_{\text{int}}|V\rangle\langle V|H_{\text{int}}|IV\rangle\langle IV|H_{\text{int}}|III\rangle\langle III|H_{\text{int}}|II\rangle\langle II|H_{\text{int}}|I\rangle\langle I|H_{\text{int}}|0\rangle}{E_{V0}E_{IV0}E_{III0}E_{II0}E_{I0}}. \tag{6.3.3}$$

Evaluation of the energy shift is facilitated by drawing Feynman diagrams depicting the possible time orderings of the interaction vertices. The initial and final states of the system are specified by

$$|0\rangle = |E_0^A, E_0^B, E_0^C; 0(\vec{k}_1,\lambda_1), 0(\vec{k}_2,\lambda_2), 0(\vec{k}_3,\lambda_3)\rangle, \tag{6.3.4}$$

corresponding to all three species in the electronic ground state and the radiation field in the vacuum state. For the present problem, 360 graphs contribute to the 3-body interaction energy. One such graph corresponds to the sequence in which virtual photon labeled 1 with mode characteristics $(\vec{k}_1,\lambda_1)$ is emitted by A and absorbed by B first of all followed by B emitting a virtual photon denoted by 2 of mode $(\vec{k}_2,\lambda_2)$ and which is absorbed by C with finally the third virtual photon designated 3 and of mode $(\vec{k}_3,\lambda_3)$ emitted by C and absorbed by A. As before, let the intermediate states of A, B, and C be labeled p, q, and r, respectively. Evaluating in the familiar way, the contribution to the energy shift due to this specific time ordering is

$$-\sum_{\vec{k}_1,\vec{k}_2,\vec{k}_3}\sum_{\lambda_1,\lambda_2,\lambda_3}\sum_{p,q,r}\left(\frac{\hbar ck_1}{2\varepsilon_0 V}\right)\left(\frac{\hbar ck_2}{2\varepsilon_0 V}\right)\left(\frac{\hbar ck_3}{2\varepsilon_0 V}\right)\bar{e}_i^{(\lambda_1)}(\vec{k}_1)e_k^{(\lambda_1)}(\vec{k}_1)\bar{e}_l^{(\lambda_2)}(\vec{k}_2)$$

$$\times e_m^{(\lambda_2)}(\vec{k}_2)\bar{e}_n^{(\lambda_3)}(\vec{k}_3)e_j^{(\lambda_3)}(\vec{k}_3)\mu_i^{p0}(A)\mu_j^{0p}(A)\mu_k^{q0}(B)\mu_l^{0q}(B)\mu_m^{r0}(C)$$

$$\times \mu_n^{0r}(C)\mathrm{e}^{i\vec{k}_1\cdot(\vec{R}_B-\vec{R}_A)}\mathrm{e}^{i\vec{k}_2\cdot(\vec{R}_C-\vec{R}_B)}\mathrm{e}^{i\vec{k}_3\cdot(\vec{R}_A-\vec{R}_C)}$$

$$\times\left[(E_{p0}+\hbar ck_1)(E_{p0}+E_{q0})(E_{p0}+\hbar ck_2)(E_{p0}+E_{r0})(E_{p0}+\hbar ck_3)\right]^{-1}. \tag{6.3.5}$$

The remaining 359 graphs may be computed similarly, noting that because the labels on the virtual photons are arbitrary, they must be permuted also. Defining the interatomic separation distance vectors $\vec{a}=\vec{R}^B-\vec{R}^C$, $\vec{b}=\vec{R}^C-\vec{R}^A$, and $\vec{c}=\vec{R}^A-\vec{R}^B$ and carrying out the polarization sum, the energy shift can be written as

$$\Delta E=-\sum_{\vec{k}_1,\vec{k}_2,\vec{k}_3}\sum_{p,q,r}\left(\frac{\hbar ck_1}{2\varepsilon_0 V}\right)\left(\frac{\hbar ck_2}{2\varepsilon_0 V}\right)\left(\frac{\hbar ck_3}{2\varepsilon_0 V}\right)\mu_i^{p0}\mu_j^{0p}\mu_k^{q0}\mu_l^{0q}\mu_m^{r0}\mu_n^{0r}$$

$$\times\left(\delta_{ik}-\hat{k}_i^{(1)}\hat{k}_k^{(1)}\right)\left(\delta_{lm}-\hat{k}_l^{(2)}\hat{k}_m^{(2)}\right)\left(\delta_{jn}-\hat{k}_j^{(3)}\hat{k}_n^{(3)}\right)\mathrm{e}^{i\vec{k}_1\cdot\vec{a}}\mathrm{e}^{i\vec{k}_2\cdot\vec{b}}\mathrm{e}^{i\vec{k}_3\cdot\vec{c}}\sum_{a=1}^{360}D_a^{-1}, \tag{6.3.6}$$

where D_a^{-1} is the denominator associated with graph (a); the molecular labels have been removed from the transition electric dipoles since the intermediate-state labels are sufficient to identify each species and $\hat{k}_i^{(n)}$, $n=1,\,2,\,3$ denotes the ith component of the unit wavevector of virtual photon of mode n. Converting the wavevector sums to integrals and performing the angular averages produces

$$\Delta E=-\sum_{p,q,r}\left(\frac{\hbar c}{4\pi\varepsilon_0}\right)^3\frac{1}{\pi^3}\mu_i^{p0}\mu_j^{0p}\mu_k^{q0}\mu_l^{0q}\mu_m^{r0}\mu_n^{0r}\int_0^\infty\int_0^\infty\int_0^\infty \mathrm{d}k_1\mathrm{d}k_2\mathrm{d}k_3$$

$$\times\left[\left(-\vec{\nabla}^2\delta_{ik}+\vec{\nabla}_i\vec{\nabla}_k\right)\frac{\sin k_1a}{a}\right]\left[\left(-\vec{\nabla}^2\delta_{lm}+\vec{\nabla}_l\vec{\nabla}_m\right)\frac{\sin k_2b}{b}\right]$$

$$\times\left[\left(-\vec{\nabla}^2\delta_{jn}+\vec{\nabla}_j\vec{\nabla}_n\right)\frac{\sin k_3c}{c}\right]\sum_{a=1}^{360}D_a^{-1}. \tag{6.3.7}$$

Performing orientational averages for each particle and carrying out the wavevector integrals results in the following formula for the retarded three-body energy shift (Power and Thirunamachandran, 1994),

$$\Delta E = -\frac{1}{64\pi^4\varepsilon_0^3\hbar^2c^2}\left(\frac{2}{3}\right)^3 \sum_{p,q,r} |\vec{\mu}^{0p}(A)|^2|\vec{\mu}^{0q}(B)|^2|\vec{\mu}^{0r}(C)|^2$$

$$\times\left(-\vec{\nabla}^2\delta_{ij}+\vec{\nabla}_i\vec{\nabla}_j\right)^a\frac{1}{a}\left(-\vec{\nabla}^2\delta_{jk}+\vec{\nabla}_j\vec{\nabla}_k\right)^b\frac{1}{b}\left(-\vec{\nabla}^2\delta_{ki}+\vec{\nabla}_k\vec{\nabla}_i\right)^c\frac{1}{c}$$

$$\times\left\{\begin{array}{l} \dfrac{k_{q0}k_{r0}}{\left(k_{q0}^2-k_{p0}^2\right)\left(k_{r0}^2-k_{p0}^2\right)}f\left[k_{p0}(a+b+c)\right] \\ +\dfrac{k_{r0}k_{p0}}{\left(k_{r0}^2-k_{q0}^2\right)\left(k_{p0}^2-k_{q0}^2\right)}f\left[k_{q0}(a+b+c)\right] \\ +\dfrac{k_{p0}k_{q0}}{\left(k_{p0}^2-k_{r0}^2\right)\left(k_{q0}^2-k_{r0}^2\right)}f\left[k_{r0}(a+b+c)\right] \end{array}\right\}, \tag{6.3.8}$$

where

$$f(x) = ci(x)\sin(x) - si(x)\cos(x), \tag{6.3.9}$$

and the superscripts on the gradient operators indicate the object on which the operator acts. Before going on to discuss the implications of the result (6.3.8), its asymptotic limits, and specialization to particular geometries, it is shown in the following section how the retarded triple-dipole dispersion energy shift may be obtained with significantly less labor using perturbation theory methods.

6.4 TRIPLE-DIPOLE DISPERSION ENERGY SHIFT VIA CRAIG–POWER HAMILTONIAN

The three-body dispersion energy shift may be obtained in a relatively facile manner using perturbation theory techniques by employing the Craig–Power interaction Hamiltonian (Craig and Power, 1969) instead of the coupling operator linear in the electric displacement field equation (6.3.2). Such a replacement was made in Section 5.4, where it was shown that the perturbation operator quadratic in the electric displacement field allowed the far-zone 2-body dispersion potential to be evaluated by summing over 2 time-ordered graphs instead of the 4 diagrams appropriate at this limit obtained from the full set of 12 graphs with each of the 2 diagrams now containing an effective 2-photon interaction vertex at each center. If the dynamic property of the electric dipole polarizability is correctly accounted for, use of the Craig–Power coupling operator $-(1/2\varepsilon_0^2)\alpha_{ij}(\xi;k)d_i^{\perp}(\xi')d_j^{\perp}(\xi')$, $\xi \neq \xi' = A, B$ leads to the Casimir–Polder potential valid for all separation distances in the case of two interacting bodies. Application of the effective 2-photon interaction Hamiltonian in the perturbation theory calculation of the 3-body dispersion energy shift now involves summation over only 6 diagrams rather than 360 (Passante et al., 1998). Another major advantage gained by the use of collapsed two-photon interaction vertices is that the order of perturbation theory required for the calculation is now the third rather than the sixth. Details of the calculation follow.

The total Hamiltonian for the three-particle system is again given by (6.3.1), but the interaction Hamiltonian is of the form

$$H_{\text{int}}(\xi) = -\frac{1}{2\varepsilon_0^2}\sum_{\text{modes}} \alpha(\xi;k)\vec{d}^{\perp}(\vec{R}_\xi, \xi'; \vec{k}, \lambda)\vec{d}^{\perp}(\vec{R}_\xi, \xi'; \vec{k}, \lambda), \qquad (6.4.1)$$

where $\alpha(\xi;k)$ is the isotropic frequency dependent polarizability of species $\xi = A, B, C$ and $\vec{d}^{\perp}(\vec{R}_\xi, \xi'; \vec{k}, \lambda)$ is the transverse electric displacement field due to species ξ', evaluated at the position of body ξ and of mode $(\vec{k}, \lambda)$. The initial and final states $|0\rangle$ are the same as (6.3.4), corresponding to all three molecules in the ground state $|E_0^{\xi}\rangle$, with no virtual photons present. Notation identical to that used in the previous section to denote excited electronic states of each species and the mode characteristics of the three virtual photons that are exchanged between any two of the three centers is again adopted.

Permutations of the three two-photon coupling vertices—with one interaction operator acting at each site, results in six possible time orderings. As before, the virtual photon exchanged between A and B is of mode $(\vec{k}_1, \lambda_1)$, between B and C is of mode $(\vec{k}_2, \lambda_2)$, and that traversing A and C is of mode $(\vec{k}_3, \lambda_3)$. In one time ordering, species A first emits spontaneously two virtual photons. One of these is then absorbed by B, which then emits spontaneously a virtual photon. The two virtual photons now in transit, that emitted by A and B, are then finally absorbed by C. In a second diagram, two virtual photon emission first occurs at B—one photon of mode $(\vec{k}_1, \lambda_1)$ that propagates to A and a second photon of mode $(\vec{k}_2, \lambda_2)$, which is first absorbed by C, which then simultaneously emits a $(\vec{k}_3, \lambda_3)$ mode virtual photon. Molecule A then finally absorbs the $(\vec{k}_1, \lambda_1)$ and $(\vec{k}_3, \lambda_3)$ mode photons. In the third graph, species C first emits two virtual photons—of modes $(\vec{k}_2, \lambda_2)$ and $(\vec{k}_3, \lambda_3)$. Next, A absorbs the $(\vec{k}_3, \lambda_3)$ photon and simultaneously emits a $(\vec{k}_1, \lambda_1)$ photon to B. Finally, B simultaneously absorbs this virtual photon and the $(\vec{k}_2, \lambda_2)$ photon first emitted by C. The three remaining diagrams are obtained from the first three on reflection. The contribution from the first graph described above is now evaluated.

On making use of the interaction Hamiltonian (6.4.1), the product of the electric displacement fields describing two different modes at the same field point $\vec{r}$, is

$$
\begin{aligned}
d_i^{\perp}(\vec{r}, \vec{k}, \lambda) d_j^{\perp}(\vec{r}, \vec{k}', \lambda') = & \sum_{\vec{k}, \vec{k}'} \sum_{\lambda, \lambda'} \left(\frac{\hbar c k \varepsilon_0}{2V} \right)^{1/2} \left(\frac{\hbar c k' \varepsilon_0}{2V} \right)^{1/2} \\
& \times \left[e_i^{(\lambda)}(\vec{k}) e_j^{(\lambda')}(\vec{k}') a^{(\lambda)}(\vec{k}) a^{(\lambda')}(\vec{k}') \mathrm{e}^{i(\vec{k}+\vec{k}')\cdot\vec{r}} \right. \\
& - e_i^{(\lambda)}(\vec{k}) \bar{e}_j^{(\lambda')}(\vec{k}') a^{(\lambda)}(\vec{k}) a^{\dagger(\lambda')}(\vec{k}') \mathrm{e}^{i(\vec{k}-\vec{k}')\cdot\vec{r}} \\
& - \bar{e}_i^{(\lambda)}(\vec{k}) e_j^{(\lambda')}(\vec{k}') a^{\dagger(\lambda)}(\vec{k}) a^{(\lambda')}(\vec{k}') \mathrm{e}^{-i(\vec{k}-\vec{k}')\cdot\vec{r}} \\
& \left. + \bar{e}_i^{(\lambda)}(\vec{k}) \bar{e}_j^{(\lambda')}(\vec{k}') a^{\dagger(\lambda)}(\vec{k}) a^{\dagger(\lambda')}(\vec{k}') \mathrm{e}^{-i(\vec{k}+\vec{k}')\cdot\vec{r}} \right],
\end{aligned}
\tag{6.4.2}
$$

which when used in the expression for the third-order perturbation theory energy shift (6.2.7), results in the contribution from the first time-ordered graph described above being

$$
\begin{aligned}
& -\left(\frac{1}{2}\right)^3 \sum_{\vec{k}_1, \vec{k}_2, \vec{k}_3} \sum_{\lambda_1, \lambda_2, \lambda_3} \left(\frac{\hbar c k_1}{2\varepsilon_0 V} \right) \left(\frac{\hbar c k_2}{2\varepsilon_0 V} \right) \left(\frac{\hbar c k_3}{2\varepsilon_0 V} \right) [\alpha_{ij}(A; k_1) + \alpha_{ij}(A; k_3)] \\
& \quad \times [\alpha_{kl}(B; k_1) + \alpha_{kl}(B; k_2)][\alpha_{mn}(C; k_2) + \alpha_{mn}(C; k_3)] \\
& \quad \times \bar{e}_i^{(\lambda_1)}(\vec{k}_1) e_k^{(\lambda_1)}(\vec{k}_1) \bar{e}_l^{(\lambda_2)}(\vec{k}_2) e_m^{(\lambda_2)}(\vec{k}_2) e_n^{(\lambda_3)}(\vec{k}_3) \bar{e}_j^{(\lambda_3)}(\vec{k}_3) \mathrm{e}^{i\vec{k}_1\cdot(\vec{R}_B - \vec{R}_A)} \\
& \quad \times \mathrm{e}^{i\vec{k}_2\cdot(\vec{R}_C - \vec{R}_B)} \mathrm{e}^{i\vec{k}_3\cdot(\vec{R}_C - \vec{R}_A)} [(\hbar c k_1 + \hbar c k_3)(\hbar c k_2 + \hbar c k_3)]^{-1}.
\end{aligned}
\tag{6.4.3}
$$

The contributions from the other five graphs may be computed similarly. On using the definitions of the distance vectors $\vec{a}$, $\vec{b}$, and $\vec{c}$ given in the previous section, the energy shift is

$$\Delta E = -\sum_{\vec{k}_1,\vec{k}_2,\vec{k}_3}\sum_{\lambda_1,\lambda_2,\lambda_3}\left(\frac{\hbar c k_1 k_2 k_3}{32\varepsilon_0^3 V^3}\right)\left[\alpha_{ij}(A;k_1)+\alpha_{ij}(A;k_3)\right]$$
$$\times\left[\alpha_{kl}(B;k_1)+\alpha_{kl}(B;k_2)\right]\left[\alpha_{mn}(C;k_2)+\alpha_{mn}(C;k_3)\right]\bar{e}_i^{(\lambda_1)}(\vec{k}_1)$$
$$\times e_k^{(\lambda_1)}(\vec{k}_1)\bar{e}_l^{(\lambda_2)}(\vec{k}_2)e_m^{(\lambda_2)}(\vec{k}_2)e_n^{(\lambda_3)}(\vec{k}_3)\bar{e}_j^{(\lambda_3)}(\vec{k}_3)\mathrm{e}^{-i\vec{k}_1\cdot\vec{c}}\mathrm{e}^{-i\vec{k}_2\cdot\vec{a}}\mathrm{e}^{i\vec{k}_3\cdot\vec{b}}$$
$$\times\left[\frac{1}{(k_1+k_3)(k_2+k_3)}+\frac{1}{(k_1+k_2)(k_1+k_3)}+\frac{1}{(k_2+k_3)(k_1+k_2)}\right]. \tag{6.4.4}$$

Performing an orientational average using the relation

$$\begin{aligned}\langle\alpha_{ij}(A)\alpha_{kl}(B)\alpha_{mn}(C)\rangle &= \langle\alpha_{ij}(A)\rangle\langle\alpha_{kl}(B)\rangle\langle\alpha_{mn}(C)\rangle\\ &= \alpha(A)\alpha(B)\alpha(C)\delta_{ij}\delta_{kl}\delta_{mn},\end{aligned} \tag{6.4.5}$$

where a factor of 1/3 has been absorbed into each of the isotropic polarizabilities, carrying out the polarization sums and converting the summations over wavevector to integrals yields

$$\Delta E = -\frac{\hbar c}{32\varepsilon_0^3}\frac{1}{(2\pi)^9}\int \mathrm{d}^3\vec{k}_1\mathrm{d}^3\vec{k}_2\mathrm{d}^3\vec{k}_3\left[\alpha(A;k_1)+\alpha(A;k_3)\right]$$
$$\times\left[\alpha(B;k_1)+\alpha(B;k_2)\right]\left[\alpha(C;k_2)+\alpha(C;k_3)\right]k_1k_2k_3\left(\delta_{ij}-\hat{k}_i^{(1)}\hat{k}_j^{(1)}\right)$$
$$\times\left(\delta_{jk}-\hat{k}_j^{(2)}\hat{k}_k^{(2)}\right)\left(\delta_{ki}-\hat{k}_k^{(3)}\hat{k}_i^{(3)}\right)\mathrm{e}^{-i\vec{k}_1\cdot\vec{c}}\mathrm{e}^{-i\vec{k}_2\cdot\vec{a}}\mathrm{e}^{i\vec{k}_3\cdot\vec{b}}$$
$$\times\left[\frac{1}{(k_1+k_3)(k_2+k_3)}+\frac{1}{(k_1+k_2)(k_1+k_3)}+\frac{1}{(k_2+k_3)(k_1+k_2)}\right]. \tag{6.4.6}$$

Transforming to spherical polar coordinates via $\mathrm{d}^3\vec{k} = k^2\mathrm{d}k\mathrm{d}\Omega$ and performing the angular integrations produces

$$\Delta E = -\frac{\hbar c}{256\pi^6\varepsilon_0^3}\left(-\vec{\nabla}^2\delta_{ij}+\vec{\nabla}_i\vec{\nabla}_j\right)^c\left(-\vec{\nabla}^2\delta_{jk}+\vec{\nabla}_j\vec{\nabla}_k\right)^a\left(-\vec{\nabla}^2\delta_{ki}+\vec{\nabla}_k\vec{\nabla}_i\right)^b$$
$$\times\frac{1}{abc}\int_0^\infty\int_0^\infty\int_0^\infty \mathrm{d}k_1\mathrm{d}k_2\mathrm{d}k_3\sin k_1c\sin k_2a\sin k_3b\left[\alpha(A;k_1)+\alpha(A;k_3)\right]$$

$$\times\left[\alpha(B;k_1)+\alpha(B;k_2)\right]\left[\alpha(C;k_2)+\alpha(C;k_3)\right]$$
$$\times\left[\frac{1}{(k_1+k_3)(k_2+k_3)}+\frac{1}{(k_1+k_2)(k_1+k_3)}+\frac{1}{(k_2+k_3)(k_1+k_2)}\right]. \tag{6.4.7}$$

Since each of the transitions from the ground state is purely virtual, the transition energies or wavevectors appearing in the energy denominator factor are all positive. To facilitate evaluation of the integrals, the wavevectors can be separated according to the following integral representation in terms of the parameter u,

$$\left[\frac{1}{(k_1+k_3)(k_2+k_3)}+\frac{1}{(k_1+k_2)(k_1+k_3)}+\frac{1}{(k_2+k_3)(k_1+k_2)}\right]$$
$$=\frac{4k_1k_2k_3}{\pi}\int_0^\infty\frac{\mathrm{d}u}{(k_1^2+u^2)(k_2^2+u^2)(k_3^2+u^2)}. \tag{6.4.8}$$

Hence, (6.4.7) becomes

$$\Delta E=-\frac{\hbar c}{64\pi^7\varepsilon_0^3}\left(-\vec{\nabla}^2\delta_{ij}+\vec{\nabla}_i\vec{\nabla}_j\right)^c\left(-\vec{\nabla}^2\delta_{jk}+\vec{\nabla}_j\vec{\nabla}_k\right)^a\left(-\vec{\nabla}^2\delta_{ki}+\vec{\nabla}_k\vec{\nabla}_i\right)^b$$
$$\times\frac{1}{abc}\int_0^\infty\int_0^\infty\int_0^\infty\int_0^\infty \mathrm{d}u\,\mathrm{d}k_1\,\mathrm{d}k_2\,\mathrm{d}k_3\,k_1k_2k_3\sin k_1c\sin k_2a\sin k_3b$$
$$\times\left[\alpha(A;k_1)+\alpha(A;k_3)\right]\left[\alpha(B;k_1)+\alpha(B;k_2)\right]$$
$$\times\left[\alpha(C;k_2)+\alpha(C;k_3)\right]\frac{1}{(k_1^2+u^2)(k_2^2+u^2)(k_3^2+u^2)}. \tag{6.4.9}$$

On noting that

$$\frac{k}{k^2+u^2}=\frac{1}{2}\left(\frac{1}{k+iu}+\frac{1}{k-iu}\right), \tag{6.4.10}$$

the wavevector integrals may be evaluated using the result

$$\int_0^\infty dk\alpha(k)\frac{k\sin kx}{k^2+u^2}=\frac{1}{4i}\int_{-\infty}^{\infty} dk\alpha(k)e^{ikx}\left(\frac{1}{k+iu}+\frac{1}{k-iu}\right)=\frac{\pi}{2}\alpha(iu)e^{-ux}, \tag{6.4.11}$$

where $\alpha(iu)$ is the isotropic polarizability at imaginary frequency icu. Therefore, the energy shift is

$$\Delta E=-\frac{\hbar c}{64\pi^4\varepsilon_0^3}\left(-\vec{\nabla}^2\delta_{ij}+\vec{\nabla}_i\vec{\nabla}_j\right)^c\left(-\vec{\nabla}^2\delta_{jk}+\vec{\nabla}_j\vec{\nabla}_k\right)^a\left(-\vec{\nabla}^2\delta_{ki}+\vec{\nabla}_k\vec{\nabla}_i\right)^b$$
$$\times\frac{1}{abc}\int_0^\infty du\alpha(A;iu)\alpha(B;iu)\alpha(C;iu)e^{-(a+b+c)u}. \tag{6.4.12}$$

The interaction energy (6.4.12) holds for all separations a, b, and c beyond overlap of charge distributions associated with the three isotropic bodies A, B, and C with arbitrary geometry. An alternative form may be written on evaluating the gradient operators (Power and Thirunamachandran, 1985). Defining

$$C(x)=1+x+x^2 \tag{6.4.13a}$$

and

$$D(x)=3+3x+x^2, \tag{6.4.13b}$$

expression (6.4.12) becomes

$$\Delta E=-\frac{\hbar c}{64\pi^4\varepsilon_0^3}\int_0^\infty du\alpha(A;iu)\alpha(B;iu)\alpha(C;iu)e^{-(a+b+c)u}\frac{1}{a^3b^3c^3}$$
$$\times\left[-3C(au)C(bu)C(cu)+C(au)C(bu)D(cu)+C(bu)C(cu)D(au)\right.$$
$$+C(cu)C(au)D(bu)-C(au)D(bu)D(cu)(\hat{b}\cdot\hat{c})^2$$
$$-C(bu)D(cu)D(au)(\hat{c}\cdot\hat{a})^2-C(cu)D(au)D(bu)(\hat{a}\cdot\hat{b})^2$$
$$\left.+D(au)D(bu)D(cu)(\hat{b}\cdot\hat{c})(\hat{c}\cdot\hat{a})(\hat{a}\cdot\hat{b})\right]. \tag{6.4.14}$$

Explicitly in terms of transition dipole moments, the energy shift is

$$\Delta E = -\frac{1}{16\pi^4\varepsilon_0^3\hbar^2c^2}\left(\frac{2}{3}\right)^3\sum_{p,q,r}k_{p0}k_{q0}k_{r0}|\vec{\mu}^{0p}(A)|^2|\vec{\mu}^{0q}(B)|^2|\vec{\mu}^{0r}(C)|^2$$
$$\times[abc(a+b+c)]^{-1}[N_I+N_{II}(\hat{b}\cdot\hat{c})(\hat{c}\cdot\hat{a})(\hat{a}\cdot\hat{b})+N_{III}(a)(\hat{b}\cdot\hat{c})^2$$
$$+N_{III}(b)(\hat{c}\cdot\hat{a})^2+N_{III}(c)(\hat{a}\cdot\hat{b})^2], \tag{6.4.15}$$

where

$$N_x = \frac{a+b+c}{4a^2b^2c^2}\int_0^\infty du\frac{N_x(u)e^{-(a+b+c)u}}{(k_{p0}^2+u^2)(k_{q0}^2+u^2)(k_{r0}^2+u^2)}, \tag{6.4.16}$$

$$N_I(u) = \frac{1}{3}\big[-9C(au)C(bu)C(cu)+3C(au)C(bu)D(cu)$$
$$+3C(bu)C(cu)D(au)+3C(cu)C(au)D(bu)-C(au)D(bu)D(cu)$$
$$-C(bu)D(cu)D(au)-C(cu)D(au)D(bu)\big], \tag{6.4.17}$$

$$N_{II}(u) = \frac{1}{3}\big[3D(au)D(bu)D(cu)-2C(au)D(bu)D(cu)$$
$$-2C(bu)D(cu)D(au)-2C(cu)D(au)D(bu)\big], \tag{6.4.18}$$

and

$$N_{III}(a,u) = \frac{1}{3}\big[-2C(au)D(bu)D(cu)+C(bu)D(cu)D(au)$$
$$+C(cu)D(au)D(bu)\big], \tag{6.4.19}$$

for $x = I, II, III$. The asymptotic limits are readily obtained from the result applicable for all separation distances. To obtain the far-zone asymptote, u^2 in the denominator of (6.4.16) is discarded, from which it is seen that the molecular part of (6.4.15) reduces to static isotropic polarizabilities. For an equilateral triangle in which $a = b = c = R$, the far-zone limit is (Aub and Zienau, 1960)

$$\Delta E_{\mathrm{FZ}} = \frac{2^4\times 79\hbar c}{3^5\pi(4\pi\varepsilon_0)^3R^{10}}\alpha(A;0)\alpha(B;0)\alpha(C;0). \tag{6.4.20}$$

For a linear arrangement of three bodies with $2a = 2b = c = R$, the potential has the asymptotic form in the far zone

$$\Delta E_{\mathrm{FZ}} = -\frac{93\hbar c}{32\pi^4\varepsilon_0^3 R^{10}}\alpha(A;0)\alpha(B;0)\alpha(C;0). \qquad (6.4.21)$$

Both limits (6.4.20) and (6.4.21) exhibit an R^{-10} dependence. It is interesting to note that the sign of the potential depends on the geometry.

At the other extreme, the displacements a, b, and c are all small relative to characteristic transition wavelengths k_{p0}^{-1}, k_{q0}^{-1}, and k_{r0}^{-1}. Hence, the near-zone potential is found by letting $a,b,c \to 0$ in the integral (6.4.16). Making use of the identity

$$\begin{aligned} I &= \int_0^\infty \frac{du}{(k_{p0}^2+u^2)(k_{q0}^2+u^2)(k_{r0}^2+u^2)} \\ &= \frac{\pi}{2}\frac{(k_{p0}+k_{q0}+k_{r0})}{(k_{p0}+k_{q0})(k_{q0}+k_{r0})(k_{r0}+k_{p0})}\frac{1}{k_{p0}k_{q0}k_{r0}}, \end{aligned} \qquad (6.4.22)$$

equations (6.4.17)–(6.4.19) reduce to

$$N_I = -\frac{3(a+b+c)}{4a^2b^2c^2}I, \qquad (6.4.23)$$

$$N_{II} = \frac{9(a+b+c)}{4a^2b^2c^2}I, \qquad (6.4.24)$$

and

$$N_{III} = 0, \qquad (6.4.25)$$

giving the near-zone result,

$$\begin{aligned} \Delta E_{\mathrm{NZ}} &= \frac{3}{2}\frac{\hbar c}{(4\pi\varepsilon_0)^3}\frac{(k_{p0}+k_{q0}+k_{r0})k_{p0}k_{q0}k_{r0}}{(k_{p0}+k_{q0})(k_{q0}+k_{r0})(k_{r0}+k_{p0})}\alpha(A;0)\alpha(B;0) \\ &\quad \times \alpha(C;0)\frac{\left[1-3(\hat{b}\cdot\hat{c})(\hat{c}\cdot\hat{a})(\hat{a}\cdot\hat{b})\right]}{a^3b^3c^3}, \end{aligned} \qquad (6.4.26)$$

which is the Axilrod–Teller–Muto triple-dipole dispersion potential derived in Section 6.2. As expected, the form of the potential obtained via static multipolar coupling coincides with the near-zone limit of the general form of the retarded energy shift derived via quantum electrodynamical theory.

Finally, it is worth noting that the Casimir–Polder potential can be written as

$$\Delta E = -\frac{\hbar c}{32\pi^3\varepsilon_0^2}\left(-\vec{\nabla}^2\delta_{ij} + \vec{\nabla}_i\vec{\nabla}_j\right)^R\left(-\vec{\nabla}^2\delta_{ij} + \vec{\nabla}_i\vec{\nabla}_j\right)^{\bar{R}}\frac{1}{R\bar{R}} \times \int_0^\infty \mathrm{d}u\, \alpha(A; iu)\alpha(B; iu)\mathrm{e}^{-u(R+\bar{R})}, \tag{6.4.27}$$

with $\bar{R}$ set equal to R after carrying out the differentiations. Comparing (6.4.27) with the form of the retarded three-body dispersion potential (6.4.12), it is easy to see how a generalized formula may be written down for the dispersion energy shift between N bodies. A derivation of this result using the alternative response theory method is given in Section 6.6.

6.5 TRIPLE-DIPOLE DISPERSION POTENTIAL VIA CORRELATIONS OF THE DRESSED VACUUM FIELD

In this section, an alternative physical viewpoint and calculational method are presented for the retarded three-body dispersion energy shift. It is similar to the induced moment method introduced in the previous chapter to calculate two-body dispersion forces between systems in both ground and excited electronic states. In obtaining the Casimir–Polder potential using that approach, the picture was one in which electric dipole moments were induced at each center by correlations of vacuum field fluctuations, with the potential given by the expectation value over the ground state of the quantum mechanical analogue of the expression for the classical interaction energy of the correlated dipoles. In a variation to be given below, the dressed spatial correlations of the zero-point field due to one atom are first calculated, followed by the evaluation of the correlation of dipole moments induced in two other atoms by this source dressed vacuum field, with the three-body potential emanating from the coupling of these induced moments (Cirone and Passante, 1997). This technique will be seen to offer a number of advantages over the conventional diagrammatic perturbation theory computation presented in the previous two sections.

Consider an atom A, in the ground electronic state and located at $\vec{R}_A$. In the electric dipole approximation, the quantum electrodynamical multipolar

Hamiltonian for such a system is

$$H = H_{\mathrm{mol}}(A) + H_{\mathrm{rad}} - \varepsilon_0^{-1}\vec{\mu}(A)\cdot\vec{d}^{\perp}(\vec{R}_A), \tag{6.5.1}$$

where $\vec{\mu}(A)$ is the electric dipole moment operator of species A and $\vec{d}^{\perp}(\vec{r})$ is the transverse electric displacement field operator. Correct to second order in the interaction Hamiltonian—the third term on the right-hand side of (6.5.1), the unnormalized perturbed wavefunction for the ground state of the system (Power and Thirunamachandran, 1993a), commonly termed a dressed state, is

$$\begin{aligned}|\tilde{0}\rangle = {} & |0;0(\vec{k},\lambda)\rangle - \frac{1}{\hbar c}\sum_{p}\sum_{\vec{k},\lambda}\frac{\langle\vec{k},\lambda;p|H_{\mathrm{int}}|0;0(\vec{k},\lambda)\rangle|p;\vec{k},\lambda\rangle}{k+k_{p0}}\\ & + \left(\frac{1}{\hbar c}\right)^2\sum_{p,p'}\sum_{\vec{k},\lambda}\sum_{\vec{k}',\lambda'}\\ & \times\frac{\langle\vec{k}',\lambda';\vec{k},\lambda;p'|H_{\mathrm{int}}|p;\vec{k},\lambda\rangle\langle\vec{k},\lambda;p|H_{\mathrm{int}}|0;0(\vec{k},\lambda)\rangle|p';\vec{k},\lambda;\vec{k}',\lambda'\rangle}{(k+k_{p0})(k+k'+k_{p'0})},\end{aligned} \tag{6.5.2}$$

where $|0;0(\vec{k},\lambda)\rangle = |0\rangle|0(\vec{k},\lambda)\rangle$ represents the product state comprised of unperturbed ground state of atom A, $|0\rangle$ and bare vacuum field state, $|0(\vec{k},\lambda)\rangle$, with p and p' denoting complete sets of energy levels of atom A. The dressed state (6.5.2) is used to calculate the expectation value of the spatial correlation function for two differing modes of the electric displacement field denoted by $\vec{k},\lambda$ and $\vec{k}',\lambda'$ due to the presence of atom A. It is given by

$$\begin{aligned}&\langle\tilde{0}|d_i^{\perp}(\vec{r};\vec{k},\lambda)d_j^{\perp}(\vec{r}';\vec{k}',\lambda')|\tilde{0}\rangle\\ &\quad = \langle 0(\vec{k},\lambda);0|d_i^{\perp}(\vec{r};\vec{k},\lambda)d_j^{\perp}(\vec{r}';\vec{k}',\lambda')|0;0(\vec{k},\lambda)\rangle\\ &\qquad + \frac{1}{V^2}kk'\sum_{p}\mu_s^{0p}(A)\mu_t^{p0}(A)e_i^{(\lambda)}(\vec{k})\bar{e}_t^{(\lambda)}(\vec{k})e_j^{(\lambda')}(\vec{k}')\bar{e}_s^{(\lambda')}(\vec{k}')\\ &\qquad \times\left\{\frac{\mathrm{e}^{-i\vec{k}\cdot(\vec{r}-\vec{R}_A)}\mathrm{e}^{i\vec{k}'\cdot(\vec{r}'-\vec{R}_A)}}{(k+k_{p0})(k'+k_{p0})} + \frac{\mathrm{e}^{i\vec{k}\cdot(\vec{r}-\vec{R}_A)}\mathrm{e}^{i\vec{k}'\cdot(\vec{r}'-\vec{R}_A)}}{(k+k')}\right.\\ &\qquad \left.\times\left(\frac{1}{(k+k_{p0})}+\frac{1}{(k'+k_{p0})}\right)+\mathrm{c.c}\right\}.\end{aligned} \tag{6.5.3}$$

The first term of (6.5.3) is the expectation value of the field-field spatial correlation function over the unperturbed ground state of the system, namely, the ground state of atom A and the vacuum state of the electromagnetic field. It is independent of species A and does not contribute to the three-body dispersion potential. Its contribution is ignored henceforth.

To leading order, the moment induced in a polarizable body by an electric displacement field is the electric dipole,

$$\mu_i^{\text{ind}}(\vec{k}) = \varepsilon_0^{-1}\alpha(k)d_i^{\perp}(\vec{r};\vec{k},\lambda), \tag{6.5.4}$$

where $\alpha(k)$ is the isotropic dynamic electric dipole polarizability. Consider two other atoms B and C that are identical to A and both in their ground electronic states and located at $\vec{R}_B$ and $\vec{R}_C$, respectively. The interaction energy between the electric dipole moments induced in atoms B and C is

$$\begin{aligned}\Delta E_{BC} &= \sum_{\vec{k},\lambda}\sum_{\vec{k}',\lambda'}\langle\tilde{0}^A|\mu_i^{\text{ind}}(B;\vec{k})\mu_j^{\text{ind}}(C;\vec{k}')|\tilde{0}^A\rangle\text{Re}V_{ij}(k,k';\vec{R}_B,\vec{R}_C)\\ &= \varepsilon_0^{-2}\sum_{\vec{k},\lambda}\sum_{\vec{k}',\lambda'}\alpha(B;k)\alpha(C;k')\langle\tilde{0}^A|d_i^{\perp}(\vec{R}_B;\vec{k},\lambda)d_j^{\perp}(\vec{R}_C;\vec{k}',\lambda')|\tilde{0}^A\rangle\\ &\quad\times\text{Re}V_{ij}(k,k';\vec{R}_B,\vec{R}_C),\end{aligned} \tag{6.5.5}$$

where the expectation value is taken over the dressed state (6.5.2) and implicitly depends on atom A. In expression (6.5.5), the two-wavevector resonant dipole interaction tensor due to two oscillating dipoles has the form

$$\text{Re}V_{ij}(k,k';\vec{R},\vec{R}') = -\frac{1}{4\pi\varepsilon_0}\frac{1}{2}\left[k^3\text{Re}F_{ij}(k|\vec{R}-\vec{R}'|) + k'^3\text{Re}F_{ij}(k'|\vec{R}-\vec{R}'|)\right], \tag{6.5.6}$$

where

$$\begin{aligned}\text{Re}F_{ij}(kR) &= \text{Re}\frac{1}{k^3}\left(-\vec{\nabla}^2\delta_{ij}+\vec{\nabla}_i\vec{\nabla}_j\right)^R\frac{e^{ikR}}{R} = \frac{1}{k^3}\left(-\vec{\nabla}^2\delta_{ij}+\vec{\nabla}_i\vec{\nabla}_j\right)^R\frac{\cos kR}{R}\\ &= \left\{(\delta_{ij}-\hat{R}_i\hat{R}_j)\frac{\cos kR}{kR} - (\delta_{ij}-3\hat{R}_i\hat{R}_j)\left(\frac{\sin kR}{k^2R^2}+\frac{\cos kR}{k^3R^3}\right)\right\}.\end{aligned} \tag{6.5.7}$$

Substituting (6.5.3) and (6.5.6) into (6.5.5) produces, for isotropic source atom A,

$$\Delta E_{BC} = -\frac{1}{3\times 2^4\pi^5\varepsilon_0^3}\sum_p |\vec{\mu}^{0p}(A)|^2 \int_0^\infty\int_0^\infty \mathrm{d}k\mathrm{d}k' k^3 k'^3 \alpha(B;k)\alpha(C;k')$$
$$\times\left[k^3\mathrm{Re}F_{ij}(ka)+k'^3\mathrm{Re}F_{ij}(k'a)\right]\mathrm{Im}F_{ik}(kc)\mathrm{Im}F_{jk}(k'b)$$
$$\times\left[\frac{1}{(k+k_{p0})(k'+k_{p0})}+\frac{1}{(k+k_{p0})(k+k')}+\frac{1}{(k'+k_{p0})(k+k')}\right], \tag{6.5.8}$$

after performing polarization sums and angular averages, where

$$\mathrm{Im}F_{ij}(kR) = \mathrm{Im}\frac{1}{k^3}\left(-\vec{\nabla}^2\delta_{ij}+\vec{\nabla}_i\vec{\nabla}_j\right)^R\frac{\mathrm{e}^{ikR}}{R} = \frac{1}{k^3}\left(-\vec{\nabla}^2\delta_{ij}+\vec{\nabla}_i\vec{\nabla}_j\right)^R\frac{\sin kR}{R}$$
$$=\left\{(\delta_{ij}-\hat{R}_i\hat{R}_j)\frac{\sin kR}{kR}+(\delta_{ij}-3\hat{R}_i\hat{R}_j)\left(\frac{\cos kR}{k^2R^2}-\frac{\sin kR}{k^3R^3}\right)\right\}. \tag{6.5.9}$$

In expression (6.5.8), the familiar lengths $a=|\vec{R}_B-\vec{R}_C|$, $b=|\vec{R}_C-\vec{R}_A|$, and $c=|\vec{R}_B-\vec{R}_A|$ have been reintroduced. It should be noted that including the contribution from the first term of (6.5.3), simply gives rise to the Casimir–Polder potential between atoms B and C as demonstrated in Section 5.8. To evaluate one of the wavevector integrals in (6.5.8), the wavevector partial fractions are rewritten as

$$\frac{1}{(k+k_{p0})(k'+k_{p0})}+\frac{1}{(k+k_{p0})(k+k')}+\frac{1}{(k'+k_{p0})(k+k')}$$
$$=\frac{1}{k+k_{p0}}\left(\frac{1}{(k+k')}-\frac{\mathrm{PV}}{(k-k')}\right)+\frac{1}{k'+k_{p0}}\left(\frac{1}{(k+k')}-\frac{\mathrm{PV}}{(k'-k)}\right), \tag{6.5.10}$$

where PV denotes the principal value. Inserting (6.5.10) into (6.5.8) and using (6.5.7) and (6.5.9), ΔE_{BC} becomes

$$\Delta E_{BC} = -\frac{1}{3\times 2^5\pi^5\varepsilon_0^3}\left(-\vec{\nabla}^2\delta_{ij}+\vec{\nabla}_i\vec{\nabla}_j\right)^a\frac{1}{a}\left(-\vec{\nabla}^2\delta_{jk}+\vec{\nabla}_j\vec{\nabla}_k\right)^b\frac{1}{b}$$
$$\times\left(-\vec{\nabla}^2\delta_{ki}+\vec{\nabla}_k\vec{\nabla}_i\right)^c\frac{1}{c}$$

$$\times \sum_p |\vec{\mu}^{0p}(A)|^2 \left\{ \begin{array}{l} \int_0^\infty dk' \alpha(C;k') \sin k'b \\ \quad \times \int_0^\infty dk \alpha(B;k) [\sin k(c+a) + \sin k(c-a)] \\ \quad + \int_0^\infty dk \alpha(B;k) \sin kc \\ \quad \times \int_0^\infty dk' \alpha(C;k') [\sin k'(b+a) + \sin k'(b-a)] \end{array} \right\}$$

$$\times \left[\frac{1}{k+k_{p0}} \left(\frac{1}{(k+k')} - \frac{\text{PV}}{(k-k')} \right) + \frac{1}{k'+k_{p0}} \left(\frac{1}{(k+k')} - \frac{\text{PV}}{(k'-k)} \right) \right]. \tag{6.5.11}$$

One type of integral occurring in (6.5.11) may be evaluated in the complex plane on making use of $\alpha(\xi;k) = \alpha(\xi;-k)$, and yields

$$\int_0^\infty \alpha(\xi;k) \sin kx \left(\frac{1}{k+k'} - \frac{\text{PV}}{k'-k} \right) dk = \text{PV} \int_{-\infty}^\infty dk \alpha(\xi;k) \frac{\sin kx}{k+k'}$$
$$= \pi \, \text{sgn}(x) \alpha(\xi;k') \cos k'x, \tag{6.5.12}$$

where $\text{sgn}(x)$ is the signum of x. Substituting (6.5.12) into (6.5.11) produces integrals of the type

$$\int_0^\infty dk \frac{\alpha(B;k)\alpha(C;k) \sin kx}{k+k_{p0}}$$
$$= \frac{1}{2i} \left\{ \int_0^\infty dk \frac{\alpha(B;k)\alpha(C;k) e^{ikx}}{k+k_{p0}} - \int_0^\infty dk \frac{\alpha(B;k)\alpha(C;k) e^{-ikx}}{k+k_{p0}} \right\}. \tag{6.5.13}$$

This may also be evaluated in the complex plane by inserting $k=iu$ in the first integral and $k=-iu$ in the second, for $x>0$, to give

$$\int_0^\infty \mathrm{d}k \frac{\alpha(B;k)\alpha(C;k)\mathrm{e}^{ikx}}{k+k_{p0}} - \int_0^\infty \mathrm{d}k \frac{\alpha(B;k)\alpha(C;k)\mathrm{e}^{-ikx}}{k+k_{p0}} = i\int_0^\infty \mathrm{d}u \frac{\alpha(B;iu)\alpha(C;iu)\mathrm{e}^{-ux}}{k_{p0}+iu} - i\int_0^\infty \mathrm{d}u \frac{\alpha(B;iu)\alpha(C;iu)\mathrm{e}^{-ux}}{k_{p0}-iu}, \tag{6.5.14}$$

with the $x<0$ case evaluated similarly. Integral expression (6.5.13) therefore becomes

$$\int_0^\infty \mathrm{d}k \frac{\alpha(B;k)\alpha(C;k)\sin kx}{k+k_{p0}} = \mathrm{sgn}(x)k_{p0}\int_0^\infty \mathrm{d}u \frac{\alpha(B;iu)\alpha(C;iu)\mathrm{e}^{-u|x|}}{k_{p0}^2+u^2}, \tag{6.5.15}$$

where in the last two relations $\alpha(\xi;iu)$ is the isotropic polarizability of species ξ at the imaginary frequency icu. Energy shift expression (6.5.11) becomes

$$\Delta E_{BC} = -\frac{\hbar c}{256\pi^4\varepsilon_0^3}\left(-\vec{\nabla}^2\delta_{ij}+\vec{\nabla}_i\vec{\nabla}_j\right)^a\frac{1}{a}\left(-\vec{\nabla}^2\delta_{jk}+\vec{\nabla}_j\vec{\nabla}_k\right)^b\frac{1}{b}$$
$$\times\left(-\vec{\nabla}^2\delta_{ki}+\vec{\nabla}_k\vec{\nabla}_i\right)^c\frac{1}{c}$$
$$\times\left\{\begin{array}{l} 2\int_0^\infty \mathrm{d}u\alpha(A;iu)\alpha(B;iu)\alpha(C;iu)\mathrm{e}^{-u(a+b+c)} \\ +\left[1+\frac{1}{2}\mathrm{sgn}(b-a)+\frac{1}{2}\mathrm{sgn}(c-a)\right] \\ \times\int_0^\infty \mathrm{d}u\alpha(A;iu)\alpha(B;iu)\alpha(C;iu)\mathrm{e}^{-u(b+c-a)} \\ +\left[\frac{1}{2}\mathrm{sgn}(b-a)-\frac{1}{2}\right]\int_0^\infty \mathrm{d}u\alpha(A;iu)\alpha(B;iu)\alpha(C;iu)\mathrm{e}^{-u(a+c-b)} \\ +\left[\frac{1}{2}\mathrm{sgn}(c-a)-\frac{1}{2}\right]\int_0^\infty \mathrm{d}u\alpha(A;iu)\alpha(B;iu)\alpha(C;iu)\mathrm{e}^{-u(a+b-c)} \end{array}\right\}. \tag{6.5.16}$$

Because the three atoms are identical, permutation of the atom labels along with separation distance vectors readily generates contributions to the three-body energy shift from the interaction of atoms A and C in the presence of B, ΔE_{AC}, which is obtained from (6.5.16) on changing B to A and b to a and from the interaction between A and B in the presence of C, ΔE_{BA}, which follows from (6.5.16) on changing C to A and c to a. The overall three-body dispersion potential is finally arrived at by averaging out the three contributions described above and accounting for all possible pairings giving

$$\Delta E = \frac{2}{3}(\Delta E_{AB} + \Delta E_{BC} + \Delta E_{CA})$$
$$= -\frac{\hbar c}{64\pi^4\varepsilon_0^3}\left(-\vec{\nabla}^2\delta_{ij} + \vec{\nabla}_i\vec{\nabla}_j\right)^a\left(-\vec{\nabla}^2\delta_{jk} + \vec{\nabla}_j\vec{\nabla}_k\right)^b\left(-\vec{\nabla}^2\delta_{ki} + \vec{\nabla}_k\vec{\nabla}_i\right)^c$$
$$\times \frac{1}{abc}\int_0^\infty du\,\alpha(A;iu)\alpha(B;iu)\alpha(C;iu)\mathrm{e}^{-(a+b+c)u}, \tag{6.5.17}$$

in agreement with result (6.4.12).

6.6 N-BODY DISPERSION POTENTIAL

A physically transparent and calculationally efficient method for the evaluation of the dispersion interaction due to N neutral polarizable molecules is response theory (Power and Thirunamachandran, 1985). The computation is a generalization of the approach presented in Section 5.7, where the response of one species through its dynamic electric dipole polarizability to the electric displacement field of a second source molecule was shown to lead to the Casimir–Polder potential. In the present case, the time-dependent Maxwell field operators for a collection of molecules is calculated, from which the response of one molecule to the fields produced by all of the others is then evaluated, leading directly to the N-body energy shift.

A characteristic of the response formalism is that molecules couple to the field via their frequency dependent polarizability. A convenient starting point in the calculation of the many-body interaction energy is the Craig– Power form of the Hamiltonian density,

$$\mathcal{H} = \mathcal{H}_{\text{rad}} + \sum_{\xi=1}^{N} \mathcal{H}_{\text{int}}(\xi)$$
$$= \frac{1}{2\varepsilon_0}\left\{\vec{d}^{\perp 2}(\vec{r}) + \varepsilon_0^2 c^2 \vec{b}^2(\vec{r})\right\} - \frac{1}{2\varepsilon_0^2}\sum_{\xi=1}^{N} \alpha_{ij}(\xi;k) d_i^{\perp}(\vec{r}) d_j^{\perp}(\vec{r}) \delta(\vec{r}-\vec{R}_\xi), \tag{6.6.1}$$

which is appropriate for the situation in which the molecules are considered sources of the radiation field. The electromagnetic fields themselves must, of course, satisfy Maxwell operator equations. The first two microscopic Maxwell equations are clearly obeyed since the dynamical fields—electric displacement and magnetic—are purely transverse. Meanwhile, use of the transverse vector potential as the canonical field variable ensures that the third Maxwell equation is satisfied. Finally, the fourth Maxwell equation is obtained from (6.6.1) as follows. Variation of the Hamiltonian density with the vector potential yields the negative time derivative of the conjugate momentum field, as in

$$\dot{\vec{\Pi}}(\vec{r}) = -\frac{\partial \mathcal{H}}{\partial \vec{a}(\vec{r})} = -\varepsilon_0 c^2 \vec{\nabla} \times \vec{\nabla} \times \vec{a}(\vec{r}). \tag{6.6.2}$$

Noting that in the multipolar framework $\vec{\Pi}(\vec{r}) = -\vec{d}^{\perp}(\vec{r})$ (equation (1.7.5)) and $\vec{b} = \vec{\nabla} \times \vec{a}(\vec{r})$, it is seen that

$$\vec{\nabla} \times \vec{b}(\vec{r}) = \frac{1}{\varepsilon_0 c^2}\frac{\partial}{\partial t}\vec{d}^{\perp}(\vec{r}), \tag{6.6.3}$$

which is a special case of equation (1.6.34) applicable when spatial variation of the vector potential is neglected. Similarly, variation of $\mathcal{H}$ with respect to $\vec{\Pi}(\vec{r})$ results in

$$\dot{\vec{a}}(\vec{r}) = \frac{\partial \mathcal{H}}{\partial \vec{\Pi}(\vec{r})} = -\varepsilon_0^{-1}\vec{d}^{\perp}(\vec{r}) + \varepsilon_0^{-1}\sum_{\xi=1}^{N} \vec{\vec{\alpha}}(\xi;k)\cdot \vec{d}^{\perp}(\vec{r})\delta(\vec{r}-\vec{R}_\xi). \tag{6.6.4}$$

Recalling that $\dot{\vec{a}}(\vec{r}) = -\vec{e}^{\perp}(\vec{r})$, relation (6.6.4) can be written as

$$\vec{e}^{\perp}(\vec{r}) = \varepsilon_0^{-1}\left[\vec{\vec{1}} - \sum_{\xi=1}^{N} \vec{\vec{\alpha}}(\xi;k)\delta(\vec{r}-\vec{R}_\xi)\right]\cdot \vec{d}^{\perp}(\vec{r}). \tag{6.6.5}$$

Defining the transverse polarization field distribution, $\vec{p}^{\perp}(\vec{r})$, in terms of the electric dipole polarizability

$$\vec{p}^{\perp}(\vec{r}) = \sum_{\xi=1}^{N} \vec{\alpha}(\xi;k)\cdot\vec{d}^{\perp}(\vec{r})\delta(\vec{r}-\vec{R}_{\xi}), \tag{6.6.6}$$

the constitutive relation between fields $\vec{e}^{\perp}(\vec{r})$ and $\vec{d}^{\perp}(\vec{r})$ is obtained,

$$\vec{d}^{\perp}(\vec{r}) = \varepsilon_0\vec{e}^{\perp}(\vec{r}) + \vec{p}^{\perp}(\vec{r}). \tag{6.6.7}$$

The next step is to derive the homogeneous equations for the components of the electric displacement field in terms of the sources. Evaluating the time derivative of (6.6.7) and substituting in (6.6.3) produces

$$\vec{\nabla}\times\vec{b}(\vec{r}) = \frac{1}{c^2}\frac{\partial\vec{e}^{\perp}(\vec{r})}{\partial t} + \frac{1}{\varepsilon_0 c^2}\frac{\partial}{\partial t}\vec{p}^{\perp}(\vec{r}) = \frac{1}{\varepsilon_0 c^2}\frac{\partial}{\partial t}\vec{d}^{\perp}(\vec{r}). \tag{6.6.8}$$

Effecting the vector cross product operator twice on (6.6.7) and using the third microscopic Maxwell equation $\vec{\nabla}\times\vec{e}^{\perp}(\vec{r}) = -(\partial/\partial t)\vec{b}(\vec{r})$ yields

$$\vec{\nabla}\times\vec{\nabla}\times\vec{d}^{\perp}(\vec{r}) = -\varepsilon_0\frac{\partial}{\partial t}\left(\vec{\nabla}\times\vec{b}(\vec{r})\right) + \vec{\nabla}\times\vec{\nabla}\times\vec{p}^{\perp}(\vec{r}), \tag{6.6.9}$$

where the order of temporal and spatial derivatives has been interchanged in achieving the first term on the right-hand side of the last equation. Inserting (6.6.3) for $\vec{\nabla}\times\vec{b}(\vec{r})$, employing the identity

$$\vec{\nabla}\times\vec{\nabla}\times = -\vec{\nabla}^2 + \vec{\nabla}\vec{\nabla}\cdot, \tag{6.6.10}$$

and noting that the divergence of the transverse displacement field vanishes outside the sources for a neutral system, (6.6.9) produces the following for the wave equation for the displacement field,

$$\left(\vec{\nabla}^2 - \frac{1}{c^2}\frac{\partial^2}{\partial t^2}\right)\vec{d}^{\perp}(\vec{r}) = -\vec{\nabla}\times\vec{\nabla}\times\vec{p}^{\perp}(\vec{r}), \tag{6.6.11}$$

or, explicitly in terms of the polarization field (6.6.6),

$$\left(\vec{\nabla}^2 - \frac{1}{c^2}\frac{\partial^2}{\partial t^2}\right)d_p^{\perp}(\vec{r}) = -\sum_{\xi=1}^{N}\alpha_{ij}(\xi;k)\varepsilon_{lmp}\varepsilon_{jmn}\vec{\nabla}_l\vec{\nabla}_n d_i^{\perp}(\vec{r})\delta(\vec{r}-\vec{R}_{\xi}). \tag{6.6.12}$$

To proceed further, the transverse displacement field is expanded as a sum of normal modes,

$$d_p^{\perp}(\vec{r}) = i\sum_{\vec{k},\lambda}\left(\frac{\hbar c k \varepsilon_0}{2V}\right)^{1/2}\left\{e_q^{(\lambda)}(\vec{k})F_{pq}(\vec{k},\vec{r})\alpha(0)\mathrm{e}^{-i\omega t} - \mathrm{H.C.}\right\}, \quad (6.6.13)$$

where $\alpha(0)$ is the initial time boson annihilation operator and H.C. is the Hermitian conjugate. In the absence of sources, the mode function for the free radiation field is of the form

$$F_{pq}^{(0)}(\vec{k},\vec{r}) = \delta_{pq}\mathrm{e}^{i\vec{k}\cdot\vec{r}}. \quad (6.6.14)$$

Inserting (6.6.14) into (6.6.12), for the spatial part of the mode function

$$\left(\vec{\nabla}^2 + k^2\right)F_{pq}(\vec{k},\vec{r}) = -\sum_{\xi=1}^{N}\alpha_{ij}(\xi;k)\left(-\vec{\nabla}^2\delta_{jp} + \vec{\nabla}_j\vec{\nabla}_p\right)F_{iq}(\vec{k},\vec{r})\delta(\vec{r}-\vec{R}_\xi). \quad (6.6.15)$$

Solutions to (6.6.15) may be obtained via Fourier transformation. Both sides of the last relation are multiplied by $\mathrm{e}^{i\vec{p}\cdot\vec{r}}$ on the left and integrated over all space to yield

$$(2\pi)^3(-p^2+k^2)G_{pq}(\vec{k},\vec{p}) = -\sum_{\xi=1}^{N}\alpha_{ij}(\xi;k)(p^2\delta_{jp} - p_j p_p)\mathrm{e}^{i\vec{p}\cdot\vec{R}_\xi}F_{iq}(\vec{k},\vec{R}_\xi), \quad (6.6.16)$$

where

$$G_{pq}(\vec{k},\vec{p}) = \frac{1}{(2\pi)^3}\int_{-\infty}^{\infty}\mathrm{e}^{i\vec{p}\cdot\vec{r}}F_{pq}(\vec{k},\vec{r})\mathrm{d}^3\vec{r}. \quad (6.6.17)$$

The inverse of (6.6.17) is

$$\begin{aligned} F_{pq}(\vec{k},\vec{r}) &= F_{pq}^{(0)}(\vec{k},\vec{r}) + \frac{1}{(2\pi)^3}\sum_{\xi=1}^{N}\int_{-\infty}^{\infty}\frac{(p^2\delta_{jp} - p_j p_p)}{p^2-k^2}\mathrm{e}^{-i\vec{p}\cdot(\vec{r}-\vec{R}_\xi)} \\ &\quad \times\alpha_{ij}(\xi;k)F_{iq}(\vec{k},\vec{R}_\xi)\mathrm{d}^3\vec{p} \\ &= F_{pq}^{(0)}(\vec{k},\vec{r}) + \frac{1}{(2\pi)^3}\left(-\vec{\nabla}^2\delta_{jp} + \vec{\nabla}_j\vec{\nabla}_p\right) \\ &\quad \times\sum_{\xi=1}^{N}\int_{-\infty}^{\infty}\frac{\mathrm{e}^{-i\vec{p}\cdot(\vec{r}-\vec{R}_\xi)}}{p^2-k^2}\alpha_{ij}(\xi;k)F_{iq}(\vec{k},\vec{R}_\xi)\mathrm{d}^3\vec{p}. \end{aligned} \quad (6.6.18)$$

Since (6.6.14) is an appropriate solution for the free field situation, it makes sense to adopt it as the complementary function. Performing the angular average using

$$\frac{1}{4\pi}\int \mathrm{e}^{-i\vec{p}\cdot(\vec{r}-\vec{R}_\xi)}\mathrm{d}\Omega = \frac{1}{2ip}\frac{1}{|\vec{r}-\vec{R}_\xi|}\left(\mathrm{e}^{ip|\vec{r}-\vec{R}_\xi|}-\mathrm{e}^{-ip|\vec{r}-\vec{R}_\xi|}\right), \tag{6.6.19}$$

and carrying out the p-integral on (6.6.18) with the pole displaced to favor outgoing waves gives

$$F_{pq}(\vec{k},\vec{r}) = F_{pq}^{(0)}(\vec{k},\vec{r}) + \frac{1}{4\pi}\sum_{\xi=1}^{N}\alpha_{ij}(\xi;k)\left(-\vec{\nabla}^2\delta_{jp}+\vec{\nabla}_j\vec{\nabla}_p\right)\frac{\mathrm{e}^{ik|\vec{r}-\vec{R}_\xi|}}{|\vec{r}-\vec{R}_\xi|}F_{iq}(\vec{k},\vec{R}_\xi). \tag{6.6.20}$$

Higher order terms of the mode function dependent on increasing powers of the polarizability may therefore be obtained on iteration,

$$F_{pq} = F_{pq}^{(0)} + F_{pq}^{(1)} + \cdots, \tag{6.6.21}$$

with

$$F_{pq}^{(n+1)}(\vec{k},\vec{r}) = \sum_{\xi=1}^{N}\left(\frac{1}{4\pi}\right)^{N-1}\alpha_{ij}(\xi;k)\left(-\vec{\nabla}^2\delta_{jp}+\vec{\nabla}_j\vec{\nabla}_p\right)\frac{\mathrm{e}^{ik|\vec{r}-\vec{R}_\xi|}}{|\vec{r}-\vec{R}_\xi|}F_{iq}^{(n)}(\vec{k},\vec{R}_\xi). \tag{6.6.22}$$

The series expansion for the electric displacement field is then obtained on substituting (6.6.22) into (6.6.13).

The response of each molecule taken in turn, to the field of all of the others gives an expression from which the N-body energy shift may be found. It is given by

$$\Delta E = -\frac{1}{2N\varepsilon_0^{N+1}}\sum_P \alpha_{i_1j_1}(P_1;k)\langle 0|d_{i_1}^{\perp}(\vec{R}_{p_1})d_{j_1}^{\perp}(\vec{R}_{p_2})|0\rangle, \tag{6.6.23}$$

where the ground-state polarizability appears and the expectation value of the product of radiation fields is taken over the vacuum state of the electromagnetic field. In formula (6.6.23), $\sum_P$ denotes the summations over all permutations corresponding to

$$P \equiv \begin{bmatrix} 1 & 2 & 3 & \dots & N \\ P_1 & P_2 & P_3 & \dots & P_N \end{bmatrix}, \tag{6.6.24}$$

where in the top line, the objects to be permuted are written in their natural order, while the bottom line signifies the order that results on carrying out the prescribed permutation (Wigner, 1959). Substituting for the displacement field from (6.6.13) and executing the polarization sum, (6.6.23) becomes

$$\begin{aligned}\Delta E &= -\frac{1}{2N\varepsilon_0^N}\sum_P\sum_{\vec{k}}\left(\frac{\hbar ck}{2V}\right)\alpha_{i_1j_1}(P_1;k)\left(\delta_{rs}-\hat{k}_r\hat{k}_s\right)F_{i_1r}\left(\vec{k},\vec{R}_{P_1}\right)\bar{F}_{j_1s}\left(\vec{k},\vec{R}_{P_1}\right)\\ &= -\frac{1}{2N\varepsilon_0}\left(\frac{1}{4\pi\varepsilon_0}\right)^{N-1}\sum_P\sum_{\vec{k}}\left(\frac{\hbar ck}{2V}\right)\alpha_{i_1j_1}(P_1;k)\left(\delta_{rs}-\hat{k}_r\hat{k}_s\right)\\ &\quad\times\sum_{n=0}^{N-1}F_{i_1r}^{(n)}(\vec{k},\vec{R}_{P_1})\bar{F}_{j_1s}^{(N-n-1)}(\vec{k},\vec{R}_{P_1}),\end{aligned} \tag{6.6.25}$$

where in the second line, summation over n is carried out of the product of F functions to ensure that the polarizability of each molecule appears only once as necessitated by the final form of the expression for the N-body energy shift. Let

$$\gamma_{i_aj_b}(k) = \left(-\vec{\nabla}^2\delta_{i_aj_b} + \vec{\nabla}_{i_a}\vec{\nabla}_{j_b}\right)\frac{e^{ikR_{P_aP_b}}}{R_{P_aP_b}}, \tag{6.6.26}$$

where the gradients act on $R_{P_aP_b} = |\vec{R}_{P_b} - \vec{R}_{P_a}|$. With the relevant effective contributions to $F_{i_1r}^{(n)}(\vec{k},\vec{R}_{p_1})$ and $\bar{F}_{j_1s}^{(N-n-1)}(\vec{k},\vec{R}_{p_1})$ being given by

$$\prod_{a=1}^{n}\alpha_{i_{a+1}j_{a+1}}(P_{a+1};k)\gamma_{i_aj_{a+1}}(k)F_{i_{n+1}r}^{(0)}(\vec{k},\vec{R}_{P_{n+1}}), \tag{6.6.27}$$

and

$$\prod_{a=n+1}^{N-1}\alpha_{i_{a+1}j_{a+1}}(P_{a+1};k)\bar{\gamma}_{i_{a+1}j_{a+2}}(k)\bar{F}_{j_{n+2}s}^{(0)}(\vec{k},\vec{R}_{P_{n+2}}), \tag{6.6.28}$$

respectively, where $j_{N+1} = j_1$, the energy shift from (6.6.25) is

$$\Delta E = -\frac{1}{2N\varepsilon_0}\left(\frac{1}{4\pi\varepsilon_0}\right)^{N-1} \sum_P \sum_{\vec{k}} \left(\frac{\hbar ck}{2V}\right) \alpha_{i_1j_1}(P_1;k)(\delta_{rs} - \hat{k}_r\hat{k}_s)$$
$$\times \sum_{n=0}^{N-1} F^{(0)}_{i_{n+1}r}(\vec{k},\vec{R}_{P_{n+1}}) \bar{F}^{(0)}_{j_{n+2}s}(\vec{k},\vec{R}_{P_{n+2}})$$
$$\times \prod_{a=1}^{n} \prod_{b=n+1}^{N-1} \alpha_{i_{a+1}j_{a+1}}(P_{a+1};k)\alpha_{i_{b+1}j_{b+1}}(P_{b+1};k)\gamma_{i_aj_{a+1}}(k)\bar{\gamma}_{i_{b+1}j_{b+2}}(k). \tag{6.6.29}$$

Inserting the zeroth-order mode function from (6.6.14), converting the mode sum to an integral and performing the angular average produces

$$\Delta E = -\frac{1}{2N}\left(\frac{1}{4\pi\varepsilon_0}\right)^{N} \frac{\hbar c}{\pi} \sum_P \sum_{n=0}^{N-1} \left(-\vec{\nabla}^2\delta_{i_{n+1}j_{n+2}} + \vec{\nabla}_{i_{n+1}}\vec{\nabla}_{j_{n+2}}\right)$$
$$\times \int_0^\infty dk\alpha_{i_1j_1}(P_1;k)\frac{\sin kR_{P_{n+1}P_{n+2}}}{R_{P_{n+1}P_{n+2}}}$$
$$\times \prod_{a=1}^{n} \prod_{b=n+1}^{N-1} \alpha_{i_{a+1}j_{a+1}}(P_{a+1};k)\alpha_{i_{b+1}j_{b+1}}(P_{b+1};k)\gamma_{i_aj_{a+1}}(k)\bar{\gamma}_{i_{b+1}j_{b+2}}(k). \tag{6.6.30}$$

Making use of the fact that

$$\left(-\vec{\nabla}^2\delta_{i_{n+1}j_{n+2}} + \vec{\nabla}_{i_{n+1}}\vec{\nabla}_{j_{n+2}}\right)\frac{\sin kR_{P_{n+1}P_{n+2}}}{R_{P_{n+1}P_{n+2}}} = \frac{1}{2i}\left(\gamma_{i_{n+1}j_{n+2}}(k) - \bar{\gamma}_{i_{n+1}j_{n+2}}(k)\right), \tag{6.6.31}$$

(6.6.30) becomes

$$\Delta E = -\frac{1}{2N}\left(\frac{1}{4\pi\varepsilon_0}\right)^{N} \frac{\hbar c}{2\pi i} \sum_P \int_0^\infty dk\alpha_{i_1j_1}(P_1;k)\alpha_{i_2j_2}(P_2;k)\ldots\alpha_{i_Nj_N}(P_N;k)$$
$$\times [\gamma_{i_1j_2}(k)\gamma_{i_2j_3}(k)\ldots\gamma_{i_Nj_1}(k) - \bar{\gamma}_{i_1j_2}(k)\bar{\gamma}_{i_2j_3}(k)\ldots\bar{\gamma}_{i_Nj_1}(k)]. \tag{6.6.32}$$

Inserting the $\gamma_{ij}(k)$ tensor from (6.6.26) and converting the integral over k to an imaginary wavevector $k = \pm iu$, the dispersion energy shift between

N bodies can be written as

$$\Delta E^{(N)} = -\left(\frac{\hbar c}{2\pi N}\right)\left(\frac{1}{4\pi\varepsilon_0}\right)^N \sum_P \left(-\vec{\nabla}^2\delta_{i_1j_2} + \vec{\nabla}_{i_1}\vec{\nabla}_{j_2}\right)^{R_{P_1P_2}} \frac{1}{R_{P_1P_2}}$$
$$\times\left(-\vec{\nabla}^2\delta_{i_2j_3} + \vec{\nabla}_{i_2}\vec{\nabla}_{j_3}\right)^{R_{P_2P_3}} \frac{1}{R_{P_2P_3}}\cdots$$
$$\times\left(-\vec{\nabla}^2\delta_{i_Nj_1} + \vec{\nabla}_{i_N}\vec{\nabla}_{j_1}\right)^{R_{P_NP_1}} \frac{1}{R_{P_NP_1}}$$
$$\times\int_0^\infty du\alpha_{i_1j_1}(P_1;iu)\alpha_{i_2j_2}(P_2;iu)\ldots\alpha_{i_Nj_N}(P_N;iu)\mathrm{e}^{-u(R_{P_1P_2}+R_{P_2P_3}+\cdots R_{P_NP_1})}. \tag{6.6.33}$$

This result holds for N molecules with anisotropic polarizabilities for all pair separation distance permutations beyond the region of overlap of electronic charge distributions.

From the general formula (6.6.33), it is a simple matter to extract results for $N=2$ and 3 and compare with previously obtained formulas. For the pair interaction energy,

$$\Delta E^{(2)} = -\left(\frac{\hbar c}{4\pi}\right)\left(\frac{1}{16\pi^2\varepsilon_0^2}\right)\sum_P \left(-\vec{\nabla}^2\delta_{i_1j_2} + \vec{\nabla}_{i_1}\vec{\nabla}_{j_2}\right)^{R_{P_1P_2}} \frac{1}{R_{P_1P_2}}$$
$$\times\left(-\vec{\nabla}^2\delta_{i_2j_1} + \vec{\nabla}_{i_2}\vec{\nabla}_{j_1}\right)^{R_{P_2P_1}} \frac{1}{R_{P_2P_1}}$$
$$\times\int_0^\infty du\alpha_{i_1j_1}(P_1;iu)\alpha_{i_2j_2}(P_2;iu)\mathrm{e}^{-u(R_{P_1P_2}+R_{P_2P_1})}, \tag{6.6.34}$$

which is identical to the Casimir–Polder potential (6.4.27) once the sum over permutations in (6.6.34) is carried out, which introduces a factor of 2!. Similarly, for the three-body energy shift,

$$\Delta E^{(3)} = -\left(\frac{\hbar c}{6\pi}\right)\left(\frac{1}{4\pi\varepsilon_0}\right)^3 \sum_P \left(-\vec{\nabla}^2\delta_{i_1j_2} + \vec{\nabla}_{i_1}\vec{\nabla}_{j_2}\right)^{R_{P_1P_2}} \frac{1}{R_{P_1P_2}}$$
$$\times(-\vec{\nabla}^2\delta_{i_2j_3} + \vec{\nabla}_{i_2}\vec{\nabla}_{j_3})^{R_{P_2P_3}} \frac{1}{R_{P_2P_3}}\left(-\vec{\nabla}^2\delta_{i_3j_1} + \vec{\nabla}_{i_3}\vec{\nabla}_{j_1}\right)^{R_{P_3P_1}} \frac{1}{R_{P_3P_1}}$$
$$\times\int_0^\infty du\alpha_{i_1j_1}(P_1;iu)\alpha_{i_2j_2}(P_2;iu)\alpha_{i_3j_3}(P_3;iu)\mathrm{e}^{-u(R_{P_1P_2}+R_{P_2P_3}+R_{P_3P_1})}, \tag{6.6.35}$$

which is seen to be equivalent to the expression derived in Section 6.4 for the retarded triple-dipole dispersion potential and given by equation (6.4.12), once a factor of 3! is accounted for on performing the sum over permutations in (6.6.35), where $a = |\vec{R}_2 - \vec{R}_3|$, $b = |\vec{R}_3 - \vec{R}_1|$, and $c = |\vec{R}_1 - \vec{R}_2|$.

The form of the asymptotic N-body dispersion potentials is easily obtained from the result applicable to all ranges of pair intermolecular separation distance (6.6.33). In the far-zone limit, all separations are much larger than characteristic reduced transition wavelengths. Hence, the polarizabilities are all static and independent of u and they can be factored outside of the integral. The resulting u-integral is elementary, leading to the result

$$\begin{aligned}\Delta E_{\mathrm{FZ}}^{(N)} = &-\left(\frac{\hbar c}{2\pi N}\right)\left(\frac{1}{4\pi\varepsilon_0}\right)^N \sum_P \alpha_{i_1j_1}(P_1;0)\alpha_{i_2j_2}(P_2;0)\ldots\alpha_{i_Nj_N}(P_N;0)\\ &\times\left(-\vec{\nabla}^2\delta_{i_1j_2}+\vec{\nabla}_{i_1}\vec{\nabla}_{j_2}\right)^{R_{P_1P_2}}\\ &\times\left(-\vec{\nabla}^2\delta_{i_2j_3}+\vec{\nabla}_{i_2}\vec{\nabla}_{j_3}\right)^{R_{P_2P_3}}\ldots\left(-\vec{\nabla}^2\delta_{i_Nj_1}+\vec{\nabla}_{i_N}\vec{\nabla}_{j_1}\right)^{R_{P_NP_1}}\\ &\times\frac{1}{R_{P_1P_2}R_{P_2P_3}\ldots R_{P_NP_1}(R_{P_1P_2}+R_{P_2P_3}+\cdots+R_{P_NP_1})},\end{aligned} \tag{6.6.36}$$

or in terms of the displacements a, b, c, and so on,

$$\begin{aligned}\Delta E_{\mathrm{FZ}}^{(N)} = &-\left(\frac{\hbar c}{2\pi N}\right)\left(\frac{1}{4\pi\varepsilon_0}\right)^N \sum_P \alpha_{i_1j_1}(P_1;0)\alpha_{i_2j_2}(P_2;0)\ldots\alpha_{i_Nj_N}(P_N;0)\\ &\times\left(-\vec{\nabla}^2\delta_{i_1j_2}+\vec{\nabla}_{i_1}\vec{\nabla}_{j_2}\right)^{a}\left(-\vec{\nabla}^2\delta_{i_2j_3}+\vec{\nabla}_{i_2}\vec{\nabla}_{j_3}\right)^{b}\\ &\times\left(-\vec{\nabla}^2\delta_{i_3j_4}+\vec{\nabla}_{i_3}\vec{\nabla}_{j_4}\right)^{c}\cdots\left(-\vec{\nabla}^2\delta_{i_Nj_1}+\vec{\nabla}_{i_N}\vec{\nabla}_{j_1}\right)^{n}\cdots\\ &\times\frac{1}{abc\ldots n(a+b+c+\cdots+n)}.\end{aligned} \tag{6.6.37}$$

At the opposite extreme, the separation distances are very much less than molecular transition wavelengths. Setting the exponential in (6.6.33) to unity leads to

$$\Delta E_{\mathrm{NZ}}^{(N)} = -\left(\frac{\hbar c}{2\pi N}\right)\left(\frac{1}{4\pi\varepsilon_0}\right)^N \sum_P \left(-\vec{\nabla}^2\delta_{i_1j_2} + \vec{\nabla}_{i_1}\vec{\nabla}_{j_2}\right)^{R_{P_1P_2}}$$

$$\times \left(-\vec{\nabla}^2\delta_{i_2j_3} + \vec{\nabla}_{i_2}\vec{\nabla}_{j_3}\right)^{R_{P_2P_3}} \ldots \left(-\vec{\nabla}^2\delta_{i_Nj_1} + \vec{\nabla}_{i_N}\vec{\nabla}_{j_1}\right)^{R_{P_NP_1}}$$

$$\times \frac{1}{R_{P_1P_2}R_{P_2P_3}\ldots R_{P_NP_1}} \int_0^\infty du\, \alpha_{i_1j_1}(P_1;iu)\alpha_{i_2j_2}(P_2;iu)\ldots\alpha_{i_Nj_N}(P_N;iu), \tag{6.6.38}$$

which can be written as

$$\Delta E_{\mathrm{NZ}}^{(N)} = -(-1)^N\left(\frac{\hbar c}{2\pi N}\right)\left(\frac{1}{4\pi\varepsilon_0}\right)^N \sum_P (\delta_{i_1j_2} - 3\hat{a}_{i_1}\hat{a}_{j_2})$$

$$\times (\delta_{i_2j_3} - 3\hat{b}_{i_2}\hat{b}_{j_3}) \cdots \frac{1}{a^3b^3\ldots}$$

$$\times \int_0^\infty du\, \alpha_{i_1j_1}(P_1;iu)\alpha_{i_2j_2}(P_2;iu)\ldots\alpha_{i_Nj_N}(P_N;iu), \tag{6.6.39}$$

which is the N-body generalization of the Axilrod–Teller result. The long- and short-range forms of the energy shift for $N=2$ and 3, obtained from limits (6.6.37) and (6.6.39), agree with formulas derived from explicit evaluation of the pair and three-body potential valid for all separations.

6.7 FOUR-BODY RETARDED DISPERSION POTENTIAL

In the previous section, a general formula was obtained for the retarded N-body dispersion interaction energy by calculating the response of one molecule at a time to the electric displacement field produced by the remaining $N-1$ atoms or molecules. Expressions for N-body dispersion potentials applicable at short- and long-range asymptotic limits were readily extracted and all three formulas were tested for $N=2$ and 3 and found to agree with results derived directly for two and three interacting bodies. In this section, the general formula is applied to calculate the quantum electrodynamical dispersion potential between four bodies

(Power and Thirunamachandran, 1994) by inserting $N=4$ into formula (6.6.33). This produces the energy shift expression

$$\Delta E = -\left(\frac{\hbar c}{16\pi}\right)\left(\frac{1}{4\pi\varepsilon_0}\right)^4 \sum_P \left(-\vec{\nabla}^2\delta_{i_1j_2} + \vec{\nabla}_{i_1}\vec{\nabla}_{j_2}\right)^{R_{P_1P_2}} \frac{1}{R_{P_1P_2}}$$
$$\times\left(-\vec{\nabla}^2\delta_{i_2j_3} + \vec{\nabla}_{i_2}\vec{\nabla}_{j_3}\right)^{R_{P_2P_3}} \frac{1}{R_{P_2P_3}}\left(-\vec{\nabla}^2\delta_{i_3j_4} + \vec{\nabla}_{i_3}\vec{\nabla}_{j_4}\right)^{R_{P_3P_4}} \frac{1}{R_{P_3P_4}}$$
$$\times\left(-\vec{\nabla}^2\delta_{i_4j_1} + \vec{\nabla}_{i_4}\vec{\nabla}_{j_1}\right)^{R_{P_4P_1}} \frac{1}{R_{P_4P_1}}$$
$$\times\int_0^\infty du\,\alpha_{i_1j_1}(P_1;iu)\alpha_{i_2j_2}(P_2;iu)\alpha_{i_3j_3}(P_3;iu)\alpha_{i_4j_4}(P_4;iu)$$
$$\times e^{-u(R_{P_1P_2}+R_{P_2P_3}+R_{P_3P_4}+R_{P_4P_1})}. \tag{6.7.1}$$

From the general formula applicable to N bodies in mutual interaction, it is seen that there are $N!/(2N)$ distinct contributions, the denominator arising when cyclic permutations and reversals are not distinguished in the ordering of particles. It is convenient to label the distinct contributions in such a case by defining a class as containing an ordering given by the ratio above and differing only by cyclic permutations and reverse ordering. For four bodies A, B, C, and D, there are three classes that are labeled *ABCD*, *ABDC*, and *ACBD*. These three groups, along with the cyclic and reverse ordering associated with them, comprise the full set of $4!=24$ possible permutations of the four bodies. The three orderings listed above form the representative labels of the three classes. Let the six displacements between the four species be defined according to $c=|\vec{R}_B-\vec{R}_A|$, $a=|\vec{R}_C-\vec{R}_B|$, $b=|\vec{R}_A-\vec{R}_C|$, $d=|\vec{R}_A-\vec{R}_D|$, $e=|\vec{R}_B-\vec{R}_D|$, and $f=|\vec{R}_C-\vec{R}_D|$. For the three classes *ABCD*, *ABDC*, and *ACBD*, the ordered interobject distances are (c, a, f, d) (i.e., $A\to B=c$, $B\to C=a$, $C\to D=f$, and $D\to A=d$), (c, e, f, b), and (b, a, e, d), respectively.

In the far zone, the potential from equation (6.6.37) after orientational averaging is given by

$$\Delta E_{\mathrm{FZ}} = -\left(\frac{\hbar c}{\pi}\right)\left(\frac{1}{4\pi\varepsilon_0}\right)^4 \alpha(A;0)\alpha(B;0)\alpha(C;0)\alpha(D;0)$$
$$\times\left[D(cafd)+D(cefb)+D(baed)\right], \tag{6.7.2}$$

where the geometric factors are defined by

$$D(abcd) = \left(-\vec{\nabla}^2\delta_{ij} + \vec{\nabla}_i\vec{\nabla}_j\right)^a \left(-\vec{\nabla}^2\delta_{jk} + \vec{\nabla}_j\vec{\nabla}_k\right)^b \left(-\vec{\nabla}^2\delta_{kl} + \vec{\nabla}_k\vec{\nabla}_l\right)^c \times \left(-\vec{\nabla}^2\delta_{li} + \vec{\nabla}_l\vec{\nabla}_i\right)^d \frac{1}{abcd(a+b+c+d)}. \tag{6.7.3}$$

For an arrangement in which the four bodies are located at the corners of a regular tetrahedron with side length R, the dispersion potential in the far zone is

$$\Delta E_{\rm FZ} = -\left(\frac{\hbar c}{\pi R^{13}}\right)\left(\frac{1}{4\pi\varepsilon_0}\right)^4 \left(\frac{3\times 41\times 2689}{2^{15}}\right)\alpha(A;0)\alpha(B;0)\alpha(C;0)\alpha(D;0), \tag{6.7.4}$$

exhibiting an inverse thirteenth power dependence on separation distance.

On the other hand, in the near zone, the potential takes the form

$$\Delta E_{\rm NZ} = -\left(\frac{\hbar c}{\pi}\right)\left(\frac{1}{4\pi\varepsilon_0}\right)^4 [F(cafd) + F(cefb) + F(baed)] \times \int_0^\infty du\,\alpha(A;iu)\alpha(B;iu)\alpha(C;iu)\alpha(D;iu), \tag{6.7.5}$$

where

$$F(abcd) = \left(-\vec{\nabla}^2\delta_{ij} + \vec{\nabla}_i\vec{\nabla}_j\right)^a \left(-\vec{\nabla}^2\delta_{jk} + \vec{\nabla}_j\vec{\nabla}_k\right)^b \left(-\vec{\nabla}^2\delta_{kl} + \vec{\nabla}_k\vec{\nabla}_l\right)^c \times \left(-\vec{\nabla}^2\delta_{li} + \vec{\nabla}_l\vec{\nabla}_i\right)^d \frac{1}{d}$$

$$= \frac{3}{a^3b^3c^3d^3}\left\{\begin{array}{l} -4 + 3[(\hat{a}\cdot\hat{b})^2 + (\hat{a}\cdot\hat{c})^2 + (\hat{a}\cdot\hat{d})^2 \\ + (\hat{b}\cdot\hat{c})^2 + (\hat{b}\cdot\hat{d})^2 + (\hat{c}\cdot\hat{d})^2] \\ -9[(\hat{b}\cdot\hat{c})(\hat{c}\cdot\hat{d})(\hat{d}\cdot\hat{b}) + (\hat{c}\cdot\hat{d})(\hat{d}\cdot\hat{a})(\hat{a}\cdot\hat{c}) \\ + (\hat{a}\cdot\hat{b})(\hat{b}\cdot\hat{d})(\hat{d}\cdot\hat{a}) + (\hat{b}\cdot\hat{c})(\hat{c}\cdot\hat{a})(\hat{a}\cdot\hat{b}) \\ + 27[(\hat{a}\cdot\hat{b})(\hat{b}\cdot\hat{c})(\hat{c}\cdot\hat{d})(\hat{d}\cdot\hat{a})] \end{array}\right\} \tag{6.7.6}$$

and the molecular factor is

$$\int_0^\infty du\alpha(A;iu)\alpha(B;iu)\alpha(C;iu)\alpha(D;iu)$$

$$= \frac{\pi}{2}\left(\frac{2}{3\hbar c}\right)^4 \sum_{p,q,r,s} |\vec{\mu}^{p0}(A)|^2|\vec{\mu}^{q0}(B)|^2|\vec{\mu}^{r0}(C)|^2|\vec{\mu}^{s0}(D)|^2$$
$$\times \left\{ \begin{array}{l} [(k_{p0}+k_{q0}+k_{r0}+k_{s0})(k_{p0}k_{q0}+k_{p0}k_{r0}+k_{p0}k_{s0}+k_{q0}k_{r0} \\ \quad +k_{q0}k_{s0}+k_{r0}k_{s0}) \\ -(k_{p0}k_{q0}k_{r0}+k_{p0}k_{q0}k_{s0}+k_{p0}k_{r0}k_{s0}+k_{q0}k_{r0}k_{s0})] \\ \times [(k_{p0}+k_{q0})(k_{p0}+k_{r0})(k_{p0}+k_{s0})(k_{q0}+k_{r0}) \\ \times (k_{q0}+k_{s0})(k_{r0}+k_{s0})]^{-1} \end{array} \right\}, \tag{6.7.7}$$

with expression (6.7.5) the four-body equivalent of the Axilrod–Teller triple-dipole dispersion potential.

6.8 THREE-BODY DISPERSION INTERACTION INVOLVING ONE EXCITED MOLECULE

In Chapter 5, three different approaches were given for the calculation of dispersion energy shifts between two molecules. Each of the methods allowed for one or both of the interacting molecules to be in electronically excited states although the degree of difficulty of the calculation varied according to the viewpoint adopted even though the final results obtained were the same in all cases. It was found that response theory and the method of induced moments greatly simplified the calculation of pair dispersion potentials when real photon emission and absorption processes occur in addition to transitions that are purely virtual in origin, as is the case when one or both species is excited. Before going on to show how a combination of the coupling of induced moments of two of the three bodies to the displacement field of an excited third molecule leads straightforwardly to the dispersion interaction between two ground-state molecules and one excited-state molecule, a time-dependent perturbation theory treatment is presented first (Power and Thirunamachandran, 1995b).

6.8.1 Time-Dependent Perturbation Theory

In Section 5.6, diagrammatic time-dependent perturbation theory was used to calculate the retarded two-body dispersion interaction between a ground-state molecule and an electronically excited species. As for the calculation between two ground-state molecules, the contribution from 12 time-ordered graphs were summed over with care being taken to account for emission of a real photon by the excited molecule when undergoing a downward transition to a lower lying state. This was dealt with by including damping factors in the energy denominator products and including pole contributions in addition to the principal value part of the integral over photon wavevector. A similar approach is now carried out for three interacting molecules, only one of which is excited.

Consider three neutral, nonpolar molecules A, B, and C situated at $\vec{R}_A$, $\vec{R}_B$, and $\vec{R}_C$, respectively. Initially, let species A be in electronically excited state $|m\rangle$, while both B and C are in the ground electronic state. Both upward and downward transitions to an intermediate-state $|p\rangle$ are allowed in A, while B and C both undergo upward transitions to excited-state $|q\rangle$ and $|r\rangle$, respectively. In the electric dipole approximation of the multipolar Hamiltonian of molecular quantum electrodynamics, the total Hamiltonian for the system is given by

$$H = H_{\text{mol}}(A) + H_{\text{mol}}(B) + H_{\text{mol}}(C) + H_{\text{rad}} + H_{\text{int}}, \tag{6.8.1}$$

where the interaction Hamiltonian is written as

$$H_{\text{int}} = -\varepsilon_0^{-1} \sum_{\xi=A,B,C} \vec{\mu}(\xi) \cdot \vec{d}^{\perp}(\vec{R}_\xi). \tag{6.8.2}$$

When this form of coupling Hamiltonian was adopted in Section 6.3 for the computation of the triple-dipole dispersion potential between three ground-state species, summation over contributions arising from 360 time orderings were necessary. Considerable simplification of the calculation was achieved by adopting an effective 2-photon interaction Hamiltonian at each center, reducing the evaluation of the energy shift to summation over 12 diagrams. A similar approach is taken in the present problem, with one important difference being that because A is excited, the linear in the electric displacement field type of coupling Hamiltonian (6.8.2) is retained for molecule A. Thus,

$$H_{\text{int}} = -\varepsilon_0^{-1}\vec{\mu}(A) \cdot \vec{d}^{\perp}(\vec{R}_A) - \frac{1}{2\varepsilon_0^2}\alpha(B)\vec{d}^{\perp 2}(\vec{R}_B) - \frac{1}{2\varepsilon_0^2}\alpha(C)\vec{d}^{\perp 2}(\vec{R}_C). \tag{6.8.3}$$

With the use of this form of interaction Hamiltonian, the leading order of perturbation theory necessary for evaluation of the energy shift is now no longer the sixth but is the fourth. When time reversals are adumbrated, 24 possible time-ordered sequences may be drawn to represent the process. The initial and final states of the system are written as

$$|0\rangle = |E_m^A, E_0^B, E_0^C; 0(\vec{k}_1, \lambda_1), 0(\vec{k}_2, \lambda_2), 0(\vec{k}_3, \lambda_3)\rangle, \tag{6.8.4}$$

where, as for the ground-state case, the photon labeled by mode $(\vec{k}_1, \lambda_1)$ is exchanged between A and B, that characterized by mode $(\vec{k}_2, \lambda_2)$ propagates between B and C, while that denoted by $(\vec{k}_3, \lambda_3)$ traverses between A and C. Evaluating the individual contributions in the usual way and adding gives, for three isotropic molecules, the expression

$$\begin{aligned} -\frac{1}{3}\frac{1}{(4\pi\varepsilon_0)^3}\sum_p |\vec{\mu}^{mp}(A)|^2 \left(-\vec{\nabla}^2\delta_{ij} + \vec{\nabla}_i\vec{\nabla}_j\right)^a \frac{1}{a}\left(-\vec{\nabla}^2\delta_{jk} + \vec{\nabla}_j\vec{\nabla}_k\right)^b \frac{1}{b} \\ \times \left(-\vec{\nabla}^2\delta_{ki} + \vec{\nabla}_k\vec{\nabla}_i\right)^c \frac{1}{c} \\ \times \frac{1}{\pi^3}\int \alpha(B;k)\alpha(C;k)\sin k_2 a \sin k_3 b \sin k_1 c \sum_{g=i}^{xii} D_g^{-1} \mathrm{d}k_1 \mathrm{d}k_2 \mathrm{d}k_3 + \text{c.c.}, \end{aligned} \tag{6.8.5}$$

where a, b, and c are the pair separation distances defined in Section 6.3, D_g^{-1} is the energy denominator product from graph g, and c.c. denotes the complex conjugate term. Since A is excited and can make downward as well as upward transitions, some of the energy denominators can vanish due to resonant excitation. It is therefore convenient to consider only downward transitions from $|m\rangle$ in A, say, to the ground-state $|0^{\mathrm{A}}\rangle$, with wavevector k_{m0}. While upward transitions clearly contribute, in this case, none of the energy denominators vanish and the contribution is identical to the ground-state triple-dipole dispersion potential (6.4.12), but with excited-state polarizability of A appearing in the expression instead of the ground-state one. As in the analogous two-body case, the poles in the denominators are handled by adding $\pm i\gamma$ using incoming and outgoing wave criteria for the choice of sign and employing the identities (5.6.5) and (5.6.7) to evaluate the wavevector integrals. Since the potential (6.8.5) is real, for denominators containing a single pole, only the principal value of (5.6.5) contributes, while for terms with two poles, both the principal value product and the contribution from the two delta functions in (5.6.7) remain. Like the situation occurring in Section 5.6 for the interaction between a ground-state molecule and an excited-state molecule, only two graphs give rise to a contribution involving a product of two delta functions, which is

$$-\frac{2}{3}\frac{1}{(4\pi\varepsilon_0)^3}|\vec{\mu}^{m0}(A)|^2\alpha(B;k_{m0})\alpha(C;k_{m0})\left(-\vec{\nabla}^2\delta_{ij}+\vec{\nabla}_i\vec{\nabla}_j\right)^a\frac{1}{a}$$
$$\times\left(-\vec{\nabla}^2\delta_{jk}+\vec{\nabla}_j\vec{\nabla}_k\right)^b\frac{1}{b}$$
$$\times\left(-\vec{\nabla}^2\delta_{ki}+\vec{\nabla}_k\vec{\nabla}_i\right)^c\frac{1}{c}\cos k_{m0}a\sin k_{m0}b\sin k_{m0}c. \qquad (6.8.6)$$

Evaluating the principal value product terms from these 2 graphs along with the 10 other graphs produces a contribution, which can be written as 2 terms. One is trigonometric and similar to (6.8.6) in that it depends only on the transition frequency of molecule A, ck_{m0},

$$-\frac{2}{3}\frac{1}{(4\pi\varepsilon_0)^3}|\vec{\mu}^{m0}(A)|^2\alpha(B;k_{m0})\alpha(C;k_{m0})\left(-\vec{\nabla}^2\delta_{ij}+\vec{\nabla}_i\vec{\nabla}_j\right)^a\frac{1}{a}$$
$$\times\left(-\vec{\nabla}^2\delta_{jk}+\vec{\nabla}_j\vec{\nabla}_k\right)^b\frac{1}{b}$$
$$\times\left(-\vec{\nabla}^2\delta_{ki}+\vec{\nabla}_k\vec{\nabla}_i\right)^c\frac{1}{c}\cos k_{m0}a\cos k_{m0}b\cos k_{m0}c. \qquad (6.8.7)$$

The other is similar to the familiar u-integral expression as found for three interacting molecules in the ground state and is

$$\frac{\hbar c}{\pi(4\pi\varepsilon_0)^3}\left(-\vec{\nabla}^2\delta_{ij}+\vec{\nabla}_i\vec{\nabla}_j\right)^a\frac{1}{a}\left(-\vec{\nabla}^2\delta_{jk}+\vec{\nabla}_j\vec{\nabla}_k\right)^b\frac{1}{b}\left(-\vec{\nabla}^2\delta_{ki}+\vec{\nabla}_k\vec{\nabla}_i\right)^c\frac{1}{c}$$
$$\times\int_0^\infty du\,\alpha(A;iu)\alpha(B;iu)\alpha(C;iu)e^{-(a+b+c)u}$$
$$=\frac{2}{3\pi}\frac{1}{(4\pi\varepsilon_0)^3}k_{m0}|\vec{\mu}^{m0}(A)|^2\left(-\vec{\nabla}^2\delta_{ij}+\vec{\nabla}_i\vec{\nabla}_j\right)^a\frac{1}{a}\left(-\vec{\nabla}^2\delta_{jk}+\vec{\nabla}_j\vec{\nabla}_k\right)^b\frac{1}{b}$$
$$\times\left(-\vec{\nabla}^2\delta_{ki}+\vec{\nabla}_k\vec{\nabla}_i\right)^c\frac{1}{c}\int_0^\infty du\frac{1}{k_{m0}^2+u^2}\alpha(B;iu)\alpha(C;iu)e^{-(a+b+c)u}. \qquad (6.8.8)$$

The total three-body dispersion potential between an excited molecule A and two ground-state species B and C is given by the sum of the last three expressions. Before examining the form of this energy shift in greater detail, it is shown how the result may be obtained using coupling of induced dipoles to their fields.

6.8.2 Coupling of Induced Dipoles

It was shown in Chapter 5 how the induced multipole moment method provided an alternative approach to the computation of dispersion pair potentials between ground- and excited-state molecules. This viewpoint is now used to calculate the dispersion interaction between three bodies, one of which, A, is excited. The physical picture is one in which A is viewed as giving rise to a dipole field, which induces dipole moments in each of B and C. These induced moments couple to the resonant dipole-dipole interaction tensor at the transition frequency of excited molecule A, resulting in an energy shift (Power and Thirunamachandran, 1995b).

As in the last subsection, let A initially be in excited electronic state $|m\rangle$, from which it makes an electric dipole allowed downward transition to the ground state with frequency ck_{m0}. In Section 2.6, the electric displacement field of such an oscillating dipole was calculated to be (equation (2.6.21))

$$\begin{aligned} d_i(\vec{\mu};\vec{r},t) &= \frac{1}{4\pi}\mu_j^{m0}(A)\left(-\vec{\nabla}^2\delta_{ij}+\vec{\nabla}_i\vec{\nabla}_j\right)\frac{e^{ik_{m0}(r-ct)}}{r} \\ &= \frac{1}{4\pi r^3}\mu_j^{m0}(A)\left[(\delta_{ij}-\hat{r}_i\hat{r}_j)k_{m0}^2r^2\right. \\ &\quad \left.+(\delta_{ij}-3\hat{r}_i\hat{r}_j)(ik_{m0}r-1)\right]e^{ik_{m0}(r-ct)}. \end{aligned} \tag{6.8.9}$$

The electric dipole moments induced in ground-state polarizable molecules B and C by this field are

$$\vec{\mu}^{\text{ind}}(B) = \varepsilon_0^{-1}\alpha(B;k_{m0})\vec{d}^{\perp}(\vec{\mu};\vec{R}_{BA},t) \tag{6.8.10a}$$

and

$$\vec{\mu}^{\text{ind}}(C) = \varepsilon_0^{-1}\alpha(C;k_{m0})\vec{d}^{\perp}(\vec{\mu};\vec{R}_{CA},t), \tag{6.8.10b}$$

where $\alpha(\xi;k)$ is the isotropic electric dipole polarizability of molecule ξ at frequency $\omega = ck$. The moments induced at B and C couple to each other via the dipole-dipole interaction tensor at the frequency of the downward transition in A, $V_{ij}(k_{m0}, \vec{R}_{BC})$ where from Chapter 4 and work relating to resonant transfer of energy between an excited and unexcited pair,

$$\begin{aligned}\mathrm{Re}V_{ij}(k,\vec{R}) &= -\frac{1}{4\pi\varepsilon_0}\mathrm{Re}\left(-\vec{\nabla}^2\delta_{ij}+\vec{\nabla}_i\vec{\nabla}_j\right)\frac{\mathrm{e}^{ikR}}{R} \\ &= \frac{1}{4\pi\varepsilon_0 R^3}\left[(\delta_{ij}-3\hat{R}_i\hat{R}_j)(\cos kR + kR\sin kR)\right. \\ &\quad \left.-(\delta_{ij}-3\hat{R}_i\hat{R}_j)k^2R^2\cos kR\right]. \end{aligned} \tag{6.8.11}$$

Thus, one part of the downward transition contribution to the three-body energy shift is computed from

$$\bar{\mu}_i^{\mathrm{ind}}(B)\mu_j^{\mathrm{ind}}(C)\mathrm{Re}V_{ij}(k_{m0},\vec{R}_{BC}) + \mathrm{c.c.} \tag{6.8.12}$$

Substituting (6.8.10) and (6.8.11) produces for (6.8.12) the expression

$$\begin{aligned}&-\frac{2}{(4\pi\varepsilon_0)^3}\mu_k^{0m}(A)\mu_l^{m0}(A)\alpha(B;k_{m0})\alpha(C;k_{m0}) \\ &\times \mathrm{Re}\left(\left[\left(-\vec{\nabla}^2\delta_{ik}+\vec{\nabla}_i\vec{\nabla}_k\right)^c\frac{\mathrm{e}^{-ik_{m0}c}}{c}\right]\left[\left(-\vec{\nabla}^2\delta_{jl}+\vec{\nabla}_j\vec{\nabla}_l\right)^b\frac{\mathrm{e}^{ik_{m0}b}}{b}\right]\right) \\ &\times \mathrm{Re}\left[\left(-\vec{\nabla}^2\delta_{ij}+\vec{\nabla}_i\vec{\nabla}_j\right)^a\frac{\mathrm{e}^{ik_{m0}a}}{a}\right]. \end{aligned} \tag{6.8.13}$$

For isotropic A, the contribution to the energy shift is

$$\begin{aligned}&-\frac{1}{3(4\pi\varepsilon_0)^3}|\vec{\mu}^{0m}(A)|^2\alpha(B;k_{m0})\alpha(C;k_{m0})\left(-\vec{\nabla}^2\delta_{ij}+\vec{\nabla}_i\vec{\nabla}_j\right)^a\frac{1}{a} \\ &\times\left(-\vec{\nabla}^2\delta_{jk}+\vec{\nabla}_j\vec{\nabla}_k\right)^b\frac{1}{b}\left(-\vec{\nabla}^2\delta_{ki}+\vec{\nabla}_k\vec{\nabla}_i\right)^c\frac{1}{c} \\ &\times\{\cos[k_{m0}(a+b-c)]+\cos[k_{m0}(a-b+c)]\}. \end{aligned} \tag{6.8.14}$$

The other factor contributing to the energy shift when A is excited has the structure of a u-integral, as occurs in the ground-state interaction between three molecules. For upward transitions in A, it has the same form and sign as the triple-dipole dispersion potential (6.4.12), but containing excited state polarizability of A. For downward transitions in A, the u-integral is of opposite sign to that found for upward transitions when A is excited. This last contribution is given by

$$\frac{1}{(4\pi\varepsilon_0)^3\pi\hbar^2c^2}\left(\frac{2}{3}\right)^3\sum_{q,r}k_{m0}|\vec{\mu}^{m0}(A)|^2k_{q0}|\vec{\mu}^{0q}(B)|^2k_{r0}|\vec{\mu}^{0r}(C)|^2$$
$$\times\left(-\vec{\nabla}^2\delta_{ij}+\vec{\nabla}_i\vec{\nabla}_j\right)^a\frac{1}{a}\left(-\vec{\nabla}^2\delta_{jk}+\vec{\nabla}_j\vec{\nabla}_k\right)^b\frac{1}{b}$$
$$\times\left(-\vec{\nabla}^2\delta_{ki}+\vec{\nabla}_k\vec{\nabla}_i\right)^c\frac{1}{c}\int_0^\infty\frac{e^{-u(a+b+c)}}{(k_{m0}^2+u^2)(k_{q0}^2+u^2)(k_{r0}^2+u^2)}du, \tag{6.8.15}$$

for transitions to the ground state in A, recalling that transitions in B and C are upward from the ground state to $|q\rangle$ and $|r\rangle$, respectively. Hence, the total energy shift is given by the sum of (6.8.14) and (6.8.15),

$$\Delta E=-\frac{1}{3(4\pi\varepsilon_0)^3}|\vec{\mu}^{0m}(A)|^2\alpha(B;k_{m0})\alpha(C;k_{m0})\left(-\vec{\nabla}^2\delta_{ij}+\vec{\nabla}_i\vec{\nabla}_j\right)^a\frac{1}{a}$$
$$\times\left(-\vec{\nabla}^2\delta_{jk}+\vec{\nabla}_j\vec{\nabla}_k\right)^b\frac{1}{b}\left(-\vec{\nabla}^2\delta_{ki}+\vec{\nabla}_k\vec{\nabla}_i\right)^c\frac{1}{c}$$
$$\times\{\cos[k_{m0}(a+b-c)]+\cos[k_{m0}(a-b+c)]\}$$
$$+\frac{1}{\pi(4\pi\varepsilon_0)^3}\left(\frac{2}{3}\right)k_{m0}|\vec{\mu}^{m0}(A)|^2\left(-\vec{\nabla}^2\delta_{ij}+\vec{\nabla}_i\vec{\nabla}_j\right)^a\frac{1}{a}$$
$$\times\left(-\vec{\nabla}^2\delta_{jk}+\vec{\nabla}_j\vec{\nabla}_k\right)^b\frac{1}{b}\left(-\vec{\nabla}^2\delta_{ki}+\vec{\nabla}_k\vec{\nabla}_i\right)^c\frac{1}{c}$$
$$\times\int_0^\infty\frac{e^{-u(a+b+c)}}{(k_{m0}^2+u^2)}\alpha(B;iu)\alpha(C;iu)du. \tag{6.8.16}$$

The result (6.8.16) is easily extended to the case where A may make both upward and downward transitions from initial excited-state $|m\rangle$ to state $|p\rangle$.

Note that only downward transitions in A contribute to (6.8.14) with $|0\rangle$ replaced by $|p\rangle$. In contrast, all transitions contribute in the u-integral term, the sign depending on whether $k_{mp} > 0$ or $k_{mp} < 0$ Hence, for multilevel A, the dispersion potential is

$$\begin{aligned}\Delta E = &-\frac{1}{3(4\pi\varepsilon_0)^3}\sum_{\substack{p\\E_m>E_p}}|\vec{\mu}^{pm}(A)|^2\alpha(B;k_{pm})\alpha(C;k_{pm})\left(-\vec{\nabla}^2\delta_{ij}+\vec{\nabla}_i\vec{\nabla}_j\right)^a\frac{1}{a}\\&\times\left(-\vec{\nabla}^2\delta_{jk}+\vec{\nabla}_j\vec{\nabla}_k\right)^b\frac{1}{b}\left(-\vec{\nabla}^2\delta_{ki}+\vec{\nabla}_k\vec{\nabla}_i\right)^c\frac{1}{c}\\&\times\{\cos\left[k_{pm}(a+b-c)\right]+\cos\left[k_{pm}(a-b+c)\right]\}\\&+\frac{\hbar c}{\pi(4\pi\varepsilon_0)^3}\left(-\vec{\nabla}^2\delta_{ij}+\vec{\nabla}_i\vec{\nabla}_j\right)^a\frac{1}{a}\left(-\vec{\nabla}^2\delta_{jk}+\vec{\nabla}_j\vec{\nabla}_k\right)^b\frac{1}{b}\\&\times\left(-\vec{\nabla}^2\delta_{ki}+\vec{\nabla}_k\vec{\nabla}_i\right)^c\frac{1}{c}\int_0^\infty\alpha(A;iu)\alpha(B;iu)\alpha(C;iu)\mathrm{e}^{-u(a+b+c)}\mathrm{d}u,\end{aligned} \tag{6.8.17}$$

where excited-state polarizability of A and ground-state polarizabilities of B and C appear, which is identical to the result obtained via diagrammatic perturbation theory at the end of the previous subsection.

From the result (6.8.17) applicable to all separation distances, the asymptotically limiting forms are obtained after the usual approximations are made for the near and far zones, namely, that characteristic transition wavelengths are greater than or less than interparticle separations distances, respectively. To simplify the structure of the limiting energy shifts, a two-level model is adopted for the three molecules, with the downward transition in A having wavevector k_A and upward transitions in B and C, k_B and k_C.

In the near zone, both real and virtual photon contributions must be included in the limiting potential. The second term of (6.8.17) gives rise to an Axilrod–Teller-type contribution

$$\begin{aligned}&-\frac{4}{9(4\pi\varepsilon_0)^3}\frac{1}{(\hbar c)^2}|\vec{\mu}(A)|^2|\vec{\mu}(B)|^2|\vec{\mu}(C)|^2\\&\times\frac{(k_A+k_B+k_C)}{(k_B+k_C)(k_C+k_A)(k_A+k_B)}\frac{[1-3(\hat{b}\cdot\hat{c})(\hat{c}\cdot\hat{a})(\hat{a}\cdot\hat{b})]}{a^3b^3c^3},\end{aligned} \tag{6.8.18}$$

while the first term of (6.8.17) produces

$$\frac{8}{9(4\pi\varepsilon_0)^3}\frac{1}{(\hbar c)^2}|\vec{\mu}(A)|^2|\vec{\mu}(B)|^2|\vec{\mu}(C)|^2$$
$$\times\frac{(k_B k_C)}{(k_B^2-k_A^2)(k_C^2-k_A^2)}\frac{[1-3(\hat{b}\cdot\hat{c})(\hat{c}\cdot\hat{a})(\hat{a}\cdot\hat{b})]}{a^3b^3c^3}. \qquad (6.8.19)$$

The addition of these two terms yields, for the near-zone energy shift, the limiting form

$$\Delta E_{\mathrm{NZ}}=\frac{4}{9(4\pi\varepsilon_0)^3(\hbar c)^2}|\vec{\mu}(A)|^2|\vec{\mu}(B)|^2|\vec{\mu}(C)|^2$$
$$\times\frac{(k_B+k_C-k_A)}{(k_B+k_C)(k_C-k_A)(k_B-k_A)}\frac{[1-3(\hat{b}\cdot\hat{c})(\hat{c}\cdot\hat{a})(\hat{a}\cdot\hat{b})]}{a^3b^3c^3}. \qquad (6.8.20)$$

It is interesting to note that (6.8.20) is the result obtained using third-order perturbation theory and static dipolar coupling potentials with molecule A excited. Moreover, result (6.8.20) is obtained by changing the sign of k_A in (6.8.18) noting that the overall sign is due to the fact that the transition being considered in A is downward.

At large separations, the first term of (6.8.17), due to downward transitions, dominates the energy shift, giving

$$\Delta E_{\mathrm{FZ}}=-\frac{4}{27(4\pi\varepsilon_0)^3(\hbar c)^2}|\vec{\mu}(A)|^2|\vec{\mu}(B)|^2|\vec{\mu}(C)|^2\frac{k_A^6 k_B k_C}{(k_B^2-k_A^2)(k_C^2-k_A^2)}$$
$$\times[\cos(k_A[a+b-c])+\cos(k_A[a-b+c])]\frac{[1+(\hat{b}\cdot\hat{c})(\hat{c}\cdot\hat{a})(\hat{a}\cdot\hat{b})]}{abc}. \qquad (6.8.21)$$

When $a=b=c=R$, corresponding to an equilateral triangle, the far-zone limit (6.8.21) reduces to

$$\Delta E_{\mathrm{FZ}}=-\frac{7}{27(4\pi\varepsilon_0)^3(\hbar c)^2R^3}|\vec{\mu}(A)|^2|\vec{\mu}(B)|^2|\vec{\mu}(C)|^2$$
$$\times\frac{k_A^6 k_B k_C}{(k_B^2-k_A^2)(k_C^2-k_A^2)}\cos(k_A R), \qquad (6.8.22)$$

exhibiting a modulated inverse cubic dependence on R.

6.9 MEDIATION OF RESONANCE ENERGY TRANSFER BY A THIRD BODY

While the retarded dispersion energy shift formula for two-, three-, and so on, *N* bodies has many similar features, for example, in each case, it can be written as an integral over an imaginary frequency *icu* of the product of the complex dynamic polarizabilities of each molecule and a geometric factor involving distinct pair separation distances, the ultimate dependence on interatomic displacements for the entire range of separations, or at the near- and far-zone asymptotic limits, varies with the number of interacting species. A similar situation applies to the resonant migration of energy, which was examined in Chapter 4 by considering the exchange of excitation between a pair of molecules. Even though pair-transfer rates may now be measured using recent advances in single- and few-body spectroscopy, transfer of energy occurs more commonly in a medium in which numerous other particles are present in addition to donor and acceptor species. In a solution, for example, the medium is comprised of solvent particles. If transfer is taking place in the gaseous phase, however, other identical systems may be present in very low concentration, but undergoing transitions nonresonant with the frequency of radiation exchanged between the donor-acceptor pair. In this section, the mediation of the resonant transfer of energy between two molecules due to the presence of a third molecule is studied (Craig and Thirunamachandran, 1989). This corresponds to the situation in which the third body is a constituent of a medium of low density and provides the leading correction to the modification of the pair-transfer rate due to the effect of many other molecules, the latter more commonly treated as a medium of uniform dielectric constant in the macroscopic limit.

Let *A* and *B* be two identical molecules positioned at $\vec{R}_A$ and $\vec{R}_B$, between which energy is transferred resonantly. Let *C*, located at $\vec{R}_C$ be a polarizable molecule that mediates the exchange of energy between *A* and *B*. The total quantum electrodynamical Hamiltonian for the system is written as

$$H = H_{\text{mol}}(A) + H_{\text{mol}}(B) + H_{\text{mol}}(C) + H_{\text{rad}} + H_{\text{int}}, \tag{6.9.1}$$

where, in the electric dipole approximation, the interaction Hamiltonian coupling radiation and matter is

$$H_{\text{int}} = -\varepsilon_0^{-1}\vec{\mu}(A)\cdot\vec{d}^{\perp}(\vec{R}_A) - \varepsilon_0^{-1}\vec{\mu}(B)\cdot\vec{d}^{\perp}(\vec{R}_B) - \varepsilon_0^{-1}\vec{\mu}(C)\cdot\vec{d}^{\perp}(\vec{R}_C). \tag{6.9.2}$$

For the problem at hand, the initial and final states of the system are specified as

$$|i\rangle = |E_n^A, E_0^B, E_0^C; 0(\vec{p}, \varepsilon)\rangle \tag{6.9.3a}$$

and

$$|f\rangle = |E_0^A, E_n^B, E_0^C; 0(\vec{p}, \varepsilon)\rangle, \tag{6.9.3b}$$

corresponding to an initial state in which species A is pre-excited to electronic state $|n^A\rangle$ of energy E_n^A, B is in the ground electronic state $|0^B\rangle$ with energy E_0^B. After transfer of energy resonantly, A returns to the ground state while B becomes excited to electronic state $|n^B\rangle$. Species C remains in the electronic ground state throughout and there is no change in the state of the radiation field, there being no photons present before and after interaction. Because transfer between A and B is mediated by C, it is appropriate to refer to this mechanism as an indirect one and insert the superscript "*in*" on the matrix element. Time-dependent perturbation theory may be used to evaluate the matrix element. Twenty-four time-ordered diagrams in which the virtual photon is exchanged between A and C and between B and C are found to contribute. Their sum gives rise to

$$\begin{aligned}M_{fi}^{\text{in}} = &-\frac{1}{4\varepsilon_0^2 V^2}\sum_{\vec{p},\varepsilon}\sum_{\vec{p}',\varepsilon'}\mu_i^{0n}(A)\mu_j^{n0}(B)\alpha_{kl}(C;k)e_i^{(\varepsilon)}(\vec{p})e_l^{(\varepsilon)}(\vec{p})\\ &\times e_k^{(\varepsilon')}(\vec{p}')e_j^{(\varepsilon')}(\vec{p}')pp'\\ &\times\left[\frac{\mathrm{e}^{-i\vec{p}\cdot\vec{R}}e^{i\vec{p}'\cdot\vec{R}'}}{(p-k)(p'-k)}+\frac{\mathrm{e}^{-i\vec{p}\cdot\vec{R}}\mathrm{e}^{-i\vec{p}'\cdot\vec{R}'}}{(p-k)(p'+k)}+\frac{\mathrm{e}^{i\vec{p}\cdot\vec{R}}\mathrm{e}^{i\vec{p}'\cdot\vec{R}'}}{(p+k)(p'-k)}+\frac{\mathrm{e}^{i\vec{p}\cdot\vec{R}}\mathrm{e}^{-i\vec{p}'\cdot\vec{R}'}}{(p+k)(p'+k)}\right],\end{aligned} \tag{6.9.4}$$

where the virtual photons are of modes $(\vec{p}, \varepsilon)$ and $(\vec{p}', \varepsilon')$, and $\alpha_{kl}(C;k)$ is the dynamic electric dipole polarizability tensor of molecule C. Species C is taken to be situated at the origin, with relative separation distance vectors with respect to A and B, $\vec{R} = \vec{R}_C - \vec{R}_A$, and $\vec{R}' = \vec{R}_C - \vec{R}_B$; ck is the resonant frequency $E_{n0}/\hbar$, and C undergoes virtual transitions to state $|r\rangle$ with energy $E_{r0} = \hbar c k_{r0}$. Performing the polarization and wavevector sums and carrying out the angular integrations produces

$$M_{fi}^{\rm in} = -\frac{1}{16\pi^4\varepsilon_0^2}\mu_i^{0n}(A)\mu_j^{n0}(B)\alpha_{kl}(C;k)\left(-\vec{\nabla}^2\delta_{il}+\vec{\nabla}_i\vec{\nabla}_l\right)^R$$
$$\times\left(-\vec{\nabla}^2\delta_{jk}+\vec{\nabla}_j\vec{\nabla}_k\right)^{R'}\frac{1}{RR'}$$
$$\times\int_0^\infty\int_0^\infty {\rm d}p{\rm d}p'\sin pR\,\sin p'R'\left(\frac{1}{(p-k)}+\frac{1}{(p+k)}\right)\left(\frac{1}{(p'-k)}+\frac{1}{(p'+k)}\right). \tag{6.9.5}$$

The integrations over p and p' are independent. They are identical to that occurring in the evaluation of the matrix element for resonant transfer of excitation between two molecules. Thus, (6.9.5) becomes

$$M_{fi}^{\rm in} = -\mu_i^{0n}(A)\mu_j^{n0}(B)\alpha_{kl}(C;k)V_{il}(k,\vec{R})V_{jk}(k,\vec{R}'), \tag{6.9.6}$$

where

$$V_{ij}(k,\vec{R}) = -\frac{1}{4\pi\varepsilon_0}\left(-\vec{\nabla}^2\delta_{ij}+\vec{\nabla}_i\vec{\nabla}_j\right)^R\frac{{\rm e}^{ikR}}{R}$$
$$= \frac{1}{4\pi\varepsilon_0 R^3}\left[(\delta_{ij}-3\hat{R}_i\hat{R}_j)(1-ikR)-(\delta_{ij}-\hat{R}_i\hat{R}_j)k^2R^2\right]{\rm e}^{ikR}. \tag{6.9.7}$$

Substituting (6.9.7) into (6.9.6) yields for the matrix element

$$M_{fi}^{\rm in} = -\frac{1}{(4\pi\varepsilon_0)^2R^3R'^3}\mu_i^{0n}(A)\mu_j^{n0}(B)\alpha_{kl}(C;k){\rm e}^{ik(R+R')}$$
$$\times[(\delta_{il}-3\hat{R}_i\hat{R}_l)(1-ikR)-(\delta_{il}-\hat{R}_i\hat{R}_l)k^2R^2]$$
$$\times[(\delta_{jk}-3\hat{R}'_j\hat{R}'_k)(1-ikR')-(\delta_{jk}-\hat{R}'_j\hat{R}'_k)k^2R'^2]. \tag{6.9.8}$$

The near-zone form of the matrix element (6.9.8) is easily obtained on noting that at this asymptotic limit kR and kR' are both significantly less than unity giving

$$M_{fi}^{\rm in}({\rm NZ}) = -\frac{1}{(4\pi\varepsilon_0)^2R^3R'^3}\mu_i^{0n}(A)\mu_j^{n0}(B)\alpha_{kl}(C;k)$$
$$\times(\delta_{il}-3\hat{R}_i\hat{R}_l)(\delta_{jk}-3\hat{R}'_j\hat{R}'_k), \tag{6.9.9}$$

and corresponds to static dipolar coupling between A and C and between B and C. It is applicable to the situation in which species C is close to both A and B.

Recalling from (4.2.18) that the near-zone matrix element for resonant transfer of energy between A and B, commonly termed the direct mechanism, is

$$M_{fi}^{\text{dir}}(\text{NZ}) = \frac{1}{4\pi\varepsilon_0|\vec{\rho}|^3}\mu_i^{0n}(A)\mu_j^{n0}(B)(\delta_{ij}-3\hat{\rho}_i\hat{\rho}_j), \tag{6.9.10}$$

where $\vec{\rho}$ is the A–B separation, $\vec{\rho} = \vec{R}_B-\vec{R}_A = \vec{R}'-\vec{R}$. The total matrix element in the near zone is therefore the sum of the direct and indirect mechanism matrix elements (6.9.10) and (6.9.9), respectively. The total rate in the near zone may be evaluated using the Fermi golden rule (1.9.33),

$$\Gamma^{\text{Tot}}(\text{NZ}) = \frac{2\pi\rho_f}{\hbar}|M_{fi}^{\text{dir}}(\text{NZ})+M_{fi}^{\text{in}}(\text{NZ})|^2 = \Gamma^{\text{dir}}(\text{NZ})+\Gamma^{\text{int}}(\text{NZ})+\Gamma^{\text{in}}(\text{NZ}), \tag{6.9.11}$$

and is a sum direct, interference, and indirect near-zone transfer rates with ρ_f the density of final states. The direct contribution is obtained straightforwardly from expression (6.9.10) and corresponds to the near-zone limit of the two-body transfer rate (4.2.21). For isotropic A and B, it is given by

$$\Gamma^{\text{dir}}(\text{NZ}) = \frac{2\pi\rho_f}{\hbar}|M_{fi}^{\text{dir}}(\text{NZ})|^2 = \frac{\rho_f}{12\pi\hbar\varepsilon_0^2\rho^6}|\vec{\mu}^{0n}(A)|^2|\vec{\mu}^{n0}(B)|^2. \tag{6.9.12}$$

Near-zone matrix elements (6.9.9) and (6.9.10) are used to calculate the interference contribution to the rate,

$$\Gamma^{\text{int}}(\text{NZ}) = \frac{4\pi\rho_f}{\hbar}\text{Re}(M_{fi}^{\text{dir}}(\text{NZ})\bar{M}_{fi}^{\text{in}}(\text{NZ})). \tag{6.9.13}$$

Thus,

$$M_{fi}^{\text{dir}}(\text{NZ})\bar{M}_{fi}^{\text{in}}(\text{NZ}) = -\frac{1}{(4\pi\varepsilon_0)^3R^3R'^3\rho^3}\mu_i^{0n}(A)\bar{\mu}_{i'}^{n0}(A)\mu_j^{0n}(B)\bar{\mu}_{j'}^{n0}(B)$$
$$\times\alpha_{kl}(C;k)(\delta_{ij}-3\hat{\rho}_i\hat{\rho}_j)(\delta_{i'l}-3\hat{R}_{i'}\hat{R}_l)(\delta_{j'k}-3\hat{R}'_{j'}\hat{R}'_k). \tag{6.9.14}$$

Carrying out a rotational average of the molecular factors using the result

$$\langle \mu_i^{0n}(A)\bar{\mu}_{i'}^{n0}(A)\mu_j^{0n}(B)\bar{\mu}_{j'}^{n0}(B)\alpha_{kl}(C;k)\rangle$$
$$= \frac{1}{9}|\vec{\mu}^{0n}(A)|^2|\vec{\mu}^{0n}(B)|^2\alpha(C;k)\delta_{ii'}\delta_{jj'}\delta_{kl}, \tag{6.9.15}$$

where a factor of 1/3 has been included in the definition of the isotropic polarizability of C. Contracting the tensors produces for (6.9.13) the expression

$$\Gamma^{\text{int}}(\text{NZ}) = \frac{4\pi\rho_f}{3\hbar\rho^3 R^3 R'^3}\frac{1}{(4\pi\varepsilon_0)^3}|\vec{\mu}^{0n}(A)|^2|\vec{\mu}^{0n}(B)|^2\alpha(C;k)$$
$$\times \{2-3[(\hat{R}\cdot\hat{R}')^2+(\hat{\rho}\cdot\hat{R})^2+(\hat{\rho}\cdot\hat{R}')^2]+9(\hat{R}\cdot\hat{R}')(\hat{\rho}\cdot\hat{R})(\hat{\rho}\cdot\hat{R}')\}. \tag{6.9.16}$$

Finally, the contribution to the total transfer rate from the indirect term is obtained from (6.9.9). Thus,

$$|M_{fi}^{\text{in}}(\text{NZ})|^2 = \frac{1}{(4\pi\varepsilon_0)^4 R^6 R'^6}\mu_i^{0n}(A)\bar{\mu}_{i'}^{n0}(A)\mu_j^{0n}(B)\bar{\mu}_{j'}^{n0}(B)$$
$$\times \alpha_{kl}(C;k)\bar{\alpha}_{k'l'}(C;k)(\delta_{il}-3\hat{R}_i\hat{R}_l)(\delta_{i'l'}-3\hat{R}_{i'}\hat{R}_{l'})$$
$$\times(\delta_{jk}-3\hat{R}'_j\hat{R}'_k)(\delta_{j'k'}-3\hat{R}'_{j'}\hat{R}'_{k'}), \tag{6.9.17}$$

which applies for the three molecules in fixed relative orientation to each other. The rotational averages for A and B are straightforward and are obtained via $\langle \mu_i^{0n}(\xi)\bar{\mu}_{i'}^{n0}(\xi)\rangle = (1/3)|\vec{\mu}^{0n}(\xi)|^2\delta_{ii'}$. The product of polarizability of C, however, requires fourth-rank Cartesian tensor averaging and is evaluated using result (B.7) of Appendix B. Hence,

$$\langle \alpha_{kl}(C;k)\bar{\alpha}_{k'l'}(C;k)\rangle = \frac{1}{30}\begin{pmatrix}\delta_{kl}\delta_{k'l'}\\ \delta_{kk'}\delta_{ll'}\\ \delta_{kl'}\delta_{k'l}\end{pmatrix}^{\text{T}}\begin{pmatrix}4 & -1 & -1\\ -1 & 4 & -1\\ -1 & -1 & 4\end{pmatrix}$$
$$\times\begin{pmatrix}\delta_{\lambda\mu}\delta_{\nu\pi}\\ \delta_{\lambda\nu}\delta_{\mu\pi}\\ \delta_{\lambda\pi}\delta_{\mu\nu}\end{pmatrix}\alpha_{\lambda\mu}(C;k)\bar{\alpha}_{\nu\pi}(C;k), \tag{6.9.18}$$

where Greek subscripts denote tensor components in the molecule-fixed frame and T designates the transpose. Evaluating the matrix product, on making use of the symmetry properties of the electric dipole polarizability tensor, produces

$$\begin{aligned}\langle \alpha_{kl}(C;k)\bar{\alpha}_{k'l'}(C;k)\rangle = \frac{1}{30}\{&\delta_{kl}\delta_{k'l'}\left[4\alpha_{\lambda\lambda}(C;k)\bar{\alpha}_{\mu\mu}(C;k)\right. \\ &-2\alpha_{\lambda\mu}(C;k)\bar{\alpha}_{\lambda\mu}(C;k)] \\ &+(\delta_{kk'}\delta_{ll'}+\delta_{kl'}\delta_{k'l})[-\alpha_{\lambda\lambda}(C;k)\bar{\alpha}_{\mu\mu}(C;k) \\ &+3\alpha_{\lambda\mu}(C;k)\bar{\alpha}_{\lambda\mu}(C;k)]\}. \end{aligned} \tag{6.9.19}$$

Substituting (6.9.19) into (6.9.17), performing the averages over species A and B and contracting yields, for the third term of (6.9.11), the near-zone rate

$$\begin{aligned}\Gamma^{\text{in}}(\text{NZ}) = \frac{2\pi\rho_f}{\hbar}|M_{fi}^{\text{in}}(\text{NZ})|^2 &= \frac{\pi\rho_f}{15\hbar R^6 R'^6}\frac{1}{(4\pi\varepsilon_0)^4}|\vec{\mu}^{0n}(A)|^2|\vec{\mu}^{0n}(B)|^2 \\ &\times\{[-1+3(\hat{R}\cdot\hat{R}')^2]\alpha_{\lambda\lambda}(C;k)\bar{\alpha}_{\mu\mu}(C;k) \\ &+[13+(\hat{R}\cdot\hat{R}')^2]\alpha_{\lambda\mu}(C;k)\bar{\alpha}_{\lambda\mu}(C;k)\}. \end{aligned} \tag{6.9.20}$$

When a large number of molecules of type C is present, the indirect mechanism with rate given by the last expression dominates the total rate, with the latter given by the sum of (6.9.12), (6.9.16), and (6.9.20). At the other extreme, when no C is present, the Förster rate limit corresponding to direct transfer between A and B dominates the overall rate in the near zone.

From the treatment given above, useful insight may be gained into the effect of one or more additional particles on the pair-transfer rate. From the microscopic point of view, all distinct couplings with the molecules of medium C are accounted for, with the result containing the permittivity of the vacuum, ε_0. In the near zone, the direct interaction is equivalent to static coupling of permanent electric dipole moments. When the pair is in an isotropic medium, expression (6.9.10) still applies, but now the vacuum permittivity is replaced by the permittivity of the medium, ε. This is the common picture on the macroscopic scale, in which only direct coupling is present along with the permittivity of the medium, which is frequency dependent. The effect of the medium, therefore, is to relay energy between A and B via transfer mediated by one, two, three, and N molecules of type C, via the frequency dependent polarizability of species C. In this section, the leading medium correction term has been evaluated by accounting explicitly for the effect of one molecule C of the medium. Clearly, as the number of C molecules rises, the contribution of the indirect mechanism increases, with the effects of higher order terms becoming more important, while the direct A–B contribution becomes less significant.

CHAPTER 7

INTERMOLECULAR INTERACTIONS IN A RADIATION FIELD

Tom's photons are not the same as Dick's photons—and as for Harry's,. . .!.
—E. A. Power, The natural line shape, in *Physics and Probability: Essays in Honour of Edwin T. Jaynes*, W. T. Grandy Jr. and P. W. Milonni (Eds), Cambridge University Press, Cambridge, 1993, p. 101.

7.1 INTRODUCTION

In Chapter 1, the classical and quantum electrodynamical theory of the interaction of a nonrelativistic charged particle with a radiation field was formulated while in Chapter 2, a completely field theoretic viewpoint was adopted and the techniques of second quantization were employed in the development. In both cases, the total Hamiltonian operator for the coupled radiation–matter system was obtained in the minimal- and multipolar-coupling schemes. It was shown in Chapters 4 and 5 how molecular quantum electrodynamics could be successfully applied to calculate and to understand the physical origin of, two fundamental intermolecular processes—the rate of resonant energy transfer and the van der Waals dispersion energy shift, respectively. It should not be forgotten, however, that the theoretical foundations detailed in the first two chapters allow processes involving the interaction of one or more photons with electrons associated with a single atomic or molecular center to be studied rigorously, forming a large part of the field known as theoretical spectroscopy. In this and allied areas, transition rates have been computed for a variety of single- and multiphoton absorption and emission processes, and cross sections calculated for a number of elastic and inelastic scattering phenomena (Mukamel, 1995; Craig and Thirunamachandran, 1998a; Andrews and Allcock, 2002). Each particular application is begun by writing down the total Hamiltonian for the system comprising the Hamiltonian for the single species, the Hamiltonian for the radiation field, and the operator coupling the two and solved for specific quantum mechanical observable quantities.

With continuing advances being made in the generation of coherent and incoherent sources of laser light, novel and esoteric experiments are being performed that confirm theoretical predictions or require theoretical interpretation and explanation. This is witnessed by the emergence of single molecule spectroscopy, development in nonlinear and quantum optics, and progress in ultracold spectroscopy. It is now possible to not only trap small particles using optomechanical forces but also control and manipulate them. In general, this relies on a particle of matter undergoing radiative attraction toward the high-intensity focal area of a laser beam. When two or more particles are present, however, the modification by light of intermolecular forces has to be reckoned with. Such phenomena are the subject of the present chapter. Firstly, time-dependent perturbation theory is employed to evaluate the change in energy shift when a pair of interacting molecules is in the presence of an intense beam of laser light. If one or both entities are polar, two terms are found to contribute to ΔE. One term is proportional to the polarizability of each body. A second depends on the product of the permanent dipole moment of one body and the molecular first hyperpolarizability of the other. It is then demonstrated how the induced multipole moment method leads straightforwardly to an expression for the radiation modified pair interaction energy. Both calculational techniques are then utilized in the computation of radiation-induced chiral discrimination. For consistency, magnetic dipole and electric quadrupole coupling terms are accounted for. The final section is devoted to higher order radiation-induced chiral discrimination in which the pair of molecules is coupled via two virtual photon exchange.

7.2 RADIATION-INDUCED DISPERSION FORCE: PERTURBATION THEORY

It is well known that the application of a constant or time-varying radiation field causes a shift in the energy levels of atomic or molecular systems. In the case in which the external field is electric, the familiar static or dynamic Stark shift ensues, while if the incident field is magnetic, the Zeeman effect results. For a free molecule subject to an oscillating electric field, the change in energy levels is easily calculated using second-order perturbation theory together with diagrams of the form shown in Fig. 7.1, which illustrate the two possible time orderings associated with scattering of a real photon of mode $(\vec{k}, \lambda)$ by a molecule in state $|E_s\rangle$. Making use of the interaction Hamiltonian in electric dipole approximation,

$$H_{\text{int}} = -\varepsilon_0^{-1}\vec{\mu} \cdot \vec{d}^{\perp}, \tag{7.2.1}$$

the energy shift for a nonpolar molecule whose energy levels are non-degenerate, is

$$\Delta E = -\frac{I}{2\varepsilon_0 c} e_i^{(\lambda)}(\vec{k})\bar{e}_j^{(\lambda)}(\vec{k})\alpha_{ij}^{ss}(\omega, -\omega), \tag{7.2.2}$$

where the dynamic polarizability is given by

$$\alpha_{ij}^{ss}(\pm\omega, \pm\omega) = \sum_r \left\{ \frac{\mu_i^{sr}\mu_j^{rs}}{E_{rs} \mp \hbar\omega} + \frac{\mu_j^{sr}\mu_i^{rs}}{E_{rs} \mp \hbar\omega} \right\}, \tag{7.2.3}$$

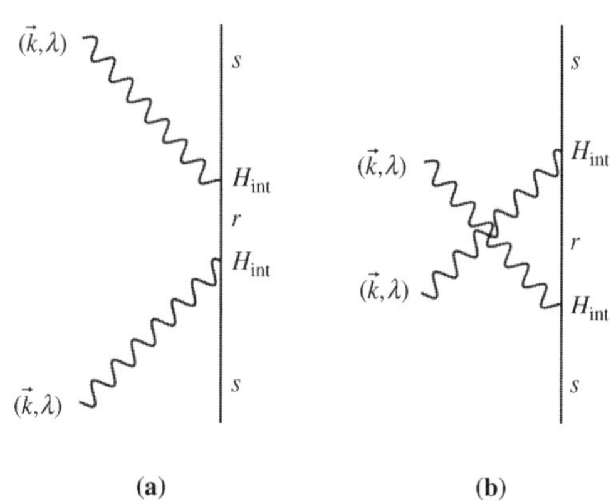

FIGURE 7.1 Time-ordered graphs illustrating dynamic Stark shift.

and the irradiance $I = N\hbar c^2 k/V$, where N specifies the number of photons in the radiation field.

Similarly, an applied field causes a change in the mutual energy of interaction between a pair of atoms or molecules (Thirunamachandran, 1980). While this interaction energy vanishes for two coupled molecules that are randomly oriented with respect to each other in the presence of an applied static electric field, it remains during the action of an oscillating field even when the pair is randomly oriented.

Consider two molecules A and B situated at $\vec{R}_A$ and $\vec{R}_B$, respectively, with internuclear separation distance $R = |\vec{R}_B - \vec{R}_A|$. Let both molecules be in ground electronic states $|E_0^A, E_0^B\rangle$ initially and finally, with the radiation field represented by a state $|N(\vec{k}, \lambda)\rangle$, corresponding to N photons of mode $(\vec{k}, \lambda)$. To leading order, the change in energy shift is given by the dynamic Stark shift (7.2.2). Since this is independent of R, it is excluded. The first contributing term to the radiation-induced intermolecular energy shift is of fourth order in perturbation theory. It corresponds to the scattering of a real photon by the molecular pair, which in turn is coupled by single virtual photon exchange.

The fourth-order contribution itself is composed of two types of terms depending on whether the real photon is scattered by the same or different centers. As these two terms have different physical origins, it is convenient to consider them separately. This is done in the following two sections. In both cases, the total Hamiltonian for the system is given by

$$H = H_{\text{mol}}(A) + H_{\text{mol}}(B) + H_{\text{rad}} + H_{\text{int}}(A) + H_{\text{int}}(B), \tag{7.2.4}$$

comprising a sum of molecular Hamiltonians for each entity, the radiation field Hamiltonian, and the electric dipole approximated form for the interaction between matter and electromagnetic field at each center,

$$H_{\text{int}}(A) + H_{\text{int}}(B) = -\varepsilon_0^{-1}\vec{\mu}(A) \cdot \vec{d}^{\perp}(\vec{R}_A) - \varepsilon_0^{-1}\vec{\mu}(B) \cdot \vec{d}^{\perp}(\vec{R}_B). \tag{7.2.5}$$

Since there is no overall change in the state of the radiation field and with both molecules remaining in the ground electronic state, identical initial and final states are used to represent the system

$$|i\rangle = |f\rangle = |E_0^A, E_0^B; N(\vec{k}, \lambda)\rangle. \tag{7.2.6}$$

Fourth-order perturbation theory for the energy shift, given by formula (5.2.4), may then be employed together with time-ordered or state sequence diagrams to calculate the change in mutual interaction energy between a pair of molecules subject to the action of an intense radiation field.

7.3 DYNAMIC MECHANISM

When the real photon of mode $(\vec{k}, \lambda)$ from the incident beam is absorbed at A and emitted at B or absorbed at B and emitted at A and a single virtual photon of mode $(\vec{p}, \varepsilon)$ propagates between the pair, the contribution to ΔE for reasons that will become apparent is commonly termed the "dynamic" mechanism. It is described completely by 48 time-ordered diagrams, which may be grouped into 4 sets of 12 graphs. One set of 12 time orderings in which absorption of a real photon occurs at A, emission at B, with the virtual photon traversing from A to B is illustrated in Fig. 7.2. From each of the 4 sets of 12 diagrams, 1 representative graph is shown in Fig. 7.3.

Concentrating for the moment on the first graph, labeled (i) from the set of 12 graphs classified as (a) and depicted in Fig. 7.2, using coupling Hamiltonian (7.2.5) and the mode expansion for the electric displacement field (1.7.17) with the initial and final states given by (7.2.6), determining the intermediate states and energy denominators from the time-ordered diagram, and substituting into the expression for ΔE from fourth-order perturbation theory (5.2.4) the contribution to the energy shift from this graph is

$$
\begin{aligned}
&-\sum_{\vec{p},\varepsilon}\sum_{r,s}\left(\frac{N\hbar ck}{2\varepsilon_0 V}\right)\left(\frac{\hbar cp}{2\varepsilon_0 V}\right)\bar{e}_i^{(\lambda)}(\vec{k})e_j^{(\lambda)}(\vec{k})e_k^{(\varepsilon)}(\vec{p})\bar{e}_l^{(\varepsilon)}(\vec{p})\\
&\times\mu_j^{0r}(A)\mu_l^{r0}(A)\mu_i^{0s}(B)\mu_k^{s0}(B)\\
&\times \mathrm{e}^{-i\vec{k}\cdot\vec{R}}\mathrm{e}^{i\vec{p}\cdot\vec{R}}[(E_{r0}+\hbar cp)(E_{r0}+E_{s0})(E_{s0}-\hbar ck)]^{-1},
\end{aligned}
\tag{7.3.1}
$$

where r and s denote the intermediate electronic states of A and B, respectively. The remaining 11 graphs may be evaluated similarly and added to (7.3.1) to yield

$$
\begin{aligned}
&-\sum_{\vec{p},\varepsilon}\sum_{r,s}\left(\frac{N\hbar ck}{2\varepsilon_0 V}\right)\left(\frac{\hbar cp}{2\varepsilon_0 V}\right)\bar{e}_i^{(\lambda)}(\vec{k})e_j^{(\lambda)}(\vec{k})e_k^{(\varepsilon)}(\vec{p})\bar{e}_l^{(\varepsilon)}(\vec{p})\\
&\times\mu_j^{0r}(A)\mu_l^{r0}(A)\mu_i^{0s}(B)\mu_k^{s0}(B)\,\mathrm{e}^{-i\vec{k}\cdot\vec{R}}\mathrm{e}^{i\vec{p}\cdot\vec{R}}\sum_{a=i}^{xii}E_a^{-1},
\end{aligned}
\tag{7.3.2}
$$

where E_a^{-1} is the energy denominator product arising from graphs $a = i$–xii. They are given explicitly in Table 7.1.

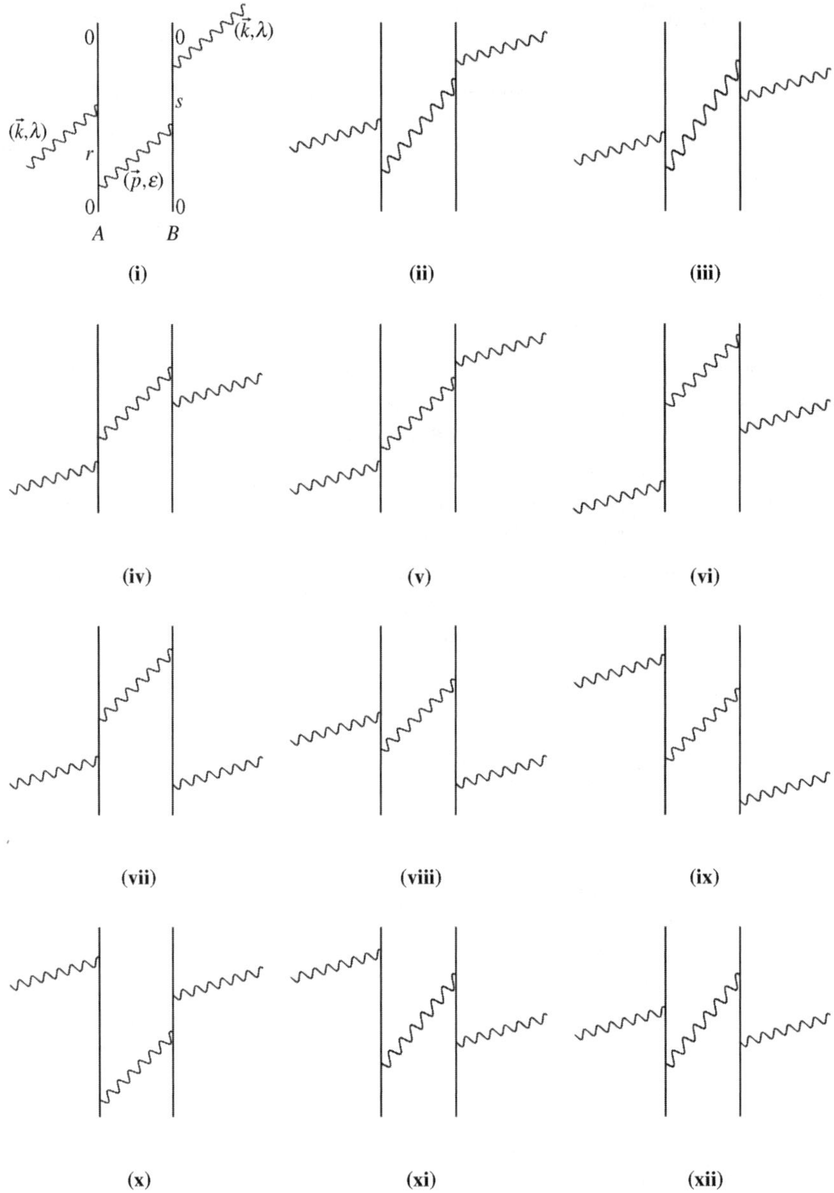

FIGURE 7.2 One set of 12 time-ordered diagrams that contribute to the dynamic mechanism of radiation-induced intermolecular energy shift. Collectively these graphs are labeled class (a).

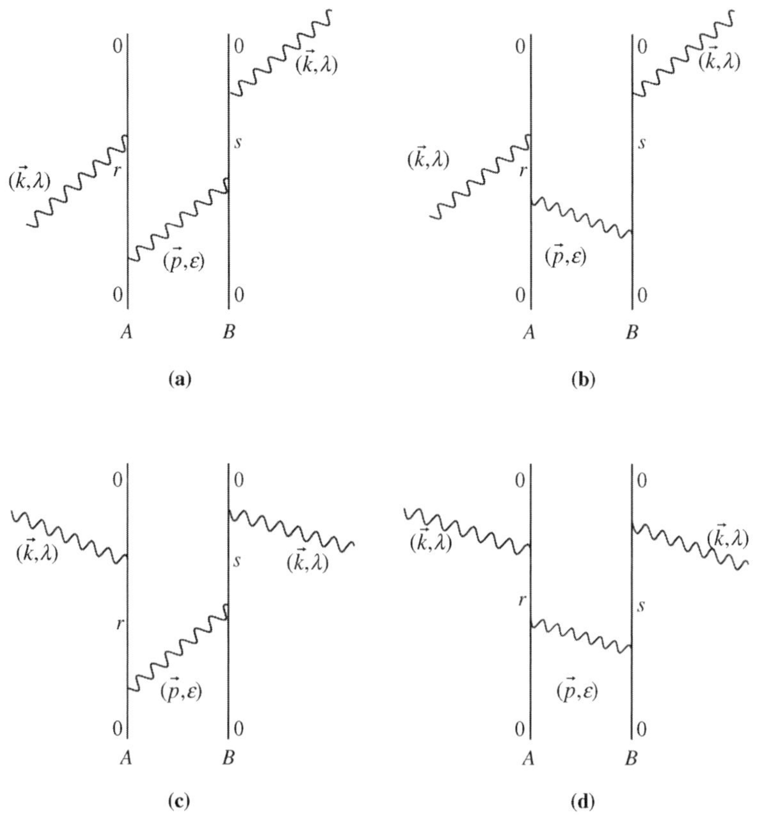

FIGURE 7.3 (a)–(d) Representative time orderings from each of the 4 sets of 12 graphs featuring in the dynamic mechanism.

To facilitate simplification of the molecular part of the energy shift, the 12 energy denominators listed in Table 7.1 may be added as follows:

$$E_{viii}^{-1} + E_{xii}^{-1} = [(E_{r0} + \hbar cp)(E_{s0} + \hbar ck)(E_{s0} + \hbar cp)]^{-1}. \qquad (7.3.3)$$

Adding E_{iii}^{-1} to the right-hand side of (7.3.3) gives

$$[(E_{r0} + \hbar cp)(E_{s0} + \hbar ck)(\hbar cp - \hbar ck)]^{-1}, \qquad (7.3.4)$$

$$E_{ix}^{-1} + E_{xi}^{-1} = [(E_{r0} + \hbar ck)(E_{r0} + \hbar cp)(E_{s0} + \hbar ck)]^{-1}, \qquad (7.3.5)$$

and

$$E_{vi}^{-1} + E_{vii}^{-1} = [(E_{r0} - \hbar ck)(E_{s0} + \hbar ck)(E_{s0} + \hbar cp)]^{-1}. \qquad (7.3.6)$$

TABLE 7.1 Energy Denominator Products Corresponding to Diagrams (i)–(xii) of Fig. 7.2

Graph	Denominator
(i)	$(E_{r0}+\hbar cp)(E_{r0}+E_{s0})(E_{s0}-\hbar ck)$
(ii)	$(E_{r0}+\hbar cp)(\hbar cp-\hbar ck)(E_{s0}-\hbar ck)$
(iii)	$(E_{r0}+\hbar cp)(\hbar cp-\hbar ck)(E_{s0}+\hbar cp)$
(iv)	$(E_{r0}-\hbar ck)(\hbar cp-\hbar ck)(E_{s0}+\hbar cp)$
(v)	$(E_{r0}-\hbar ck)(\hbar cp-\hbar ck)(E_{s0}-\hbar ck)$
(vi)	$(E_{r0}-\hbar ck)(E_{r0}+E_{s0})(E_{s0}+\hbar cp)$
(vii)	$(E_{s0}+\hbar ck)(E_{r0}+E_{s0})(E_{s0}+\hbar cp)$
(viii)	$(E_{s0}+\hbar ck)(E_{r0}+E_{s0}+\hbar ck+\hbar cp)(E_{s0}+\hbar cp)$
(ix)	$(E_{s0}+\hbar ck)(E_{r0}+E_{s0}+\hbar ck+\hbar cp)(E_{r0}+\hbar ck)$
(x)	$(E_{r0}+\hbar cp)(E_{r0}+E_{s0})(E_{r0}+\hbar ck)$
(xi)	$(E_{r0}+\hbar cp)(E_{r0}+E_{s0}+\hbar ck+\hbar cp)(E_{r0}+\hbar ck)$
(xii)	$(E_{r0}+\hbar cp)(E_{r0}+E_{s0}+\hbar ck+\hbar cp)(E_{s0}+\hbar cp)$

Adding E_{iv}^{-1} to the right-hand side of (7.3.6) produces

$$[(E_{r0}-\hbar ck)(E_{s0}+\hbar ck)(\hbar cp-\hbar ck)]^{-1}. \tag{7.3.7}$$

Now,

$$E_i^{-1}+E_x^{-1}=[(E_{r0}+\hbar ck)(E_{r0}+\hbar cp)(E_{s0}-\hbar ck)]^{-1}. \tag{7.3.8}$$

Adding E_{ii}^{-1} to the right-hand side of (7.3.8) yields

$$[(E_{r0}+\hbar ck)(E_{s0}-\hbar ck)(\hbar cp-\hbar ck)]^{-1}. \tag{7.3.9}$$

Adding (7.3.4) and (7.3.5) gives

$$[(E_{r0}+\hbar ck)(E_{s0}+\hbar ck)(\hbar cp-\hbar ck)]^{-1}. \tag{7.3.10}$$

The denominator from graph (v) remains as

$$E_v^{-1}=[(E_{r0}-\hbar ck)(E_{s0}-\hbar ck)(\hbar cp-\hbar ck)]^{-1}. \tag{7.3.11}$$

The four terms (7.3.7), (7.3.9), (7.3.10), and (7.3.11) may be factored as

$$\left[\frac{1}{(E_{r0}+\hbar ck)}+\frac{1}{(E_{r0}-\hbar ck)}\right]\left[\frac{1}{(E_{s0}+\hbar ck)}+\frac{1}{(E_{s0}-\hbar ck)}\right]\frac{1}{\hbar cp-\hbar ck}. \tag{7.3.12}$$

Hence, (7.3.2) becomes

$$
\begin{aligned}
&-\sum_{\vec{p},\varepsilon}\sum_{r,s}\left(\frac{N\hbar ck}{2\varepsilon_0 V}\right)\left(\frac{\hbar cp}{2\varepsilon_0 V}\right)\bar{e}_i^{(\lambda)}(\vec{k})e_j^{(\lambda)}(\vec{k})e_k^{(\varepsilon)}(\vec{p})\bar{e}_l^{(\varepsilon)}(\vec{p})\\
&\times\mu_j^{0r}(A)\mu_l^{r0}(A)\mu_i^{0s}(B)\mu_k^{s0}(B)\\
&\times e^{-i\vec{k}\cdot\vec{R}}e^{i\vec{p}\cdot\vec{R}}\left[\frac{1}{(E_{r0}+\hbar ck)}+\frac{1}{(E_{r0}-\hbar ck)}\right]\\
&\times\left[\frac{1}{(E_{s0}+\hbar ck)}+\frac{1}{(E_{s0}-\hbar ck)}\right]\frac{1}{\hbar cp-\hbar ck}.
\end{aligned}
\tag{7.3.13}
$$

Recognizing that the molecular polarizability of A is defined as

$$
\alpha_{jl}(A;k)=\sum_r \mu_j^{0r}(A)\mu_l^{r0}(A)\left[\frac{1}{(E_{r0}+\hbar ck)}+\frac{1}{(E_{r0}-\hbar ck)}\right] \tag{7.3.14}
$$

with an analogous formula for the corresponding quantity of species B, (7.3.13) can be written as

$$
\begin{aligned}
&-\sum_{\vec{p},\varepsilon}\left(\frac{N\hbar ck}{2\varepsilon_0 V}\right)\left(\frac{\hbar cp}{2\varepsilon_0 V}\right)\alpha_{jl}(A;k)\alpha_{ik}(B;k)\bar{e}_i^{(\lambda)}(\vec{k})e_j^{(\lambda)}(\vec{k})e_k^{(\varepsilon)}(\vec{p})\bar{e}_l^{(\varepsilon)}(\vec{p})\\
&\times e^{-i\vec{k}\cdot\vec{R}}e^{i\vec{p}\cdot\vec{R}}\frac{1}{\hbar cp-\hbar ck}.
\end{aligned}
\tag{7.3.15}
$$

The remaining 3 sets of 12 diagrams can be evaluated similarly and added to (7.3.15) to give, for the change in energy shift, the expression

$$
\begin{aligned}
\Delta E_{\text{dyn}}=&\left(\frac{N\hbar ck}{2\varepsilon_0 V}\right)\alpha_{jl}(A;k)\alpha_{ik}(B;k)\bar{e}_i^{(\lambda)}(\vec{k})e_j^{(\lambda)}(\vec{k})\frac{1}{2\varepsilon_0 V}\\
&\times\left\{\begin{aligned}&e^{i\vec{k}\cdot\vec{R}}\sum_{\vec{p},\varepsilon}\hbar cpe_k^{(\varepsilon)}(\vec{p})\bar{e}_l^{(\varepsilon)}(\vec{p})\left[\frac{e^{i\vec{p}\cdot\vec{R}}}{-\hbar ck-\hbar cp}+\frac{e^{-i\vec{p}\cdot\vec{R}}}{\hbar ck-\hbar cp}\right]\\
&+e^{-i\vec{k}\cdot\vec{R}}\sum_{\vec{p},\varepsilon}\hbar cp\bar{e}_k^{(\varepsilon)}(\vec{p})e_l^{(\varepsilon)}(\vec{p})\left[\frac{e^{i\vec{p}\cdot\vec{R}}}{\hbar ck-\hbar cp}+\frac{e^{-i\vec{p}\cdot\vec{R}}}{-\hbar ck-\hbar cp}\right]\end{aligned}\right\}.
\end{aligned}
\tag{7.3.16}
$$

The summation over virtual photon $(\vec{p},\varepsilon)$ appearing in (7.3.16) is identical to that carried out in Section 4.2 in the context of resonant transfer of excitation. Its result may be written down immediately in terms of the retarded coupling tensor $V_{ij}(k,\vec{R})$ (4.2.17), as in

$$\sum_{\vec{p},\varepsilon}\left(\frac{\hbar cp}{2\varepsilon_0 V}\right)e_i^{(\varepsilon)}(\vec{p})\bar{e}_j^{(\varepsilon)}(\vec{p})\left\{\frac{\mathrm{e}^{i\vec{p}\cdot\vec{R}}}{\hbar ck-\hbar cp}+\frac{\mathrm{e}^{-i\vec{p}\cdot\vec{R}}}{-\hbar ck-\hbar cp}\right\}=V_{ij}(k,\vec{R}). \tag{7.3.17}$$

Hence, the energy shift (7.3.16) between a pair of molecules in fixed relative orientation to each other and to the wavevector of the incident radiation beam is

$$\Delta E_{\mathrm{dyn}}=\left(\frac{N\hbar ck}{2\varepsilon_0 V}\right)\alpha_{jl}(A;k)\alpha_{ik}(B;k)\bar{e}_i^{(\lambda)}(\vec{k})e_j^{(\lambda)}(\vec{k})\mathrm{Re}V_{kl}(k,\vec{R})(\mathrm{e}^{i\vec{k}\cdot\vec{R}}+\mathrm{e}^{-i\vec{k}\cdot\vec{R}}), \tag{7.3.18}$$

which is seen to be proportional to the polarizability of each molecule. The designation of this contribution as being dynamic is now clear in that a definite amount of energy migrates between the pair accompanying the exchange of a single virtual photon. This is manifested through the appearance in ΔE of the coupling tensor $V_{kl}(k,\vec{R})$ whose real part is taken since the energy shift is real.

A state sequence representation of the dynamic contribution to the radiation-induced intermolecular energy shift may also be effected. It bears a close resemblance to the depiction of the Casimir–Polder potential via state sequences and leads to the expression (7.3.16) for the energy shift. In the present case, the hyperspace dimension n is also four since there are four distinct photon creation–destruction events. For the contribution being considered, they correspond to emission and absorption of a real photon at different centers and single virtual photon exchange. The 48 possible time orderings map onto two state sequence diagrams illustrated in Figs. 7.4 and 7.5, each containing 24 unique paths due to the 4! possible permutations of the four unique interaction vertices or vector coefficients used to depict them. Therefore, the basis set used for the construction of the state sequence diagram for the retarded dispersion potential, $I=\{1\vec{i}_1,1\vec{i}_2,1\vec{i}_3,1\vec{i}_4\}$, can be used in the present problem. A consequence of this identification is that the structure coefficients calculated using (1.10.14) are the binomial coefficients 1 4 6 4 1 as occurred in Section 5.5. In Fig. 7.4, the state sequences represent time orderings in which absorption of a real photon at either center occurs before emission of the real photon with the virtual photon propagating in either direction. Meanwhile in Fig. 7.5, the 24

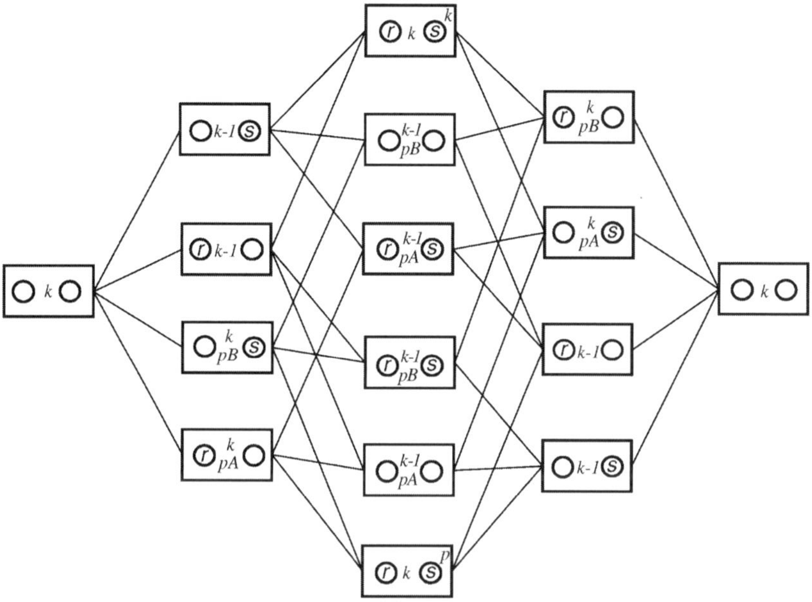

FIGURE 7.4 One of the two state sequence diagrams used in the calculation of the dynamic mechanism of the radiation-induced intermolecular interaction. This set is associated with the absorption of the real photon $(\vec{k}, \lambda)$occurring before its time-ordered emission at the other center.

pathways depicted in the state sequences correspond to time orderings in which emission of a real photon, at either site, takes place before absorption of the real photon at the other species, with the virtual photon again traversing from A to B or from B to A. Hence, in each of the two state sequence diagrams, six pathways come from each of the 4 sets of 12 diagrams shown in Fig. 7.3a–d. The superscript appearing on the virtual photon label inside of a state sequence box denotes the site of virtual emission. Furthermore, a k or a p appearing in the upper right-hand corner of a cell signifies the type of photon that has already been emitted and absorbed in some order.

7.4 STATIC MECHANISM

When either one or both of the species is polar, there is an additional contribution to the radiation-induced intermolecular energy shift (Bradshaw and Andrews, 2005). It is also of fourth order in perturbation

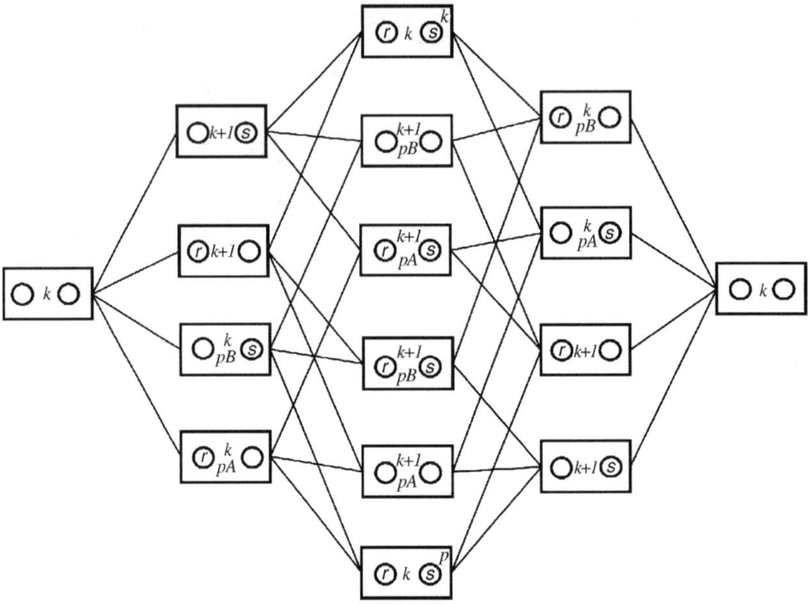

FIGURE 7.5 One of the two state sequence diagrams used in the calculation of the dynamic mechanism of the radiation-induced intermolecular interaction. This set is associated with the emission of the real photon occurring before the time-ordered absorption of the real photon at the other interaction site.

theory with initial and final states that are identical to one another and equal to those used to represent the dynamic mechanism and given by (7.2.6). As in the dynamic mechanism, the two molecules interact via the exchange of a single virtual photon, but unlike the dynamic contribution, the real photon is scattered by the same molecular center. This extra contribution to ΔE is termed as the "static" mechanism as no net energy is relayed between the pair on migration of a virtual photon. Pictorially, this process may be represented by 48 time-ordered diagrams, which may also be grouped into 4 sets of 12 graphs. A representative graph from each of the four sets is illustrated in Fig. 7.6. As for the dynamic mechanism, a state sequence diagram may be drawn to depict the process and used to compute the energy shift. Because there are again four unique photonic processes, corresponding to emission and absorption of a real photon at one site, either A or B, and single virtual photon exchange, the hyperspace dimension $n = 4$. Two state sequence diagrams each containing 24 pathways are generated with structure coefficients given by the fourth row of Pascal's

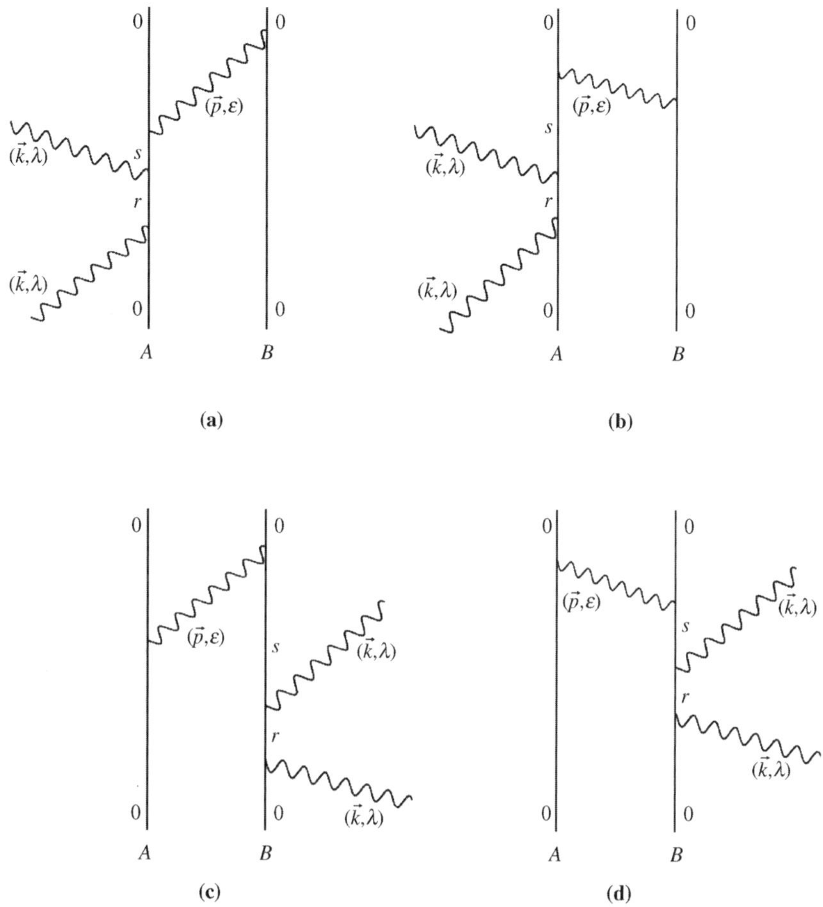

FIGURE 7.6 (a)–(d) Representative time-ordered diagrams from each of the 4 sets of 12 graphs that contribute to the static mechanism of the radiation-induced energy shift.

triangle. One of them is shown in Fig. 7.7 in which scattering of the real photon occurs at molecule A. The other diagram is easily obtained on interchanging A and B.

Again the starting point for the computation is the total Hamiltonian (7.2.4), with electric dipole coupling Hamiltonian (7.2.5). It is now shown how the first set of 12 time-ordered diagrams, explicitly drawn in Fig. 7.8, of the type shown in Fig. 7.6a, are added and simplified. Listed in Table 7.2 are the energy denominator products corresponding to the 12 graphs of Fig. 7.8. Adding the contribution from the 12

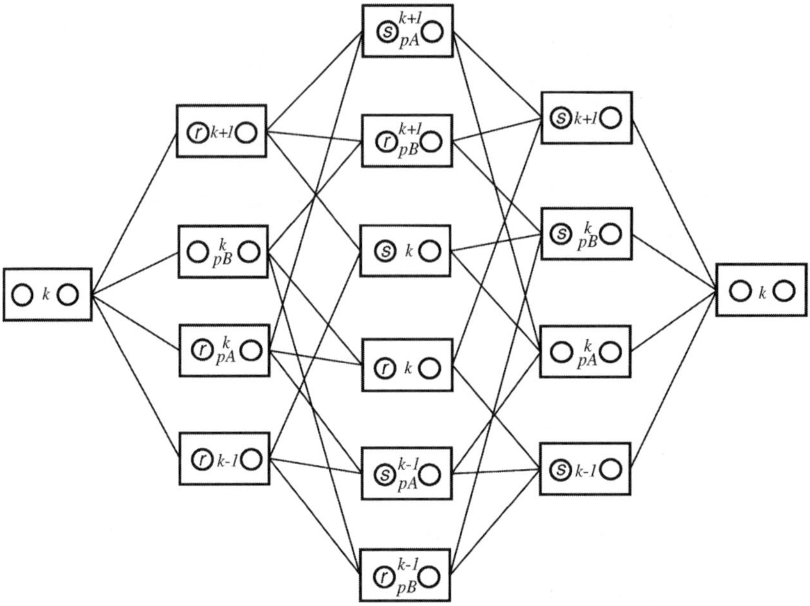

FIGURE 7.7 One of the two state sequence diagrams representing the static contribution to the radiation-induced intermolecular interaction energy. In this picture, scattering of a real photon occurs at molecule A. The other diagram is obtained on interchanging A and B.

graphs produces

$$-\sum_{\vec{p},\varepsilon}\sum_{r,s}\left(\frac{N\hbar ck}{2\varepsilon_0 V}\right)\left(\frac{\hbar cp}{2\varepsilon_0 V}\right)e_i^{(\lambda)}(\vec{k})\bar{e}_j^{(\lambda)}(\vec{k})\bar{e}_k^{(\varepsilon)}(\vec{p})e_l^{(\varepsilon)}(\vec{p})\mathrm{e}^{i\vec{p}\cdot\vec{R}}\mu_l^{00}(B)$$
$$\times\left\{\begin{array}{l}\dfrac{\mu_k^{0s}(A)\mu_j^{sr}(A)\mu_i^{r0}(A)}{E_i}+\dfrac{\mu_j^{0s}(A)\mu_k^{sr}(A)\mu_i^{r0}(A)}{E_{ii}}+\dfrac{\mu_j^{0s}(A)\mu_k^{sr}(A)\mu_i^{r0}(A)}{E_{iii}}\\ +\dfrac{\mu_j^{0s}(A)\mu_i^{sr}(A)\mu_k^{r0}(A)}{E_{iv}}+\dfrac{\mu_j^{0s}(A)\mu_i^{sr}(A)\mu_k^{r0}(A)}{E_{v}}+\dfrac{\mu_j^{0s}(A)\mu_i^{sr}(A)\mu_k^{r0}(A)}{E_{vi}}\\ +\dfrac{\mu_k^{0s}(A)\mu_i^{sr}(A)\mu_j^{r0}(A)}{E_{vii}}+\dfrac{\mu_i^{0s}(A)\mu_k^{sr}(A)\mu_j^{r0}(A)}{E_{viii}}+\dfrac{\mu_i^{0s}(A)\mu_k^{sr}(A)\mu_j^{r0}(A)}{E_{ix}}\\ +\dfrac{\mu_i^{0s}(A)\mu_j^{sr}(A)\mu_k^{r0}(A)}{E_{x}}+\dfrac{\mu_i^{0s}(A)\mu_j^{sr}(A)\mu_k^{r0}(A)}{E_{xi}}+\dfrac{\mu_i^{0s}(A)\mu_j^{sr}(A)\mu_k^{r0}(A)}{E_{xii}}\end{array}\right\}. \tag{7.4.1}$$

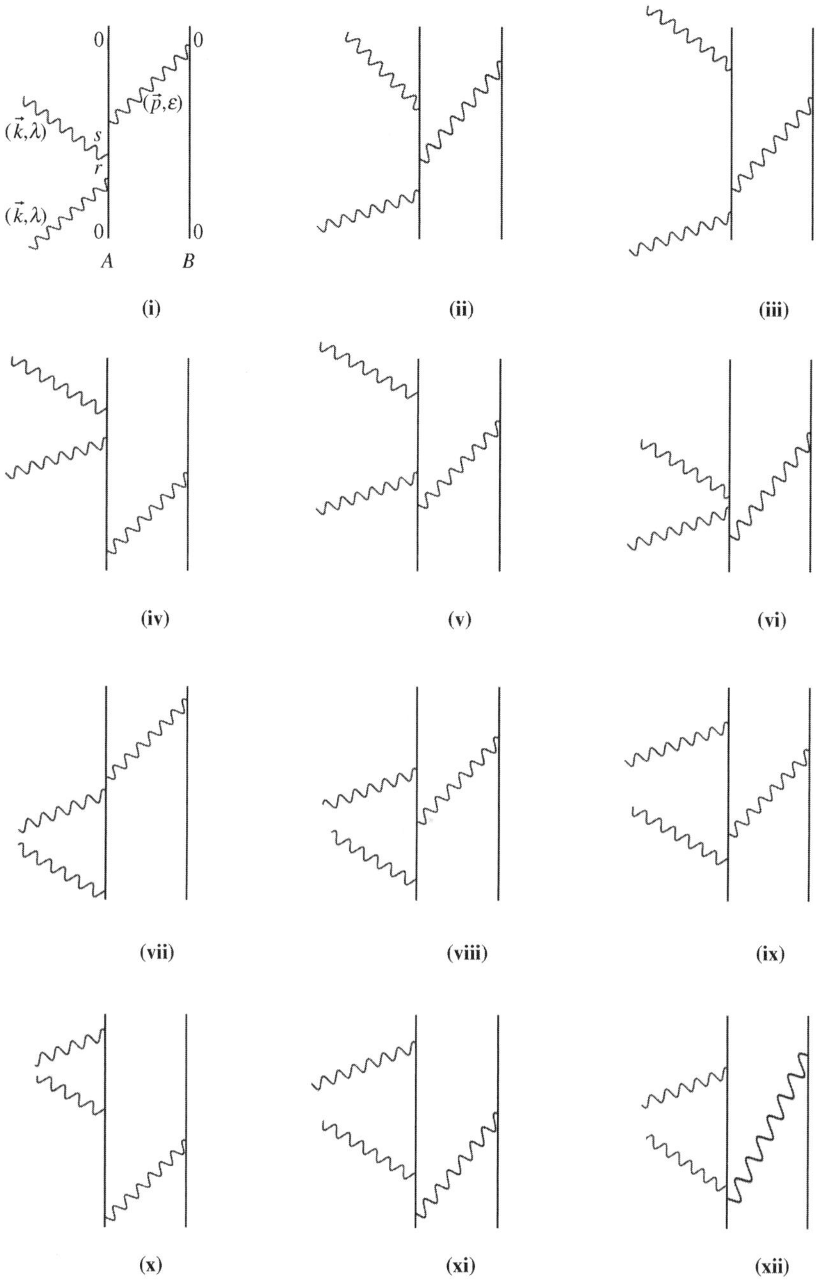

FIGURE 7.8 Twelve time-ordered graphs contributing to the static mechanism in which scattering of a real photon occurs at A with virtual photon traveling from A to B.

TABLE 7.2 Energy Denominators Corresponding to Time-Ordered Diagrams of Fig. 7.8

Graph	Energy Denominator
(i)	$E_i = (E_{r0}-\hbar ck)E_{s0}\hbar cp$
(ii)	$E_{ii} = (E_{r0}-\hbar ck)(E_{s0}-\hbar ck+\hbar cp)\hbar cp$
(iii)	$E_{iii} = (E_{r0}-\hbar ck)(E_{s0}-\hbar ck+\hbar cp)(E_{s0}-\hbar ck)$
(iv)	$E_{iv} = (E_{r0}+\hbar cp)(E_{s0}-\hbar ck)E_{r0}$
(v)	$E_v = (E_{r0}+\hbar cp)(E_{s0}-\hbar ck+\hbar cp)(E_{s0}-\hbar ck)$
(vi)	$E_{vi} = (E_{r0}+\hbar cp)(E_{s0}-\hbar ck+\hbar cp)\hbar cp$
(vii)	$E_{vii} = (E_{r0}+\hbar ck)E_{s0}\hbar cp$
(viii)	$E_{viii} = (E_{r0}+\hbar ck)(E_{s0}+\hbar ck+\hbar cp)\hbar cp$
(ix)	$E_{ix} = (E_{r0}+\hbar ck)(E_{s0}+\hbar ck+\hbar cp)(E_{s0}+\hbar ck)$
(x)	$E_x = (E_{r0}+\hbar cp)(E_{s0}+\hbar ck)E_{r0}$
(xi)	$E_{xi} = (E_{r0}+\hbar cp)(E_{s0}+\hbar ck+\hbar cp)(E_{s0}+\hbar ck)$
(xii)	$E_{xii} = (E_{r0}+\hbar cp)(E_{s0}+\hbar ck+\hbar cp)\hbar cp$

In the expression above, $\mu_i^{00}(\xi)$ is the ground-state permanent electric dipole moment of species ξ, $\langle E_0^{\xi}|\mu_i(\xi)|E_0^{\xi}\rangle$, and E_α, $\alpha = i$–xii, are the energy denominators displayed in Table 7.2. It is now shown how the terms within braces, including the sums over r and s, may be simplified considerably. Adding the fifth and sixth terms gives

$$\mu_j^{0s}\mu_i^{sr}\mu_k^{r0}\left(\frac{1}{E_v}+\frac{1}{E_{vi}}\right) = \frac{\mu_j^{0s}\mu_i^{sr}\mu_k^{r0}}{(E_{r0}+\hbar cp)(E_{s0}-\hbar ck)\hbar cp}, \tag{7.4.2}$$

which when added to the fourth term results in

$$\mu_j^{0s}\mu_i^{sr}\mu_k^{r0}\left(\frac{1}{(E_{r0}+\hbar cp)(E_{s0}-\hbar ck)\hbar cp}+\frac{1}{E_{iv}}\right) = \frac{\mu_j^{0s}\mu_i^{sr}\mu_k^{r0}}{(E_{s0}-\hbar ck)E_{r0}\hbar cp}. \tag{7.4.3}$$

Addition of second and third terms produces

$$\mu_j^{0s}\mu_k^{sr}\mu_i^{r0}\left(\frac{1}{E_{ii}}+\frac{1}{E_{iii}}\right) = \frac{\mu_j^{0s}\mu_k^{sr}\mu_i^{r0}}{(E_{r0}-\hbar ck)(E_{s0}-\hbar ck)\hbar cp}. \tag{7.4.4}$$

Terms eight and nine sum to

$$\mu_i^{0s}\mu_k^{sr}\mu_j^{r0}\left(\frac{1}{E_{viii}}+\frac{1}{E_{ix}}\right) = \frac{\mu_i^{0s}\mu_k^{sr}\mu_j^{r0}}{(E_{r0}+\hbar ck)(E_{s0}+\hbar ck)\hbar cp}. \tag{7.4.5}$$

Meanwhile, adding the tenth term to the sum of terms 11 and 12 yields

$$\mu_i^{0s}\mu_j^{sr}\mu_k^{r0}\left(\frac{1}{E_{xi}}+\frac{1}{E_{xii}}+\frac{1}{E_x}\right)$$
$$=\mu_i^{0s}\mu_j^{sr}\mu_k^{r0}\left(\frac{1}{(E_{r0}+\hbar cp)(E_{s0}+\hbar ck)\hbar cp}+\frac{1}{E_x}\right)=\frac{\mu_i^{0s}\mu_j^{sr}\mu_k^{r0}}{(E_{s0}+\hbar ck)E_{r0}\hbar cp}. \tag{7.4.6}$$

Terms one and seven remain unaltered. Adding these six terms, namely, equations (7.4.3)–(7.4.6) to $\frac{1}{E_i}+\frac{1}{E_{vii}}$, results in

$$\frac{1}{\hbar cp}\sum_{r,s}\left\{\begin{array}{l}\dfrac{\mu_i^{0s}(A)\mu_j^{sr}(A)\mu_k^{r0}(A)}{(E_{s0}+\hbar ck)E_{r0}}+\dfrac{\mu_i^{0s}(A)\mu_k^{sr}(A)\mu_j^{r0}(A)}{(E_{r0}+\hbar ck)(E_{s0}+\hbar ck)}\\ +\dfrac{\mu_j^{0s}(A)\mu_i^{sr}(A)\mu_k^{r0}(A)}{(E_{s0}-\hbar ck)E_{r0}}+\dfrac{\mu_j^{0s}(A)\mu_k^{sr}(A)\mu_i^{r0}(A)}{(E_{r0}-\hbar ck)(E_{s0}-\hbar ck)}\\ +\dfrac{\mu_k^{0s}(A)\mu_i^{sr}(A)\mu_j^{r0}(A)}{(E_{r0}+\hbar ck)E_{s0}}+\dfrac{\mu_k^{0s}(A)\mu_j^{sr}(A)\mu_i^{r0}(A)}{(E_{r0}-\hbar ck)E_{s0}}\end{array}\right\} \tag{7.4.7}$$
$$=\frac{1}{\hbar cp}\beta_{ijk}(A;k),$$

which is defined to be $(\hbar cp)^{-1}$ multiplied by the molecular first hyperpolarizability of molecule A. This enables the energy shift (7.4.1) to be written as

$$-\sum_{\vec{p},\varepsilon}\left(\frac{N\hbar ck}{2\varepsilon_0 V}\right)\left(\frac{1}{2\varepsilon_0 V}\right)e_i^{(\lambda)}(\vec{k})\bar{e}_j^{(\lambda)}(\vec{k})\bar{e}_k^{(\varepsilon)}(\vec{p})e_l^{(\varepsilon)}(\vec{p})\mathrm{e}^{i\vec{p}\cdot\vec{R}}\beta_{ijk}(A;k)\mu_l^{00}(B). \tag{7.4.8}$$

For the second set of 12 graphs, one of which is shown in Fig. 7.6b, the virtual photon propagates from B to A. The contribution to the energy shift from this set of diagrams is then identical to (7.4.8) except for replacement of the factor $\mathrm{e}^{i\vec{p}\cdot\vec{R}}$ by $\mathrm{e}^{-i\vec{p}\cdot\vec{R}}$. The molecular factor can again be summed to yield the hyperpolarizability tensor (7.4.7). The contribution from the remaining 24 graphs is then easily obtained from the expression calculated from the first 24 diagrams by interchanging $A\leftrightarrow B$. Hence, the contribution to the energy shift due to the static mechanism is

$$\Delta E_{\mathrm{stat}}=-\sum_{\vec{p},\varepsilon}\left(\frac{N\hbar ck}{2\varepsilon_0 V}\right)\left(\frac{1}{2\varepsilon_0 V}\right)e_i^{(\lambda)}(\vec{k})\bar{e}_j^{(\lambda)}(\vec{k})e_k^{(\varepsilon)}(\vec{p})\bar{e}_l^{(\varepsilon)}(\vec{p})$$
$$\times[\beta_{ijk}(A;k)\mu_l^{00}(B)+\mu_l^{00}(A)\beta_{ijk}(B;k)](\mathrm{e}^{i\vec{p}\cdot\vec{R}}+\mathrm{e}^{-i\vec{p}\cdot\vec{R}}). \tag{7.4.9}$$

To proceed further, the virtual photon polarization sum is executed using the identity

$$\sum_{\varepsilon} e_k^{(\varepsilon)}(\vec{p})\bar{e}_l^{(\varepsilon)}(\vec{p}) = \delta_{kl} - \hat{p}_k\hat{p}_l \tag{7.4.10}$$

and the $\vec{p}$-sum converted to an integral via

$$\sum_{\vec{p}} \frac{1}{V} \rightarrow \frac{1}{(2\pi)^3}\int d^3\vec{p}. \tag{7.4.11}$$

Equation (7.4.9) therefore becomes

$$\Delta E_{\text{stat}} = -\left(\frac{N\hbar ck}{4\varepsilon_0^2 V}\right) e_i^{(\lambda)}(\vec{k})\bar{e}_j^{(\lambda)}(\vec{k})\left[\beta_{ijk}(A;k)\mu_l^{00}(B) + \mu_l^{00}(A)\beta_{ijk}(B;k)\right]$$
$$\times \frac{1}{(2\pi)^3}\int (\delta_{kl} - \hat{p}_k\hat{p}_l)e^{i\vec{p}\cdot\vec{R}}d^3\vec{p}, \tag{7.4.12}$$

on noting that the right-hand side of (7.4.10) is even in $\vec{p}$. Since for $\vec{p} \neq 0$ (Power, 1964),

$$\frac{1}{(2\pi)^3\varepsilon_0}\int (\delta_{kl} - \hat{p}_k\hat{p}_l)e^{i\vec{p}\cdot\vec{R}}d^3\vec{p} = \vec{\nabla}_k\vec{\nabla}_l\frac{1}{4\pi\varepsilon_0 R} = -\frac{1}{4\pi\varepsilon_0 R^3}(\delta_{kl} - 3\hat{R}_k\hat{R}_l)$$
$$= -V_{kl}(0,\vec{R}), \tag{7.4.13}$$

ΔE_{stat} can be written in terms of the static coupling tensor $V_{kl}(0,\vec{R})$ as (Bradshaw and Andrews, 2005)

$$\Delta E_{\text{stat}} = \left(\frac{N\hbar ck}{2\varepsilon_0 V}\right) e_i^{(\lambda)}(\vec{k})\bar{e}_j^{(\lambda)}(\vec{k})\left[\beta_{ijk}(A;k)\mu_l^{00}(B) + \mu_l^{00}(A)\beta_{ijk}(B;k)\right]V_{kl}(0,\vec{R}). \tag{7.4.14}$$

The total modification in intermolecular energy shift due to the action of an intense laser field is then given by the sum of dynamic and static terms, equations (7.3.18) and (7.4.14), respectively,

$$\Delta E = \Delta E_{\text{dyn}} + \Delta E_{\text{stat}}$$
$$= \frac{I}{\varepsilon_0 c} e_i^{(\lambda)}(\vec{k})\bar{e}_j^{(\lambda)}(\vec{k})\left\{\alpha_{jl}(A;k)\alpha_{ik}(B;k)\cos(\vec{k}\cdot\vec{R})\text{Re}V_{kl}(k,\vec{R})\right.$$
$$\left. + \frac{1}{2}\left[\beta_{ijk}(A;k)\mu_l^{00}(B) + \mu_l^{00}(A)\beta_{ijk}(B;k)\right]V_{kl}(0,\vec{R})\right\}, \tag{7.4.15}$$

where I is the irradiance of the incident beam. It is worth pointing out that the first term of (7.4.15) permits A and B to be transposed, as evident by i,j- and k, l-index symmetry. In contrast, A and B are distinct species in the second, static contribution. Further, both the permanent moment and the linear hyperpolarizability are equal to zero if either A or B is centrosymmertric resulting in the ΔE_{stat} contribution vanishing.

It should, of course, not be forgotten that the dominant contribution to the interaction energy between two polar species is the electrostatic interaction between two permanent dipoles, which is larger than both ΔE_{dyn} and ΔE_{stat}. In the multipolar formalism, this energy shift is calculated using second-order perturbation theory via the formula

$$\Delta E = -\sum_{I} \frac{\langle 0|H_{\text{int}}|I\rangle\langle I|H_{\text{int}}|0\rangle}{E_I - E_0}. \tag{7.4.16}$$

The total and interaction Hamiltonians are once again given by (7.2.4) and (7.2.5), with the initial and final states given by the ground state of the total system,

$$|0\rangle = |E_0^A, E_0^B; 0(\vec{p}, \varepsilon)\rangle. \tag{7.4.17}$$

Two time-ordered diagrams represent the coupling and they are depicted in Fig. 7.9 in which a single virtual photon propagates between the two sites.

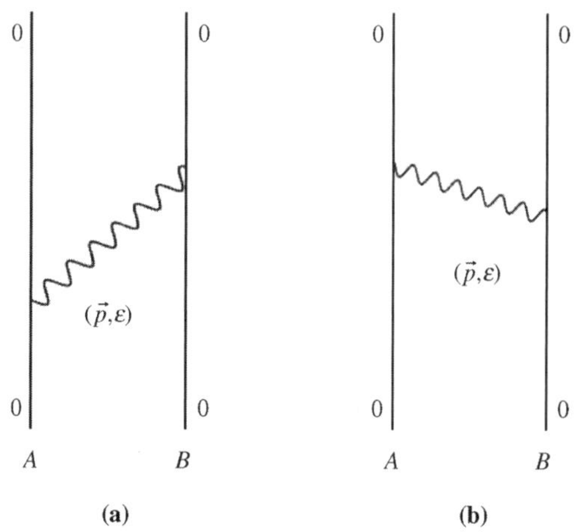

FIGURE 7.9 Electrostatic Coulomb interaction.

Standard evaluation leads directly to

$$\Delta E_{\text{Coul}} = -\frac{1}{2\varepsilon_0 V}\sum_{\vec{p},\varepsilon} \bar{e}_i^{(\varepsilon)}(\vec{p}) e_j^{(\varepsilon)}(\vec{p}) \times\left\{\mu_i^{00}(A)\mu_j^{00}(B)\mathrm{e}^{i\vec{p}\cdot\vec{R}} + \mu_j^{00}(A)\mu_i^{00}(B)\mathrm{e}^{-i\vec{p}\cdot\vec{R}}\right\}. \tag{7.4.18}$$

Carrying out the polarization sum using (7.4.10), taking advantage of i, j-index symmetry, and converting the $\vec{p}$-sum to an integral using (7.4.11) yields

$$\Delta E_{\text{Coul}} = -\frac{1}{(2\pi)^3 \varepsilon_0}\mu_i^{00}(A)\mu_j^{00}(B)\int(\delta_{ij}-\hat{p}_i\hat{p}_j)\mathrm{e}^{i\vec{p}\cdot\vec{R}}\mathrm{d}^3\vec{p}. \tag{7.4.19}$$

Comparing (7.4.19) with expression (7.4.12) for ΔE_{stat}, it is seen that the $\vec{p}$−dependent part is identical in both cases. Hence, using the result (7.4.13), the interaction energy between two ground-state permanent dipoles is

$$\Delta E_{\text{Coul}} = \frac{1}{4\pi\varepsilon_0 R^3}\mu_i^{00}(A)\mu_j^{00}(B)(\delta_{ij}-3\hat{R}_i\hat{R}_j), \tag{7.4.20}$$

which is the familiar static dipolar coupling energy. It is interesting to note that use of multipolar coupling that involves the emission and absorption of transverse photons contains the static interaction term.

7.5 MOLECULAR AND PAIR ORIENTATIONAL AVERAGING

The result (7.4.15) for the radiation-induced energy shift between two interacting molecules obtained in the previous section applies to an A–B pair in which the internuclear separation $\vec{R}$ is fixed relative to the wave-vector $\vec{k}$ of the impinging laser and with the orientation of each species fixed relative to each other. For a molecular pair in the gaseous or liquid state, an average over all $\vec{R}$ relative to $\vec{k}$ and over the relative orientations of A and B is required. The former is called the pair orientational average while the latter is known as the molecular average. Carrying out this second average on ΔE_{stat}, (7.4.14), results in this contribution vanishing. This is easy to see since an orientational average of the first hyperpolarizability using result (B.5) from Appendix B introduces a factor $\langle\beta_{ijk}(\xi;k)\rangle = \frac{1}{6}\varepsilon_{ijk}\varepsilon_{\lambda\mu\nu}\beta_{\lambda\mu\nu}(\xi;k)$, where Latin subscripts refer to Cartesian components in the space-fixed frame, while Greek suffixes refer to molecule-fixed frame axes. When contracting this factor with other tensor components featuring in ΔE_{stat}, it is evident that the product of polarization factors is symmetric in the indices i and j, while the alternating tensor ε_{ijk} is antisymmetric in this pair. Hence, $\langle \Delta E_{\mathrm{stat}} \rangle = 0$, where the angular brackets denote an averaged result. Performing a similar molecular average on the dynamic term (7.3.18) using the result $\langle \alpha_{jl}(A;k)\alpha_{ik}(B;k) \rangle = \delta_{jl}\delta_{ik}\alpha(A;k)\alpha(B;k)$ where a factor of 1/3 has been included in each of the isotropic polarizabilities leads to

$$\Delta E_{\mathrm{dyn}} = \left(\frac{I}{\varepsilon_0 c}\right) e_i^{(\lambda)}(\vec{k})\bar{e}_j^{(\lambda)}(\vec{k})\alpha(A;k)\alpha(B;k)\cos(\vec{k}\cdot\vec{R})\mathrm{Re}V_{ij}(k,\vec{R}). \tag{7.5.1}$$

A pair orientational average is now carried out on the dynamic term (7.5.1). Replacing the polarization vector product $e_i^{(\lambda)}(\vec{k})\bar{e}_j^{(\lambda)}(\vec{k})$ by $\frac{1}{2}(\delta_{ij}-\hat{k}_i\hat{k}_j)$, and rewriting $\cos(\vec{k}\cdot\vec{R})$ in exponential form produces

$$\Delta E_{\mathrm{dyn}} = \left(\frac{I}{4\varepsilon_0 c}\right)\alpha(A;k)\alpha(B;k)(\delta_{ij}-\hat{k}_i\hat{k}_j)\mathrm{Re}V_{ij}(k,\vec{R})(\mathrm{e}^{i\vec{k}\cdot\vec{R}}+\mathrm{e}^{-i\vec{k}\cdot\vec{R}}). \tag{7.5.2}$$

The tumbling average is performed using the result

$$\begin{aligned}\langle(\delta_{ij}-\hat{k}_i\hat{k}_j)\mathrm{e}^{\pm i\vec{k}\cdot\vec{R}}\rangle &= \frac{1}{4\pi}\int(\delta_{ij}-\hat{k}_i\hat{k}_j)\mathrm{e}^{\pm i\vec{k}\cdot\vec{R}}\mathrm{d}\Omega = \frac{1}{k^3}(-\vec{\nabla}^2\delta_{ij}+\vec{\nabla}_i\vec{\nabla}_j)\frac{\sin kR}{R}\\ &= \left\{(\delta_{ij}-\hat{R}_i\hat{R}_j)\frac{\sin kR}{kR}+(\delta_{ij}-3\hat{R}_i\hat{R}_j)\left(\frac{\cos kR}{k^2R^2}-\frac{\sin kR}{k^3R^3}\right)\right\}.\end{aligned} \tag{7.5.3}$$

Inserting the right-hand most side of (7.5.3) along with $\mathrm{Re}V_{ij}(k,\vec{R})$ from expression (4.2.17) into (7.5.2), contracting the tensors and expressing the trigonometric factors in terms of double angles produces, for the dynamic term contributing to the change in mutual interaction energy caused by an external radiation field, the result for a freely tumbling pair

$$\begin{aligned}\langle \Delta E_{\mathrm{dyn}} \rangle &= -\frac{I}{8\pi\varepsilon_0^2 cR^3}\alpha(A;k)\alpha(B;k)\\ &\quad\times\left[kR\sin 2kR+2\cos 2kR-5\frac{\sin 2kR}{kR}-6\frac{\cos 2kR}{k^2R^2}+3\frac{\sin 2kR}{k^3R^3}\right].\end{aligned} \tag{7.5.4}$$

The energy shift is seen to be linearly proportional to the laser irradiance, and to the polarizability of each molecule, and is independent of the polarization characteristics of the incident field. From the expression valid for all R beyond wavefunction overlap (7.5.4), it is a simple matter to obtain the asymptotic form of the energy shift in the limit of far and near zones. For the former, in which $kR \gg 1$,

$$\langle \Delta E_{\text{dyn}}^{\text{FZ}} \rangle = -\frac{Ik}{8\pi\varepsilon_0^2 cR^2}\alpha(A;k)\alpha(B;k)\sin 2kR, \tag{7.5.5}$$

exhibiting a modulated inverse square law behavior. In the near zone, where $kR \ll 1$, expanding the sine and cosine functions as McLaurin series leads to the limiting form

$$\langle \Delta E_{\text{dyn}}^{\text{NZ}} \rangle = -\frac{11Ik^2}{60\pi\varepsilon_0^2 cR}\alpha(A;k)\alpha(B;k), \tag{7.5.6}$$

which has R^{-1} dependence on separation distance. Comparing with the London dispersion energy (5.2.22), which has inverse sixth power dependence and a much more rapid falloff, (7.5.6) can be appreciable for a large number of pairs as in a molecular assembly. Another method by which $\langle \Delta E_{\text{dyn}}^{\text{NZ}} \rangle$ may be increased in magnitude is if the frequency of the incident laser is tuned to near resonance with an atomic or molecular transition frequency, thereby resonantly enhancing the dynamic polarizability.

7.6 POLARIZATION ANALYSIS

It is instructive to return to the result (7.5.1) when the orientation of the molecular pair is kept fixed relative to the direction of propagation of the incident beam and examine the polarization characteristics of the applied radiation field and its effect on the energy shift. The incoming beam is taken to have one of the two different polarizations—either linear polarization or circular (left-hand/right-hand) polarization. Moreover, the incident laser is taken to propagate in one of the two different directions with respect to the intermolecular separation distance vector, $\vec{R}$, being either parallel (||) or perpendicular ($\perp$). Expression (7.5.1) is to be analyzed in detail for each of the four possible combinations (Thirunamachandran, 1980), on making use of the fact that $\vec{e}$, $\vec{b}$, and $\hat{k}$ form a right-handed set of vectors.

7.6.1 Parallel Propagation

When the incident field travels parallel to $\vec{R}$, namely, $\vec{k}||\vec{R}$, then $\vec{e}$ is perpendicular to $\vec{R}$, that is, $\vec{e} \perp \vec{R}$, and $\cos(\vec{k}\cdot\vec{R}) = \cos kR$. Substituting for $\mathrm{Re}V_{ij}(k,\vec{R})$ in (7.5.1) produces

$$\Delta E_{\mathrm{dyn}}^{||} = \frac{I}{4\pi\varepsilon_0^2 cR^3} e_i^{(\lambda)}(\vec{k})\bar{e}_j^{(\lambda)}(\vec{k})\alpha(A;k)\alpha(B;k)$$
$$\times [(\delta_{ij}-3\hat{R}_i\hat{R}_j)(\cos kR + kR\sin kR) - (\delta_{ij}-\hat{R}_i\hat{R}_j)k^2R^2\cos kR]\cos(kR). \tag{7.6.1}$$

7.6.1.1 Linear Polarization Noting that for linearly polarized light, $\delta_{ij}e_i^{(\lambda)}(\vec{k})\bar{e}_j^{(\lambda)}(\vec{k}) = 1$ and $e_i^{(\lambda)}(\vec{k})\bar{e}_j^{(\lambda)}(\vec{k})\hat{R}_i\hat{R}_j = 0$, equation (7.6.1) becomes

$$\Delta E_{\mathrm{dyn}}^{||(\mathrm{lin})} = \frac{I}{4\pi\varepsilon_0^2 cR^3}\alpha(A;k)\alpha(B;k)$$
$$\times \left[\cos^2 kR + kR\sin\ kR\cos\ kR) - k^2R^2\cos^2 kR\right]. \tag{7.6.2}$$

In the far zone, $kR \gg 1$ so that (7.6.2) tends to

$$\Delta E_{\mathrm{dyn}}^{||(\mathrm{lin})}(\mathrm{FZ}) = -\frac{Ik^2}{4\pi\varepsilon_0^2 cR}\alpha(A;k)\alpha(B;k)\cos^2 kR, \tag{7.6.3}$$

while in the near zone, after expanding all three terms within square brackets of (7.6.2), which approximates to unity,

$$\Delta E_{\mathrm{dyn}}^{||(\mathrm{lin})}(\mathrm{NZ}) = \frac{I}{4\pi\varepsilon_0^2 cR^3}\alpha(A;k)\alpha(B;k). \tag{7.6.4}$$

7.6.1.2 Circular Polarization When the incident radiation field is circularly polarized, the energy shift (7.5.1) is

$$\Delta E_{\mathrm{dyn}}^{||(\mathrm{L/R})} = \frac{I}{\varepsilon_0 c} e_i^{(\mathrm{L/R})}(\vec{k})\bar{e}_j^{(\mathrm{L/R})}(\vec{k})\alpha(A;k)\alpha(B;k)\mathrm{Re}V_{ij}(k,\vec{R})\cos kR. \tag{7.6.5}$$

Using the identity

$$e_i^{(\mathrm{L/R})}(\vec{k})\bar{e}_j^{(\mathrm{L/R})}(\vec{k}) = \frac{1}{2}\left[(\delta_{ij}-\hat{k}_i\hat{k}_j) \mp i\varepsilon_{ijk}\hat{k}_k\right], \tag{7.6.6}$$

where the upper and lower signs refer to L- and R-circular polarization, respectively, and noting that $V_{ij}(k,\vec{R})$ is symmetric in i, j so that only the i, j-symmetric part of (7.6.6) contributes, (7.6.5) becomes

$$\Delta E_{\mathrm{dyn}}^{||(\mathrm{L/R})} = \frac{I}{8\pi\varepsilon_0^2 cR^3}\alpha(A;k)\alpha(B;k)(\delta_{ij}-\hat{k}_i\hat{k}_j)$$
$$\times [(\delta_{ij}-3\hat{R}_i\hat{R}_j)(\cos^2 kR + kR\sin kR\cos kR)$$
$$-(\delta_{ij}-\hat{R}_i\hat{R}_j)k^2R^2\cos^2 kR], \tag{7.6.7}$$

on inserting $\mathrm{Re}V_{ij}(k,\vec{R})$. Contracting the tensors and making use of the fact that $\hat{k}_i\hat{k}_j\hat{R}_i\hat{R}_j = 1$, $\Delta E_{\mathrm{dyn}}^{||(\mathrm{L/R})}$ is found to be identical to the result (7.6.2) obtained for linear polarization, that is,

$$\Delta E_{\mathrm{dyn}}^{||(\mathrm{L/R})} = \Delta E_{\mathrm{dyn}}^{||(\mathrm{lin})}. \tag{7.6.8}$$

Identical limiting forms (7.6.3) and (7.6.4) therefore follow.

7.6.2 Perpendicular Propagation

If the incident field travels in a direction orthogonal to $\vec{R}$, that is, $\vec{k} \perp \vec{R}$, then $\vec{e}||\vec{R}$, and $\cos \vec{k}\cdot\vec{R} = 1$. Thus, (7.5.1) can be written as

$$\Delta E_{\mathrm{dyn}}^{\perp} = \frac{I}{4\pi\varepsilon_0^2 cR^3} e_i^{(\lambda)}(\vec{k})\bar{e}_j^{(\lambda)}(\vec{k})\alpha(A;k)\alpha(B;k) \times [(\delta_{ij}-3\hat{R}_i\hat{R}_j)(\cos kR + kR\sin kR) - (\delta_{ij}-\hat{R}_i\hat{R}_j)k^2R^2\cos kR]. \tag{7.6.9}$$

7.6.2.1 *Linear Polarization*

Contracting the tensor products for linearly polarized light in (7.6.9), for which $(\delta_{ij}-3\hat{R}_i\hat{R}_j)e_i^{(\lambda)}(\vec{k})\bar{e}_j^{(\lambda)}(\vec{k}) = -2$ and $(\delta_{ij}-\hat{R}_i\hat{R}_j)e_i^{(\lambda)}(\vec{k})\bar{e}_j^{(\lambda)}(\vec{k}) = 0$, yields

$$\Delta E_{\mathrm{dyn}}^{\perp(\mathrm{lin})} = -\frac{I}{2\pi\varepsilon_0^2 cR^3}\alpha(A;k)\alpha(B;k)[\cos kR + kR\sin kR]. \tag{7.6.10}$$

The asymptotic limits of this result are

$$\Delta E_{\mathrm{dyn}}^{\perp(\mathrm{lin})}(\mathrm{FZ}) = -\frac{Ik}{2\pi\varepsilon_0^2 cR^2}\alpha(A;k)\alpha(B;k)\sin kR \tag{7.6.11}$$

and

$$\Delta E_{\mathrm{dyn}}^{\perp(\mathrm{lin})}(\mathrm{NZ}) = -\frac{I}{2\pi\varepsilon_0^2 cR^3}\alpha(A;k)\alpha(B;k). \tag{7.6.12}$$

7.6.2.2 *Circular Polarization*

For circularly polarized radiation propagating perpendicularly to $\vec{R}$, substituting the first term of (7.6.6) and $\cos \vec{k}\cdot\vec{R} = 1$ into (7.5.1) gives

$$\Delta E_{\mathrm{dyn}}^{\perp(\mathrm{L/R})} = \frac{I}{8\pi\varepsilon_0^2 cR^3}\alpha(A;k)\alpha(B;k)(\delta_{ij}-\hat{k}_i\hat{k}_j) \times [(\delta_{ij}-3\hat{R}_i\hat{R}_j)(\cos kR + kR\sin kR) - (\delta_{ij}-\hat{R}_i\hat{R}_j)k^2R^2\cos kR]. \tag{7.6.13}$$

With tensor contraction producing $(\delta_{ij}-\hat{k}_i\hat{k}_j)(\delta_{ij}-3\hat{R}_i\hat{R}_j) = -1$ and $(\delta_{ij}-\hat{k}_i\hat{k}_j)(\delta_{ij}-\hat{R}_i\hat{R}_j) = 1$ since now $\hat{k}_i\hat{k}_j\hat{R}_i\hat{R}_j = 0$, (7.6.13) becomes

$$\Delta E_{\text{dyn}}^{\perp(\text{L/R})} = -\frac{I}{8\pi\varepsilon_0^2 cR^3}\alpha(A;k)\alpha(B;k)\left[\cos kR + kR\sin kR + k^2R^2\cos kR\right]. \tag{7.6.14}$$

In the far zone, (7.6.14) reduces to

$$\Delta E_{\text{dyn}}^{\perp(\text{L/R})}(\text{FZ}) = -\frac{Ik^2}{8\pi\varepsilon_0^2 cR}\alpha(A;k)\alpha(B;k)\cos kR, \tag{7.6.15}$$

while at small separations, the limit is

$$\Delta E_{\text{dyn}}^{\perp(\text{L/R})}(\text{NZ}) = -\frac{I}{8\pi\varepsilon_0^2 cR^3}\alpha(A;k)\alpha(B;k). \tag{7.6.16}$$

7.7 COLLAPSED GRAPHS AND EFFECTIVE INTERACTION HAMILTONIAN

In Section 5.4, it was shown how second-order perturbation theory together with an effective coupling Hamiltonian that is quadratic in the electric displacement field, could be used to calculate the far-zone limit of the Casimir–Polder potential more efficiently than using an interaction Hamiltonian that is first order in $\vec{d}^{\perp}(\vec{r})$. Pictorially, this amounted to collapsing the linear interaction vertices occurring at each center to a two-photon coupling vertex. This reduced the number of time-ordered graphs to be summed over from four to two. A similar approach may be adopted for the computation of the change in intermolecular interaction energy caused by an applied radiation field. It is again convenient to consider the two contributions to the energy shift, the dynamic and static mechanisms, separately. Beginning with the former term, the 48 Feynman diagrams used to visualize the interaction, with representatives shown in Figs. 7.2 and 7.3, may be reduced to four time orderings with each containing collapsed two-photon interaction vertices at each site. They are illustrated in Fig. 7.10, the respective classes representing 12 time orderings are shown in Fig. 7.10a–d.

Instead of the interaction Hamiltonian (7.2.5), the effective coupling operator (5.4.12) is employed,

$$\begin{aligned} H_{\text{dyn}}^{\text{eff, int}} &= -\varepsilon_0^{-2}\alpha_{ik}(A;k)d_i^{\perp}(\vec{k},\lambda;\vec{R}_A)d_k^{\perp}(\vec{p},\varepsilon;\vec{R}_A) \\ &\quad -\varepsilon_0^{-2}\alpha_{jl}(B;k)d_j^{\perp}(\vec{k},\lambda;\vec{R}_B)d_l^{\perp}(\vec{p},\varepsilon;\vec{R}_B), \end{aligned} \tag{7.7.1}$$

which is proportional to the molecular polarizability, $\alpha_{ij}(\xi;k)$, and is bilinear in the electric displacement field, with real and virtual photons characterized by modes $(\vec{k},\lambda)$ and $(\vec{p},\varepsilon)$, respectively. Hence, the effective

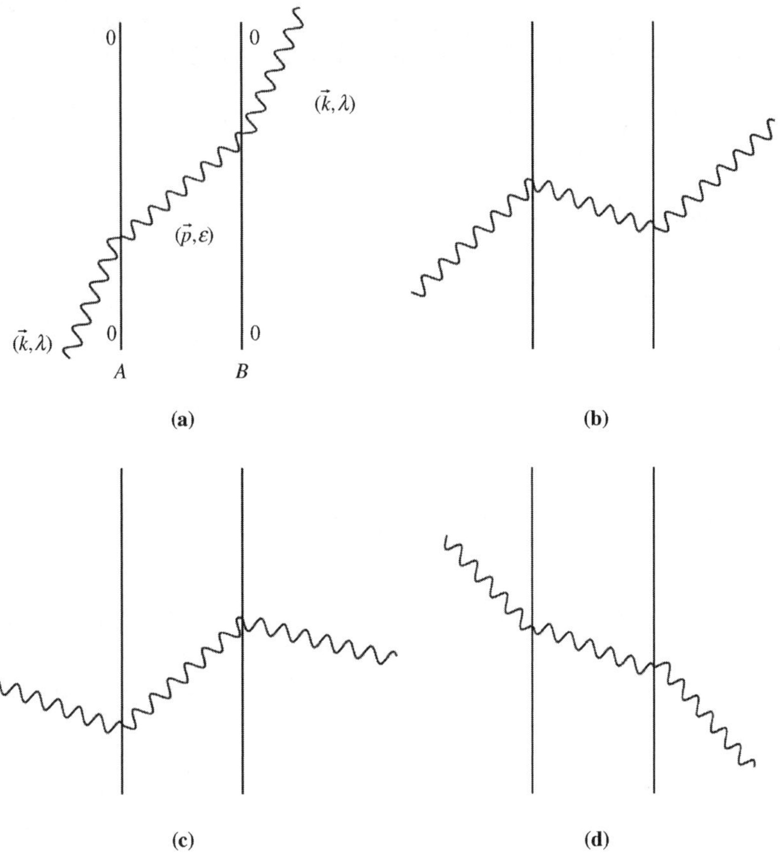

FIGURE 7.10 Time-ordered diagrams for dynamic mechanism containing collapsed interaction vertices.

two-photon interaction Hamiltonian can be interpreted as each molecule responding to the incident field via its frequency dependent polarizability and radiating a virtual photon. The product of the displacement field is easily written as

$$
\begin{aligned}
d_i^{\perp}(\vec{k})d_j^{\perp}(\vec{p}) = &-\left(\frac{\hbar c k \varepsilon_0}{2V}\right)^{1/2}\left(\frac{\hbar c p \varepsilon_0}{2V}\right)^{1/2} \\
&\times \Big[e_i^{(\lambda)}(\vec{k})e_j^{(\varepsilon)}(\vec{p})a^{(\lambda)}(\vec{k})a^{(\varepsilon)}(\vec{p})\mathrm{e}^{i(\vec{k}+\vec{p})\cdot\vec{R}} - e_i^{(\lambda)}(\vec{k})\bar{e}_j^{(\varepsilon)}(\vec{p}) \\
&\times a^{(\lambda)}(\vec{k})a^{\dagger(\varepsilon)}(\vec{p})\mathrm{e}^{i(\vec{k}-\vec{p})\cdot\vec{R}} \\
&-\bar{e}_i^{(\lambda)}(\vec{k})e_j^{(\varepsilon)}(\vec{p})a^{\dagger(\lambda)}(\vec{k})a^{(\varepsilon)}(\vec{p})\mathrm{e}^{-i(\vec{k}-\vec{p})\cdot\vec{R}} + \bar{e}_i^{(\lambda)}(\vec{k})\bar{e}_j^{(\varepsilon)}(\vec{p}) \\
&\times a^{\dagger(\lambda)}(\vec{k})a^{\dagger(\varepsilon)}(\vec{p})\mathrm{e}^{-i(\vec{k}+\vec{p})\cdot\vec{R}}\Big].
\end{aligned}
\tag{7.7.2}
$$

The energy shift is calculated using the second-order perturbation theory formula (5.4.13). Because there is no overall change in the state of the radiation field, only the second and third terms in the field operator expansion (7.7.2) are needed in the evaluation of matrix elements. With initial and final states given by (7.2.6), the sum of the contributions from the four graphs of Fig. 7.10 is found to be

$$\Delta E_{\text{dyn}} = \left(\frac{N\hbar ck}{2\varepsilon_0 V}\right)\bar{e}_i^{(\lambda)}(\vec{k})e_j^{(\lambda)}(\vec{k})\alpha_{ik}(A;k)\alpha_{jl}(B;k)$$
$$\times\left\{\begin{array}{l} e^{-i\vec{k}\cdot\vec{R}}\sum_{\vec{p},\varepsilon}\left(\frac{\hbar cp}{2\varepsilon_0 V}\right) \\ \times\left[e_k^{(\varepsilon)}(\vec{p})\bar{e}_l^{(\varepsilon)}(\vec{p})\frac{e^{i\vec{p}\cdot\vec{R}}}{\hbar ck-\hbar cp}+\bar{e}_k^{(\varepsilon)}(\vec{p})e_l^{(\varepsilon)}(\vec{p})\frac{e^{-i\vec{p}\cdot\vec{R}}}{-\hbar ck-\hbar cp}\right] \\ +e^{i\vec{k}\cdot\vec{R}}\sum_{\vec{p},\varepsilon}\left(\frac{\hbar cp}{2\varepsilon_0 V}\right) \\ \times\left[\bar{e}_k^{(\varepsilon)}(\vec{p})e_l^{(\varepsilon)}(\vec{p})\frac{e^{-i\vec{p}\cdot\vec{R}}}{\hbar ck-\hbar cp}+e_k^{(\varepsilon)}(\vec{p})\bar{e}_l^{(\varepsilon)}(\vec{p})\frac{e^{i\vec{p}\cdot\vec{R}}}{-\hbar ck-\hbar cp}\right] \end{array}\right\}, \tag{7.7.3}$$

on letting $N + 1$ equal N, which is appropriate for an intense laser. Expression (7.7.3) is identical to result (7.3.16).

As detailed in Section 7.4 and illustrated in Fig. 7.6, an extra contribution to the radiation-induced intermolecular energy shift occurs if A and B are polar, which according to perturbation theory is understood as arising from traversal of a virtual photon between the pair and scattering of a real incident photon exclusively at one center or the other. As in the computation of ΔE_{dyn}, the evaluation of the static contribution to the energy shift can be simplified considerably by collapsing the interaction vertices at the site at which the real photon is first absorbed/emitted and then emitted/absorbed and employing an effective nonlinear interaction Hamiltonian. Again, the 48 time-ordered diagrams that are required to be summed over when field operators that can only change the number of photons by one are used are reduced by a factor of 12 on collapsing interaction vertices. The 4 graphs, each of which represents the 12 time orderings exemplified by diagrams shown in Fig. 7.6a–d, are now drawn as in Fig. 7.11 and feature a 3-photon interaction vertex.

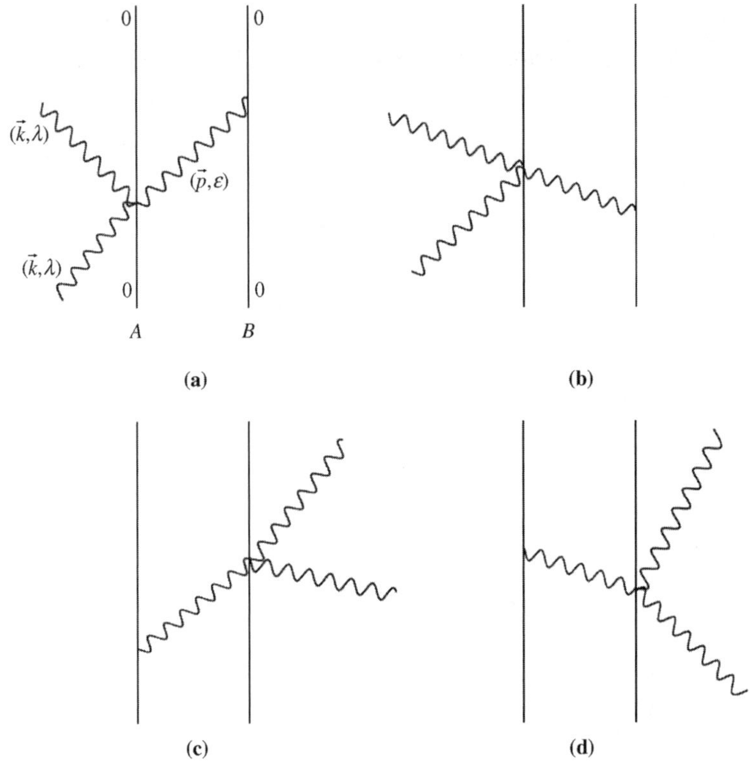

FIGURE 7.11 Static mechanism graphs containing collapsed interaction vertices.

It is sufficient to use second-order perturbation theory to compute the energy shift via formula (5.4.13). The appropriate effective interaction Hamiltonian takes the form

$$H_{\text{stat}}^{\text{eff, int}}(\xi) = -\varepsilon_0^{-1}\mu_i(\xi)d_i^{\perp}(\vec{p},\varepsilon;\vec{R}_\xi) \\ -\varepsilon_0^{-3}\beta_{ijk}(\xi;k)d_i^{\perp}(\vec{k},\lambda;\vec{R}_\xi)d_j^{\perp}(\vec{p},\varepsilon;\vec{R}_\xi)d_k^{\perp}(\vec{k},\lambda;\vec{R}_\xi), \\ \xi = A, B. \tag{7.7.4}$$

The second term of (7.7.4) represents the three-photon coupling and is interpreted as molecule ξ responding to the incident radiation field of mode $(\vec{k},\lambda)$ through its hyperpolarizability tensor, $\beta_{ijk}(\xi;k)$, by elastic scattering of a photon in the forward direction and by the radiation of a virtual photon of mode $(\vec{p},\varepsilon)$. An identical coupling term has been used to treat optical rotation in the two-group model (Craig and Thirunamachandran, 1998a,

Section 8.10). Computation of the energy shift is facilitated by calculating the trilinear field product

$$\begin{aligned} & d_i^{\perp}(\vec{k},\lambda;\vec{r})d_j^{\perp}(\vec{p},\varepsilon;\vec{r})d_k^{\perp}(\vec{k},\lambda;\vec{r}) \\ & \quad = i^3\left(\frac{\hbar ck\varepsilon_0}{2V}\right)\left(\frac{\hbar cp\varepsilon_0}{2V}\right)^{1/2} \\ & \qquad \times \left[e_i^{(\lambda)}(\vec{k})a^{(\lambda)}(\vec{k})\mathrm{e}^{i\vec{k}\cdot\vec{r}} - \bar{e}_i^{(\lambda)}(\vec{k})a^{\dagger(\lambda)}(\vec{k})\mathrm{e}^{-i\vec{k}\cdot\vec{r}}\right] \\ & \qquad \times \left[e_j^{(\varepsilon)}(\vec{p})a^{(\varepsilon)}(\vec{p})\mathrm{e}^{i\vec{p}\cdot\vec{r}} - \bar{e}_j^{(\varepsilon)}(\vec{p})a^{\dagger(\varepsilon)}(\vec{p})\mathrm{e}^{-i\vec{p}\cdot\vec{r}}\right] \\ & \qquad \times \left[e_k^{(\lambda)}(\vec{k})a^{(\lambda)}(\vec{k})\mathrm{e}^{i\vec{k}\cdot\vec{r}} - \bar{e}_k^{(\lambda)}(\vec{k})a^{\dagger(\lambda)}(\vec{k})\mathrm{e}^{-i\vec{k}\cdot\vec{r}}\right]. \end{aligned} \tag{7.7.5}$$

With the initial and final states given by (7.2.6), the sum of the contributions from the four graphs of Fig. 7.11 is found to be

$$\begin{aligned} \Delta E_{\text{stat}} = & -\sum_{\vec{p},\varepsilon}\left(\frac{N\hbar ck}{2\varepsilon_0 V}\right)\left(\frac{1}{2\varepsilon_0 V}\right)e_i^{(\lambda)}(\vec{k})\bar{e}_k^{(\lambda)}(\vec{k})e_j^{(\varepsilon)}(\vec{p})\bar{e}_l^{(\varepsilon)}(\vec{p}) \\ & \times [\mu_l^{00}(A)\beta_{ijk}(B;k) + \beta_{ijk}(A;k)\mu_l^{00}(B)](e^{i\vec{p}\cdot\vec{R}} + e^{-i\vec{p}\cdot\vec{R}}), \end{aligned} \tag{7.7.6}$$

which is equivalent to expression (7.4.9) and leads to the result (7.4.14) for the static contribution to the energy shift.

7.8 RADIATION-INDUCED INTERMOLECULAR INTERACTION VIA THE METHOD OF INDUCED MOMENTS

In Section 5.8, the induced multipole moment approach was introduced as an alternative physical viewpoint and calculational method for the evaluation of dispersion energy shifts. Not only the interactions between ground-state species were easily obtained but also coupling energies involving molecules in electronically excited states were derived more readily relative to diagrammatic perturbation theory techniques. The versatility of the method is now demonstrated by applying it to the computation of the radiation-induced change in intermolecular interaction energy (Craig and Thirunamachandran, 1999).

As in the calculation of dispersion forces, the central concept remains that fluctuations in the electromagnetic field induce multipole moments in polarizable molecules, which in turn couple via the resonant interaction tensor. Instead of calculating the expectation value of the interaction energy over the vacuum state of the radiation field, as was done for both ground- and excited-state contributions to the dispersion potential, for the computation of the modification of ΔE for an interacting molecular pair by electromagnetic radiation, the radiation field is now represented by the state $|N(\vec{k},\lambda)\rangle$ corresponding to an intense laser containing N photons. Apart from a few subtle changes, the formulas presented in Section 5.8 are applicable to the current problem. In this section, it is shown how the induced moment approach enables ΔE_{dyn} and ΔE_{stat} to be calculated in a facile manner (Salam, 2006b, 2007).

Let the incident laser be of mode $(\vec{k},\lambda)$. In a polarizable species ξ, the leading contribution to the induced electric dipole moment is given by

$$\mu_i^{\mathrm{ind}}(\xi) = \varepsilon_0^{-1}\alpha_{ij}(\xi;k)d_j^{\perp}(\vec{k},\lambda;\vec{R}_{\xi}), \tag{7.8.1}$$

where $\alpha_{ij}(\xi;k)$ is the anisotropic frequency dependent electric dipole polarizability, defined by (5.8.2). Coupling of the moments induced at each site through the resonant interaction tensor $V_{ij}(k,\vec{R})$, the latter given by (5.8.3), gives rise to the dynamic contribution to the energy shift,

$$\begin{aligned}\Delta E_{\mathrm{dyn}} &= \mu_i^{\mathrm{ind}}(A)\mu_j^{\mathrm{ind}}(B)\mathrm{Re}V_{ij}(k,\vec{R})\\ &= \varepsilon_0^{-2}\alpha_{ik}(A;k)\alpha_{jl}(B;k)d_k^{\perp}(\vec{k},\lambda;\vec{R}_A)d_l^{\perp}(\vec{k},\lambda;\vec{R}_B)\mathrm{Re}V_{ij}(k,\vec{R}),\end{aligned} \tag{7.8.2}$$

where the second line of (7.8.2) has been obtained on inserting (7.8.1). The expectation value of (7.8.2) is taken over the state $|0^A,0^B;N(\vec{k},\lambda)\rangle$. As earlier, the molecular part results in the ground-state electric dipole polarizability of each species. In contrast to the calculation of the dispersion interaction, where for the radiation field, the expectation value was taken over the spatial correlation function of the vacuum field, the product of transverse electric displacement fields at spatially different points is evaluated over a state of the field containing N photons in the present case. This quantity was evaluated previously and is given by expression (5.9.38). For an intense beam of laser light, $N+1 \approx N$. After making this approximation and inserting (5.9.38) into (7.8.2), the change in energy shift is found to be

$$\begin{aligned}\Delta E_{\mathrm{dyn}} = &\left(\frac{N\hbar ck}{2\varepsilon_0 V}\right)e_k^{(\lambda)}(\vec{k})\bar{e}_l^{(\lambda)}(\vec{k})\alpha_{ik}(A;k)\alpha_{jl}(B;k)\\ &\times(\mathrm{e}^{i\vec{k}\cdot\vec{R}}+\mathrm{e}^{-i\vec{k}\cdot\vec{R}})\mathrm{Re}V_{ij}(k,\vec{R}),\end{aligned} \tag{7.8.3}$$

which is identical to result (7.3.18) obtained using diagrammatic perturbation theory for a pair of anisotropic molecules in fixed relative orientation with respect to the incoming laser.

When either one or both of A and B are polar, it was shown in Section 7.4 that there is a contribution to the change in mutual interaction energy for oriented systems that depends on static intermolecular coupling. To calculate this term using the induced moment method, additional multipole moments induced by the external radiation field have to be accounted for. These extra terms are

$$\mu_i^{\text{ind}}(\xi) = \mu_i^s(\xi) + \varepsilon_0^{-1}\alpha_{ij}(\xi;k)d_j^{\perp}(\vec{k},\lambda;\vec{R}_\xi) + \varepsilon_0^{-2}\beta_{ijk}(\xi;k)d_j^{\perp}(\vec{k},\lambda;\vec{R}_\xi)d_k^{\perp}(\vec{k},\lambda;\vec{R}_\xi), \tag{7.8.4}$$

where $\mu_i^s(\xi)$ is the ith Cartesian component of the static electric dipole moment operator of molecule ξ and $\beta_{ijk}(\xi;k)$ is the first hyperpolarizability tensor. Substituting formula (7.8.4) for the induced dipole moment of each species into the expression for the interaction energy, retaining terms appropriate for the static contribution to the energy shift, namely, the contributions involving μ_i^s and β_{ijk}, and neglecting the term proportional to the molecular polarizability since this contribution was already accounted for in the computation of the dynamic term of the energy shift, ΔE_{stat} is obtained from

$$\Delta E_{\text{stat}} = [\mu_i^s(A) + \varepsilon_0^{-2}\beta_{ipq}(A;k)d_p^{\perp}(\vec{k},\lambda;\vec{R}_A)d_q^{\perp}(\vec{k},\lambda;\vec{R}_A)] \times [\mu_j^s(B) + \varepsilon_0^{-2}\beta_{jrs}(B;k)d_r^{\perp}(\vec{k},\lambda;\vec{R}_B)d_s^{\perp}(\vec{k},\lambda;\vec{R}_B)]V_{ij}(0,\vec{R}). \tag{7.8.5}$$

Since no energy is transferred between centers in the static mechanism, the $\omega \to 0$ limit of the resonant coupling tensor, $V_{ij}(k,\vec{R})$, (5.8.3)

$$V_{ij}(0,\vec{R}) = \frac{1}{4\pi\varepsilon_0 R^3}(\delta_{ij} - 3\hat{R}_i\hat{R}_j), \tag{7.8.6}$$

appears in expression (7.8.5). As expected, taking the expectation value over the ground state of the first term of (7.8.5) yields the Coulomb interaction energy between two ground-state permanent moments (7.4.20), evaluated in Section 7.4 using diagrammatic perturbation theory techniques.

To derive the field-induced energy shift, the expectation value of the cross terms in (7.8.5) is evaluated over the state $|0^A,0^B;N(\vec{k},\lambda)\rangle$,

$$\Delta E_{\text{stat}} = \langle N(\vec{k},\lambda);0^A,0^B|\frac{1}{2}\varepsilon_0^{-2}\{\mu_i^s(A)\beta_{jrs}(B;k)d_r^{\perp}(\vec{k},\lambda;\vec{R}_B)d_s^{\perp}(\vec{k},\lambda;\vec{R}_B) + \beta_{ipq}(A;k)\mu_j^s(B)d_p^{\perp}(\vec{k},\lambda;\vec{R}_A)d_q^{\perp}(\vec{k},\lambda;\vec{R}_A)\}V_{ij}(0,\vec{R})|0^B,0^A;N(\vec{k},\lambda)\rangle. \tag{7.8.7}$$

Index symmetry introduces the factor of one-half and enables the identity of A and B, which may be identical or different, to be distinguished. Interestingly, the average value for the radiation field part of the energy shift (7.8.7) involves the product of the electric displacement field at the same point in space. Using the mode representation for $d_i^{\perp}(\vec{k}, \lambda; \vec{r})$, this expectation value is easily calculated to be

$$\begin{aligned}&\langle N(\vec{k}, \lambda)| d_i^{\perp}(\vec{k}, \lambda; \vec{R}_{\xi}) d_j^{\perp}(\vec{k}, \lambda; \vec{R}_{\xi}) | N(\vec{k}, \lambda)\rangle \\ &= \left(\frac{\hbar c k \varepsilon_0}{2V}\right)\left[(N+1) e_i^{(\lambda)}(\vec{k}) \bar{e}_j^{(\lambda)}(\vec{k}) + N \bar{e}_i^{(\lambda)}(\vec{k}) e_j^{(\lambda)}(\vec{k})\right].\end{aligned} \tag{7.8.8}$$

Comparison of (7.8.8) with the field–field spatial correlation function (5.9.38) shows that the former may be obtained from the latter on letting the two points in space coincide. For an intense beam of incident laser light, it is justifiable to assume that $N + 1 \sim N$. In that case, the two terms within square brackets of (7.8.8) are seen to be complex conjugates of each other. After evaluating the expectation value of the molecular part using the matter states, (7.8.7) becomes

$$\Delta E_{\text{stat}} = \frac{I}{2\varepsilon_0 c} e_r^{(\lambda)}(\vec{k}) \bar{e}_s^{(\lambda)}(\vec{k}) \left[\mu_i^{00}(A) \beta_{jrs}(B; k) + \beta_{jrs}(A; k) \mu_i^{00}(B)\right] V_{ij}(0, \vec{R}), \tag{7.8.9}$$

where the definition of the irradiance of the laser $I = N\hbar c^2 k/V$ has been used, $\mu_i^{00}(\xi)$ is the ground-state permanent electric dipole moment, and the hyperpolarizability tensor is given explicitly by (7.4.7). Expression (7.8.9) is in agreement with the perturbative result (7.4.14).

7.9 DISCRIMINATORY INTERMOLECULAR INTERACTION IN A RADIATION FIELD: PERTURBATION THEORY

Intermolecular interactions between optically active species are discriminatory. They are dependent on the handedness of each molecule, chromophore or functional group. Examples were given of two such fundamental interactions between chiral entities in each of Chapters 4 and 5. One was the resonant transfer of electronic excitation energy, which was understood to arise from single virtual photon exchange between the pair of molecules. A second was the retarded van der Waals dispersion potential, which according to perturbation theory, was interpreted as arising from the exchange of two virtual photons. In each of these cases, the source of discrimination was due to inclusion of magnetic dipole coupling to the radiation field and the resulting interference of this interaction term with the leading electric dipole contribution to coupling. Thus far, the treatment of the change in intermolecular interaction due to the presence of an external radiation field has been restricted to the electric dipole approximation and the resulting modification of the energy shift—both static and dynamic terms—are independent of the chirality of either molecule. By taking account of the effects of magnetic dipole coupling, the shift in interaction energy between a pair of coupled chiral molecules when subject to an applied electromagnetic field is evaluated in this and the following two sections. This is first carried out using the techniques of diagrammatic time-dependent perturbation theory. Next, it is demonstrated how an identical result may be obtained with significantly reduced labor and technical sophistication by using the method of induced multipole moments. The approach is a straightforward extension of the calculation performed in the electric dipole approximation and presented in the previous section, and the application of this method to the computation of the chiral discrimination dispersion potential detailed in Section 5.9.3. A complete polarization analysis is then carried out for circularly polarized incoming radiation.

To evaluate the leading contribution to the change in interaction energy between optically active systems, the electric dipole coupling terms of the interaction Hamiltonian (7.2.5) are no longer sufficient. Magnetic dipole interaction terms must be added to give

$$\begin{aligned} H_{\text{int}} = & -\varepsilon_0^{-1}\vec{\mu}(A)\cdot\vec{d}^{\perp}(\vec{R}_A) - \vec{m}(A)\cdot\vec{b}(\vec{R}_A) \\ & -\varepsilon_0^{-1}\vec{\mu}(B)\cdot\vec{d}^{\perp}(\vec{R}_B) - \vec{m}(B)\cdot\vec{b}(\vec{R}_B) \end{aligned} \tag{7.9.1}$$

and the contribution proportional to the product of electric and magnetic dipole moments at each molecule is extracted. Again, the process involves scattering of a real photon at different centers and single virtual photon exchange. The initial and final states of the total matter–field system are given by (7.2.6), and the energy shift is computed using the expression for the fourth-order perturbation theory correction. Evaluation is aided by drawing of time-ordered diagrams. In total, 192 graphs contribute to ΔE. They may be grouped into 4 sets of 48 diagrams. The four sets of diagrams are identical to those that feature in the dynamic mechanism to the laser-induced intermolecular energy shift when working in the electric dipole

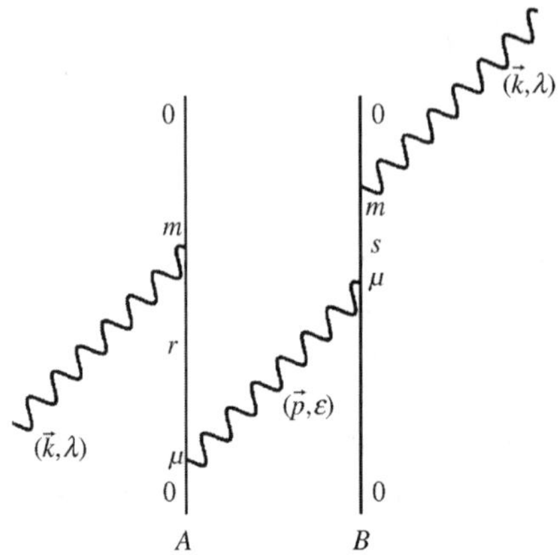

FIGURE 7.12 One of the 192 graphs for single virtual photon radiation-induced chiral discrimination.

approximation, with representatives of the quartet displayed in Figs. 7.2 and 7.3. The fourfold increase in the number of time orderings in the present case is due to replacement of one electric dipole interaction vertex by a magnetic dipole one simultaneously at each center. A typical graph is depicted in Fig. 7.12, one of the four possible time orderings associated with graph (i) of Fig. 7.2.

Evaluating in the usual way, the four contributions associated with permuting electric and magnetic interaction vertices in Fig. 7.12 whose energy denominator in each case is given by the entry for graph (i) in Table 7.1 produces the term

$$\begin{aligned}
&-\sum_{\vec{p},\varepsilon}\sum_{r,s}\left(\frac{N\hbar k}{2\varepsilon_0 V}\right)\left(\frac{\hbar p}{2\varepsilon_0 V}\right)\mu_i^{0r}(A)m_k^{r0}(A)\mu_j^{0s}(B)m_l^{s0}(B)\mathrm{e}^{-i\vec{k}\cdot\vec{R}}\mathrm{e}^{i\vec{p}\cdot\vec{R}}\\
&\times\Big[e_i^{(\lambda)}(\vec{k})\bar{e}_j^{(\lambda)}(\vec{k})\bar{b}_k^{(\varepsilon)}(\vec{p})b_l^{(\varepsilon)}(\vec{p})-e_i^{(\lambda)}(\vec{k})\bar{b}_l^{(\lambda)}(\vec{k})e_j^{(\varepsilon)}(\vec{p})\bar{b}_k^{(\varepsilon)}(\vec{p})\\
&\quad-\bar{e}_j^{(\lambda)}(\vec{k})b_k^{(\lambda)}(\vec{k})\bar{e}_i^{(\varepsilon)}(\vec{p})b_l^{(\varepsilon)}(\vec{p})+b_k^{(\lambda)}(\vec{k})\bar{b}_l^{(\lambda)}(\vec{k})\bar{e}_i^{(\varepsilon)}(\vec{p})e_j^{(\varepsilon)}(\vec{p})\Big]\\
&\times\left[(E_{r0}+\hbar cp)(E_{r0}+E_{s0})(E_{s0}-\hbar ck)\right]^{-1},
\end{aligned} \tag{7.9.2}$$

where use has been made of the relations $m_k^{0r}\mu_i^{r0}=-\mu_i^{0r}m_k^{r0}$ and $m_l^{0s}\mu_j^{s0}=-\mu_j^{0s}m_l^{s0}$. The remaining 44 graphs of this set, obtained from the 11 time orderings illustrated by graphs (ii)–(xii) of Fig. 7.2, with denominators given in Table 7.1 may be computed similarly and added to (7.9.2) to give

$$-\sum_{\vec{p},\varepsilon}\sum_{r,s}\left(\frac{N\hbar k}{2\varepsilon_0 V}\right)\left(\frac{\hbar p}{2\varepsilon_0 V}\right)\mu_i^{0r}(A)m_k^{r0}(A)\mu_j^{0s}(B)m_l^{s0}(B)\mathrm{e}^{-i\vec{k}\cdot\vec{R}}\mathrm{e}^{i\vec{p}\cdot\vec{R}}$$
$$\times\Big[b_k^{(\lambda)}(\vec{k})\bar{b}_l^{(\lambda)}(\vec{k})\bar{e}_i^{(\varepsilon)}(\vec{p})e_j^{(\varepsilon)}(\vec{p})+e_i^{(\lambda)}(\vec{k})\bar{e}_j^{(\lambda)}(\vec{k})\bar{b}_k^{(\varepsilon)}(\vec{p})b_l^{(\varepsilon)}(\vec{p})$$
$$-e_i^{(\lambda)}(\vec{k})\bar{b}_l^{(\lambda)}(\vec{k})e_j^{(\varepsilon)}(\vec{p})\bar{b}_k^{(\varepsilon)}(\vec{p})-\bar{e}_j^{(\lambda)}(\vec{k})b_k^{(\lambda)}(\vec{k})\bar{e}_i^{(\varepsilon)}(\vec{p})b_l^{(\varepsilon)}(\vec{p})\Big]$$
$$\times\left[\frac{1}{E_i}+\frac{1}{E_{ii}}-\frac{1}{E_{iii}}+\frac{1}{E_{iv}}-\frac{1}{E_v}+\frac{1}{E_{vi}}+\frac{1}{E_{vii}}-\frac{1}{E_{viii}}-\frac{1}{E_{ix}}\right.$$
$$\left.+\frac{1}{E_x}-\frac{1}{E_{xi}}-\frac{1}{E_{xii}}\right], \tag{7.9.3}$$

where $E_a^{-1}, a=i-xii$ refer to energy denominators corresponding to graphs (i)–(xii) of Fig. 7.2 and listed in Table 7.1. It is now shown how the denominator sum may be simplified.

$$\frac{1}{E_i}+\frac{1}{E_x}+\frac{1}{E_{ii}}=\frac{1}{(E_{r0}+\hbar ck)(E_{s0}-\hbar ck)(E_{r0}+\hbar cp)}$$
$$+\frac{1}{(E_{r0}+\hbar cp)(\hbar cp-\hbar ck)(E_{s0}-\hbar ck)}$$
$$=\frac{1}{(E_{r0}+\hbar ck)(E_{s0}-\hbar ck)(\hbar cp-\hbar ck)}. \tag{7.9.4}$$

$$-\left(\frac{1}{E_{ix}}+\frac{1}{E_{xi}}\right)=-\frac{1}{(E_{r0}+\hbar ck)(E_{s0}+\hbar ck)(E_{r0}+\hbar cp)}. \tag{7.9.5}$$

$$-\left(\frac{1}{E_{viii}}+\frac{1}{E_{xii}}+\frac{1}{E_{iii}}\right)=-\frac{1}{(E_{r0}+\hbar cp)(E_{s0}+\hbar ck)(E_{s0}+\hbar cp)}$$
$$-\frac{1}{(E_{r0}+\hbar cp)(\hbar cp-\hbar ck)(E_{s0}+\hbar cp)}$$
$$=-\frac{1}{(E_{r0}+\hbar cp)(E_{s0}+\hbar ck)(\hbar cp-\hbar ck)}. \tag{7.9.6}$$

$$\frac{1}{E_{vi}} + \frac{1}{E_{vii}} + \frac{1}{E_{iv}} = \frac{1}{(E_{r0} - \hbar ck)(E_{s0} + \hbar ck)(E_{s0} + \hbar cp)} + \frac{1}{(E_{r0} - \hbar ck)(\hbar cp - \hbar ck)(E_{s0} + \hbar cp)} = \frac{1}{(E_{r0} - \hbar ck)(E_{s0} + \hbar ck)(\hbar cp - \hbar ck)}. \quad (7.9.7)$$

Adding the right-hand side of (7.9.6) to the right-hand side of (7.9.5) gives

$$-\frac{1}{(E_{r0} + \hbar ck)(E_{s0} + \hbar ck)(\hbar cp - \hbar ck)}. \quad (7.9.8)$$

Finally, adding (7.9.4), (7.9.7), and (7.9.8) to $-E_v^{-1}$ results in the sum of energy denominators being given by

$$\frac{1}{(\hbar ck - \hbar cp)}\left[\frac{1}{(E_{r0} - \hbar ck)(E_{s0} - \hbar ck)} - \frac{1}{(E_{r0} - \hbar ck)(E_{s0} + \hbar ck)} - \frac{1}{(E_{r0} + \hbar ck)(E_{s0} - \hbar ck)} + \frac{1}{(E_{r0} + \hbar ck)(E_{s0} + \hbar ck)}\right]. \quad (7.9.9)$$

From the definition of the anisotropic dynamic mixed electric–magnetic dipole polarizability tensor $G_{ij}(\xi;k)$(5.9.16), multiplying the molecular factor appearing in (7.9.3) by (7.9.9) yields

$$\sum_{r,s} \mu_i^{0r}(A) m_k^{r0}(A) \mu_j^{0s}(B) m_l^{s0}(B) \left\{\frac{1}{(E_{r0} - \hbar ck)} - \frac{1}{(E_{r0} + \hbar ck)}\right\} \times \left\{\frac{1}{(E_{s0} - \hbar ck)} - \frac{1}{(E_{s0} + \hbar ck)}\right\} \frac{1}{(\hbar ck - \hbar cp)} = G_{ik}(A;k) G_{jl}(B;k)(\hbar ck - \hbar cp)^{-1}. \quad (7.9.10)$$

Hence, (7.9.3) can be written as

$$-\sum_{\vec{p},\varepsilon} \left(\frac{N\hbar k}{2\varepsilon_0 V}\right)\left(\frac{\hbar p}{2\varepsilon_0 V}\right) G_{ik}(A;k) G_{jl}(B;k) \mathrm{e}^{-i\vec{k}\cdot\vec{R}} \mathrm{e}^{i\vec{p}\cdot\vec{R}} \frac{1}{(\hbar ck - \hbar cp)} \times \Big[b_k^{(\lambda)}(\vec{k}) \bar{b}_l^{(\lambda)}(\vec{k}) \bar{e}_i^{(\varepsilon)}(\vec{p}) e_j^{(\varepsilon)}(\vec{p}) + e_i^{(\lambda)}(\vec{k}) \bar{e}_j^{(\lambda)}(\vec{k}) \bar{b}_k^{(\varepsilon)}(\vec{p}) b_l^{(\varepsilon)}(\vec{p}) - e_i^{(\lambda)}(\vec{k}) \bar{b}_l^{(\lambda)}(\vec{k}) e_j^{(\varepsilon)}(\vec{p}) \bar{b}_k^{(\varepsilon)}(\vec{p}) - \bar{e}_j^{(\lambda)}(\vec{k}) b_k^{(\lambda)}(\vec{k}) \bar{e}_i^{(\varepsilon)}(\vec{p}) b_l^{(\varepsilon)}(\vec{p})\Big]. \quad (7.9.11)$$

The contribution from the remaining 3 sets of 48 graphs may be similarly evaluated and summed and then added to (7.9.11) to give

$$\Delta E = -\sum_{\vec{p},\varepsilon}\left(\frac{N\hbar k}{2\varepsilon_0 V}\right)\left(\frac{\hbar p}{2\varepsilon_0 V}\right)G_{ik}(A;k)G_{jl}(B;k)$$
$$\times\left[b_k^{(\lambda)}(\vec{k})\bar{b}_l^{(\lambda)}(\vec{k})\bar{e}_i^{(\varepsilon)}(\vec{p})e_j^{(\varepsilon)}(\vec{p})+e_i^{(\lambda)}(\vec{k})\bar{e}_j^{(\lambda)}(\vec{k})\bar{b}_k^{(\varepsilon)}(\vec{p})b_l^{(\varepsilon)}(\vec{p})\right.$$
$$\left.-e_i^{(\lambda)}(\vec{k})\bar{b}_l^{(\lambda)}(\vec{k})e_j^{(\varepsilon)}(\vec{p})\bar{b}_k^{(\varepsilon)}(\vec{p})-\bar{e}_j^{(\lambda)}(\vec{k})b_k^{(\lambda)}(\vec{k})\bar{e}_i^{(\varepsilon)}(\vec{p})b_l^{(\varepsilon)}(\vec{p})\right]$$
$$\times\left[\frac{e^{i\vec{k}\cdot\vec{R}}e^{i\vec{p}\cdot\vec{R}}}{-\hbar ck-\hbar cp}+\frac{e^{i\vec{k}\cdot\vec{R}}e^{-i\vec{p}\cdot\vec{R}}}{\hbar ck-\hbar cp}+\frac{e^{-i\vec{k}\cdot\vec{R}}e^{i\vec{p}\cdot\vec{R}}}{\hbar ck-\hbar cp}+\frac{e^{-i\vec{k}\cdot\vec{R}}e^{-i\vec{p}\cdot\vec{R}}}{-\hbar ck-\hbar cp}\right]. \tag{7.9.12}$$

To proceed further, the familiar steps associated with the summation of virtual photon variables are carried out. The various polarization sums are executed with the aid of identities (1.4.56) to (1.4.58). After converting the wavevector sum to an integral, the angular averages are performed using relations (4.2.12) and (4.4.7). The ensuing p-integrals are evaluated using the results

$$\frac{1}{2\pi^2\varepsilon_0}\int_0^\infty\frac{p^4}{(k^2-p^2)}\left\{(\delta_{ij}-\hat{R}_i\hat{R}_j)\frac{\sin pR}{pR}+(\delta_{ij}-3\hat{R}_i\hat{R}_j)\left(\frac{\cos pR}{p^2R^2}-\frac{\sin pR}{p^3R^3}\right)\right\}\mathrm{d}p$$
$$= -\mathrm{Re}V_{ij}(k,\vec{R}), \tag{7.9.13}$$

and

$$\frac{k}{2\pi^2\varepsilon_0 c}\varepsilon_{ijk}\hat{R}_k\int_0^\infty\frac{1}{(k^2-p^2)}\left(\frac{p^2\cos pR}{R}-\frac{p\sin pR}{R^2}\right)\mathrm{d}p = \mathrm{Im}U_{ij}(k,\vec{R}), \tag{7.9.14}$$

where $V_{ij}(k,\vec{R})$ and $U_{ij}(k,\vec{R})$ are defined by (4.2.17) and (4.4.11), respectively. In the last two relations, the $\pm$ superscripts have been dropped from the interaction tensors as the operators preceding $V_{ij}(k,\vec{R})$ and $U_{ij}(k,\vec{R})$ render the distinction in signs superfluous. The resulting change in energy shift between a pair of interacting oriented chiral molecules due to the presence of an intense electromagnetic field is

$$\Delta E = \left(\frac{N\hbar k}{\varepsilon_0 V}\right) G_{ik}(A;k)G_{jl}(B;k)\left\{\left[b_k^{(\lambda)}(\vec{k})\bar{b}_l^{(\lambda)}(\vec{k})\,\mathrm{Re}V_{ij}(k,\vec{R})\right.\right.$$
$$\left.+e_i^{(\lambda)}(\vec{k})\bar{e}_j^{(\lambda)}(\vec{k})\,\mathrm{Re}V_{kl}(k,\vec{R})\right]-c\left[e_i^{(\lambda)}(\vec{k})\bar{b}_l^{(\lambda)}(\vec{k})\,\mathrm{Im}\,U_{jk}(k,\vec{R})\right.$$
$$\left.\left.-e_j^{(\lambda)}(\vec{k})\bar{b}_k^{(\lambda)}(\vec{k})\,\mathrm{Im}\,U_{il}(k,\vec{R})\right]\right\}\cos(\vec{k}\cdot\vec{R}). \qquad (7.9.15)$$

Before going on to derive results for randomly oriented and freely tumbling molecular species and carrying out a polarization analysis, it is shown how the energy shift (7.9.15) may be obtained using the method of induced multipole moments.

7.10 RADIATION-INDUCED CHIRAL DISCRIMINATION: INDUCED MOMENT METHOD

In Section 5.9.3, it was shown how the method of induced multipole moments could be applied to calculate the discriminatory retarded dispersion potential between a pair of optically active molecules. To correctly account for higher multipole allowed transitions in such systems, the electric dipole approximation was relaxed and magnetic dipole moments induced by fluctuating electromagnetic fields were included in the formalism, a consequence of the fact that the leading electric dipole polarizability is no longer sufficient to describe the characteristics of chiral species. It was shown that discriminatory effects in the energy shift for isotropic systems arose from electric dipole–dipole and magnetic dipole–dipole terms, as well as from the interference of electric dipole–magnetic dipole couplings. On taking the expectation value of the ground state of the matter–field system of the interaction of these moments with the appropriate resonant coupling tensor, the dispersion energy shift resulted. The method is now extended to treat the change in the mutual energy of interaction of a pair of chiral molecules when subject to an external radiation field (Salam, 2006a). As demonstrated in Section 7.8, the computation of the leading contribution to the laser-induced intermolecular energy shift via the method of induced moments involves evaluating the expectation value of the dipole–dipole coupling term over a state of the radiation field containing $N(\vec{k},\lambda)$ photons with both molecules in the electronic ground state. This method is now applied to a pair of chiral molecules.

Consider two interacting optically active molecules each possessing mixed electric–magnetic dipole polarizability $G_{ij}(\xi;k)$, $\xi = A, B$ given by (5.9.16). Application of an electromagnetic field induces both electric and magnetic dipole moments as expressed in relations (5.9.32) and (5.9.33), respectively. If fluctuations of the field are in the same mode, the interaction between the moments induced at each center occurs in resonance, enabling the energy shift to be written as

$$\Delta E = \left[\mu_i^{\mathrm{ind}}(A)\mu_j^{\mathrm{ind}}(B) + \frac{1}{c^2} m_i^{\mathrm{ind}}(A) m_j^{\mathrm{ind}}(B)\right] \mathrm{Re}V_{ij}(k,\vec{R}) + [\mu_i^{\mathrm{ind}}(A) m_j^{\mathrm{ind}}(B) + m_i^{\mathrm{ind}}(A)\mu_j^{\mathrm{ind}}(B)]\mathrm{Im}U_{ij}(k,\vec{R}), \qquad (7.10.1)$$

where the resonant coupling tensors $V_{ij}(k,\vec{R})$ and $U_{ij}(k,\vec{R})$ are given by (4.2.17) and (4.4.11), respectively. An expression for the energy shift in terms of $G_{ij}(\xi;k)$ may be obtained by substituting for the induced moments (5.9.32) and (5.9.33) producing

$$\begin{aligned}\Delta E = {} & [G_{ik}(A;k)G_{jl}(B;k)b_k(\vec{R}_A)b_l(\vec{R}_B) \\ & + \frac{1}{\varepsilon_0^2 c^2} G_{ki}(A;k)G_{lj}(B;k)d_k^{\perp}(\vec{R}_A)d_l^{\perp}(\vec{R}_B)]\mathrm{Re}V_{ij}(k,\vec{R}) \\ & + \varepsilon_0^{-1}[G_{ik}(A;k)G_{lj}(B;k)b_k(\vec{R}_A)d_l^{\perp}(\vec{R}_B) \\ & + G_{ki}(A;k)G_{jl}(B;k)d_k^{\perp}(\vec{R}_A)b_l(\vec{R}_B)]\mathrm{Im}U_{ij}(k,\vec{R}). \qquad (7.10.2)\end{aligned}$$

Next, the expectation value is taken of (7.10.2) for the state $|0^A, 0^B; N(\vec{k},\lambda)\rangle$. As previously, the molecular part results in ground-state mixed electric–magnetic dipole dynamic polarizabilities. For the radiation field part, use is made of the expectation value over the field state $|N(\vec{k},\lambda)\rangle$ of the four combinations of field–field spatial correlation functions given by (5.9.38) to (5.9.41). Examining, for instance, the second term occurring within the first set of square brackets in (7.10.2), its expectation value on using (5.9.38) and assuming high photon occupation number is

$$\begin{aligned}& \langle N(\vec{k},\lambda); E_0^B; E_0^A|\varepsilon_0^{-2}c^{-2}G_{ki}(A;k)G_{lj}(B;k)d_k^{\perp}(\vec{R}_A)d_l^{\perp}(\vec{R}_B) \\ & \quad \times \mathrm{Re}V_{ij}(k,\vec{R})|E_0^A; E_0^B; N(\vec{k},\lambda)\rangle \\ & = \left(\frac{N\hbar k}{2\varepsilon_0 cV}\right) G_{ki}(A;k)G_{lj}(B;k) \\ & \quad \times \left\{ e_k^{(\lambda)}(\vec{k})\bar{e}_l^{(\lambda)}(\vec{k})\mathrm{e}^{-i\vec{k}\cdot\vec{R}} + \bar{e}_k^{(\lambda)}(\vec{k})e_l^{(\lambda)}(\vec{k})\mathrm{e}^{i\vec{k}\cdot\vec{R}} \right\} \mathrm{Re}V_{ij}(k,\vec{R}).\end{aligned} \qquad (7.10.3)$$

Noting that the term within braces are complex conjugates of each other, allowing twice the real part to be taken, (7.10.3) becomes

$$\left(\frac{N\hbar k}{\varepsilon_0 cV}\right) G_{ki}(A;k)G_{lj}(B;k)e_k^{(\lambda)}(\vec{k})\bar{e}_l^{(\lambda)}(\vec{k})\mathrm{Re}V_{ij}(k,\vec{R})\cos(\vec{k}\cdot\vec{R}). \quad (7.10.4)$$

Similar evaluation of the three remaining terms of (7.10.2) when added to (7.10.4) results in the energy shift

$$\begin{aligned}\Delta E = &\left(\frac{N\hbar k}{\varepsilon_0 cV}\right)\left\{\left[G_{ik}(A;k)G_{jl}(B;k)b_k^{(\lambda)}(\vec{k})\bar{b}_l^{(\lambda)}(\vec{k})\right.\right. \\ &\left. + G_{ki}(A;k)G_{lj}(B;k)e_k^{(\lambda)}(\vec{k})\bar{e}_l^{(\lambda)}(\vec{k})\right] \mathrm{Re}V_{ij}(k,\vec{R}) \\ &+ c\left[G_{ik}(A;k)G_{lj}(B;k)b_k^{(\lambda)}(\vec{k})\bar{e}_l^{(\lambda)}(\vec{k})\right. \\ &\left.\left. + G_{ki}(A;k)G_{jl}(B;k)e_k^{(\lambda)}(\vec{k})\bar{b}_l^{(\lambda)}(\vec{k})\right]\mathrm{Im}U_{ij}(k,\vec{R})\right\} \cos(\vec{k}\cdot\vec{R}),\end{aligned} \quad (7.10.5)$$

which holds for A and B oriented relative to each other and to the direction of the incident laser. Expression (7.10.5), after index manipulation, is seen to be identical to the result (7.9.15) obtained using perturbative techniques. Carrying out the pair orientational average via

$$\langle G_{ik}(A;k)G_{jl}(B;k)\rangle = \delta_{ik}\delta_{jl}G(A;k)G(B;k) \quad (7.10.6)$$

produces for the energy shift the formula

$$\begin{aligned}\Delta E = &\left(\frac{N\hbar k}{\varepsilon_0 cV}\right) G(A;k)G(B;k)\left\{\left[b_i^{(\lambda)}(\vec{k})\bar{b}_j^{(\lambda)}(\vec{k}) + e_i^{(\lambda)}(\vec{k})\bar{e}_j^{(\lambda)}(\vec{k})\right]\right. \\ &\times\mathrm{Re}V_{ij}(k,\vec{R}) + c\left[b_i^{(\lambda)}(\vec{k})\bar{e}_j^{(\lambda)}(\vec{k}) + e_i^{(\lambda)}(\vec{k})\bar{b}_j^{(\lambda)}(\vec{k})\right] \\ &\left.\times\mathrm{Im}U_{ij}(k,\vec{R})\right\} \cos(\vec{k}\cdot\vec{R}),\end{aligned} \quad (7.10.7)$$

which holds for isotropic A and B. The energy shift is dependent on the chirality of each molecule through the polarizability $G(\xi;k)$, which changes sign when one molecule is replaced by its enantiomer. Expression (7.10.7) forms a convenient starting point for carrying out a polarization analysis of the incident laser and its effect on the energy shift. As in the case of the energy shift within the electric dipole approximation when an identical analysis was performed, the incoming field is taken to be either linearly or circularly polarized and propagating either parallel or perpendicular to $\vec{R}$. When $\vec{k}||\vec{R}$, $\vec{b} \perp \vec{R}$, and $\cos(\vec{k}\cdot\vec{R}) = \cos kR$ while if $\vec{k} \perp \vec{R}$, then $\vec{b}||\vec{R}$ and $\cos \vec{k}\cdot\vec{R} = 1$ with $\vec{e}$, $\vec{b}$, and $\hat{k}$ forming a right-handed frame of vectors.

7.10.1 Linearly Polarized Radiation

For $\vec{k}||\vec{R}$, inserting $\mathrm{Re}V_{ij}(k,\vec{R})$ and $\mathrm{Im}\,U_{ij}(k,\vec{R})$ into (7.10.7) produces for the energy shift

$$\begin{aligned}\Delta E_{\text{lin}}^{||} = \left(\frac{N\hbar k}{4\pi\varepsilon_0^2 cR^3 V}\right) G(A;k)G(B;k)\Bigg\{ &\left[b_i^{(\lambda)}(\vec{k})\bar{b}_j^{(\lambda)}(\vec{k}) + e_i^{(\lambda)}(\vec{k})\bar{e}_j^{(\lambda)}(\vec{k})\right] \\ &\times \left[(\delta_{ij}-3\hat{R}_i\hat{R}_j)(\cos kR + kR\sin kR) - (\delta_{ij}-\hat{R}_i\hat{R}_j)k^2R^2\cos kR\right] \\ &+ \varepsilon_{ijk}\hat{R}_k\left[e_i^{(\lambda)}(\vec{k})\bar{b}_j^{(\lambda)}(\vec{k}) + b_i^{(\lambda)}(\vec{k})\bar{e}_j^{(\lambda)}(\vec{k})\right] \\ &\times \left[kR\cos kR + k^2R^2\sin kR\right]\Bigg\}\cos kR. \end{aligned} \tag{7.10.8}$$

Using $b_j = (\hat{k}\times\vec{e})_j = \varepsilon_{jmn}\hat{k}_m e_n$ in the second term within braces with $\varepsilon_{jki}\varepsilon_{jmn} = \delta_{km}\delta_{in} - \delta_{kn}\delta_{im}$ and contracting results in

$$\begin{aligned}\Delta E_{\text{lin}}^{||} = \frac{I}{2\pi\varepsilon_0^2 c^3 R^3} G(A;k)G(B;k)[&\cos kR + kR(\cos kR + \sin kR) \\ &+ k^2R^2(\sin kR - \cos kR)]\cos kR,\end{aligned} \tag{7.10.9}$$

on using the definition of the irradiance $I = N\hbar c^2 k/V$. In the far zone, $kR \gg 1$ and the limiting form is

$$\Delta E_{\text{lin}}^{||}(\text{FZ}) = \frac{Ik^2}{2\pi\varepsilon_0^2 c^3 R} G(A;k)G(B;k)(\sin kR - \cos kR)\cos kR, \tag{7.10.10}$$

having an R^{-1} dependence on separation distance. At short distances, $kR \ll 1$ and the energy shift (7.10.9) exhibits inverse cube behavior,

$$\Delta E_{\text{lin}}^{||}(\text{NZ}) = \frac{I}{2\pi\varepsilon_0^2 c^3 R^3} G(A;k)G(B;k). \tag{7.10.11}$$

For $\vec{k} \perp \vec{R}$, $\cos \vec{k}\cdot\vec{R} = 1$ and the energy shift is given by (7.10.7) on inserting $\cos \vec{k}\cdot\vec{R} = 1$. Since $\varepsilon_{jki}\varepsilon_{jmn}e_i\bar{e}_n\hat{R}_k\hat{R}_m = 0$ for this particular configuration, the energy shift is

$$\Delta E_{\text{lin}}^{\perp} = -\frac{I}{\pi\varepsilon_0^2 c^3 R^3} G(A;k)G(B;k)(\cos kR + kR\sin kR). \tag{7.10.12}$$

Its asymptotic limits are

$$\Delta E_{\mathrm{lin}}^{\perp}(\mathrm{FZ}) = -\frac{Ik}{\pi\varepsilon_0^2 c^3 R^2} G(A;k)G(B;k)\sin kR, \tag{7.10.13}$$

which has a modulated inverse square dependence on R and

$$\Delta E_{\mathrm{lin}}^{\perp}(\mathrm{NZ}) = -\frac{I}{\pi\varepsilon_0^2 c^3 R^3} G(A;k)G(B;k) \tag{7.10.14}$$

exhibiting R^{-3} behavior.

7.10.2 Circularly Polarized Radiation

To examine the effect of circular polarization on the energy shift, use is made of identity (7.6.6) involving the product of circularly polarized electric polarization vectors. Also required are magnetic–magnetic and electric–magnetic combinations. These are derived from (7.6.6) together with use of the relation $b_i^{(\mathrm{L/R})}(\vec{k}) = \mp i e_i^{(\mathrm{L/R})}(\vec{k})$. Hence,

$$\begin{aligned} b_i^{(\mathrm{L/R})}(\vec{k})\bar{b}_j^{(\mathrm{L/R})}(\vec{k}) &= \left[\mp i e_i^{(\mathrm{L/R})}(\vec{k})\right] * \left[\pm i \bar{e}_j^{(\mathrm{L/R})}(\vec{k})\right] \\ &= \frac{1}{2}\left[(\delta_{ij} - \hat{k}_i\hat{k}_j) \mp i\varepsilon_{ijk}\hat{k}_k\right] \end{aligned} \tag{7.10.15}$$

and

$$e_i^{(\mathrm{L/R})}(\vec{k})\bar{b}_j^{(\mathrm{L/R})}(\vec{k}) = \pm i e_i^{(\mathrm{L/R})}(\vec{k})\bar{e}_j^{(\mathrm{L/R})}(\vec{k}) = \frac{1}{2}\left[\pm i(\delta_{ij} - \hat{k}_i\hat{k}_j) + \varepsilon_{ijk}\hat{k}_k\right]. \tag{7.10.16}$$

From (7.10.7), for $\vec{k}||\vec{R}$ for which $\cos(\vec{k}\cdot\vec{R}) = \cos kR$,

$$\begin{aligned} \Delta E_{\mathrm{L/R}}^{||} = \left(\frac{N\hbar k}{\varepsilon_0 c V}\right) G(A;k)G(B;k) \\ \times\Bigg\{ \left[b_i^{(\mathrm{L/R})}(\vec{k})\bar{b}_j^{(\mathrm{L/R})}(\vec{k}) + e_i^{(\mathrm{L/R})}(\vec{k})\bar{e}_j^{(\mathrm{L/R})}(\vec{k})\right] \mathrm{Re}V_{ij}(k,\vec{R}) \\ + c\left[e_i^{(\mathrm{L/R})}(\vec{k})\bar{b}_j^{(\mathrm{L/R})}(\vec{k}) + b_i^{(\mathrm{L/R})}(\vec{k})\bar{e}_j^{(\mathrm{L/R})}(\vec{k})\right] \mathrm{Im}U_{ij}(k,\vec{R})\Bigg\}\cos(kR). \end{aligned} \tag{7.10.17}$$

Noting that the first term within braces of (7.10.17) is symmetric in the indices i and j, while the second term is i,j-antisymmetric, only the i,j-symmetric parts of identities (7.6.6) and (7.10.15) and the i,j-antisymmetric part of (7.10.16) contribute, respectively. On tensor contraction, after substituting for the resonant coupling tensors and using the relation $\varepsilon_{ijk}\varepsilon_{ijl} = 2\delta_{kl}$, it is found that the energy shift is equal to the expression obtained when linearly polarized light propagates parallel to $\vec{R}$, equation (7.10.9), that is, $\Delta E^{||}_{\mathrm{L/R}} = \Delta E^{||}_{\mathrm{lin}}$.

For perpendicular propagation of circularly polarized light, $\vec{k} \perp \vec{R}$, the energy shift is given by (7.10.17) on inserting $\cos kR = 1$. After substituting the appropriate identities involving circular polarization vectors and contracting, with the second term within braces of (7.10.17) vanishing, the interaction energy is given by

$$\Delta E^{\perp}_{\mathrm{L/R}} = -\frac{I}{4\pi\varepsilon_0^2 c^3 R^3} G(A;k)G(B;k)\left[\cos kR + kR\sin kR + k^2R^2\cos kR\right]. \tag{7.10.18}$$

The limiting forms at the extremes of intermolecular separation are

$$\Delta E^{\perp}_{\mathrm{L/R}}(\mathrm{FZ}) = -\frac{Ik^2}{4\pi\varepsilon_0^2 c^3 R} G(A;k)G(B;k)\cos kR \tag{7.10.19}$$

and

$$\Delta E^{\perp}_{\mathrm{L/R}}(\mathrm{NZ}) = -\frac{I}{4\pi\varepsilon_0^2 c^3 R^3} G(A;k)G(B;k). \tag{7.10.20}$$

A common feature in all of the results obtained involving interacting chiral molecules is the discriminatory nature of the energy shift for chemically identical species, changing sign when one of the pairs is exchanged for its optical isomer. Also worthy of remark is that the inspection of results shows that for both types of polarization, the energy shift is repulsive for parallel propagation but is attractive for a perpendicular arrangement of $\hat{k}$ and $\vec{R}$.

7.11 FREELY TUMBLING CHIRAL PAIR IN THE PRESENCE OF CIRCULARLY POLARIZED LIGHT

For a pair of coupled chiral molecules subject to circularly polarized radiation in the fluid phase, not only are the orientations of the two species relative to each other random but also all possible directions of the A–B separation distance vector $\vec{R}$ are allowed relative to the laser propagation direction, $\vec{k}$. An exact expression for the energy shift can be calculated for the molecular and tumble averaged situations without having to make the high photon occupation number approximation.

Substituting the field–field spatial correlation functions (5.9.38) to (5.9.41) for circularly polarized radiation into the energy shift expression (7.10.2) yields

$$\Delta E = \left(\frac{\hbar k}{2\varepsilon_0 cV}\right)\Bigg\{G_{ik}(A;k)G_{jl}(B;k)\Big[(N+1)b_k^{(\mathrm{L/R})}(\vec{k})\bar{b}_l^{(\mathrm{L/R})}(\vec{k})\mathrm{e}^{-i\vec{k}\cdot\vec{R}}$$
$$+N\bar{b}_k^{(\mathrm{L/R})}(\vec{k})b_l^{(\mathrm{L/R})}(\vec{k})\mathrm{e}^{i\vec{k}\cdot\vec{R}}\Big]+G_{ki}(A;k)G_{lj}(B;k)$$
$$\times\Big[(N+1)e_k^{(\mathrm{L/R})}(\vec{k})\bar{e}_l^{(\mathrm{L/R})}(\vec{k})\mathrm{e}^{-i\vec{k}\cdot\vec{R}}+N\bar{e}_k^{(\mathrm{L/R})}(\vec{k})e_l^{(\mathrm{L/R})}(\vec{k})\mathrm{e}^{i\vec{k}\cdot\vec{R}}\Big]\Bigg\}$$
$$\times\mathrm{Re}V_{ij}(k,\vec{R})+\left(\frac{\hbar k}{2\varepsilon_0 V}\right)\Bigg\{G_{ik}(A;k)G_{lj}(B;k)$$
$$\times\Big[(N+1)b_k^{(\mathrm{L/R})}(\vec{k})\bar{e}_l^{(\mathrm{L/R})}(\vec{k})\mathrm{e}^{-i\vec{k}\cdot\vec{R}}+N\bar{b}_k^{(\mathrm{L/R})}(\vec{k})e_l^{(\mathrm{L/R})}(\vec{k})\mathrm{e}^{i\vec{k}\cdot\vec{R}}\Big]$$
$$+G_{ki}(A;k)G_{jl}(B;k)\Big[(N+1)e_k^{(\mathrm{L/R})}(\vec{k})\bar{b}_l^{(\mathrm{L/R})}(\vec{k})\mathrm{e}^{-i\vec{k}\cdot\vec{R}}$$
$$+N\bar{e}_k^{(\mathrm{L/R})}(\vec{k})b_l^{(\mathrm{L/R})}(\vec{k})\mathrm{e}^{i\vec{k}\cdot\vec{R}}\Big]\Bigg\}\,\mathrm{Im}U_{ij}(k,\vec{R}). \tag{7.11.1}$$

Concentrating on the second term present within the first set of braces above, carrying out the molecular average via (7.10.6) and substituting the relation (7.6.6) produces

$$\frac{1}{2}\left(\frac{\hbar k}{2\varepsilon_0 cV}\right)G(A;k)G(B;k)\Big\{(\delta_{ij}-\hat{k}_i\hat{k}_j)\mp i\varepsilon_{ijs}\hat{k}_s\Big\}$$
$$\times\Big[(N+1)\mathrm{e}^{-i\vec{k}\cdot\vec{R}}+N\mathrm{e}^{i\vec{k}\cdot\vec{R}}\Big]\,\mathrm{Re}V_{ij}(k,\vec{R}). \tag{7.11.2}$$

Judicious use is again made of index symmetry: since $V_{ij}(k,\vec{R})$ is *i,j*-symmetric, while the Levi-Civita tensor is antisymmetric in this pair of suffixes, only the first term within braces of (7.11.2) survives. Carrying out the pair orientational average using formula (7.5.3) and substituting for $V_{ij}(k,\vec{R})$ gives

$$\frac{\hbar k}{16\pi\varepsilon_0^2 cVR^3}(2N+1)G(A;k)G(B;k)\big[(\delta_{ij}-3\hat{R}_i\hat{R}_j)(\cos kR+kR\sin kR)$$
$$-(\delta_{ij}-\hat{R}_i\hat{R}_j)k^2R^2\cos kR\big]$$
$$\times\left[(\delta_{ij}-\hat{R}_i\hat{R}_j)\frac{\sin kR}{kR}+(\delta_{ij}-3\hat{R}_i\hat{R}_j)\left(\frac{\cos kR}{k^2R^2}-\frac{\sin kR}{k^3R^3}\right)\right], \tag{7.11.3}$$

which on tensor reduction results in

$$-\frac{I}{16\pi\varepsilon_0^2 c^3 R^3} G(A;k)G(B;k) \times \left[kR \sin 2kR + 2\cos 2kR - 5\frac{\sin 2kR}{kR} - 6\frac{\cos 2kR}{k^2R^2} + 3\frac{\sin 2kR}{k^3R^3}\right], \tag{7.11.4}$$

the irradiance of the incoming field now being defined as $I = (2N+1)\hbar c^2 k/V$. Due to the equality of identities (7.6.6) and (7.10.15), the first term within the first set of braces of expression (7.11.1) produces a contribution identical to equation (7.11.4). Going back to expression (7.11.1) and examining the second term appearing within the second set of braces, substituting (7.10.16) gives

$$\frac{1}{2}\left(\frac{\hbar k}{2\varepsilon_0 V}\right) G(A;k)G(B;k) \left\{ \left[\pm i(\delta_{ij} - \hat{k}_i\hat{k}_j) + \varepsilon_{ijs}\hat{k}_s\right](N+1)e^{-i\vec{k}\cdot\vec{R}} + \left[\mp i(\delta_{ij} - \hat{k}_i\hat{k}_j) - \varepsilon_{ijs}\hat{k}_s\right] N e^{i\vec{k}\cdot\vec{R}} \right\} \mathrm{Im}U_{ij}(k,\vec{R}), \tag{7.11.5}$$

for isotropic A and B. Since $U_{ij}(k,\vec{R})$ is antisymmetric in i and j, only the i,j-antisymmetric part of (7.11.5) remains. With the tumbling average given by

$$\langle \hat{k}_k e^{\pm i\vec{k}\cdot\vec{R}} \rangle = \frac{1}{4\pi}\int \hat{k}_k e^{\pm i\vec{k}\cdot\vec{R}} d\Omega = \mp i\left(\frac{\cos kR}{kR} - \frac{\sin kR}{k^2R^2}\right)\hat{R}_k \tag{7.11.6}$$

and substituting for $U_{ij}(k,\vec{R})$, (7.11.5) becomes

$$\frac{I}{16\pi\varepsilon_0^2 c^3 R^3} G(A;k)G(B;k) \left[kR \sin 2kR + 2\cos 2kR - \frac{\sin 2kR}{kR}\right]. \tag{7.11.7}$$

An identical contribution to (7.11.7) arises from the first term in the second set of braces of (7.11.1). Hence, the pair averaged energy shift is obtained from twice the sum of (7.11.4) and (7.11.7),

$$\Delta E = \frac{I}{8\pi\varepsilon_0^2 c^3 R^3} G(A;k)G(B;k) \left[4\frac{\sin 2kR}{kR} + 6\frac{\cos 2kR}{k^2R^2} - 3\frac{\sin 2kR}{k^3R^3}\right]. \tag{7.11.8}$$

The energy shift is linearly dependent on the irradiance of the incident beam, but is independent of its polarization. Circular polarization does not produce discriminatory effects. The latter arises solely from the mixed electric–magnetic polarizability $G(\xi; k)$. In the far zone, the energy shift exhibits a modulated inverse fourth power dependence on R,

$$\Delta E(\mathrm{FZ}) = \frac{I}{2\pi\varepsilon_0^2 c^3 k R^4} G(A; k) G(B; k) \sin 2kR, \tag{7.11.9}$$

while having inverse R dependence at short range,

$$\Delta E(\mathrm{NZ}) = -\frac{4Ik^2}{15\pi\varepsilon_0^2 c^3 R} G(A; k) G(B; k). \tag{7.11.10}$$

7.12 RADIATION-INDUCED INTERMOLECULAR ENERGY SHIFTS INVOLVING MAGNETIC DIPOLE AND ELECTRIC QUADRUPOLE POLARIZABLE MOLECULES

The change in mutual interaction energy between a pair of chiral molecules in the presence of an intense electromagnetic field was shown to be proportional to the chiroptical response tensor $G_{ij}(\xi; \omega)$, the dynamic mixed electric–magnetic dipole polarizability, which is equal and opposite for two identical optical isomers. For the sake of consistency, the change in energy shift between an electric dipole polarizable molecule and a magnetically susceptible molecule due to external radiation should also be computed, it being the same order of magnitude as the radiation-induced discriminatory interaction, containing two electric dipole and two magnetic dipole interaction vertices overall, but each now occurring at the same molecular center. Since the electric quadrupole is of comparable order of magnitude to the magnetic dipole, the field modified energy shift between an electric dipole polarizable molecule and an electric quadrupole polarizable molecule is also evaluated (Salam, 2006b). Neglected is the interaction between two electric dipole–quadrupole polarizable molecules in a radiation field, also of a similar order, but which vanishes for isotropic A and B. Due to the calculational simplicity of the induced moment approach, this method will be employed in the work of this section.

Consider an electric dipole polarizable molecule A and a magnetically susceptible one B. Application of an electromagnetic field induces electric and magnetic dipole moments, respectively,

$$\mu_i^{\mathrm{ind}}(A) = \varepsilon_0^{-1} \alpha_{ik}(A; k) d_k^{\perp}(\vec{R}_A), \tag{7.12.1}$$

and

$$m_j^{\text{ind}}(B) = \chi_{jl}(B;k)b_l(\vec{R}_B), \tag{7.12.2}$$

where $\alpha_{ik}(A;k)$ and $\chi_{jl}(B;k)$ are dynamic electric dipole and magnetic dipole polarizability tensors. The two induced moments interact via the resonant coupling tensor, $U_{ij}(k,\vec{R})$, giving rise to an interaction energy

$$\begin{aligned}\Delta E &= \text{Im}\mu_i^{\text{ind}}(A)m_j^{\text{ind}}(B)U_{ij}(k,\vec{R}) \\ &= \varepsilon_0^{-1}\,\text{Im}\alpha_{ik}(A;k)\chi_{jl}(B;k)d_k^{\perp}(\vec{R}_A)b_l(\vec{R}_B)U_{ij}(k,\vec{R}).\end{aligned} \tag{7.12.3}$$

The expectation value of (7.12.3) is taken over the state $|E_0^A, E_0^B; N(\vec{k},\lambda)\rangle$, the molecular part resulting in ground-state polarizability tensors of A and B. For the radiation field factor, use is made of the expectation value of the field–field spatial correlation function (5.9.40). Hence,

$$\begin{aligned}\Delta E = \text{Im}\left(\frac{\hbar k}{2\varepsilon_0 V}\right)&\alpha_{ik}(A;k)\chi_{jl}(B;k) \\ &\times\left[(N+1)e_k^{(\lambda)}(\vec{k})\bar{b}_l^{(\lambda)}(\vec{k})\mathrm{e}^{-i\vec{k}\cdot\vec{R}} + N\bar{e}_k^{(\lambda)}(\vec{k})b_l^{(\lambda)}(\vec{k})\mathrm{e}^{i\vec{k}\cdot\vec{R}}\right]U_{ij}(k,\vec{R}).\end{aligned} \tag{7.12.4}$$

For a pair of isotropic molecules, rotational averaging can be carried out using

$$\langle\alpha_{ik}(A;k)\chi_{jl}(B;k)\rangle = \delta_{ik}\delta_{jl}\alpha(A;k)\chi(B;k). \tag{7.12.5}$$

Next, a pair orientational average is performed. Assuming the incident laser is circularly polarized, the product of polarization vectors may be re-expressed via the identity (7.10.16). Since $U_{ij}(k,\vec{R})$ is antisymmetric in i,j, only the i,j-antisymmetric part of (7.10.16) contributes. Utilizing (7.12.5), the energy shift (7.12.4) becomes

$$\Delta E = \text{Im}\left(\frac{\hbar k}{4\varepsilon_0 V}\right)\alpha(A;k)\chi(B;k)\varepsilon_{ijk}\hat{k}_k\left[(N+1)\mathrm{e}^{-i\vec{k}\cdot\vec{R}} - N\mathrm{e}^{i\vec{k}\cdot\vec{R}}\right]U_{ij}(k,\vec{R}). \tag{7.12.6}$$

With the tumbling average calculated using (7.11.6), the relevant terms in (7.12.6) yield

$$\langle(N+1)\hat{k}_k\mathrm{e}^{-i\vec{k}\cdot\vec{R}}\rangle - \langle N\hat{k}_k\mathrm{e}^{i\vec{k}\cdot\vec{R}}\rangle = i(2N+1)\left(\frac{\cos kR}{kR} - \frac{\sin kR}{k^2R^2}\right)\hat{R}_k \tag{7.12.7}$$

so that the energy shift becomes

$$\Delta E = \mathrm{Im}\, i\left(\frac{(2N+1)\hbar k}{4\varepsilon_0 V}\right)\alpha(A;k)\chi(B;k)\varepsilon_{ijk}\hat{R}_k\left(\frac{\cos kR}{kR}-\frac{\sin kR}{k^2R^2}\right)U_{ij}(k,\vec{R}). \tag{7.12.8}$$

Substituting for $U_{ij}(k,\vec{R})$, defining the irradiance as $I=(2N+1)\hbar c^2k/V$, and contracting results in

$$\Delta E = \frac{I}{16\pi\varepsilon_0^2c^3R^3}\alpha(A;k)\chi(B;k)\left[kR\sin 2kR+2\cos 2kR-\frac{\sin 2kR}{kR}\right], \tag{7.12.9}$$

which holds for all R outside the charge overlap region. In the far zone, the dominant term is given by the first in square brackets resulting in a modulated R^{-2} asymptote

$$\Delta E_{\mathrm{FZ}} = \frac{Ik}{16\pi\varepsilon_0^2c^3R^2}\alpha(A;k)\chi(B;k)\sin 2kR, \tag{7.12.10}$$

while at very close range, McLaurin series expansion of all trigonometric terms produces

$$\Delta E_{\mathrm{NZ}} = -\frac{Ik^2}{24\pi\varepsilon_0^2c^3R}\alpha(A;k)\chi(B;k), \tag{7.12.11}$$

having an inverse dependence on separation distance and with sign opposite to that found in expression (7.12.9) and (7.12.10).

If molecule B is electric quadrupole polarizable, an electric quadrupole moment is induced by an electric field as in

$$Q_{pq}^{\mathrm{ind}}(B) = \varepsilon_0^{-1}\Theta_{pqrs}(B;k)\vec{\nabla}_s d_r^{\perp}(\vec{R}_B), \tag{7.12.12}$$

where the dynamic electric quadrupole polarizability tensor is defined as

$$\Theta_{pqrs}(B;k) = \sum_t\left\{\frac{Q_{pq}^{0t}(B)Q_{rs}^{t0}(B)}{E_{t0}-\hbar ck}+\frac{Q_{rs}^{0t}(B)Q_{pq}^{t0}(B)}{E_{t0}+\hbar ck}\right\}, \tag{7.12.13}$$

where $Q_{pq}^{0t}(B)$ is the $0t$th matrix element of the electric quadrupole moment operator,

$$\langle 0|Q_{pq}(B)|t\rangle = \left\langle 0\left|-\frac{e}{2!}(\vec{q}-\vec{R}_B)_p(\vec{q}-\vec{R}_B)_q\right|t\right\rangle. \tag{7.12.14}$$

The induced electric dipole moment (7.12.1) and $Q_{pq}^{\text{ind}}(B)$ interact via the resonant electric dipole–quadrupole coupling tensor, $V_{ipq}(k, \vec{R})$, resulting in an energy shift

$$\begin{aligned}\Delta E &= \mu_i^{\text{ind}}(A)Q_{pq}^{\text{ind}}(B)\,\mathrm{Re}V_{ipq}(k,\vec{R})\\ &= \varepsilon_0^{-2}\alpha_{ik}(A;k)\Theta_{pqrs}(B;k)d_k^{\perp}(\vec{R}_A)\vec{\nabla}_s d_r^{\perp}(\vec{R}_B)\,\mathrm{Re}V_{ipq}(k,\vec{R}),\end{aligned} \tag{7.12.15}$$

where

$$\begin{aligned}V_{ipq}(k,\vec{R}) &= -\frac{1}{4\pi\varepsilon_0}\left(-\vec{\nabla}^2\delta_{ip}+\vec{\nabla}_i\vec{\nabla}_p\right)\vec{\nabla}_q\frac{\mathrm{e}^{ikR}}{R}\\ &= -\frac{1}{4\pi\varepsilon_0}\Bigg[(\delta_{ip}-\hat{R}_i\hat{R}_p)\hat{R}_q\left(\frac{ik^3}{R}-\frac{k^2}{R^2}\right)+(\delta_{ip}\hat{R}_q+\delta_{iq}\hat{R}_p\\ &\qquad +\delta_{pq}\hat{R}_i-5\hat{R}_i\hat{R}_p\hat{R}_q)\left(-\frac{k^2}{R^2}-\frac{3ik}{R^3}+\frac{3}{R^4}\right)\Bigg]\mathrm{e}^{ikR}.\end{aligned} \tag{7.12.16}$$

Using the mode expansion for the electric displacement field and its gradient, the expectation value of the field gradient–field spatial correlation function for an N-photon state is

$$\begin{aligned}&\langle N(\vec{k},\lambda)|d_i^{\perp}(\vec{R}_A)\vec{\nabla}_k d_j^{\perp}(\vec{R}_B)|N(\vec{k},\lambda)\rangle\\ &\quad= i\left(\frac{\hbar ck\varepsilon_0}{2V}\right)k_k\left[N\bar{e}_i^{(\lambda)}(\vec{k})e_j^{(\lambda)}(\vec{k})\mathrm{e}^{i\vec{k}\cdot\vec{R}}-(N+1)e_i^{(\lambda)}(\vec{k})\bar{e}_j^{(\lambda)}(\vec{k})\mathrm{e}^{-i\vec{k}\cdot\vec{R}}\right].\end{aligned} \tag{7.12.17}$$

When inserted in the right-hand side of (7.12.15), the energy shift is

$$\begin{aligned}\Delta E &= i\left(\frac{\hbar ck}{2\varepsilon_0 V}\right)\alpha_{ik}(A;k)\Theta_{pqrs}(B;k)k_s\\ &\quad\times\left[N\bar{e}_k^{(\lambda)}(\vec{k})e_r^{(\lambda)}(\vec{k})\mathrm{e}^{i\vec{k}\cdot\vec{R}}-(N+1)e_k^{(\lambda)}(\vec{k})\bar{e}_r^{(\lambda)}(\vec{k})\mathrm{e}^{-i\vec{k}\cdot\vec{R}}\right]\\ &\quad\times\mathrm{Re}V_{ipq}(k,\vec{R}).\end{aligned} \tag{7.12.18}$$

To examine the effect of linearly polarized incident light on the energy shift, use is made of the relation $e_k^{(\lambda)}(\vec{k})\bar{e}_r^{(\lambda)}(\vec{k}) = \frac{1}{2}(\delta_{kr}-\hat{k}_k\hat{k}_r)$. For freely tumbling molecules, a pair orientational average is required. This is carried out via

$$\left\langle (\delta_{ij}-\hat{k}_i\hat{k}_j)k_k e^{\pm i\vec{k}\cdot\vec{R}} \right\rangle = \frac{1}{4\pi}\int (\delta_{ij}-\hat{k}_i\hat{k}_j)k_k e^{\pm i\vec{k}\cdot\vec{R}} d\Omega$$
$$= \mp\frac{i}{k^3}\left(-\vec{\nabla}^2\delta_{ij}+\vec{\nabla}_i\vec{\nabla}_j\right)\vec{\nabla}_k\frac{\sin kR}{R}. \quad (7.12.19)$$

Substituting for $\mathrm{Re}V_{ipq}(k,\vec{R})$ from (7.12.16) and using (7.12.19) gives for the energy shift

$$\Delta E = -\left(\frac{I}{16\pi\varepsilon_0^2 c}\right)\alpha_{ik}(A;k)\Theta_{pqrs}(B;k)\left[\frac{1}{k^3}\left(-\vec{\nabla}^2\delta_{kr}+\vec{\nabla}_k\vec{\nabla}_r\right)\vec{\nabla}_s\frac{\sin kR}{R}\right]$$
$$\times\left[\left(-\vec{\nabla}^2\delta_{ip}+\vec{\nabla}_i\vec{\nabla}_p\right)\vec{\nabla}_q\frac{\cos kR}{R}\right]. \quad (7.12.20)$$

The first factor in square brackets is readily obtained from (7.12.16), being $(-4\pi\varepsilon_0/k^3)\,\mathrm{Im}V_{krs}(k,\vec{R})$, while the second term in square brackets is simply $-4\pi\varepsilon_0\,\mathrm{Re}V_{ipq}(k,\vec{R})$. Thus,

$$\Delta E = -\left(\frac{I}{16\pi\varepsilon_0^2 c}\right)\alpha_{ik}(A;k)\Theta_{pqrs}(B;k)$$
$$\times\left[(\delta_{kr}-\hat{R}_k\hat{R}_r)\hat{R}_s\left(\frac{\cos kR}{R}-\frac{\sin kR}{kR^2}\right)\right.$$
$$\left.+(\delta_{kr}\hat{R}_s+\delta_{ks}\hat{R}_r+\delta_{rs}\hat{R}_k-5\hat{R}_k\hat{R}_r\hat{R}_s)\left(-\frac{\sin kR}{kR^2}-\frac{3\cos kR}{k^2R^3}+\frac{3\sin kR}{k^3R^4}\right)\right]$$
$$\times\left[(\delta_{ip}-\hat{R}_i\hat{R}_p)\hat{R}_q\left(-\frac{k^3\sin kR}{R}-\frac{k^2\cos kR}{R^2}\right)\right.$$
$$\left.+(\delta_{ip}\hat{R}_q+\delta_{iq}\hat{R}_p+\delta_{pq}\hat{R}_i-5\hat{R}_i\hat{R}_p\hat{R}_q)\left(-\frac{k^2\cos kR}{R^2}+\frac{3k\sin kR}{R^3}+\frac{3\cos kR}{R^4}\right)\right].$$
$$(7.12.21)$$

Finally, the molecular average is performed. This is done using the tensor quantity for the average of a second-rank tensor (electric dipole polarizability) multiplied by the average of a fourth-rank tensor (electric quadrupole polarizability) on making use of the fact that the electric quadrupole moment is symmetric in its component indices and is traceless. The pertinent results are given in Appendix B by equations (B.4) and (B.7). Hence, contracting (7.12.21) with

$$\langle \alpha_{ik}(A;k)\Theta_{pqrs}(B;k)\rangle = \frac{1}{10}\delta_{ik}(\delta_{pr}\delta_{qs} + \delta_{ps}\delta_{qr})\alpha(A;k)\Theta_{\lambda\mu\lambda\mu}(B;k) \tag{7.12.22}$$

results in the energy shift applicable to a pair of isotropic and freely tumbling electric dipole and quadrupole polarizable molecules in the presence of a laser field,

$$\Delta E = \frac{I}{80\pi\varepsilon_0^2 cR^5}\alpha(A;k)\Theta_{\lambda\mu\lambda\mu}(B;k)\left[k^3R^3\sin 2kR + 6k^2R^2\cos 2kR \right.$$
$$\left. -27kR\sin 2kR - 84\cos 2kR + 162\frac{\sin 2kR}{kR} + 180\frac{\cos 2kR}{k^2R^2} - 90\frac{\sin 2kR}{k^3R^3}\right]. \tag{7.12.23}$$

Asymptotically, energy shift (7.12.23) tends to

$$\Delta E(\mathrm{FZ}) = \frac{Ik^3}{80\pi\varepsilon_0^2 cR^2}\alpha(A;k)\Theta_{\lambda\mu\lambda\mu}(B;k)\sin 2kR \tag{7.12.24}$$

in the far zone and tends to

$$\Delta E(\mathrm{NZ}) = -\frac{9Ik^4}{1400\pi\varepsilon_0^2 cR}\alpha(A;k)\Theta_{\lambda\mu\lambda\mu}(B;k) \tag{7.12.25}$$

in the near zone, exhibiting modulated R^{-2} and R^{-1} behavior, respectively.

7.13 HIGHER ORDER RADIATION-INDUCED DISCRIMINATORY INTERMOLECULAR INTERACTION

In Section 7.9, it was shown that the action of circularly polarized electromagnetic radiation on a pair of interacting optically active molecules resulted in a change in the energy shift relative to the absence of light and that this modification of the interaction energy was discriminatory. The source of the discrimination was the handedness of the individual molecules as characterized by the molecular chiroptical response tensor, $G_{ij}(\xi;\omega)$, which is bilinear in the transition electric and magnetic dipole moments. Interestingly, in this case, the field-induced intermolecular energy shift is, in fact, independent of the chirality of the incident radiation. Thus far, in this chapter, the effect of an intense source of external radiation on a pair of molecules coupled via single virtual photon exchange has been examined. When perturbative techniques have been used to solve this problem, it has required implementation of fourth-order theory. For a pair of neutral, polar electric dipole polarizable molecules, two distinct mechanisms were identified as contributing to the field modified energy shift. One was a dynamic mechanism, proportional to the electric dipole polarizability of each molecule. A second was a static mechanism that depended on the product of the permanent electric dipole moment of one species and the molecular first hyperpolarizability of another. Both terms were nondiscriminatory.

Energy shifts between interacting molecules that are induced by a radiation field, which exhibit discriminatory behavior even within the electric dipole approximation can occur in higher order. This first manifests itself when molecules are coupled via two- virtual photon exchange. A full treatment therefore necessitates the use of sixth-order theory in a perturbation calculation. Considerable simplification is possible, without loss of the essential features of the phenomenon, if the interacting pair of molecules are taken to be in proximity to one another. This enables the coupling occurring between the two molecules to be static in origin. As a consequence, fourth-order perturbation theory can be used to evaluate the change in mutual interaction energy in the near zone (Taylor and Thirunamachandran, 1983). Details of the calculation are as follows.

Consider two neutral molecules A and B, situated at $\vec{R}_A$ and $\vec{R}_B$, respectively, with internuclear separation distance, $R = |\vec{R}_B - \vec{R}_A|$. In the near zone, the two entities are coupled via the static dipolar interaction potential $V_{ij}(\vec{R}) = (4\pi\varepsilon_0)^{-1}\mu_i(A)\mu_j(B)(\delta_{ij} - 3\hat{R}_i\hat{R}_j)R^{-3}$. In the higher order approximation being considered, two such couplings take place between A and B. In the electric dipole approximation, the interaction Hamiltonian used to describe the effect of external radiation on the system is the familiar $-\varepsilon_0^{-1}\mu_i(\xi)d_i^{\perp}(\vec{R}_\xi)$ coupling. Let the applied field be of mode $(\vec{k}, \lambda)$. As for the case, when only one virtual photon was exchanged between the pair, two types of contributions are found in the present situation. They correspond to the scattering of a real photon occurring at the same site or at different centers. Each contribution is examined separately. The initial and final states are equal to each other and are identical for both types of contributions. They are represented by $|E_0^A, E_0^B; N(\vec{k}, \lambda)\rangle$, corresponding to both molecules in the ground electronic state $|0^\xi\rangle$, with energy E_n^ξ, $\xi = A, B$ and the incident field containing N real photons of mode $(\vec{k}, \lambda)$. Incidentally, the initial- and final-state specifications of the system are identical to (7.2.6), the state used when the pairs are coupled via single virtual photon exchange.

In the case where absorption and emission of a real photon takes place at species A, 12 distinct time orderings of the pertinent interactions are possible. Another 12 diagrams may be drawn when scattering of the real photon occurs exclusively at B. Taken together, these two sets give rise to one type of contribution. Ultimately, this contribution will be proportional to the product of a fourth-rank hyperpolarizability tensor of one molecule and the polarizability of the other, the former containing four transition electric dipole moments, and the latter two. Due to the even number of couplings at each site, the parity of the transitions imposes no restrictions on selection rules so that the interaction energy holds for both centrosymmetric and noncentrosymmetric systems. Evaluating in the usual way, the sum of the 24 graphs gives rise to

$$
\begin{aligned}
\Delta E = &-\left(\frac{N\hbar ck}{2\varepsilon_0 VR^6}\right)(\delta_{kl}-3\hat{R}_k\hat{R}_l)(\delta_{mn}-3\hat{R}_m\hat{R}_n)e_i^{(\lambda)}(\vec{k})\bar{e}_j^{(\lambda)}(\vec{k}) \\
&\times\sum_{m,n,p}\sum_{r}\left\{\frac{[\mu_j^{0p}\mu_m^{pn}\mu_k^{nm}\mu_i^{m0}][\mu_n^{0r}\mu_l^{r0}]}{(E_{p0}-\hbar ck)(E_{n0}+E_{r0}-\hbar ck)(E_{m0}-\hbar ck)}\right. \\
&+\frac{[\mu_m^{0p}\mu_j^{pn}\mu_k^{nm}\mu_i^{m0}][\mu_n^{0r}\mu_l^{r0}]}{(E_{p0}+E_{r0})(E_{n0}+E_{r0}-\hbar ck)(E_{m0}-\hbar ck)} \\
&+\frac{[\mu_m^{0p}\mu_k^{pn}\mu_j^{nm}\mu_i^{m0}][\mu_n^{0r}\mu_l^{r0}]}{(E_{p0}+E_{r0})(E_{m0}-\hbar ck)E_{n0}}+\frac{[\mu_j^{0p}\mu_m^{pn}\mu_i^{nm}\mu_k^{m0}][\mu_n^{0r}\mu_l^{r0}]}{(E_{p0}-\hbar ck)(E_{n0}+E_{r0}-\hbar ck)(E_{m0}+E_{r0})} \\
&+\frac{[\mu_m^{0p}\mu_j^{pn}\mu_i^{nm}\mu_k^{m0}][\mu_n^{0r}\mu_l^{r0}]}{(E_{p0}+E_{r0})(E_{n0}+E_{r0}-\hbar ck)(E_{m0}+E_{r0})}+\frac{[\mu_j^{0p}\mu_i^{pn}\mu_m^{nm}\mu_k^{m0}][\mu_n^{0r}\mu_l^{r0}]}{(E_{p0}-\hbar ck)(E_{m0}+E_{r0})E_{n0}} \\
&+\frac{[\mu_i^{0p}\mu_m^{pn}\mu_k^{nm}\mu_j^{m0}][\mu_n^{0r}\mu_l^{r0}]}{(E_{p0}+\hbar ck)(E_{n0}+E_{r0}+\hbar ck)(E_{m0}+\hbar ck)} \\
&+\frac{[\mu_m^{0p}\mu_i^{pn}\mu_k^{nm}\mu_j^{m0}][\mu_n^{0r}\mu_l^{r0}]}{(E_{p0}+E_{r0})(E_{n0}+E_{r0}+\hbar ck)(E_{m0}+\hbar ck)} \\
&+\frac{[\mu_m^{0p}\mu_k^{pn}\mu_i^{nm}\mu_j^{m0}][\mu_n^{0r}\mu_l^{r0}]}{(E_{p0}+E_{r0})(E_{m0}+\hbar ck)E_{n0}}+\frac{[\mu_i^{0p}\mu_m^{pn}\mu_j^{nm}\mu_k^{m0}][\mu_n^{0r}\mu_l^{r0}]}{(E_{p0}+\hbar ck)(E_{n0}+E_{r0}+\hbar ck)(E_{m0}+E_{r0})} \\
&+\frac{[\mu_m^{0p}\mu_i^{pn}\mu_j^{nm}\mu_k^{m0}][\mu_n^{0r}\mu_l^{r0}]}{(E_{p0}+E_{r0})(E_{n0}+E_{r0}+\hbar ck)(E_{m0}+E_{r0})} \\
&\left.+\frac{[\mu_i^{0p}\mu_j^{pn}\mu_m^{nm}\mu_k^{m0}][\mu_n^{0r}\mu_l^{r0}]}{(E_{p0}+\hbar ck)(E_{m0}+E_{r0})E_{n0}}\right\}+A\leftrightarrow B. \qquad (7.13.1)
\end{aligned}
$$

As written in expression (7.13.1), the intermediate states of the species at which scattering of a real photon occurs are designated by m, n, and p, while those of the other molecule are labeled by n. This is further emphasized by placing square brackets around the product of transition electric dipole moments, one factor containing four such moments, the other two, with the former reflecting the two additional electric dipole couplings associated with absorption and emission of the $(\vec{k},\lambda)$ mode photon. Also, apparent from (7.13.1) are the symmetry properties of Cartesian tensor components. The geometric factor is invariant to interchange of $m \leftrightarrow k$ and $n \leftrightarrow l$ when taken in conjunction. Noting that $\mu_n^{0r}\mu_l^{r0}$ remains the same on interchanging $n \leftrightarrow l$, it is clear that only the k,m-symmetric part of the molecular term in (7.13.1) will contribute to the energy shift.

To proceed further, the energy denominators are separated into integral products of one center terms. A similar step was carried out when the London dispersion energy was expressed in terms of dynamic polarizabilities at imaginary frequency using the identity (5.2.21). Its generalization to an arbitrary number of factors in the denominator each of which contains a common term is

$$\frac{1}{(a+b)(c+b)\cdots(n+b)} = \frac{1}{2\pi}\int_{-\infty}^{\infty}\left[\frac{1}{(a+iu)(c+iu)\cdots(n+iu)}+\frac{1}{(a-iu)(c-iu)\cdots(n-iu)}\right] \times \frac{b}{(b^2+u^2)}\mathrm{d}u; \quad a,b,c,\ldots,n>0. \tag{7.13.2}$$

To be able to apply relation (7.13.2), the approximation is made in (7.13.1) that the frequency of the incident radiation field is less than molecular transition frequencies. Thus, (7.13.1) becomes

$$\Delta E = -\left(\frac{N\hbar ck}{2\varepsilon_0 VR^6}\right)(\delta_{kl}-3\hat{R}_k\hat{R}_l)(\delta_{mn}-3\hat{R}_m\hat{R}_n)e_i^{(\lambda)}(\vec{k})\bar{e}_j^{(\lambda)}(\vec{k}) \times \frac{\hbar}{8\pi}\int_{-\infty}^{\infty}\{T_{ijkm}(A;\omega;icu)+T_{ijmk}(A;\omega;icu)\}\alpha_{nl}(B;icu)\mathrm{d}u + A \leftrightarrow B. \tag{7.13.3}$$

Appearing in (7.13.3) are the familiar dynamic electric dipole polarizability tensor at imaginary frequency $\alpha_{nl}(\xi;icu)$ and the fourth-rank hyperpolarizability tensor $T_{ijkm}(\omega; icu)$. The latter is defined as

$$
\begin{aligned}
T_{ijkm}(\omega; icu) = \frac{1}{\hbar^3}\sum_{m,n,p}\Bigg\{ & \frac{\mu_j^{0p}\mu_m^{pn}\mu_k^{nm}\mu_i^{m0}}{(\omega_{p0}-\omega)(\omega_{n0}-\omega-icu)(\omega_{m0}-\omega)} \\
& + \frac{\mu_j^{0p}\mu_m^{pn}\mu_i^{nm}\mu_k^{m0}}{(\omega_{p0}-\omega)(\omega_{n0}-\omega-icu)(\omega_{m0}-icu)} \\
& + \frac{\mu_j^{0p}\mu_i^{pn}\mu_m^{nm}\mu_k^{m0}}{(\omega_{p0}-\omega)(\omega_{m0}-icu)\omega_{n0}} + \frac{\mu_i^{0p}\mu_j^{pn}\mu_m^{nm}\mu_k^{m0}}{(\omega_{p0}+\omega)(\omega_{m0}-icu)\omega_{n0}} \\
& + \frac{\mu_m^{0p}\mu_j^{pn}\mu_k^{nm}\mu_i^{m0}}{(\omega_{p0}-icu)(\omega_{n0}-\omega-icu)(\omega_{m0}-\omega)} + \frac{\mu_m^{0p}\mu_j^{pn}\mu_i^{nm}\mu_k^{m0}}{(\omega_{p0}-icu)(\omega_{n0}-\omega-icu)(\omega_{m0}-icu)} \\
& + \frac{\mu_m^{0p}\mu_i^{pn}\mu_j^{nm}\mu_k^{m0}}{(\omega_{p0}-icu)(\omega_{n0}+\omega-icu)(\omega_{m0}-icu)} + \frac{\mu_i^{0p}\mu_m^{pn}\mu_j^{nm}\mu_k^{m0}}{(\omega_{p0}+\omega)(\omega_{n0}+\omega-icu)(\omega_{m0}-icu)} \\
& + \frac{\mu_m^{0p}\mu_k^{pn}\mu_j^{nm}\mu_i^{m0}}{(\omega_{p0}-icu)(\omega_{m0}-\omega)\omega_{n0}} + \frac{\mu_m^{0p}\mu_k^{pn}\mu_i^{nm}\mu_j^{m0}}{(\omega_{p0}-icu)(\omega_{m0}+\omega)\omega_{n0}} \\
& + \frac{\mu_m^{0p}\mu_i^{pn}\mu_k^{nm}\mu_j^{m0}}{(\omega_{p0}-icu)(\omega_{n0}+\omega-icu)(\omega_{m0}+\omega)} + \frac{\mu_i^{0p}\mu_m^{pn}\mu_k^{nm}\mu_j^{m0}}{(\omega_{p0}+\omega)(\omega_{n0}+\omega-icu)(\omega_{m0}+\omega)}\Bigg\} \\
& + \text{terms from } k \leftrightarrow m \text{ and } u \to -u.
\end{aligned} \tag{7.13.4}
$$

Making use of k, m-index symmetry and the invariance to simultaneous exchange of virtual-state labels, it is seen that the molecular part of (7.13.3) is i,j-symmetric. Hence, only the i,j-symmetric part of the polarization factor contributes. Inserting identity (7.6.6) into (7.13.3) results in

$$
\begin{aligned}
\Delta E = & -\left(\frac{N\hbar ck}{4\varepsilon_0 VR^6}\right)(\delta_{ij}-\hat{k}_i\hat{k}_j)(\delta_{kl}-3\hat{R}_k\hat{R}_l)(\delta_{mn}-3\hat{R}_m\hat{R}_n) \\
& \times \frac{\hbar}{8\pi}\int_{-\infty}^{\infty}\{T_{ijkm}(A;\omega;icu)+T_{ijmk}(A;\omega;icu)\}\alpha_{nl}(B;icu)\mathrm{d}u + A \leftrightarrow B.
\end{aligned} \tag{7.13.5}
$$

Energy shift (7.13.5) is independent of the helicity of the incident radiation as well as being independent of the chirality of each molecule. Hence, (7.13.5) is nondiscriminatory.

In the second possible mechanism, the two molecules are again coupled by the exchange of two virtual photons—which are taken to propagate instantaneously in the near zone, but scattering of the real photon occurs at different centers. Thus, if A absorbs a $(\vec{k}, \lambda)$-mode photon, B emits one and vice versa. As in the case where scattering of the real photon occurs at the same center, 24 distinct time orderings of the photonic creation and destruction events are possible when both A and B are associated with three electric dipole coupling vertices. This contribution to the interaction energy clearly holds only for noncentrosymmetric molecules. Summing over the 24 terms produces

$$
\begin{aligned}
\Delta E = {} & -\left(\frac{N\hbar ck}{2\varepsilon_0 VR^6}\right)(\delta_{kl}-3\hat{R}_k\hat{R}_l)(\delta_{mn}-3\hat{R}_m\hat{R}_n)e_i^{(\lambda)}(\vec{k})\bar{e}_j^{(\lambda)}(\vec{k}) \\
& \times \sum_{m,n}\sum_{r,s}\left\{\frac{[\mu_m^{0n}\mu_k^{nm}\mu_i^{m0}][\mu_n^{0s}\mu_l^{sr}\mu_j^{r0}]}{(E_{n0}+E_{s0})(E_{m0}+E_{r0})(E_{r0}+\hbar ck)}\right. \\
& + \frac{[\mu_m^{0n}\mu_k^{nm}\mu_i^{m0}][\mu_n^{0s}\mu_l^{sr}\mu_j^{r0}]}{(E_{n0}+E_{s0})(E_{m0}+E_{r0})(E_{m0}-\hbar ck)} \\
& + \frac{[\mu_m^{0n}\mu_k^{nm}\mu_i^{m0}][\mu_n^{0s}\mu_j^{sr}\mu_l^{r0}]}{(E_{n0}+E_{s0})(E_{n0}+E_{r0}-\hbar ck)(E_{m0}-\hbar ck)} \\
& + \frac{[\mu_m^{0n}\mu_k^{nm}\mu_i^{m0}][\mu_j^{0s}\mu_n^{sr}\mu_l^{r0}]}{(E_{s0}-\hbar ck)(E_{n0}+E_{r0}-\hbar ck)(E_{m0}-\hbar ck)} \\
& + \frac{[\mu_m^{0n}\mu_i^{nm}\mu_k^{m0}][\mu_n^{0s}\mu_l^{sr}\mu_j^{r0}]}{(E_{n0}+E_{s0})(E_{m0}+E_{s0}-\hbar ck)(E_{r0}+\hbar ck)} \\
& + \frac{[\mu_m^{0n}\mu_i^{nm}\mu_k^{m0}][\mu_n^{0s}\mu_j^{sr}\mu_l^{r0}]}{(E_{n0}+E_{s0})(E_{m0}+E_{s0}+\hbar ck)(E_{m0}+E_{r0})} \\
& + \frac{[\mu_m^{0n}\mu_i^{nm}\mu_k^{m0}][\mu_n^{0s}\mu_j^{sr}\mu_l^{r0}]}{(E_{n0}+E_{s0})(E_{n0}+E_{r0}-\hbar ck)(E_{m0}+E_{r0})} \\
& + \frac{[\mu_m^{0n}\mu_i^{nm}\mu_k^{m0}][\mu_j^{0s}\mu_n^{sr}\mu_l^{r0}]}{(E_{s0}-\hbar ck)(E_{n0}+E_{r0}-\hbar ck)(E_{m0}+E_{r0})} \\
& + \frac{[\mu_i^{0n}\mu_m^{nm}\mu_k^{m0}][\mu_n^{0s}\mu_l^{sr}\mu_j^{r0}]}{(E_{n0}+\hbar ck)(E_{m0}+E_{s0}+\hbar ck)(E_{r0}+\hbar ck)} \\
& + \frac{[\mu_i^{0n}\mu_m^{nm}\mu_k^{m0}][\mu_n^{0s}\mu_j^{sr}\mu_l^{r0}]}{(E_{n0}+\hbar ck)(E_{m0}+E_{s0}+\hbar ck)(E_{m0}+E_{r0})} \\
& + \frac{[\mu_i^{0n}\mu_m^{nm}\mu_k^{m0}][\mu_j^{0s}\mu_n^{sr}\mu_l^{r0}]}{(E_{n0}+\hbar ck)(E_{n0}+E_{s0})(E_{m0}+E_{r0})} \\
& \left. + \frac{[\mu_i^{0n}\mu_m^{nm}\mu_k^{m0}][\mu_j^{0s}\mu_n^{sr}\mu_l^{r0}]}{(E_{s0}-\hbar ck)(E_{n0}+E_{s0})(E_{m0}+E_{r0})}\right\} \mathrm{e}^{-i\vec{k}\cdot(\vec{R}_B-\vec{R}_A)} + A \leftrightarrow B.
\end{aligned}
\tag{7.13.6}
$$

Square brackets are again used to enclose molecular moments associated with each molecule whose intermediate states are labeled by m and n and r and s. Because the potential coupling the two bodies is applicable to the near zone, the exponential factor in (7.13.6) may be taken to be unity since $kR \ll 1$. Employing identity (7.13.2) after choosing the frequency of the incident field to be lower than that of molecular transition frequencies enables (7.13.6) to be written as

$$\Delta E = -\left(\frac{N\hbar ck}{2\varepsilon_0 VR^6}\right)(\delta_{kl}-3\hat{R}_k\hat{R}_l)(\delta_{mn}-3\hat{R}_m\hat{R}_n)e_i^{(\lambda)}(\vec{k})\bar{e}_j^{(\lambda)}(\vec{k})$$
$$\times\frac{\hbar}{4\pi}\int_{-\infty}^{\infty}\{\beta_{ikm}(A;\omega;icu)\beta_{jln}(B;-\omega;-icu)$$
$$+\beta_{jln}(A;-\omega;-icu)\beta_{ikm}(B;\omega;icu)\}\mathrm{d}u, \tag{7.13.7}$$

which is seen to be proportional to the product of one hyperpolarizability tensor of rank three from each molecule. This tensor is given explicitly by

$$\beta_{ikm}(\omega;icu) = \frac{1}{\hbar^2}\sum_{m,n}\left\{\frac{\mu_i^{0n}\mu_k^{nm}\mu_m^{m0}}{(\omega_{m0}+icu)(\omega_{n0}+\omega)}\right.$$
$$+\frac{\mu_k^{0n}\mu_i^{nm}\mu_m^{m0}}{(\omega_{m0}+icu)(\omega_{n0}-\omega+icu)}+\frac{\mu_k^{0n}\mu_m^{nm}\mu_i^{m0}}{(\omega_{m0}-\omega)(\omega_{n0}-\omega+icu)}$$
$$+\frac{\mu_i^{0n}\mu_m^{nm}\mu_k^{m0}}{(\omega_{m0}+\omega-icu)(\omega_{n0}+\omega)}+\frac{\mu_m^{0n}\mu_i^{nm}\mu_k^{m0}}{(\omega_{m0}+\omega-icu)(\omega_{n0}-icu)}$$
$$\left.+\frac{\mu_m^{0n}\mu_k^{nm}\mu_i^{m0}}{(\omega_{m0}-\omega)(\omega_{n0}-icu)}\right\}. \tag{7.13.8}$$

A more compact expression of (7.13.7) is possible by interchanging intermediate-state labels of the hyperpolarizability tensor to give

$$\beta_{ikm}(-\omega;-icu) = \beta_{ikm}(\omega;icu) \tag{7.13.9}$$

so that

$$\Delta E = -\left(\frac{N\hbar ck}{2\varepsilon_0 VR^6}\right)(\delta_{kl}-3\hat{R}_k\hat{R}_l)(\delta_{mn}-3\hat{R}_m\hat{R}_n)$$
$$\times\left\{e_i^{(\lambda)}(\vec{k})\bar{e}_j^{(\lambda)}(\vec{k})+\bar{e}_i^{(\lambda)}(\vec{k})e_j^{(\lambda)}(\vec{k})\right\}$$
$$\times\frac{\hbar}{4\pi}\int_{-\infty}^{\infty}\beta_{ikm}(A;\omega;icu)\beta_{jln}(B;-\omega;-icu)\mathrm{d}u. \tag{7.13.10}$$

Since $\vec{\mu}$ is a polar vector, the molecular hyperpolarizability changes sign when one molecule is substituted by its mirror image form. Hence, the energy shift (7.13.10) is discriminatory in contrast to result (7.13.5) arising from the first mechanism. Recalling that the irradiance of the incident field is given by $I = N\hbar c^2 k/V$, energy shifts (7.13.5) and (7.13.10) are seen to be linearly proportional to I.

The results obtained thus far hold for the situation in which A and B are in a fixed orientation relative to each other and as a pair in fixed orientation relative to the wavevector of the incoming beam. If the molecules are held rigid due to strong intermolecular coupling, but are allowed to rotate freely as a pair, the energy shifts (7.13.5) and (7.13.10), respectively, become

$$\Delta E = -\left(\frac{I}{2\varepsilon_0 c R^6}\right)(\delta_{kl} - 3\hat{R}_k\hat{R}_l)(\delta_{mn} - 3\hat{R}_m\hat{R}_n) \times \frac{\hbar}{24\pi}\int_{-\infty}^{\infty} [\{T_{iikm}(A;\omega;icu) + T_{iimk}(A;\omega;icu)\}\alpha_{nl}(B;icu) + \{T_{iikm}(B;\omega;icu) + T_{iimk}(B;\omega;icu)\}\alpha_{nl}(A;icu)]\mathrm{d}u \tag{7.13.11}$$

and

$$\Delta E = -\left(\frac{I}{2\varepsilon_0 c R^6}\right)(\delta_{kl} - 3\hat{R}_k\hat{R}_l)(\delta_{mn} - 3\hat{R}_m\hat{R}_n)\frac{\hbar}{6\pi} \times \int_{-\infty}^{\infty} \beta_{ikm}(A;\omega;icu)\beta_{iln}(B;-\omega;-icu)\mathrm{d}u. \tag{7.13.12}$$

For random relative orientations of molecules within the pair, a molecular average is performed. This corresponds to carrying out a Boltzmann-weighted average at the limit $T \to \infty$, where T is the temperature. Thus, (7.13.11) and (7.13.12) become

$$\Delta E = -\left(\frac{I}{2\varepsilon_0 c R^6}\right)\frac{\hbar}{36\pi}\int_{-\infty}^{\infty} [\{T_{\lambda\lambda\mu\mu}(A;\omega;icu) + T_{\lambda\lambda\mu\mu}(A;\omega;icu)\}\alpha_{\nu\nu}(B;icu) + \{T_{\lambda\lambda\mu\mu}(B;\omega;icu) + T_{\lambda\lambda\mu\mu}(B;\omega;icu)\}\alpha_{\nu\nu}(A;icu)]\mathrm{d}u \tag{7.13.13}$$

and

$$\Delta E = \left(\frac{I}{2\varepsilon_0 c R^6}\right)\frac{\hbar}{36\pi}\varepsilon_{\lambda\nu\pi}\varepsilon_{\mu\rho\sigma}\int\limits_{-\infty}^{\infty}\beta_{\lambda\nu\pi}(A;\omega;icu)\beta_{\mu\rho\sigma}(B;-\omega;-icu)\mathrm{d}u, \tag{7.13.14}$$

both results being independent of the polarization of the incident beam. Energy shift (7.13.13) is nondiscriminatory while result (7.13.14) depends on the chirality of each molecule and changes sign when one species is replaced by its mirror image form.

APPENDIX A

HIGHER MULTIPOLE-DEPENDENT SECOND-ORDER MAXWELL FIELD OPERATORS

In Section 2.7, the Maxwell field operators linearly dependent on the electric dipole, quadrupole, and magnetic dipole moments were obtained, while Section 2.6 contained results for the quadratic fields within the electric dipole approximation. Following the procedure outlined in Chapter 2, the electric displacement and magnetic field operators bilinear and quadratic in these first three multipole moments can be evaluated. Explicit expressions are given below. Also included for completeness are the second-order electric dipole-dependent fields. For all of the fields listed below, the fermion operators act on the electronic state $|m\rangle$ and the source is taken to be situated at the origin, that is, $\vec{R} = 0$.

Molecular Quantum Electrodynamics, by Akbar Salam

$$d_i^{(2)}(\vec{\mu}\vec{\mu};\vec{r},t)=\frac{i}{4\pi}\sum_{\vec{k},\lambda}\sum_{m}\left(\frac{\hbar ck}{2\varepsilon_0 V}\right)^{1/2}$$

$$\times\left\{\begin{array}{l} e_k\alpha(0)\beta_m^{\dagger}(0)\beta_m(0)\\ \times\left[\begin{array}{l}\sum_n\left\{\dfrac{\mu_j^{mn}\mu_k^{nm}}{E_{nm}-\hbar\omega}+\dfrac{\mu_k^{mn}\mu_j^{nm}}{E_{nm}+\hbar\omega}\right\}k^3f_{ij}(kr)\mathrm{e}^{ik(r-ct)}\\ -\sum_n\dfrac{\mu_j^{mn}\mu_k^{nm}}{E_{nm}-\hbar\omega}k_{nm}^3\,f_{ij}(k_{nm}r)\mathrm{e}^{ik_{nm}(r-ct)}\\ -\sum_n\dfrac{\mu_k^{mn}\mu_j^{nm}}{E_{nm}+\hbar\omega}k_{mn}^3\,f_{ij}(k_{mn}r)\mathrm{e}^{ik_{mn}(r-ct)}\end{array}\right]\end{array}\right\}+\mathrm{H.C.}\tag{A.1}$$

$$d_i^{(2)}(\vec{\mu}\vec{m};\vec{r},t)=\frac{i}{4\pi}\sum_{\vec{k},\lambda}\sum_{m}\left(\frac{\hbar k}{2\varepsilon_0 cV}\right)^{1/2}$$

$$\times\left\{\begin{array}{l} b_k\alpha(0)\beta_m^{\dagger}(0)\beta_m(0)\\ \times\left[\begin{array}{l}\sum_n\left\{\dfrac{\mu_j^{mn}m_k^{nm}}{E_{nm}-\hbar\omega}+\dfrac{m_k^{mn}\mu_j^{nm}}{E_{nm}+\hbar\omega}\right\}k^3f_{ij}(kr)\mathrm{e}^{ik(r-ct)}\\ -\sum_n\dfrac{\mu_j^{mn}m_k^{nm}}{E_{nm}-\hbar\omega}k_{nm}^3\,f_{ij}(k_{nm}r)\mathrm{e}^{ik_{nm}(r-ct)}\\ -\sum_n\dfrac{m_k^{mn}\mu_j^{nm}}{E_{nm}+\hbar\omega}k_{mn}^3\,f_{ij}(k_{mn}r)\mathrm{e}^{ik_{mn}(r-ct)}\end{array}\right]\end{array}\right\}$$

$$-\frac{i}{4\pi}\sum_{\vec{k},\lambda}\sum_{m}\left(\frac{\hbar k}{2\varepsilon_0 cV}\right)^{1/2}$$

$$\times\left\{\begin{array}{l} e_k\alpha(0)\beta_m^{\dagger}(0)\beta_m(0)\\ \times\left[\begin{array}{l}\sum_n\left\{\dfrac{m_j^{mn}\mu_k^{nm}}{E_{nm}-\hbar\omega}+\dfrac{\mu_k^{mn}m_j^{nm}}{E_{nm}+\hbar\omega}\right\}k^3g_{ij}(kr)\mathrm{e}^{ik(r-ct)}\\ -\sum_n\dfrac{m_j^{mn}\mu_k^{nm}}{E_{nm}-\hbar\omega}k_{nm}^3g_{ij}(k_{nm}r)\mathrm{e}^{ik_{nm}(r-ct)}\\ -\sum_n\dfrac{\mu_k^{mn}m_j^{nm}}{E_{nm}+\hbar\omega}k_{mn}^3g_{ij}(k_{mn}r)\mathrm{e}^{ik_{mn}(r-ct)}\end{array}\right]\end{array}\right\}+\mathrm{H.C.}\tag{A.2}$$

$$d_i^{(2)}(\vec{m}\vec{m};\vec{r},t) = \frac{-i}{4\pi c}\sum_{\vec{k},\lambda}\sum_{m}\left(\frac{\hbar k}{2\varepsilon_0 cV}\right)^{1/2} \times \left\{ \begin{array}{l} b_k \alpha(0)\beta_m^{\dagger}(0)\beta_m(0) \\ \times \left[\begin{array}{l} \sum_n \left\{ \frac{m_j^{mn} m_k^{nm}}{E_{nm}-\hbar\omega} + \frac{m_k^{mn} m_j^{nm}}{E_{nm}+\hbar\omega} \right\} k^3 g_{ij}(kr) e^{ik(r-ct)} \\ -\sum_n \frac{m_j^{mn} m_k^{nm}}{E_{nm}-\hbar\omega} k_{nm}^3 g_{ij}(k_{nm} r) e^{ik_{nm}(r-ct)} \\ -\sum_n \frac{m_k^{mn} m_j^{nm}}{E_{nm}+\hbar\omega} k_{mn}^3 g_{ij}(k_{mn} r) e^{ik_{mn}(r-ct)} \end{array} \right] \end{array} \right\} + \text{H.C.} \tag{A.3}$$

$$\begin{aligned} d_i^{(2)}(\vec{\mu}\vec{\vec{Q}};\vec{r},t) = {} & \frac{-1}{4\pi}\sum_{\vec{k},\lambda}\sum_{m}\left(\frac{\hbar ck}{2\varepsilon_0 V}\right)^{1/2} \times \left\{ \begin{array}{l} k_l e_k \alpha(0)\beta_m^{\dagger}(0)\beta_m(0) \\ \times \left[\begin{array}{l} \sum_n \left\{ \frac{\mu_j^{mn} Q_{kl}^{nm}}{E_{nm}-\hbar\omega} + \frac{Q_{kl}^{mn} \mu_j^{nm}}{E_{nm}+\hbar\omega} \right\} k^3 f_{ij}(kr) e^{ik(r-ct)} \\ -\sum_n \frac{\mu_j^{mn} Q_{kl}^{nm}}{E_{nm}-\hbar\omega} k_{nm}^3 f_{ij}(k_{nm} r) e^{ik_{nm}(r-ct)} \\ -\sum_n \frac{Q_{kl}^{mn} \mu_j^{nm}}{E_{nm}+\hbar\omega} k_{mn}^3 f_{ij}(k_{mn} r) e^{ik_{mn}(r-ct)} \end{array} \right] \end{array} \right\} \\ & -\frac{i}{4\pi}\sum_{\vec{k},\lambda}\sum_{m}\left(\frac{\hbar ck}{2\varepsilon_0 V}\right)^{1/2} \times \left\{ \begin{array}{l} e_j \alpha(0)\beta_m^{\dagger}(0)\beta_m(0) \\ \times \left[\begin{array}{l} \sum_n \left\{ \frac{Q_{kl}^{mn} \mu_j^{nm}}{E_{nm}-\hbar\omega} + \frac{\mu_j^{mn} Q_{kl}^{nm}}{E_{nm}+\hbar\omega} \right\} k^4 h_{ikl}(kr) e^{ik(r-ct)} \\ -\sum_n \frac{Q_{kl}^{mn} \mu_j^{nm}}{E_{nm}-\hbar\omega} k_{nm}^4 h_{ikl}(k_{nm} r) e^{ik_{nm}(r-ct)} \\ -\sum_n \frac{\mu_j^{mn} Q_{kl}^{nm}}{E_{nm}+\hbar\omega} k_{mn}^4 h_{ikl}(k_{mn} r) e^{ik_{mn}(r-ct)} \end{array} \right] \end{array} \right\} + \text{H.C.} \end{aligned} \tag{A.4}$$

$$d_i^{(2)}(\vec{m}\vec{Q};\vec{r},t) = \frac{1}{4\pi c}\sum_{\vec{k},\lambda}\sum_{m}\left(\frac{\hbar ck}{2\varepsilon_0 V}\right)^{1/2}$$

$$\times\left\{\begin{array}{l} k_l e_k \alpha(0)\beta_m^{\dagger}(0)\beta_m(0) \\ \times\left[\begin{array}{l}\sum_n\left\{\dfrac{m_j^{mn}Q_{kl}^{nm}}{E_{nm}-\hbar\omega}+\dfrac{Q_{kl}^{mn}m_j^{nm}}{E_{nm}+\hbar\omega}\right\}k^3 g_{ij}(kr)e^{ik(r-ct)} \\ -\sum_n\dfrac{m_j^{mn}Q_{kl}^{nm}}{E_{nm}-\hbar\omega}k_{nm}^3 g_{ij}(k_{nm}r)e^{ik_{nm}(r-ct)} \\ -\sum_n\dfrac{Q_{kl}^{mn}m_j^{nm}}{E_{nm}+\hbar\omega}k_{mn}^3 g_{ij}(k_{mn}r)e^{ik_{mn}(r-ct)}\end{array}\right]\end{array}\right\}$$

$$-\frac{i}{4\pi c}\sum_{\vec{k},\lambda}\sum_{m}\left(\frac{\hbar ck}{2\varepsilon_0 V}\right)^{1/2}$$

$$\times\left\{\begin{array}{l} b_j \alpha(0)\beta_m^{\dagger}(0)\beta_m(0) \\ \times\left[\begin{array}{l}\sum_n\left\{\dfrac{Q_{kl}^{mn}m_j^{nm}}{E_{nm}-\hbar\omega}+\dfrac{m_j^{mn}Q_{kl}^{nm}}{E_{nm}+\hbar\omega}\right\}k^4 h_{ikl}(kr)e^{ik(r-ct)} \\ -\sum_n\dfrac{Q_{kl}^{mn}m_j^{nm}}{E_{nm}-\hbar\omega}k_{nm}^4 h_{ikl}(k_{nm}r)e^{ik_{nm}(r-ct)} \\ -\sum_n\dfrac{m_j^{mn}Q_{kl}^{nm}}{E_{nm}+\hbar\omega}k_{mn}^4 h_{ikl}(k_{mn}r)e^{ik_{mn}(r-ct)}\end{array}\right]\end{array}\right\}+\text{H.C.} \tag{A.5}$$

$$d_i^{(2)}(\vec{Q}\vec{Q};\vec{r},t) = \frac{1}{4\pi}\sum_{\vec{k},\lambda}\sum_{m}\left(\frac{\hbar ck}{2\varepsilon_0 V}\right)^{1/2}$$

$$\times\left\{\begin{array}{l} k_m e_l \alpha(0)\beta_m^{\dagger}(0)\beta_m(0) \\ \times\left[\begin{array}{l}\sum_n\left\{\dfrac{Q_{jk}^{mn}Q_{lm}^{nm}}{E_{nm}-\hbar\omega}+\dfrac{Q_{lm}^{mn}Q_{jk}^{nm}}{E_{nm}+\hbar\omega}\right\}k^4 h_{ijk}(kr)e^{ik(r-ct)} \\ -\sum_n\dfrac{Q_{jk}^{mn}Q_{lm}^{nm}}{E_{nm}-\hbar\omega}k_{nm}^4 h_{ijk}(k_{nm}r)e^{ik_{nm}(r-ct)} \\ -\sum_n\dfrac{Q_{lm}^{mn}Q_{jk}^{nm}}{E_{nm}+\hbar\omega}k_{mn}^4 h_{ijk}(k_{mn}r)e^{ik_{mn}(r-ct)}\end{array}\right]\end{array}\right\}+\text{H.C.} \tag{A.6}$$

$$b_i^{(2)}(\vec{\mu}\vec{\mu};\vec{r},t) = \frac{i}{4\pi\varepsilon_0}\sum_{\vec{k},\lambda}\sum_m \left(\frac{\hbar k}{2\varepsilon_0 cV}\right)^{1/2} \times \left\{ \begin{array}{l} e_k \alpha(0)\beta_m^\dagger(0)\beta_m(0) \\ \times \left[\begin{array}{l} \sum_n \left\{ \frac{\mu_j^{mn}\mu_k^{nm}}{E_{nm}-\hbar\omega} + \frac{\mu_k^{mn}\mu_j^{nm}}{E_{nm}+\hbar\omega} \right\} k^3 g_{ij}(kr)\mathrm{e}^{ik(r-ct)} \\ -\sum_n \frac{\mu_j^{mn}\mu_k^{nm}}{E_{nm}-\hbar\omega} k_{nm}^3 g_{ij}(k_{nm}r)\mathrm{e}^{ik_{nm}(r-ct)} \\ -\sum_n \frac{\mu_k^{mn}\mu_j^{nm}}{E_{nm}+\hbar\omega} k_{mn}^3 g_{ij}(k_{mn}r)\mathrm{e}^{ik_{mn}(r-ct)} \end{array} \right] \end{array} \right\} + \mathrm{H.C.} \tag{A.7}$$

$$\begin{aligned} b_i^{(2)}(\vec{\mu}\vec{m};\vec{r},t) = {} & \frac{i}{4\pi\varepsilon_0 c}\sum_{\vec{k},\lambda}\sum_m \left(\frac{\hbar k}{2\varepsilon_0 cV}\right)^{1/2} \\ & \times \left\{ \begin{array}{l} b_k \alpha(0)\beta_m^\dagger(0)\beta_m(0) \\ \times \left[\begin{array}{l} \sum_n \left\{ \frac{\mu_j^{mn}m_k^{nm}}{E_{nm}-\hbar\omega} + \frac{m_k^{mn}\mu_j^{nm}}{E_{nm}+\hbar\omega} \right\} k^3 g_{ij}(kr)\mathrm{e}^{ik(r-ct)} \\ -\sum_n \frac{\mu_j^{mn}m_k^{nm}}{E_{nm}-\hbar\omega} k_{nm}^3 g_{ij}(k_{nm}r)\mathrm{e}^{ik_{nm}(r-ct)} \\ -\sum_n \frac{m_k^{mn}\mu_j^{nm}}{E_{nm}+\hbar\omega} k_{mn}^3 g_{ij}(k_{mn}r)\mathrm{e}^{ik_{mn}(r-ct)} \end{array} \right] \end{array} \right\} \\ & + \frac{i}{4\pi\varepsilon_0 c}\sum_{\vec{k},\lambda}\sum_m \left(\frac{\hbar k}{2\varepsilon_0 cV}\right)^{1/2} \\ & \times \left\{ \begin{array}{l} e_k \alpha(0)\beta_m^\dagger(0)\beta_m(0) \\ \times \left[\begin{array}{l} \sum_n \left\{ \frac{m_j^{mn}\mu_k^{nm}}{E_{nm}-\hbar\omega} + \frac{\mu_k^{mn}m_j^{nm}}{E_{nm}+\hbar\omega} \right\} k^3 f_{ij}(kr)\mathrm{e}^{ik(r-ct)} \\ -\sum_n \frac{m_j^{mn}\mu_k^{nm}}{E_{nm}-\hbar\omega} k_{nm}^3 f_{ij}(k_{nm}r)\mathrm{e}^{ik_{nm}(r-ct)} \\ -\sum_n \frac{\mu_k^{mn}m_j^{nm}}{E_{nm}+\hbar\omega} k_{mn}^3 f_{ij}(k_{mn}r)\mathrm{e}^{ik_{mn}(r-ct)} \end{array} \right] \end{array} \right\} + \mathrm{H.C.} \end{aligned} \tag{A.8}$$

$$b_i^{(2)}(\vec{m}\vec{m};\vec{r},t)=\frac{i}{4\pi\varepsilon_0 c^2}\sum_{\vec{k},\lambda}\sum_m\left(\frac{\hbar k}{2\varepsilon_0 cV}\right)^{1/2}$$

$$\times\left\{\begin{array}{l} b_k\alpha(0)\beta_m^{\dagger}(0)\beta_m(0) \\ \times\left[\begin{array}{l}\sum_n\left\{\frac{m_j^{mn}m_k^{nm}}{E_{nm}-\hbar\omega}+\frac{m_k^{mn}m_j^{nm}}{E_{nm}+\hbar\omega}\right\}k^3 f_{ij}(kr)\mathrm{e}^{ik(r-ct)} \\ -\sum_n\frac{m_j^{mn}m_k^{nm}}{E_{nm}-\hbar\omega}k_{nm}^3 f_{ij}(k_{nm}r)\mathrm{e}^{ik_{nm}(r-ct)} \\ -\sum_n\frac{m_k^{mn}m_j^{nm}}{E_{nm}+\hbar\omega}k_{mn}^3 f_{ij}(k_{mn}r)\mathrm{e}^{ik_{mn}(r-ct)}\end{array}\right]\end{array}\right\}+\mathrm{H.C.} \tag{A.9}$$

$$b_i^{(2)}(\vec{\mu}\vec{\vec{Q}};\vec{r},t)=-\frac{1}{4\pi\varepsilon_0}\sum_{\vec{k},\lambda}\sum_m\left(\frac{\hbar k}{2\varepsilon_0 cV}\right)^{1/2}$$

$$\times\left\{\begin{array}{l} k_m e_l\alpha(0)\beta_m^{\dagger}(0)\beta_m(0) \\ \times\left[\begin{array}{l}\sum_n\left\{\frac{\mu_j^{mn}Q_{lm}^{nm}}{E_{nm}-\hbar\omega}+\frac{Q_{lm}^{mn}\mu_j^{nm}}{E_{nm}+\hbar\omega}\right\}k^3 g_{ij}(kr)\mathrm{e}^{ik(r-ct)} \\ -\sum_n\frac{\mu_j^{mn}Q_{lm}^{nm}}{E_{nm}-\hbar\omega}k_{nm}^3 g_{ij}(k_{nm}r)\mathrm{e}^{ik_{nm}(r-ct)} \\ -\sum_n\frac{Q_{lm}^{mn}\mu_j^{nm}}{E_{nm}+\hbar\omega}k_{mn}^3 g_{ij}(k_{mn}r)\mathrm{e}^{ik_{mn}(r-ct)}\end{array}\right]\end{array}\right\}$$

$$+\frac{1}{4\pi\varepsilon_0}\sum_{\vec{k},\lambda}\sum_m\left(\frac{\hbar k}{2\varepsilon_0 cV}\right)^{1/2}$$

$$\times\left\{\begin{array}{l} e_l\alpha(0)\beta_m^{\dagger}(0)\beta_m(0) \\ \times\left[\begin{array}{l}\sum_n\left\{\frac{Q_{jk}^{mn}\mu_l^{nm}}{E_{nm}-\hbar\omega}+\frac{\mu_l^{mn}Q_{jk}^{nm}}{E_{nm}+\hbar\omega}\right\}k^4 j_{ijk}(kr)\mathrm{e}^{ik(r-ct)} \\ -\sum_n\frac{Q_{jk}^{mn}\mu_l^{nm}}{E_{nm}-\hbar\omega}k_{nm}^4 j_{ijk}(k_{nm}r)\mathrm{e}^{ik_{nm}(r-ct)} \\ -\sum_n\frac{\mu_l^{mn}Q_{jk}^{nm}}{E_{nm}+\hbar\omega}k_{mn}^4 j_{ijk}(k_{mn}r)\mathrm{e}^{ik_{mn}(r-ct)}\end{array}\right]\end{array}\right\}+\mathrm{H.C.} \tag{A.10}$$

$$b_i^{(2)}(\vec{m}\vec{\vec{Q}};\vec{r},t) = -\frac{1}{4\pi\varepsilon_0 c}\sum_{\vec{k},\lambda}\sum_m\left(\frac{\hbar k}{2\varepsilon_0 cV}\right)^{1/2}$$

$$\times\left\{\begin{array}{l} k_m e_l \alpha(0)\beta_m^\dagger(0)\beta_m(0) \\ \times\left[\begin{array}{l}\sum_n\left\{\dfrac{m_j^{mn}Q_{lm}^{nm}}{E_{nm}-\hbar\omega}+\dfrac{Q_{lm}^{mn}m_j^{nm}}{E_{nm}+\hbar\omega}\right\}k^3 f_{ij}(kr)\mathrm{e}^{ik(r-ct)} \\ -\sum_n\dfrac{m_j^{mn}Q_{lm}^{nm}}{E_{nm}-\hbar\omega}k_{nm}^3 f_{ij}(k_{nm}r)\mathrm{e}^{ik_{nm}(r-ct)} \\ -\sum_n\dfrac{Q_{lm}^{mn}m_j^{nm}}{E_{nm}+\hbar\omega}k_{mn}^3 f_{ij}(k_{mn}r)\mathrm{e}^{ik_{mn}(r-ct)}\end{array}\right]\end{array}\right\}$$

$$+\frac{1}{4\pi\varepsilon_0 c}\sum_{\vec{k},\lambda}\sum_m\left(\frac{\hbar k}{2\varepsilon_0 cV}\right)^{1/2}$$

$$\times\left\{\begin{array}{l} b_l \alpha(0)\beta_m^\dagger(0)\beta_m(0) \\ \times\left[\begin{array}{l}\sum_n\left\{\dfrac{Q_{jk}^{mn}m_l^{nm}}{E_{nm}-\hbar\omega}+\dfrac{m_l^{mn}Q_{jk}^{nm}}{E_{nm}+\hbar\omega}\right\}k^4 j_{ijk}(kr)\mathrm{e}^{ik(r-ct)} \\ -\sum_n\dfrac{Q_{jk}^{mn}m_l^{nm}}{E_{nm}-\hbar\omega}k_{nm}^4 j_{ijk}(k_{nm}r)\mathrm{e}^{ik_{nm}(r-ct)} \\ -\sum_n\dfrac{m_l^{mn}Q_{jk}^{nm}}{E_{nm}+\hbar\omega}k_{mn}^4 j_{ijk}(k_{mn}r)\mathrm{e}^{ik_{mn}(r-ct)}\end{array}\right]\end{array}\right\}+\mathrm{H.C.} \tag{A.11}$$

$$b_i^{(2)}(\vec{\vec{Q}}\vec{\vec{Q}};\vec{r},t) = \frac{i}{4\pi\varepsilon_0}\sum_{\vec{k},\lambda}\sum_m\left(\frac{\hbar k}{2\varepsilon_0 cV}\right)^{1/2}$$

$$\times\left\{\begin{array}{l} k_m e_l \alpha(0)\beta_m^\dagger(0)\beta_m(0) \\ \times\left[\begin{array}{l}\sum_n\left\{\dfrac{Q_{jk}^{mn}Q_{lm}^{nm}}{E_{nm}-\hbar\omega}+\dfrac{Q_{lm}^{mn}Q_{jk}^{nm}}{E_{nm}+\hbar\omega}\right\}k^4 j_{ijk}(kr)\mathrm{e}^{ik(r-ct)} \\ -\sum_n\dfrac{Q_{jk}^{mn}Q_{lm}^{nm}}{E_{nm}-\hbar\omega}k_{nm}^4 j_{ijk}(k_{nm}r)\mathrm{e}^{ik_{nm}(r-ct)} \\ -\sum_n\dfrac{Q_{lm}^{mn}Q_{jk}^{nm}}{E_{nm}+\hbar\omega}k_{mn}^4 j_{ijk}(k_{mn}r)\mathrm{e}^{ik_{mn}(r-ct)}\end{array}\right]\end{array}\right\}+\mathrm{H.C.} \tag{A.12}$$

The tensor fields $f_{ij}(kr)$ and $g_{ij}(kr)$ have been defined in equations (2.9.4) and (2.9.34), respectively. The new ones appearing in the fields above are

$$\begin{aligned} H_{ijk}(kr) &= \frac{1}{k}\vec{\nabla}_k F_{ij}(kr) = \frac{1}{k^4}(-\vec{\nabla}^2\delta_{ij} + \vec{\nabla}_i\vec{\nabla}_j)\vec{\nabla}_k\frac{\mathrm{e}^{ikr}}{r} \\ &= \left\{(\delta_{ij} - \hat{r}_i\hat{r}_j)\hat{r}_k\left(\frac{i}{kr} - \frac{1}{k^2r^2}\right) + (\delta_{ij}\hat{r}_k + \delta_{ik}\hat{r}_j + \delta_{jk}\hat{r}_i - 5\hat{r}_i\hat{r}_j\hat{r}_k)\right. \\ &\quad \left.\times\left(-\frac{1}{k^2r^2} - \frac{3i}{k^3r^3} + \frac{3}{k^4r^4}\right)\right\}\mathrm{e}^{ikr} = h_{ijk}(kr)\mathrm{e}^{ikr} \end{aligned} \tag{A.13}$$

and

$$\begin{aligned} J_{ijl}(kr) &= \frac{i}{k}\vec{\nabla}_l G_{ij}(kr) = -\frac{1}{k^3}\varepsilon_{ijk}\vec{\nabla}_k\vec{\nabla}_l\frac{\mathrm{e}^{ikr}}{r} \\ &= \varepsilon_{ijk}\left\{\hat{r}_k\hat{r}_l\frac{1}{kr} - (\delta_{kl} - 3\hat{r}_k\hat{r}_l)\left(\frac{i}{k^2r^2} - \frac{1}{k^3r^3}\right)\right\}\mathrm{e}^{ikr} = j_{ijl}(kr)\mathrm{e}^{ikr}. \end{aligned} \tag{A.14}$$

APPENDIX B

ROTATIONAL AVERAGING OF CARTESIAN TENSORS

Observable quantities calculated using quantum electrodynamical theory, be they transition rates as a result of interaction of electromagnetic radiation with one or more atoms or molecules, or energy shifts arising from species in mutual interaction, or both, often apply to bodies in fixed orientation relative to each other and to the direction of propagation of the radiation field if present. To calculate expectation values applicable to species in the fluid phase therefore require performing a rotational average of the molecule or molecules, which are characterized by a multipole moment of a particular order and that are typically expressed in terms of components of a Cartesian tensor. This is usually followed by contraction with polarization vectors of the radiation field or some other geometric factors, both of which are also written in terms of Cartesian tensor components. This Appendix outlines the procedure by which Cartesian tensors may be orientationally averaged, the results presented being able to be used to yield measurables for isotropic systems.

Consider a molecular tensorial property $\boldsymbol{T}$ of rank r whose Cartesian components in a space-fixed frame of reference are $T_{i_1 i_2 \ldots i_r}$ and in a molecule-fixed frame are $T_{\lambda_1 \lambda_2 \ldots \lambda_r}$. The components in the two frames are related through

$$T_{i_1 i_2 \ldots i_r} = l_{i_1 \lambda_1} l_{i_2 \lambda_2} \ldots l_{i_r \lambda_r} T_{\lambda_1 \lambda_2 \ldots \lambda_r}, \tag{B.1}$$

Molecular Quantum Electrodynamics, by Akbar Salam

where $l_{i_n\lambda_n}$ is the direction cosine between the space-fixed axis i_n and the molecule-fixed axis λ_n. A rotational average of $T_{i_1 i_2 \ldots i_r}$ therefore requires a rotational average of the factor $l_{i_1\lambda_1} l_{i_2\lambda_2} \ldots l_{i_r\lambda_r}$, which for convenience is designated by $I^{(r)}$. This last quantity is expressible as a linear combination of isotropic tensors, each member being a product of two isotropic tensors—one for the space-fixed frame, the other for the molecule-fixed axes. The two fundamental isotropic tensors in three dimensions being the Kronecker delta δ_{ij} and the Levi-Civita epsilon ε_{ijk}. Isotropic tensor products are formed by permuting the indices $i_1, i_2, \ldots, i_n$ in a particular factor, and are called isomers. Results are given below for $I^{(r)}$ applicable for tensors up to rank six (Andrews and Thirunamachandran, 1977).

$$I^{(0)} = 1 \tag{B.2}$$

and

$$I^{(1)} = 0. \tag{B.3}$$

For $r = 2$, the sole isomer is $\delta_{i_1 i_2}$ and

$$I^{(2)} = \frac{1}{3}\, \delta_{i_1 i_2} \delta_{\lambda_1 \lambda_2}. \tag{B.4}$$

For $r = 3$, there is also a single isomer, $\varepsilon_{i_1 i_2 i_3}$, and

$$I^{(3)} = \frac{1}{6}\, \varepsilon_{i_1 i_2 i_3} \varepsilon_{\lambda_1 \lambda_2 \lambda_3}. \tag{B.5}$$

When $r = 4$, there are three linearly independent isomers

$$\delta_{i_1 i_2}\delta_{i_3 i_4}; \quad \delta_{i_1 i_3}\delta_{i_2 i_4}; \quad \delta_{i_1 i_4}\delta_{i_2 i_3} \tag{B.6}$$

and

$$I^{(4)} = \frac{1}{30} \begin{pmatrix} \delta_{i_1 i_2}\delta_{i_3 i_4} \\ \delta_{i_1 i_3}\delta_{i_2 i_4} \\ \delta_{i_1 i_4}\delta_{i_2 i_3} \end{pmatrix}^{\mathrm{T}} \begin{pmatrix} 4 & -1 & -1 \\ -1 & 4 & -1 \\ -1 & -1 & 4 \end{pmatrix} \begin{pmatrix} \delta_{\lambda_1 \lambda_2}\delta_{\lambda_3 \lambda_4} \\ \delta_{\lambda_1 \lambda_3}\delta_{\lambda_2 \lambda_4} \\ \delta_{\lambda_1 \lambda_4}\delta_{\lambda_2 \lambda_3} \end{pmatrix}, \tag{B.7}$$

where the superscript T denotes the transpose. Although there are 10 different isomers for $r = 5$, only 6 are linearly independent. The ones used are

$$\varepsilon_{i_1 i_2 i_3}\delta_{i_4 i_5}; \quad \varepsilon_{i_1 i_2 i_4}\delta_{i_3 i_5}; \quad \varepsilon_{i_1 i_2 i_5}\delta_{i_3 i_4}; \quad \varepsilon_{i_1 i_3 i_4}\delta_{i_2 i_5}; \quad \varepsilon_{i_1 i_3 i_5}\delta_{i_2 i_6}; \quad \varepsilon_{i_1 i_4 i_5}\delta_{i_2 i_3}, \tag{B.8}$$

resulting in

$$I^{(5)} = \frac{1}{30}\begin{bmatrix} \varepsilon_{i_1 i_2 i_3}\delta_{i_4 i_5} \\ \varepsilon_{i_1 i_2 i_4}\delta_{i_3 i_5} \\ \varepsilon_{i_1 i_2 i_5}\delta_{i_3 i_4} \\ \varepsilon_{i_1 i_3 i_4}\delta_{i_2 i_5} \\ \varepsilon_{i_1 i_3 i_5}\delta_{i_2 i_4} \\ \varepsilon_{i_1 i_4 i_5}\delta_{i_2 i_3} \end{bmatrix}^T \begin{bmatrix} 3 & -1 & -1 & 1 & 1 & 0 \\ -1 & 3 & -1 & -1 & 0 & 1 \\ -1 & -1 & 3 & 0 & -1 & -1 \\ 1 & -1 & 0 & 3 & -1 & 1 \\ 1 & 0 & -1 & -1 & 3 & -1 \\ 0 & 1 & -1 & 1 & -1 & 3 \end{bmatrix} \begin{bmatrix} \varepsilon_{\lambda_1\lambda_2\lambda_3}\delta_{\lambda_4\lambda_5} \\ \varepsilon_{\lambda_1\lambda_2\lambda_4}\delta_{\lambda_3\lambda_5} \\ \varepsilon_{\lambda_1\lambda_2\lambda_5}\delta_{\lambda_3\lambda_4} \\ \varepsilon_{\lambda_1\lambda_3\lambda_4}\delta_{\lambda_2\lambda_5} \\ \varepsilon_{\lambda_1\lambda_3\lambda_5}\delta_{\lambda_2\lambda_4} \\ \varepsilon_{\lambda_1\lambda_4\lambda_5}\delta_{\lambda_2\lambda_3} \end{bmatrix}. \tag{B.9}$$

There are 15 distinct isomers for $r = 6$ and they comprise a linearly independent set. $I^{(6)}$ is given by

$$I^{(6)} = \frac{1}{210}$$

$$\times \begin{pmatrix} \delta_{ij}\delta_{kl}\delta_{mn} \\ \delta_{ij}\delta_{km}\delta_{nl} \\ \delta_{ij}\delta_{kn}\delta_{lm} \\ \delta_{ik}\delta_{jl}\delta_{mn} \\ \delta_{ik}\delta_{jm}\delta_{nl} \\ \delta_{ik}\delta_{jn}\delta_{lm} \\ \delta_{il}\delta_{jk}\delta_{mn} \\ \delta_{il}\delta_{jm}\delta_{kn} \\ \delta_{il}\delta_{jn}\delta_{km} \\ \delta_{im}\delta_{jk}\delta_{nl} \\ \delta_{im}\delta_{jl}\delta_{kn} \\ \delta_{im}\delta_{jn}\delta_{kl} \\ \delta_{in}\delta_{jk}\delta_{lm} \\ \delta_{in}\delta_{jl}\delta_{km} \\ \delta_{in}\delta_{jm}\delta_{kl} \end{pmatrix}^{\mathrm{T}} \begin{pmatrix} 16 & -5 & -5 & -5 & 2 & 2 & -5 & 2 & 2 & 2 & 2 & -5 & 2 & 2 & -5 \\ -5 & 16 & -5 & 2 & -5 & 2 & 2 & 2 & -5 & -5 & 2 & 2 & 2 & -5 & 2 \\ -5 & -5 & 16 & 2 & 2 & -5 & 2 & -5 & 2 & 2 & -5 & 2 & -5 & 2 & 2 \\ -5 & 2 & 2 & 16 & -5 & -5 & -5 & 2 & 2 & 2 & -5 & 2 & 2 & -5 & 2 \\ 2 & -5 & 2 & -5 & 16 & -5 & 2 & -5 & 2 & -5 & 2 & 2 & 2 & 2 & -5 \\ 2 & 2 & -5 & -5 & -5 & 16 & 2 & 2 & -5 & 2 & 2 & -5 & -5 & 2 & 2 \\ -5 & 2 & 2 & -5 & 2 & 2 & 16 & -5 & -5 & -5 & 2 & 2 & -5 & 2 & 2 \\ 2 & 2 & -5 & 2 & -5 & 2 & -5 & 16 & -5 & 2 & -5 & 2 & 2 & 2 & -5 \\ 2 & -5 & 2 & 2 & 2 & -5 & -5 & -5 & 16 & 2 & 2 & -5 & 2 & -5 & 2 \\ 2 & -5 & 2 & 2 & -5 & 2 & -5 & 2 & 2 & 16 & -5 & -5 & -5 & 2 & 2 \\ 2 & 2 & -5 & -5 & 2 & 2 & 2 & -5 & 2 & -5 & 16 & -5 & 2 & -5 & 2 \\ -5 & 2 & 2 & 2 & 2 & -5 & 2 & 2 & -5 & -5 & -5 & 16 & 2 & 2 & -5 \\ 2 & 2 & -5 & 2 & 2 & -5 & -5 & 2 & 2 & -5 & 2 & 2 & 16 & -5 & -5 \\ 2 & -5 & 2 & -5 & 2 & 2 & 2 & 2 & -5 & 2 & -5 & 2 & -5 & 16 & -5 \\ -5 & 2 & 2 & 2 & -5 & 2 & 2 & -5 & 2 & 2 & 2 & -5 & -5 & -5 & 16 \end{pmatrix} \begin{pmatrix} \delta_{\lambda\mu}\delta_{\nu\rho}\delta_{\sigma\tau} \\ \delta_{\lambda\mu}\delta_{\nu\sigma}\delta_{\rho\tau} \\ \delta_{\lambda\mu}\delta_{\nu\tau}\delta_{\rho\sigma} \\ \delta_{\lambda\nu}\delta_{\mu\rho}\delta_{\sigma\tau} \\ \delta_{\lambda\nu}\delta_{\mu\sigma}\delta_{\rho\tau} \\ \delta_{\lambda\nu}\delta_{\mu\tau}\delta_{\rho\sigma} \\ \delta_{\lambda\rho}\delta_{\mu\nu}\delta_{\sigma\tau} \\ \delta_{\lambda\rho}\delta_{\mu\sigma}\delta_{\nu\tau} \\ \delta_{\lambda\rho}\delta_{\mu\tau}\delta_{\nu\sigma} \\ \delta_{\lambda\sigma}\delta_{\mu\nu}\delta_{\rho\tau} \\ \delta_{\lambda\sigma}\delta_{\mu\rho}\delta_{\nu\tau} \\ \delta_{\lambda\sigma}\delta_{\mu\tau}\delta_{\nu\rho} \\ \delta_{\lambda\tau}\delta_{\mu\nu}\delta_{\rho\sigma} \\ \delta_{\lambda\tau}\delta_{\mu\rho}\delta_{\nu\sigma} \\ \delta_{\lambda\tau}\delta_{\mu\sigma}\delta_{\nu\rho} \end{pmatrix}. \tag{B.10}$$

REFERENCES

Abrikosova, I. I. and Derjaguin, B. V. (1957) Direct measurement of molecular attraction of solid bodies. II. Method for measuring the gap. Results of experiments. *Sov. Phys. – JETP* **4**, 2.

Alligood, B. W. and Salam, A. (2007) On the application of state sequence diagrams to the calculation of the Casimir–Polder potential. *Mol. Phys.* **105**, 395.

Andrews, D. L. (1989) A unified theory of radiative and radiationless molecular energy transfer. *Chem. Phys.* **135**, 195.

Andrews, D. L. and Allcock, P. (2002) *Optical Harmonics in Molecular Systems.* Wiley-VCH, Weinheim.

Andrews, D. L. and Sherborne, B. S. (1987) Resonant excitation transfer: a quantum electrodynamical study. *J. Chem. Phys.* **86**, 4011.

Andrews, D. L. and Thirunamachandran, T. (1977) On three-dimensional rotational averages. *J. Chem. Phys.* **67**, 5026.

Aoyama, T., Hayakawa, M., Kinoshita, T., and Nio, M. (2007) Revised value of the eighth-order contribution to the electron *g*-2. *Phys. Rev. Lett.* **99**, 110406.

Au, C. K., Feinberg, G., and Sucher, J. (1984) Retarded long-range interaction in He Rydberg states. *Phys. Rev. Lett.* **53**, 1145.

Aub, M. R. and Zienau, S. (1960) Studies on the retarded interaction between neutral atoms. I. Three-body London–van der Waals interaction of neutral atoms. *Proc. R. Soc. Lond. A* **257**, 464.

Avery, J. S. (1966) Resonance energy transfer and spontaneous photon emission. *Proc. Phys. Soc. Lond.* **88**, 1.

Axilrod, B. M. and Teller, E. (1943) Interaction of the van der Waals type between three atoms. *J. Chem. Phys.* **11**, 299.

Babb, J. F. and Spruch, L. (1988) Evaluation of retardation energy shifts in a Rydberg helium atom. *Phys. Rev. A* **38**, 13.

Belinfante, F. J. (1946) On the longitudinal and the transverse delta-functions, with some applications. *Physica* **12**, 1.

Bordag, M., Mohideen, U., and Mostapanenko, V. M. (2001) New developments in the Casimir effect. *Phys. Rep.* **353**, 1.

Bordag, M., Klimchitskaya, G. L., Mohideen, U., and Mostepanenko, V. M. (2009) Advances in the Casimir Effect. Oxford University Press, New York.

Born, M. and Oppenheimer, J. R. (1927) Zur Quantentheorie der Molekeln. *Ann. Phys.* **84**, 457.

Born, M., Heisenberg, W., and Jordan, P. (1926) Zur Quantenmechanik II. *Z. Phys.* **35**, 557.

Bradshaw, D. S. and Andrews, D. L. (2005) Optically induced forces and torques: interactions between nanoparticles in a laser beam. *Phys. Rev. A* **72**, 033816.

Buckingham, A. D. (1967) Permanent and induced molecular moments and long-range intermolecular forces. *Adv. Chem. Phys.* **12**, 107.

Casimir, H. B. G. (1948) On the attraction between two perfectly conducting plates. *Proc. K. Ned. Akad. Wet. B* **51**, 793.

Casimir, H. B. G. and Polder, D. (1948) The influence of retardation on the London van der Waals forces. *Phys. Rev.* **73**, 360.

Cirone, M. and Passante, R. (1997) Dressed zero-point field correlations and the non-additive three-body van der Waals potential. *J. Phys. B: At. Mol. Opt. Phys.* **30**, 5579.

Claverie, P. (1978) Elaboration of approximate formulas for the interaction between large molecules: applications in organic chemistry. In: Pullman, B. (ed.), *Intermolecular Interactions: From Diatomics to Biopolymers*. Wiley, New York, p. 69.

Craig, D. P. and Power, E. A. (1969) The asymptotic Casimir–Polder potential from second-order perturbation theory and its generalization for anisotropic polarizabilities. *Int. J. Quantum Chem.* **3**, 903.

Craig, D. P. and Thirunamachandran, T. (1989) Third-body mediation of resonance coupling between identical molecules. *Chem. Phys.* **135**, 37.

Craig, D. P. and Thirunamachandran, T. (1992) An analysis of models for resonant transfer of excitation using quantum electrodynamics. *Chem. Phys.* **167**, 229.

Craig, D. P. and Thirunamachandran, T. (1998a) *Molecular Quantum Electrodynamics*. Dover, New York.

Craig, D. P. and Thirunamachandran, T. (1998b) Chiral discrimination in molecular excitation transfer. *J. Chem. Phys.* **109**, 1259.

Craig, D. P. and Thirunamachandran, T. (1999) New approaches to chiral discrimination in coupling between molecules. *Theor. Chem. Acc.* **102**, 112.

Daniels, G. J., Jenkins, R. D., Bradshaw, D. S., and Andrews, D. L. (2003) Resonance energy transfer: the unified theory revisited. *J. Chem. Phys.* **119**, 2264.

Dexter, D. L. (1953) A theory of sensitized luminescence in solids. *J. Chem. Phys.* **21**, 836.

Dirac, P. A. M. (1927) The quantum theory of the emission and absorption of radiation. *Proc. R. Soc. Lond. A* **114**, 243.

Dirac, P. A. M. (1958) *The Principles of Quantum Mechanics*. Clarendon, Oxford.

Dyson, F. J. (1949) The radiation theories of Tomonaga, Schwinger and Feynman. *Phys. Rev.* **75**, 486.

Fermi, E. (1932) Quantum theory of radiation. *Rev. Mod. Phys.* **4**, 87.

Feynman, R. P. (1948) Space–time approach to non-relativistic quantum mechanics. *Rev. Mod. Phys.* **20**, 367.

Feynman, R. P. (1949a) The theory of positrons. *Phys. Rev.* **76**, 749.

Feynman, R. P. (1949b) Space–time approach to quantum electrodynamics. *Phys. Rev.* **76**, 769.

Feynman, R. P. (1985) *QED: The Strange Theory of Light and Matter*. Princeton University Press, Princeton, NJ.

Fock, V. (1932) Konfigurationsraum und zweite Quantelung. *Z. Phys.* **75**, 622.

Förster, T. (1948) Zwischenmolekulare Energiewanderung und Fluoreszenz. *Ann. Phys.* **6**, 55.

Goldstein, H. (1960) *Classical Mechanics*. Addison-Wesley, Reading, MA.

Göppert-Mayer, M. (1931) Uber Elementarakte mit Zwei Quantensprüngen. *Ann. Phys.* **9**, 273.

Hagley, E. W. and Pipkin, F. M. (1994) Separated oscillatory field measurement of hydrogen $2S_{1/2}$–$2P_{3/2}$ fine structure interval. *Phys. Rev. Lett.* **72**, 1172.

Hanneke, D., Fogwell, S., and Gabrielse, G. (2008) New measurement of the electron magnetic moment and the fine structure constant. *Phys. Rev. Lett.* **100**, 120801.

Healy, W. P. (1982) *Non-Relativistic Quantum Electrodynamics*. Academic Press, London.

Heitler, W. (1954) *The Quantum Theory of Radiation*. Clarendon, Oxford.

Hessels, E. A., Deck, F. J., Arcuni, P. W., and Lundeen, S. R. (1990) Precision spectroscopy of high-L, $n = 10$ Rydberg helium: an important test of relativistic, radiative, and retardation effects. *Phys. Rev. Lett.* **65**, 2765.

Hirschfelder, J. O. (ed.) (1967) Intermolecular forces. *Adv. Chem. Phys.* **12**.

Jackson, J. D. (1963) *Classical Electrodynamics*. Wiley, New York.

Jeans, J. H. (1905) On the partition of energy between matter and aether. *Philos. Mag.* **10**, 91.

Jenkins, J. K., Salam, A., and Thirunamachandran, T. (1994a) Discriminatory dispersion interactions between chiral molecules. *Mol. Phys.* **82**, 835.

Jenkins, J. K., Salam, A., and Thirunamachandran, T. (1994b) Retarded dispersion interaction energies between chiral molecules. *Phys. Rev. A* **50**, 4767.

Jenkins, R. D., Andrews, D. L., and Dávila-Romero, L. C. (2002) A new diagrammatic methodology for non-relativistic quantum electrodynamics. *J. Phys. B: At. Mol. Opt. Phys.* **35**, 445.

Lamoreaux, S. K. (1997) Demonstration of the Casimir force in the 0.6 to 6 μm range. *Phys. Rev. Lett.* **78**, 5.

Lamoreaux, S. K. (2005) The Casimir force: background, experiments and applications. *Rep. Prog. Phys.* **68**, 201.

Landau, L. D. and Lifshitz, E. M. (1972) *Mechanics and Electrodynamics*. Pergamon, Oxford.

Levine, I. N. (2000) *Quantum Chemistry*. Prentice-Hall, New Jersey.

Lifshitz, E. M. (1956) The theory of molecular attractive forces between solids. *Sov. Phys. – JETP* **2**, 73.

London, F. (1930) Zur Theorie und Systematik der Molekularkräfte. *Z. Phys.* **63**, 245.

Lundeen, S. R. (1993) Experimental studies of high-*L* Rydberg states in helium. In: Levin, F. S. and Micha, D. A. (eds), *Long-Range Casimir Forces*, Plenum, New York, p. 73.

Maitland, G. C., Rigby, M., Smith, E. B., and Wakeham, W. A. (1981) *Intermolecular Forces*. Clarendon, Oxford.

Mandl, F. (1959) *Introduction to Quantum Field Theory*. Interscience, New York.

Margenau, H. and Kestner, N. R. (1969) *Theory of Intermolecular Forces*. Pergamon, Oxford.

Masiello, D., Deumens, E., and Öhrn, Y. (2005) On the canonical formulation of electrodynamics and wave mechanics. *Adv. Quantum Chem.* **49**, 249.

Mattuck, R. D. (1976) *A Guide to Feynman Diagrams in the Many-Body Problem*. McGraw-Hill, New York.

McLone, R. R. and Power, E. A. (1964) On the interaction between two identical neutral dipole systems, one in an excited state and the other in the ground state. *Mathematika* **11**, 91.

Milonni, P. W. (1994) *The Quantum Vacuum*. Academic Press, San Diego, CA.

Mohideen, U. and Roy, A. (1998) Precision measurement of the Casimir force from 0.1 to 0.9 μm. *Phys. Rev. Lett.* **81**, 4549.

Mukamel, S. (1995) *Principles of Nonlinear Optical Spectroscopy*. Oxford University Press, New York.

Muto, Y. (1943) Force between nonpolar molecules. *J. Phys. Math. Soc. Japan.* **17**, 629.

Pachucki, K. (1994) Complete two-loop binding correction to the lamb shift. *Phys. Rev. Lett.* **72**, 3154.

Palfrey, S. L. and Lundeen, S. R. (1984) Measurement of high-angular momentum fine structure in helium: an experimental test of long-range electromagnetic forces. *Phys. Rev. Lett.* **53**, 1141.

Passante, R., Power, E. A., and Thirunamachandran, T. (1998) Radiation–molecule coupling using dynamic polarizabilities: application to many-body forces. *Phys. Lett. A* **249**, 77.

Power, E. A. (1964) *Introductory Quantum Electrodynamics*. Longmans, London.

Power, E. A. (1978) A review of canonical transformations as they affect multiphoton processes. In: Eberly, J. H. and Lambropoulos, P. (eds), *Multiphoton Processes*, Wiley, New York. p. 11.

Power, E. A. (1993) The natural line shape. In: Grandy, W. J., Jr. and Milonni, P. W. (eds), *Physics and Probability*, Cambridge University Press, Cambridge, p. 101.

Power, E. A. and Thirunamachandran, T. (1971) Three distribution identities and Maxwell's atomic field equations including the Röntgen current. *Mathematika* **18**, 240.

Power, E. A. and Thirunamachandran, T. (1978) On the nature of the Hamiltonian for the interaction of radiation with atoms and molecules: $(e/mc)\vec{\mathrm{p}}\cdot\vec{\mathrm{A}}$, $-\vec{\mu}\cdot\vec{\mathrm{E}}$, and all that. *Am. J. Phys.* **46**, 370.

Power, E. A. and Thirunamachandran, T. (1983a) Quantum electrodynamics with nonrelativistic sources. I. Transformation to the multipolar formalism for second-quantized electron and Maxwell interacting fields. *Phys. Rev. A* **28**, 2649.

Power, E. A. and Thirunamachandran, T. (1983b) Quantum electrodynamics with nonrelativistic sources. II. Maxwell fields in the vicinity of an atom. *Phys. Rev. A* **28**, 2649.

Power, E. A. and Thirunamachandran, T. (1983c) Quantum electrodynamics with nonrelativistic sources. III. Intermolecular interactions. *Phys. Rev. A* **28**, 2671.

Power, E. A. and Thirunamachandran, T. (1985) The non-additive dispersion energies for *N* molecules: a quantum electrodynamical theory. *Proc. R. Soc. Lond. A* **401**, 267

Power, E. A. and Thirunamachandran, T. (1992) Quantum electrodynamics with nonrelativistic sources. IV. Poynting vector, energy densities, and other quadratic operators of the electromagnetic field. *Phys. Rev. A* **45**, 54.

Power, E. A. and Thirunamachandran, T. (1993a) Quantum electrodynamics with nonrelativistic sources. V. Electromagnetic field correlations and intermolecular interactions between molecules in either ground or excited states. *Phys. Rev. A* **47**, 2539.

Power, E. A. and Thirunamachandran, T. (1993b) Casimir–Polder potential as an interaction between induced dipoles. *Phys. Rev. A* **48**, 4761

Power, E. A. and Thirunamachandran, T. (1994) Zero-point energy differences and many-body dispersion forces. *Phys. Rev. A* **50**, 3929.

Power, E. A. and Thirunamachandran, T. (1995a) Dispersion forces between molecules with one or both molecules excited. *Phys. Rev. A* **51**, 3660.

Power, E. A. and Thirunamachandran, T. (1995b) Two- and three-body dispersion forces with one excited molecule. *Chem. Phys.* **198**, 5.

Power, E. A. and Thirunamachandran, T. (1997) Analysis of the causal behavior in energy transfer between atoms. *Phys. Rev. A* **56**, 3395.

Power, E. A. and Thirunamachandran, T. (1999a) Time dependence of operators in minimal and multipolar nonrelativistic quantum electrodynamics. I. Electromagnetic fields in the neighborhood of an atom. *Phys. Rev. A* **60**, 4927.

Power, E. A. and Thirunamachandran, T. (1999b) Time dependence of operators in minimal and multipolar nonrelativistic quantum electrodynamics. II. Analysis of the functional forms of operators in the two frameworks. *Phys. Rev. A* **60**, 4936.

Power, E. A. and Zienau, S. (1959) Coulomb gauge in non-relativistic quantum electrodynamics and the shape of spectral lines. *Philos. Trans. R. Soc. Lond. A* **251**, 427.

Rayleigh, Lord. (1900) Remarks upon the law of complete radiation. *Philos. Mag.* **49**, 539.

Salam, A. (1996) Intermolecular energy shifts between two chiral molecules in excited electronic states. *Mol. Phys.* **87**, 919.

Salam, A. (2000a) On the contribution of the diamagnetic coupling term to the two-body retarded dispersion interaction. *J. Phys. B: At. Mol. Opt. Phys.* **33**, 2181.

Salam, A. (2000b) Retarded intermolecular interactions involving diamagnetic molecules. *Int. J. Quantum Chem.* **78**, 437.

Salam, A. (2005a) A general formula for the rate of resonant transfer of energy between two electric multipole moments of arbitrary order using molecular quantum electrodynamics. *J. Chem. Phys.* **122**, 044112.

Salam, A. (2005b) Resonant transfer of excitation between two molecules using Maxwell fields. *J. Chem. Phys.* **122**, 044113.

Salam, A. (2006a) On the effect of a radiation field in modifying the intermolecular interaction between two chiral molecules. *J. Chem. Phys.* **124**, 014302.

Salam, A. (2006b) Intermolecular interactions in a radiation field via the method of induced moments. *Phys. Rev. A* **73**, 013406.

Salam, A. (2007) Two alternative derivations of the static contribution to the radiation-induced intermolecular energy shift. *Phys. Rev. A* **76**, 063402.

Salam, A. (2008) Molecular quantum electrodynamics in the Heisenberg picture: a field theoretic viewpoint. *Int. Rev. Phys. Chem.* **27**, 405.

Salam, A. and Thirunamachandran, T. (1994) Maxwell fields and Poynting vector in the proximity of a chiral molecule. *Phys. Rev. A* **50**, 4755.

Schiff, L. I. (1955) *Quantum Mechanics*. McGraw-Hill, New York.

Scholes, G. D. (2003) Long range resonance energy transfer in molecular systems. *Annu. Rev. Phys. Chem.* **54**, 57.

Scholes, G. D., Harcourt, R. D., and Fleming, G. R. (1997) Electronic interactions in photosynthetic light-harvesting complexes: the role of carotenoids. *J. Phys. Chem. B* **101**, 7302.

Schweber, S. S. (1994) *QED and the Men Who Made It*. Princeton University Press, Princeton, NJ.

Schwinger, J. (ed.) (1958) *Selected Papers on Quantum Electrodynamics*. Dover, New York.

Sparnaay, M. J. (1958) Measurements of attractive forces between flat plates. *Physica* **24**, 751.

Spruch, L. and Kelsey, E. J. (1978) Vacuum fluctuation and retardation effects on long-range potentials. *Phys. Rev. A* **18**, 845.

Sukenik, C. I., Boshier, M. G., Cho, D., Sandoghdar, V., and Hinds, E. A. (1993) Measurement of the Casimir–Polder force. *Phys. Rev. Lett.* **70**, 560.

Szalewicz, K., Patkowski, K., and Jeziorski, B. (2005) Intermolecular interactions via perturbation theory: from diatomics to biomolecules. In: Wales, D. J. (ed.), *Intermolecular Forces and Clusters. II. Structure and Bonding*, Vol. 116, Springer-Verlag, Berlin, pp. 43–117.

Tabor, D. and Winterton, R. H. S. (1969) The direct measurement of normal and retarded van der Waals forces. *Proc. R. Soc. Lond. A* **312**, 435.

Taylor, M. D. and Thirunamachandran, T. (1983) Discriminatory interactions between chiral molecules in the presence of an intense radiation field. *Mol. Phys.* **49**, 881.

Thirunamachandran, T. (1980) Intermolecular interactions in the presence of an intense radiation field. *Mol. Phys.* **40**, 393.

Thirunamachandran, T. (1988) Vacuum fluctuations and intermolecular interactions. *Phys. Scr.* **T21**, 123.

van Blokland, P. H. G. M. and Overbeek, J. T. G. (1978) van der Waals forces between objects covered with a chromium layer. *J. Chem. Soc. Faraday Trans.* **74**, 2637.

Verwey, E. J. W. and Overbeek, J. T. G. (1948) *Theory of the Stability of Lyophobic Colloids*. Elsevier, Amsterdam.

Wigner, E. P. (1959) *Group Theory and Its Applications to Quantum Mechanics of Atomic Spectra*. Academic Press, New York.

Woolley, R. G. (1971) Molecular quantum electrodynamics. *Proc. R. Soc. Lond. A* **321**, 557.

Ziman, J. M. (1969) *Elements of Advanced Quantum Theory*. Cambridge University Press, London.

INDEX